Mechanik, Werkstoffe und Konstruktion im Bauwesen

Band 65

Reihe herausgegeben von

Ulrich Knaack, Darmstadt, Deutschland

Jens Schneider, Darmstadt, Deutschland

Johann-Dietrich Wörner, Darmstadt, Deutschland

Stefan Kolling, Gießen, Deutschland

Institutsreihe zu Fortschritten bei Mechanik, Werkstoffen, Konstruktionen, Gebäudehüllen und Tragwerken. Das Institut für Statik und Konstruktion der TU Darmstadt sowie das Institut für Mechanik und Materialforschung der TH Mittelhessen in Gießen bündeln die Forschungs- und Lehraktivitäten in den Bereichen Mechanik, Werkstoffe im Bauwesen, Statik und Dynamik, Glasbau und Fassadentechnik, um einheitliche Grundlagen für werkstoffgerechtes Entwerfen und Konstruieren zu erreichen. Die Institute sind national und international sehr gut vernetzt und kooperieren bei grundlegenden theoretischen Arbeiten und angewandten Forschungsprojekten mit Partnern aus Wissenschaft, Industrie und Verwaltung. Die Forschungsaktivitäten finden sich im gesamten Ingenieurbereich wieder. Sie umfassen die Modellierung von Tragstrukturen zur Erfassung des statischen und dynamischen Verhaltens, die mechanische Modellierung und Computersimulation des Deformations-, Schädigungs- und Versagensverhaltens von Werkstoffen, Bauteilen und Tragstrukturen, die Entwicklung neuer Materialien, Produktionsverfahren und Gebäudetechnologien sowie deren Anwendung im Bauwesen unter Berücksichtigung sicherheitstheoretischer Überlegungen und der Energieeffizienz, konstruktive Aspekte des Umweltschutzes sowie numerische Simulationen von komplexen Stoßvorgängen und Kontaktproblemen in Statik und Dynamik.

Andrei Firus

A Contribution to Moving Force Identification in Bridge Dynamics

Springer Vieweg

Andrei Firus
Horgenzell, Baden-Württemberg, Germany

Vom Fachbereich 13 – Bau- und Umweltingenieurwissenschaften
der Technischen Universität Darmstadt zur Erlangung
des akademischen Grades eines Doktor-Ingenieurs (Dr.-Ing.)
genehmigte Dissertation von Master of Engineering Andrei Firus
aus Roşiori de Vede, Rumänien

1. Gutachten: Prof. Dr.-Ing. Jens Schneider
2. Gutachten: Prof. Dr. Ir. Geert Lombaert
3. Gutachten: Prof. Dr.-Ing. Roman Kemmler

Tag der Einreichung: 27.10.2021
Tag der mündlichen Prüfung: 17.12.2021
Darmstadt 2022

ISSN 2512-3238 ISSN 2512-3246 (electronic)
Mechanik, Werkstoffe und Konstruktion im Bauwesen
ISBN 978-3-658-39837-8 ISBN 978-3-658-39838-5 (eBook)
https://doi.org/10.1007/978-3-658-39838-5

This Springer Vieweg imprint is published by the registered company Springer Fachmedien Wiesbaden GmbH, part of Springer Nature.
The registered company address is: Abraham-Lincoln-Str. 46, 65189 Wiesbaden, Germany

Acknowledgement

This dissertation was elaborated as part of my work as a scientific assistant at the Institute of Structural Mechanics and Design at the Technical University of Darmstadt. At this point, I would like to express my warmest thanks to all the people who have supported and accompanied me during this time and thus contributed to the success of this work.

First of all, I wish to express my deepest appreciation to my doctoral supervisor, Prof. Dr.-Ing. Jens Schneider, for his trust in me and the great opportunity to pursue my doctorate at his institute. I would also like to thank him for his continuous support, enthusiastic guidance and encouragement, as well as for the constructive and valuable advice provided during the preparation of the dissertation. I am very grateful to have worked with him all these years and to have learned so much, which has provided me a great deal of confidence for my future professional and personal life.

I would also like to express my sincere gratitude to my co-supervisor, Prof. Dr.-Ing. Roman Kemmler (HTWG Konstanz - University of Applied Sciences), for his intensive support since the early conception phase of my dissertation. His extensive knowledge, motivation, competent advice and bright ideas, which he provided in countless and always fruitful discussions, have helped me stay on course and overcome the stumbling blocks of the past years.

I am also deeply grateful to Prof. Dr. Ir. Geert Lombaert (KU Leuven), who agreed to step in as co-supervisor during the last year of elaborating the dissertation and who contributed greatly to the success of this work through his valuable comments and suggestions.

In addition, I would also like to express my deep appreciation to my former professor, Prof. Dr.-Ing. Horst Werkle (HTWG Konstanz - University of Applied Sciences), who awakened my passion for structural dynamics and thus laid the groundwork for my professional career. The time spent under his guidance and the numerous projects we worked on together made an enormous contribution to my professional and personal development. An additional thanks goes to Professor Werkle also for his support in reading the manuscript of the dissertation and for the numerous hints.

I would also like to thank all my colleagues at the Institute of Structural Mechanics and Design for the very warm and friendly atmosphere over all the years we have spent together. I am truly honored to have been part of this group, which has not only helped me gain new insights into various technical topics, but also allowed me to make some real friends. A special thanks goes to my longtime office mate, Hagen Berthold, M.Sc., for his continuous tireless support in all experimental investigations and for struggling together in the early days of the Structural Dynamics Group.

Special thanks also go to the DB Netz AG, in particular to Mr. Günther Grunert, for the pleasant and fruitful collaboration over the past years. I would also like to thank Mr. Kai Wittig and the entire team *Fachstelle Brückenmessung* of the DB Netz AG for their intensive support during the experimental investigations on the railway bridge in operation. In this context, I would also like to extend my thanks to Prof. Dr.-Ing. Wolfgang Francke (HTWG Konstanz - University of Applied Sciences) for his personal support of the experimental part of my work. In fact, Prof. Francke firstly introduced me to the subject of structural measurements during my master's studies, for which I will be forever grateful.

The preparation of a dissertation also requires the support of the family. Therefore, I would like to thank my parents, Ileana and Daniel Firus, my brother Alexandru Firus as well as my parents in law, Maria and Viorel Spirchez, who are always by my side and have supported me in every situation. My sincerest personal thanks go to my beloved wife, Diana-Elena Firus, for bringing happiness, wisdom and tenderness into my life. Her patience, unconditional understanding for my work and the constant motivational boosts have been an indispensable support. Thank you!

Darmstadt, October 2021 *Andrei Firus*

Abstract

In the actual context of an aging infrastructure and a clear trend towards ever faster and heavier vehicles, the reassessment of existing bridges becomes an ever more challenging task for civil engineers, especially when crucial decisions regarding costly replacement or rehabilitation measures have to be made. An accurate knowledge of the real forces acting on the structure offers the engineers the possibility of a less conservative, but still reliable reassessment process, which can lead to important extensions of the remaining service life and thus to considerable financial benefits for both the bridge owners and the society. However, the direct measurement of dynamic forces acting on bridges is in most cases either not possible or very cumbersome. On the contrary, time histories of the structural responses can be recorded with reasonable effort.

The present work presents a novel approach for identification of dynamic moving forces acting on a bridge structure. It seeks to find the optimal force values that minimize the difference between the computed and measured displacement and acceleration time histories for a limited number of sensor locations. For this, an optimization problem is formulated in time domain. It considers the magnitude of each moving force in each time instant as an unknown variable. The regularization of the inverse problem is addressed by introducing two penalty functions that define physical a priori information on the solution. The problem is solved by means of the sensitivity-based trust region optimization algorithm, which requires the exact gradient and Hessian matrix of the objective function. These are determined by analytical differentiation, whereas the time stepping method of Newmark is used to solve a set of differential equations occurring in the implicit differentiation process. Moreover, an efficient solution of problems with a large number of unknown force values is possible by using a sliding window approach. The proposed moving identification method is applicable to any 3D Finite Element model with arbitrary level of detail. This is achieved by employing the principle of modal superposition.

A comprehensive study with simulated measurement data was firstly conducted to analyze the isolated effects of various parameters on the identification accuracy of the proposed method. In addition, an experimental validation was performed based on two investigations: an existing railway bridge in operation subjected to the passages of an ICE 4 vehicle and an experimental full-scale pedestrian bridge. They underline the wide range of possible applications for the moving force identification method developed within this work: from small structures excited by rather low pedestrian forces up to the "heavy category" of a complete train passing a railway bridge in operation. The results are very satisfactory, consistently indicating that the proposed method allows a sufficiently accurate identification of the moving forces. In this context, a set of particularities and limitations that arise in the practical the application of the method on real structures are also discussed. These are mostly related to modeling inaccuracies, which have been identified as a crucial factor influencing the solution behavior.

Zusammenfassung

Vor dem Hintergrund einer alternden Infrastruktur und eines eindeutigen Trends zu immer schnelleren und schwereren Fahrzeugen wird die Nachrechnung bestehender Brücken zu einer immer anspruchsvolleren Aufgabe für Bauingenieure, insbesondere wenn wichtige Entscheidungen über kostspielige Ersatz- oder Sanierungsmaßnahmen getroffen werden müssen. Eine genaue Kenntnis der tatsächlichen Kräfte, die auf das Bauwerk wirken, bietet den Ingenieuren die Möglichkeit einer weniger konservativen, aber dennoch zuverlässigen Nachrechnung, die zu einer beträchtlichen Verlängerung der Restlebensdauer und damit zu erheblichen finanziellen Vorteilen sowohl für die Infrastrukturbetreiber als auch für die Gesellschaft führen kann. Die direkte Messung der dynamischen auf Brückenbauwerke wirkenden bewegten Kräfte ist jedoch in den meisten Fällen entweder nicht möglich oder sehr umständlich. Hingegen lassen sich Zeitverläufe der Bauwerksantworten mit vertretbarem Aufwand aufzeichnen.

In der vorliegenden Arbeit wird ein neuartiger Ansatz zur Identifikation von dynamischen, auf ein Brückenbauwerk wirkenden bewegten Kräften vorgestellt. Die Grundidee ist, die optimalen Kraftwerte zu finden, die die Differenz zwischen den für eine begrenzte Anzahl an Punkten berechneten und gemessenen Zeitverläufen der Verschiebungen und Beschleunigungen minimieren. Hierfür wird ein Optimierungsproblem im Zeitbereich formuliert. Es betrachtet die Magnitude jeder bewegten Kraft in jedem Zeitschritt als unbekannte Variable. Die Regularisierung des inversen Problems wird durch die Einführung von zwei Straffunktionen addressiert, die physikalische a priori Informationen über die Lösung definieren. Das Problem wird mit Hilfe des gradientenbasierten Trust Region Optimierungsalgorithmus gelöst. Dieser erfordert den genauen Gradienten und die genaue Hesse-Matrix der Zielfunktion. Diese werden durch analytische Ableitungen bestimmt, während das numerische Integrationsverfahren von Newmark zur Lösung einer Reihe von Differenzialgleichungen verwendet wird, die im impliziten Ableitungsprozess auftreten. Darüber hinaus wird die effiziente Lösung von Problemen mit einer großen Anzahl von unbekannten Kraftwerten durch die Verwendung eines gleitenden Fensters ermöglicht. Die vorgestellte Kraftidentifikationsmethode ist auf jedes 3D Finite-Elemente-Modell mit beliebigem Detaillierungsgrad anwendbar. Dies wird durch die Verwendung der Modalanalyse ermöglicht.

Zuerst wurde eine umfassende Studie mit simulierten Messdaten durchgeführt, um die isolierten Auswirkungen verschiedener Parameter auf die Genauigkeit der vorgestellten Kraftidentifikationsmethode zu analysieren. Darüber hinaus wurde eine experimentelle Validierung anhand von zwei Untersuchungen durchgeführt: eine bestehende Eisenbahnbrücke im Betrieb, die den Überfahrten eines ICE 4 Fahrzeugs ausgesetzt wurde, und eine experimentelle Fußgängerbrücke. Sie unterstreichen den weiten Anwendungsbereich der im Rahmen dieser Arbeit entwickelten Methode zur Ermittlung von dynamischen bewegten Kräften: von kleinen Strukturen, die durch eher geringe Fußgängerkräfte angeregt werden, bis hin zur "schweren Kategorie" eines kompletten Zuges, der über eine Eisenbahnbrücke im Betrieb fährt. Die Ergebnisse sind sehr zufriedenstellend und zeigen durchweg, dass die vorgeschlagene Methode eine ausreichend genaue Identifikation der bewegten Kräfte ermöglicht. In diesem Zusammenhang wird auch auf eine Reihe von Besonderheiten und Einschränkungen eingegangen, die sich bei der praktischen Anwendung der Methode auf reale Bauwerke ergeben. Diese hängen vor allem mit Modellierungsungenauigkeiten zusammen, die als ein entscheidender Einflussfaktor des Lösungsverhaltens identifiziert wurden.

Contents

List of Figures

List of Tables

Glossaries

Abbreviations

1D	one-dimensional
2D	two-dimensional
3D	three-dimensional
AI	artificial intelligence
B-WIM	Bridge Weigh-In-Motion
CG	conjugate gradient
CSI/ref	Reference-Based Combined Deterministic-Stochastic Subspace Identification
DB AG	Deutsche Bahn AG
DFS-SAV	discrete finite signal with stable average value
DOF	degree of freedom
DOFs	degrees of freedom
FE	finite element
FEA	finite element analysis
FEM	finite element method
FTDM	frequency and time domain method
GCV	generalized cross-validation
GRF	ground reaction force
HSI	human-structure-interaction
HSLM-A	High-Speed-Load-Model A
IMI	interpretive method I
IMII	interpretive method II
LVDTs	Linear Variable Displacement Transducers
MAC	Modal Assurance Criterion
MDOF	multiple degree of freedom
MFI	moving force identification

MI	microwave interferometry
MP	measurement point
RMSE	root-mean-square error
SDOF	single degree of freedom
SNR	signal-to-noise ratio
SVD	singular value decomposition
TDM	time domain method
TSVD	truncated singular value decomposition
WIM	Weigh-In-Motion

Symbols: bridge dynamics

a_1, a_2	weighting factors of the Rayleigh damping
$A_{\mathrm{I}j}$	amplitude of the j-th impulse
α_j	j-th modal (generalized) coordinate
β, γ	integration parameters of the Newmark method
$\boldsymbol{C}$	damping matrix
c	damping coefficient
C_j	modal damping of the j-th mode
Δt	time step
d_j	coordinate of the j-th axle of a vehicle with respect to its front axle
δ	Dirac delta function
E	Young's modulus
$E_{\mathrm{inv},j}$	perturbed Young's modulus of the j-th element for the inverse computation
f_j	j-th computed natural frequency of the structure
$\bar{f}_j$	j-th experimentally identified natural frequency of the structure
$\tilde{f}_j$	j-th natural frequency of the perturbed structural model
F_j	j-th axle force passing over the structure
$F_{\mathrm{st},j}$	static component of the j-th moving force
I	moment of inertia
$I_{\mathrm{inv},j}$	perturbed moment of inertia of the j-th element for the inverse computation
$k_{\mathrm{h},i,x}$	horizontal spring defined at both ends of substructure i
$\boldsymbol{K}$	stifness matrix
$\hat{\boldsymbol{K}}_{\mathrm{tt}}$	condensed stifness matrix
K_j	modal stifness of the j-th mode
$k_{\mathrm{v},12,z}$	vertical spring defined along the longitudinal ballast joint

l	length of the bridge
λ	wave length
$L_{\mathrm{ob},k}$	length over buffers of the train k
$\boldsymbol{M}$	mass matrix
m_j	nodal mass of the j-th node
$\bar{m}$	mass per unit length
$m_{\mathrm{inv},j}$	perturbed nodal mass of the j-th node for the inverse computation
M_j	modal mass of the j-th mode
n	number of modes considered in the analysis
n_{dof}	number of degrees of freedom
n_{F}	number of forces acting on the bridge
n_{ma}	number of acceleration measurement points
n_{md}	number of displacement measurement points
$n_{\mathrm{tp},l}$	number of time steps, in which the l-th force is applied on the bridge
ν	Poisson's ratio
ω_j	j-th natural angular frequency of the structure
$\omega_{\mathrm{d},j}$	j-th damped natural angular frequency of the structure
$\boldsymbol{P}$	force vector
ϕ_j	j-th mode shape
P_j	modal load of the j-th mode
ρ	density
σ_{a}	standard deviation of the computed acceleration
σ_{d}	standard deviation of the computed displacement
T_j	j-th natural vibration period of the structure
$t_{\mathrm{aI}j}$	time point corresponding to the begin of the j-th impulse
$t_{\mathrm{I}j}$	duration of the j-th impulse
$t_{j\mathrm{s}}$	time point at which the j-th force enters the bridge
$t_{j\mathrm{e}}$	time point at which the j-th force leaves the bridge
t_{out}	time needed by a complete vehicle to cross the bridge
T_{u}	time needed by a single force to cross the bridge
$\boldsymbol{u}$	(computed) displacement vector
$\bar{\boldsymbol{u}}$	measured displacement vector
v	vehicle speed
$v_{\mathrm{res},i,j,k}$	i-th resonance speed of the j-th mode for the train k
ξ_j	modal damping ratio of the j-th mode
(x,y,z)	Cartesian coordinates

Symbols: inverse problems

$\boldsymbol{A}$	coefficient matrix of a linear system of equations
α	Tikhonov regularization parameter
$\boldsymbol{b}$	right side of a linear system of equations
Δx_i, Δx_k	perturbation factors for computation of numerical sensitivities
$\Delta^{(k)}$	trust region radius in the k-th iteration
$\boldsymbol{e}_\square$	unity vector in direction of the coordinate $\square$
e_p	noise level
$\mathcal{F}$	total objective function
$\mathcal{F}_\mathrm{mc}$	main optimization criterion
$\mathcal{F}_\mathrm{p1}$	first penalty function
$\mathcal{F}_\mathrm{p2}$	second penalty function
$\boldsymbol{\Gamma}$	Tikhonov matrix
$\boldsymbol{H}$	Hessian
$\boldsymbol{I}$	identity matrix
$\boldsymbol{k}$	vector of design variables in the model updating
$\mathcal{L}$	L-curve
$m^{(k)}$	approximation of the objective function in the k-th iteration
n_var	number of unknown force values
$\boldsymbol{N}_\mathrm{rand}$	normal distribution vector with zero mean and unit standard deviation
n_w	number of time windows used in the identification process
$\boldsymbol{Q}_\mathrm{glo}$	global smoothing matrix
$\boldsymbol{Q}_\mathrm{loc}$	local smoothing matrix
r_a	smoothing radius of the local smoothing matrix
$\sigma_\mathrm{err,st,avg}$	standard deviation of the individual errors of the static axle forces
σ_j	j-th singular value of a matrix
$\sigma_\mathrm{rmse,avg}$	average standard deviation of the RMSEs for the identified axle forces
w_a	weighting factor of the acceleration term of the objective function
w_d	weighting factor of the displacement term of the objective function
$w_{\mathrm{f},j}$	weighting factor of the j-th frequency in the model updating
w_p1	weighting factor of the first penalty function
w_p2	weighting factor of the second penalty function
$w_{\mathrm{u}j}$	weighting factor of the static displacement at point j in the model updating

$\boldsymbol{x}$	vector of unknown variables
x_{err}	relative error of the identified forces with respect to the reference value
$x_{\text{err,f}}$	relative error at the end of the forward identification
$x_{\text{err,b}}$	relative error at the end of the backward identification
$x_{\text{err,st,tot}}$	error of the identified total vehicle weight
$x_{\text{err,st},l}$	error of the static force component of the l-th axle
$x_{\text{err,st,avg}}$	average error of the static force components of all axles
$\boldsymbol{x}^*$	optimal solution vector
$x_{\text{rmse,avg}}$	average RMSE values for all axle forces of a vehicle
$x_{\text{rmse},l}$	RMSE of the l-th dynamic axle force with respect to its static component
$x_{\text{identified,st},l}$	identified static force component of the l-th axle
$x_{\text{reference,st},l}$	reference static force component of the l-th axle

Symbols: microwave interferometry

α	projection angle
B	band width of the transmitted wave
c	speed of light
δ_{r}	range resolution
Δr	displacement of target object in line of sight
Δx	projected displacement
λ	wave length of the electromagnetic waves
n	phase ambiguity
ϕ	interferometric phase change
ϕ_{atm}	interferometric phase change due to atmospheric effects
ϕ_{disp}	main component of the interferometric phase change
ϕ_{noise}	interferometric phase change due to noise
s_{r}	width of the antenna beam
θ	opening angle of the antenna beam

Mathematical Operators

$\cos\square$	cosine function
$\cosh\square$	hyperbolic cosine function
$\nabla\square$	gradient of the function $\square$

$\square^{-1}$	inverse of a matrix $\square$
$\square^{\dagger}$	pseudoinverse of a matrix $\square$
$\square^{\top}$	transpose of a matrix $\square$
$\det(\square)$	determinant of the matrix $\square$
$\Vert\square\Vert_1$	l_1 norm of the vector $\square$
$\Vert\square\Vert_2$	l_2 (Euclidean) norm of the vector $\square$
$\mu(\square)$	condition number of the matrix $\square$
$\mathrm{rank}(\square)$	rank of the matrix $\square$
$\sin\square$	sine function
$\sinh\square$	hyperbolic sine function
$\partial/\partial\square$	first order partial derivative with respect to the variable $\square$
$\partial^2/\partial^2\square$	second order partial derivative with respect to the variable $\square$
$\mathrm{d}/\mathrm{d}\square$	first order total derivative with respect to the variable $\square$
$\mathrm{d}^2/\mathrm{d}^2\square$	second order total derivative with respect to the variable $\square$
$\dot{\square}$	first order derivative of the variable $\square$ with respect to the time
$\ddot{\square}$	second order derivative of the variable $\square$ with respect to the time
$\square'$	first order derivative of the variable $\square$ with respect to the length coordinate

1 Introduction

1.1 Motivation

The infrastructure network of a country must be – regardless of age and type of its components – safe, efficient, durable, environmentally friendly, fully functional at any time and able to withstand the current volume of traffic. This facilitates mobility and transportation for the citizens of a nation and contributes to their quality of life and economic strength. This holds especially in case of countries with intensive trade relations and highly developed industrial and economic activities. As a matter of fact, the railway network of the Deutsche Bahn AG (DB AG)[1] consists of approximately 33 500 line-kilometers including around 25 200 bridges (DEUTSCHE BAHN AG, 2019). Considering the average age of the existing bridges of about 74 years and the fact that the regular high-speed rail traffic in Germany started about 30 years ago, it becomes clear that the design of many existing bridges did not consider any dynamic effects induced by high-speed railway transport. Moreover, the railway industry of the last decades is characterized by an increased traffic volume and a clear trend towards ever faster and heavier vehicles with innovative axle arrangements. At the same time, the high-speed railway networks are continuously expanding in many countries all over the world, leading thus to an increasingly internationalized railway traffic. In this context and under consideration of a continuously aging and degrading infrastructure, the reassessment of existing bridges becomes an ever more challenging task for civil engineers. This holds especially when crucial decisions regarding costly replacement or rehabilitation measures have to be made.

Within the reassessment process, structural engineers are not only concerned with the structural capacity but also with the fatigue safety and ultimately with the remaining service life of the analyzed structure (RIL 805). In case of steel bridges, the latter essentially comprises two components that provide different assertions: the remaining service life and the operating time intervals between two inspections. While the first one employs the concept of the Wöhler[2] lines (S-N curves), the second one is based on the principles of elastic fracture mechanics (RIL 805, QUOOS et al., 2000, MENSINGER et al., 2014). In this context, it should be noted that besides the annual tonnage, also the traffic composition plays

[1] Deutsche Bahn AG is the largest German railway company. It is organized as a private joint-stock company (AG, *Aktiengesellschaft*), with the Federal Republic of Germany as its single shareholder.

[2] August Wöhler (1819-1914) was a German railway engineer involved in research of steel and iron. He developed the so called *Wöhler lines*, which establish the relationship between the amplitude of an alternating stress and the number of load cycles to failure for a given material.

A. Firus, *A Contribution to Moving Force Identification in Bridge Dynamics*, Mechanik, Werkstoffe und Konstruktion im Bauwesen 65, https://doi.org/10.1007/978-3-658-39838-5_1

a relevant role within the process of estimation of the remaining service life. Considering that partly very little detailed knowledge about the traffic loads on bridges in the past is available, the analysis of the remaining service life for railway bridges of the DB AG is currently performed according to idealized load models, as established by QUOOS (1996). They consider important development trends of the railway traffic, divided in nine characteristic time periods, starting with the reference year 1876. Based on the idealized load models, a procedure for estimating the remaining service life was formulated in GEISSLER et al. (1998), then elaborated more precisely by QUOOS et al. (2000) and finally adopted by RIL 805, which is still in use[3]. It assumes linear accumulation of the fatigue damage in the past, according to the Palmgren-Miner[4] damage accumulation hypothesis (MINER, 1945). The yearly fatigue damage for the future is subsequently estimated as a function of the total accumulated fatigue damage of the past. The remaining fatigue life can be then determined according to the allowable limit damage provided in DIN EN 1992-1-1 for prestressing and reinforcing steel and in DIN EN 1993-1-9 for structural steel.

Considering that traffic volumes and compositions vary greatly from site to site, and that the two previously mentioned components of the fatigue safety analysis are very sensitive to the (load-dependent) stress range, it is very likely that the estimated remaining service life will be very inaccurate in many cases (LYDON et al., 2016). On the one hand, this could imply excessive conservatism, e.g. bridges may be recommended for unnecessary and costly repair; on the other hand, the procedure could overestimate bridge safety in certain cases[5]. It can be thus stated that one of the key factors in estimating the remaining service life of existing railway bridges is represented by the load assumption used in the calculation of damage accumulation. It is obvious that a precise knowledge of the real forces acting on the structure allows for a less conservative, but still reliable reassessment process. This could lead to important extensions of the remaining service life of existing structures and thus to considerable financial benefits for both bridge owners and society. This also applies to road and highway bridges.

The preceding discussion emphasizes the importance of detailed knowledge of the loads acting on a bridge structure in terms of improved condition assessment, especially for (steel) railway bridges. It mainly refers to the static axle loads, which are in fact relevant for the current approach of condition assessment (RIL 805). However, it is to be noted that the total contact force between wheel and surface can be considerably influenced by several factors, such as surface roughness, vehicle acceleration, out-of-round wheels, dynamically unbalanced wheels, tire inflation pressure (not relevant for railway vehicles), vehicle suspension characteristics, aerodynamic properties and wind (ASTM E1318). All these generate more or less pronounced dynamic force components, which open up a much

---

[3] as of July 2021

[4] The *Palmgren-Miner linear damage hypothesis* is widely known as the *Miner rule*, since it was popularized by M.A. Miner in 1945. However, it had been firstly proposed by A. Palmgren in 1924.

[5] It is known, for example, that during the Second World War a considerable, improperly documented amount of heavy material was transported on the European railway network (MENSINGER et al., 2014).

wider application field for the realistic load estimation. For example, it is well known that the dynamic components of the vehicle forces can increase the responses and thus the road surface damage by a factor of 2 to 4 (CEBON, 1987). In this context, accurate information regarding the real dynamic forces could possibly lead to an improved design method for road bridge pavements.

The understanding of the interaction effects between traffic load elements (vehicles or pedestrians) and the bridge structure is another aspect, which could profit from a precise knowledge of the real dynamic loads acting at the surface of the bridge. In this context, the accurate knowledge of the dynamic interaction forces between a pedestrian and a vibrating footbridge could lead to an improved understanding of the human-structure-interaction (HSI) effect and thus to less conservative design load assumptions for dynamic analysis of pedestrian induced vibrations (ŽIVANOVIĆ, 2016; SHAHABPOOR et al., 2016; CAPRANI et al., 2016; VAN NIMMEN et al., 2017; AHMADI et al., 2018a; FIRUS et al., 2018b). In case of railway bridges, the dynamic interaction forces can also show considerable differences compared to the static axle loads. This is particularly relevant for a resonance excitation, which can significantly influence the global structural behavior (ZHANG et al., 2008; LIU et al., 2009a). A better knowledge of the interaction forces and their positive influence regarding the reduction of structural responses could lead to improved, more reliable approaches for the bridge design codes (DOMÉNECH et al., 2014; CANTERO et al., 2015; YAU et al., 2019; GLATZ et al., 2021).

As previously elaborated, the accurate knowledge of both static and dynamic forces is of great importance for improved condition assessment and enhanced bridge design approaches. In this respect, several methods for direct or indirect force measurement have been developed[6]. For example, different systems based on the Weigh-In-Motion (WIM) technology are already commercially available (AL-QADI et al., 2016; VAN LOO et al., 2019). However, they are usually very expensive, require the installation at track/pavement level (implying thus temporary track/road closure) and are only able to retrieve static axle loads. An alternative to the classical WIM solution are the so called Bridge Weigh-In-Motion (B-WIM) systems (OBRIEN et al., 2001; OBRIEN et al., 2008; YU et al., 2016; LYDON et al., 2016; VAN LOO et al., 2019). They do not require an instrumentation at track level, since the whole bridge is used as a scale for the axle forces passing at regular speed. Their functioning principle is based on strain measurements in combination with theoretically or experimentally determined influence lines of the structure (BRAUNE, 1977; SNYDER et al., 1986). However, even in this case, only static forces can be retrieved, whereas these might be subjected to high inaccuracies if the measured signals contain considerable dynamic components (GONZÁLES et al., 2008).

With respect to the measurement of the dynamic forces induced by vehicles into the bridge structure, very few approaches are reported in the literature. They are based either on infrared measurements of the tire deflection (CANTIENI, 1992) or on strain and acceleration

[6] Both the direct and indirect measurement methods for moving forces will be discussed in detail in Chap. 2.

measurements at the axle level (LEBLANC et al., 1992; JACOB et al., 1998; CANTIENI et al., 2000; KOULOCHERIS, 2014). It is thus clear that all these methods require the instrumentation of the vehicle at axle level. As a consequence, it can be stated that a direct measurement of the dynamic forces for all axles of a large number of vehicles is not possible with reasonable effort and cost.

In order to overcome the previously mentioned drawbacks of the classical measurement methods for moving axle loads, several researchers around the world are concerned with the so called moving force identification (MFI) techniques. They are based on measured structural responses at a few discrete points and seek to retrieve the set of dynamic forces, which generated the measured behavior. Thus, it is a matter of determining the cause under assumption of known effects. This type of formulation is widely denoted as an inverse problem, which may also be ill-posed and ill-conditioned (LAW et al., 2011). This fact typically requires the use of a regularization procedure, in order to provide bounds to the solution. Extensive overviews on MFI are provided by YU et al. (2007), LAW et al. (2011), ZHU et al. (2016) and YU et al. (2016), whereas a detailed discussion of the current state of research in this field is given in Chap. 2 of this thesis. The MFI methods can be basically divided in three categories: deterministic methods (PINKAEW, 2006; LAW et al., 2011; FENG et al., 2015; PAN et al., 2017), stochastic methods (WU et al., 2010; WU et al., 2012) and approaches based on artificial intelligence (AI) (JIANG et al., 2003; OBATA et al., 2006; WANG et al., 2011; CHEN et al., 2021). The deterministic methods are based on establishing a linear dependency between the input and output of a system, the stochastic methods provide a statistical relationship between the random excitation force and the random structural response, while the AI based approaches require a learning process in order to find the relation between the inputs and outputs. However, the vast majority of the methods reported in the literature deal with the deterministic methods. In this regard, two main techniques can be distinguished: the time domain methods (LAW et al., 1997; LAW et al., 2000; ZHU et al., 2001; NORDSTRÖM, 2006; PAN et al., 2017) and the combined frequency and time domain methods (LAW et al., 1999; YU et al., 2003). In this context, a great part of the research was concerned with finding proper regularization approaches (ZHU et al., 2000; LAW et al., 2001a; NORDSTRÖM, 2006; PINKAEW, 2006; ASNACHINDA et al., 2008; PAN et al., 2017; FENG et al., 2015). In order to reduce the error sources related to the model uncertainty, some theoretical attempts were also made towards the joint estimation of external forces and system parameters (ZHU et al., 2007; LAW et al., 2008; LAW et al., 2010; FENG et al., 2015).

Despite the significant progress made in the field of MFI over the last two decades, there are still several issues that need to be addressed prior to a robust and reliable practical implementation, especially for the long-term application in the field of railway bridges. This concerns, for example, the solution of large problems characterized by a large numbers of axles that require several thousand time steps to pass the structure. In this respect, it is to be noted that the number of load values, which can be reconstructed by most approaches available in the literature, is still limited. The reason for this is that usual time domain based MFI methods imply the solution of an inverse problem, which requires computationally expensive matrix manipulations (LAW et al., 2011; WANG et al., 2011; ZHU et al., 2016; LAI et al., 2016). This issue is even more relevant if the ill-conditioned problem is regularized by the classical Tikhonov regularization (TIKHONOV et al., 1995), which usually requires a computationally very demanding determination of the regularization parameter (PINKAEW, 2006; FENG et al., 2015). Moreover, most of the available MFI procedures were developed using very simplified bridge models, such as analytically or numerically formulated 1D beam models or 2D orthotropic plate models (LAW et al., 2011; ZHU et al., 2016). While these approaches are well suited for demonstrating the concepts, they would become unrealistic when applied to real 3D structures. Therefore, in order to extend the area of application to real bridges in operation, new general and reliable MFI techniques, which allow the force identification based on 3D structural models with arbitrary level of detail, are required. In addition, probably the most serious drawback of the current MFI methods is that the previous works are either of pure numerical nature or are only concerned with experimental validations in the laboratory (LAW et al., 1997; LAW et al., 2000; LAW et al., 2004; YU et al., 2008b; PAN et al., 2017). A systematic experimental validation on real structures in operation, however, is generally missing. This fact is closely related to the complete lack of discussion regarding other difficulties that may arise when dealing with real practical applications.

1.2 Objectives

Given the aforementioned drawbacks of current MFI methods and in the context of a clear need for accurate load monitoring, the ultimate goal of this thesis is:

The development of a novel MFI approach, which opens the way towards a robust and reliable practical application of the MFI concept to real bridges in operation.

In order to achieve this goal, the following particular objectives will be pursued, which attempt to fill – to the greatest extent possible – the previously mentioned gaps of the currently available MFI methods:

(1) To develop an efficient solution algorithm for MFI that enables an accurate identification of dynamic moving forces in case of large problems[7].

(2) To formulate the novel MFI method in such a way, so that it can be applied to 3D finite element (FE) models with arbitrary level of detail.

[7] As large problem is understood, for example, the identification of the axle forces for a complete railway vehicle with several tens of axles, which need thousands of time steps to cross the bridge. This are usual characteristics of the MFI problem in case of railway bridge applications.

(3) To ensure that the developed method has a robust behavior to noisy signals and that it achieves a sufficient accuracy with a low number of acceleration and displacement sensors[8].

(4) To quantify the sensitivity of the developed method with respect to different computational parameters.

(5) To perform an experimental validation of the novel MFI method on a real railway bridge in operation and on a real full-scale pedestrian bridge, in order to ensure that a broad band of possible applications is covered.

(6) To identify and discuss the potential difficulties and challenges arising in the practical applications of the MFI method on real structures in operation.

1.3 Original contributions

An outline of the main original contributions of the present work is given in the following:

- A novel MFI method has been developed, which enables the identification of dynamic moving forces for arbitrarily large practical problems with reasonable computational effort.
- The solution method relies on the sensitivity-based trust region optimization algorithm. This requires the exact gradient and Hessian matrix of the objective function, which were determined by analytical differentiation in combination with the numerical solution of the equations of motion using the time integration method of Newmark.
- Solving large problems is made possible by implementing a forward/backward sliding window approach. A first identification result is obtained by the forward identification procedure. A subsequent backward identification process refines the results by taking into account the additional information from the decay phase.
- The proposed MFI approach is applicable to any 3D FE model with arbitrary level of detail. This is achieved by employing the principle of modal superposition. This implies that the FE model is generated with any FE software and the modal equations of motion are formulated externally (e.g. using the computer algebra software Matlab), using only few regular outputs of the FE modal analysis.

[8] The requirement of using displacement sensors in the identification process is motivated by a previous research project, in which the author of this thesis had a major contribution. The project *STRAMIK - Strukturanalyse mit Mikrowelleninterferometrie* (English: *Structural analysis using microwave interferometry*) dealt with non-contact displacement measurements on railway bridges by means of the microwave interferometry (MI) principle, which is briefly presented in Section 7.3.2.1. The project *STRAMIK* triggered to some extent the idea of the present work, which arose in fact as the project associates were looking for further possible applications of the non-contact dynamic displacement measurements of railway bridges.

- The MFI problem is regularized using two separate penalty functions with physical meanings. The first one ensures that the dynamic components of the solution vary around the corresponding static components. The second one minimizes the effect of noisy data on the solution accuracy.
- A set of solvability conditions for the MFI problem were formulated. They indicate in which conditions an exact solution can be achieved, if noise-free data and no regularization are used.
- A comprehensive numerical parametric study was performed to elucidate the effect of different computational parameters on the identification accuracy of the developed MFI method.
- The proposed MFI method was experimentally validated by means of two investigations: a real railway bridge in operation and an experimental full-scale pedestrian bridge.
- Measurement data recorded by a microwave interferometer are used – to the best knowledge of the author – for the first time within a MFI process.
- A set of issues occurring when dealing with real structures in operation and real measurement data are identified and discussed.

1.4 Outline of the text

Chap. 1 is the introductory chapter, which mainly focuses on the motivation of the research work and the formulation of its objectives. In addition, it highlights the main contributions and clarifies the organization of the text.

Chap. 2 gives a detailed insight into the state of the art and research related to moving force measurement and moving force identification methods. In addition to the discussion on moving forces, this chapter also covers some aspects of the state of the art for identifying dynamic non-moving forces.

Chap. 3 briefly presents the theoretical background on bridge dynamics under moving loads, which is necessary for understanding the developments presented in the further course of the thesis. The chapter starts with a short introduction into the analytical solutions of a beam under moving loads and then moves on to more detailed insights into the numerical treatment of the more general case of discrete FE systems. In this context, both the solution based on mode superposition and direct integration are discussed.

Chap. 4 is mainly concerned with the fundamentals of inverse problems. It defines the concept of inverse problems and discusses different solution approaches for general inverse problems. In addition to the classical direct solution methods, the chapter includes also the approach based on gradient-based optimization algorithms, which was adopted for the method developed in this work. In this context, the basics of unconstrained optimization and the formulation of some relevant optimization algorithms are briefly presented.

Chap. 5 represents the theoretical core of the present dissertation. It presents the formulation of the novel MFI problem, which is solved using a gradient-based optimization algorithm. This approach requires exact sensitivity information of the problem, i.e. the gradients and the Hessian matrices. Their analytical derivation and all other particularities of the problem formulation are also presented.

Chap. 6 presents the numerical validation process based on investigations with simulated measurement data. In this context, two solvability conditions are also presented. They provide the necessary conditions for achieving the exact solution in the noise-free and regularization-free case. In addition, the influence of various computational parameters on the performance of the method is highlighted in a comprehensive parametric study.

Chap. 7 describes a comprehensive experimental validation process of the proposed MFI method by means of investigations on a real railway bridge. The chapter starts with the presentation of the measurement campaign and the measurement results. In this context, a short introduction of the MI measurement principle is also provided. Subsequently, the definition of the structural model and its calibration process are briefly described. The core of this chapter is represented by the application of the developed MFI approach with real measurement data. The results obtained for five different passages are presented, whereas the main challenges occurring when dealing with applications of real bridges in operation are highlighted.

Chap. 8 presents an additional validation procedure based on real measurements of a full-scale experimental pedestrian bridge. After a short introduction into the topic of ground reaction force (GRF) measurement, the framework of the experimental investigation is presented. Subsequently, the structural model and the corresponding calibration process are discussed. The identification results of the GRFs of a pedestrian are in the foreground of this chapter. However, the application of the proposed MFI method to real pedestrian bridges requires some particular data processing steps, which are also described. Finally, a set of specific influence factors, which are particularly relevant for applications on pedestrian bridges are discussed.

Chap. 9 summarizes the most important findings of this work and gives recommendations for possible further developments.

2 Force identification methods – state of the art and research

In the context of a clear need for accurate load histories of existing bridges, emphasized in Chap. 1, several methods for moving force reconstruction have been developed. Basically, a distinction can be made between direct and indirect methods. The direct methods imply the direct measurement of the forces at vehicle or at track level, whereas the indirect force identification is based on structural response measurements of a bridge structure, which is subjected to the action of moving forces. The following sections provide insights into both direct and indirect force identification methods.

2.1 Direct measurement

First of all, it should be mentioned that the direct measurement of the dynamic forces acting on bridges is usually very cumbersome. The classical way for acquisition of the dynamic axle loads of a moving vehicle usually implies the instrumentation of the vehicle at the axle level (LEBLANC et al., 1992). Fig. 2.1 shows an example of a measurement system developed by CANTIENI (1992), which uses the tires as measurement springs. The time history of the tire deflection is measured contactless, using an infrared reflection method.

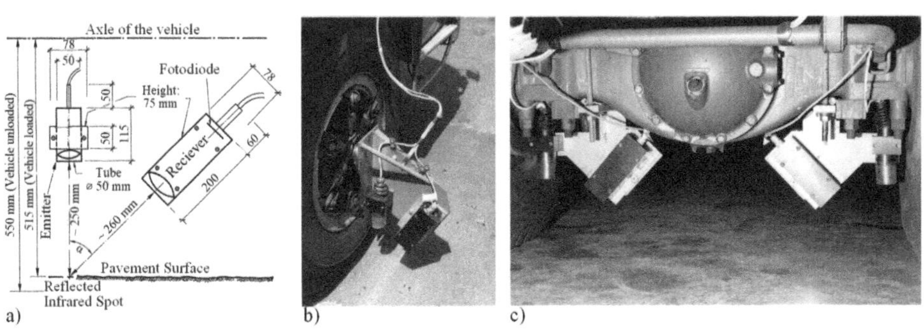

Figure 2.1 Opto-electronic device to measure dynamic wheel loads (CANTIENI, 1992): a) schematic representation of the measurement system; b) installation on a front wheel of a truck (source: Reto Cantieni); c) installation on the two wheels of a rear axle of a truck (source: Reto Cantieni)

A. Firus, *A Contribution to Moving Force Identification in Bridge Dynamics*, Mechanik, Werkstoffe und Konstruktion im Bauwesen 65, https://doi.org/10.1007/978-3-658-39838-5_2

Considering the (non-linear) spring characteristic of the tire, determined in a separate experiment, the total force transmitted to the bridge can be determined.

Another method for measuring dynamic wheel forces implies the acquisition of strains and accelerations at axle level, very close to the wheel hubs (JACOB et al., 1998; CANTIENI et al., 2000; KOULOCHERIS, 2014). Fig. 2.2 schematically shows the measurement setup for a truck axle. The strain gauge rosettes are used to record the principal strains induced by shear in the stub axle and thus the shear force in the axle. Acceleration sensors are used to estimate the inertial effects of the mass between the shear transducer and the point of application of the dynamic wheel force. The shear and inertial forces are then analytically combined to determine the total dynamic wheel force:

$$F_P = F_V + m_o \cdot a \tag{2.1}$$

where F_P is the total dynamic wheel force applied to the underlying structure, F_V is the dynamic shear force measured in the axle stub, m_o denotes the mass outboard the strain gauges and a is the acceleration of the outboard mass m_o. A similar method based on the combination of bending strains and acceleration measurements is also available (MIDDLETON et al., 1994; KOULOCHERIS, 2014). Similar application examples on railway vehicles are provided in GUPTA et al. (2009) and AUERSCH (2015).

Although both measurement principles described above are widely accepted as repeatable and reliable systems, they require a high level of installation effort and have the disadvantage that each axle has to be instrumented, in order to get a complete load pattern of the entire vehicle. It is obvious that long time monitoring of all axle loads for a large number of vehicles would not be possible with reasonable cost and justifiable effort.

In order to tackle the issue of instrumenting each vehicle axle, several Weigh-In-Motion (WIM) systems have been developed in the last decades (AL-QADI et al., 2016; VAN LOO et al., 2019). WIM describes a measuring system that is able to record the axle loads of a

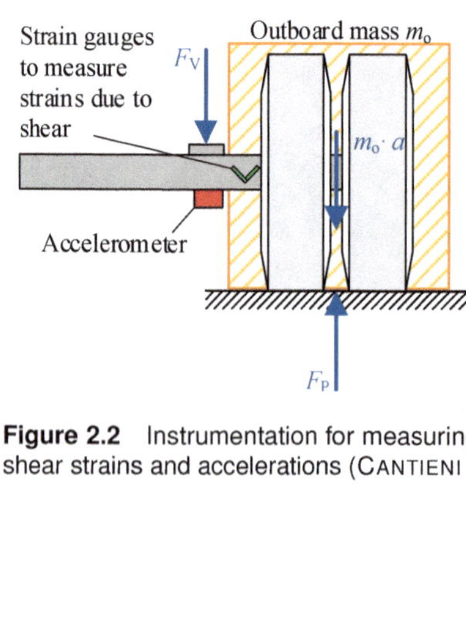

Figure 2.2 Instrumentation for measuring dynamic wheel forces of a truck based on acquisition of shear strains and accelerations (CANTIENI et al., 2000)

vehicle while it is driving over a measurement site with integrated weight sensors. The main sensor types used for WIM systems include bending plates, load cells and piezoelectric sensors or fiber optic devices (AL-QADI et al., 2016; VAN LOO et al., 2019).

The principal advantage of the WIM technology is that the measurement is carried out at regular traffic speed, so that the vehicle does not have to come to a stop[1]. Moreover, WIM systems are not only well suited for use on railways, but are also capable of handling any number of vehicles and axles. This is possible due to their functioning principle, which requires the instrumentation of the track instead of the vehicle. Basically, it can be distinguished between two WIM technologies for railway specific applications (VAN LOO et al., 2019):

- Recovering the forces from strain measurements on the rail;
- Direct measurements of the forces transmitted by the wheel-set through the rail and the sleepers to the embankment.

VAN LOO et al. (2019) states that the WIM systems for railway applications deliver in general better results than in case of road applications. This effect occurs on the one side due to the fact that there is no lateral variation of loading path, since the train strictly follows the rails. On the other side, the railway WIM systems allow a more reliable calibration of the measurement set-up due to the mostly fixed and very well known weight of the locomotives.

However, the main drawback of the WIM systems is that they are all designed to retrieve static axle loads (ASTM E1318). In fact, the WIM sensors detect dynamic components of the forces during the short sensor passage, which are generated by bouncing, rolling or pitching movements of the vehicles. However, these are only used to calculate estimates of other variables, e.g. static loads or gross vehicle weight (VAN LOO et al., 2019). In addition, WIM systems usually require a complicated installation at track level, which in any case also entails temporary track closures. Fig. 2.3 shows an example of the WIM system *DafuR*[2] used by the DB AG. Besides the measurement of axle loads, *DafuR* also enables the detection of flat spots, build-up welding spots, damages on wheel treads, polygons, unsymmetrical loads as well as parameters for vehicle identification.

2.2 Indirect measurement: Bridge Weigh-In-Motion

An alternative to WIM systems is the newly developed Bridge Weigh-In-Motion (B-WIM) technology (OBRIEN et al., 2001; OBRIEN et al., 2008; YU et al., 2016; LYDON et al.,

[1] It is to be mentioned that there are also low speed WIM systems available, denoted als LS-WIM. They are used to weight vehicles in a controlled area under defined traffic conditions, in order to minimize dynamic effects (VAN LOO et al., 2019). However, they are not relevant for this thesis and will not be discussed in detail.

[2] *DafuR* comes from the german "Detektionsanlagen für unrunde Räder" which stands for "detection system for out-of-round wheels".

Figure 2.3 WIM measurement system *DafuR* (source: Matthias Krüger, DB AG)

2016; VAN LOO et al., 2019). It eliminates the need of instrumentation at track level by using the whole bridge as scale to weigh the vehicles as they pass the structure at regular speed. The method is based on strain measurements of the bridge superstructure (girders or slabs) in combination with theoretically or experimentally determined influence lines (BRAUNE, 1977; SNYDER et al., 1986). It uses the assumption that the measured strains are in proportion to the product between the load magnitude and the influence line ordinate of the bridge (MOSES, 1979). In this respect, it is generally recommended to determine the influence line experimentally, since the theoretical values may provide unrealisic results (OBRIEN et al., 2006; YU et al., 2016). The weight of the vehicle is obtained by minimizing an objective function defined as the sum of the squared differences between the measured and the computed strain records. The singular value decomposition (SVD) and the Tikhonov regularization method are typical tools used in the solution process (OBRIEN et al., 2009; ŽNIDARIČ et al., 2016).

Although the classical B-WIM method offers significant logistical advantages, e.g. no track closure is needed for the system installation, its major disadvantage with regard to high-speed railway bridges is that it is not capable of retrieving dynamic loads, i.e. interaction forces between vehicle and structure. The reason for this is most likely that the developments of B-WIM have mainly concentrated on applications for highway bridges. In this case, the duration of the impact forces is usually too short to overcome the large inertia of the – rather massive – structures and to generate significant transient responses (SNYDER et al., 1986). Nevertheless, dynamic loads can play an important role even in case of highway bridges, but especially in terms of road damage and not global structural behavior. For instance, it has been reported, that the dynamic component of the moving forces may contribute to an increased road damage by a factor of two to four compared with static axle loads (CEBON, 1987). In this respect, VAN LOO et al. (2019) states that the B-WIM systems deliver accurate loads only if the road surface is smooth and does not present major irregularities, which can generate excessive dynamic behavior of the bridge.

To the best knowledge of the author, very few publications deal with application of B-WIM on railway bridges. LILJENCRANTZ et al. (2007) made probably one of the first attempts in this direction by adapting a common B-WIM algorithm for road bridges. The main adjustment was the consideration of the vehicle acceleration in the computations. This was required due to the partly significant speed differences between the first and last wagon of a train, whose length can reach up to several hundred of meters. After a calibration using a locomotive with known loads, the investigation revealed promising results with respect to the identification of the static loads. In addition, it highlighted some practical difficulties regarding the axle detection within a full-scale experiment, e.g. the two axles of a bogie could not be separated using only measured strain signals. A few years later, the investigation of ŽNIDARIČ et al. (2016), performed on an old truss-bridge, reported similar difficulties both with respect to the relative speed between the first and the last wagon and to the axle detection. While the first one required an adaptation of the computation software, the issue of the axle detection was solved by moving the strain sensors from the bridge superstructre to the bottom flange of the rail, between two slippers. However, this step required access to the track and implicitly a temporary track closure. Nevertheless, the investigation of ŽNIDARIČ et al. (2016) revealed a good match between the results of the static weights recovered by the B-WIM algorithm and the values recorded by a classical quasi-static low speed WIM system.

2.3 Dynamic load identification

In order to overcome the above mentioned drawbacks of the WIM and B-WIM methods, some alternative techniques for identification of dynamic moving loads have been developed in the last years. They are mainly based on dynamic measurements of different structural responses and seek to retrieve the dynamic forces passing the structure[3]. YU et al. (2007), UHL (2007), LAW et al. (2011), SANCHEZ et al. (2014), ZHU et al. (2016) and YU et al. (2016) provide extensive overviews on the current methods for force reconstruction, both in context of moving and non-moving loads. In the following, a brief overview of available dynamic load identification methods will be given.

2.3.1 Moving force identification

One of the pioneering works on moving force identification (MFI) was done by O'CONNOR et al. (1987), which developed the so called interpretive method I (IMI). It considers a bridge structure modeled as a set of lumped masses, which are connected by massless elastic beam elements. The force reconstruction procedure consists of a predictive analysis, used for generating the structural response based on the equilibrium of each lumped mass, and an interpretive analysis, in which the loads are identified from measured responses.

[3] In the broader sense, the methods for dynamic moving load identification could also be assigned to the category of indirect measurements (Section 2.2), since they are based on real measurement data of the bridge behavior. YU et al. (2016) and LYDON et al. (2016) even include the MFI in the category of B-WIM. However, in order to emphasize that dynamic force identification represents an essential aspect of this work and to provide a better clarity of the text, it is addressed in a separate section.

Basically, the method requires the numerical determination of velocities and accelerations from deflections using the central difference procedure. The force can be subsequently directly calculated by substituting all response data into a set of FE equations. Later on, CHAN et al. (1999) developed interpretive method II (IMII), which represents an extension of the IMI. The main improvements are the analytical formulation of the bridge model using Euler's equation for beams and the adopted solution approach in modal coordinates.

LAW et al. (1997) proposed an approach for MFI based on system identification methods in time domain, denoted as the time domain method (TDM). It is based on the analytical modeling of a simply supported beam with viscous damping subjected to moving loads, whereas the structural response is determined using the modal superposition method. The linear relationship between the modal structural responses and the moving axle forces is established using the convolution integral expressed in matrix form, taking into account a time discretization with a small time step. The time-varying moving forces are determined based on measured accelerations and bending moments. This is achieved by direct inversion in case of full rank matrices or using a least-square method for rank-deficient matrices. The problem formulation of LAW et al. (1997) was later extended by a Tikhonov regularization term (TIKHONOV et al., 1995), in order to provide bounds to the solution of the ill-conditioned inverse problem (LAW et al., 2001a). In the meanwhile, the Tikhonov regularization has established itself as the most popular regularization method in this field[4]. However, it has been reported that the determination of the Tikhonov regularization parameters usually generates serious issues with respect to the computational performance of the MFI, especially in case applications on large problems (PINKAEW, 2006; FENG et al., 2015). In order to better cope with the regularization issues, ASNACHINDA et al. (2008) developed a force identification method based on strain measurements and an iterative updating process for the static force component, originally proposed by PINKAEW (2006) and PINKAEW et al. (2007). It implies the decomposition of the axle loads in their static and dynamic components. The static component is updated iteratively with the help of the regularized dynamic component until convergence is achieved. This approach is accurate and robust with respect to the ill-conditioning and noise polluted signals and overcomes the issue of selecting a proper regularization parameter. The work of ASNACHINDA et al. (2008) addresses also the dependency of the solution error on the vehicle parameters. In general, a lower error is expected for vehicles with larger axle spacing than for shorter vehicles. Moreover, vehicles with larger mass and lower speed resulted in a better accuracy than in case of lighter and faster vehicles.

The regularized method of LAW et al. (2001a) was later applied for an analytically modeled multi-span continuous bridge (ZHU et al., 2001), revealing that the measurement noise leads to higher errors than in the case of a single-span bridge. ZHU et al. (2002a) used an improved method for the force reconstruction of a multi-span bridge. It is based on

[4] The theoretical background of the Tikhonov regularization method is briefly presented in Section 4.3.2 of this thesis.

splitting the system matrix into a set of matrices of lower dimensions, in order to increase the computational efficiency. Moreover, it also uses the Tikhonov approach and singular value decomposition (SVD) in the regularization process, whereas the regularization parameter is determined using the L-curve criterion and generalized cross validation (HANSEN, 1998). The method used by ZHU et al. (2002a) emphasizes that the identification of forces located in the neighborhood of the supports presents large errors and that the modeling errors have a significant influence on the MFI results. It postulates therefore that an accurate model is generally required within the force identification process. A further investigation deals with the reconstruction of two forces moving over multi-span continuous bridges with elastic restraints at the supports (ZHU et al., 2006a). The problem is based on a Tikhonov regularized least-squares problem with analytical formulation of the structure. Acceleration and strain measurements are considered in the analysis, whereas the latter category presents a lower identification accuracy. The investigation concluded that the sampling frequency should not be too high and that the vertical stiffness of the supports has to be large in comparison to the bridge stiffness, in order to ensure good quality of the results. Moreover, it is stated as a general rule for all the force identification methods based on modal superposition, that the frequency of the highest mode considered in the identification has to be higher than the highest frequency expected in the identified force time history.

ZHU et al. (2000) used the regularized method of LAW et al. (2001a) for the identification of moving forces acting on a bridge deck, which is modeled as a rectangular orthotropic plate. The investigation reported a similar identification accuracy as in case of the beam models, even in case of eccentric load path. Moreover, it is generally postulated that acceleration measurements provide better results than strain recordings. The latter remark is generally valid also for the beam models (LAW et al., 1997; LAW et al., 2001a). A further investigation on a bridge deck modeled as a rectangular orthotropic plate is given by ZHU et al. (2006b), which formulates the problem in the state space with Tikhonov regularization. CHEN et al. (2008) presents another approach for MFI in the state space, which is based on the Kalman filter and a recursive least-squares estimator.

PAN et al. (2017) addressed the aforementioned issue of a low accuracy for the force values located in the proximity of the supports by formulating a novel Tikhonov regularization approach based on the moving average technique. It makes use of a general physical feature of moving loads on bridges, which states that they always fluctuate around the axle weight (e.g. due to inertia of the vehicle and the surface roughness). A signal presenting this characteristic is denoted as a discrete finite signal with stable average value (DFS-SAV). This feature is accounted for in the inverse problem of PAN et al. (2017) by formulating an additional penalty term based on a smoothing matrix. This prevents the moving averages of the identified force values to lie too far from the global average. The proposed Tikhonov regularization approach revealed better results than the classical Tikhonov regularization or the truncated singular value decomposition (TSVD) methods, showing significant improvements of the force reconstruction around the supports. It should be noted that the results of PAN et al. (2017) are supported by an experimental investigation in the laboratory, implying a single span bridge model and a two-axled model car.

NORDSTRÖM (2006) presented a force reconstruction method for linear systems with time-dependent parameters. It is based on the dynamic programming method (TRUJILLO, 1978), while the system is formulated in the state space. The ill-posedness of the problem is solved by the classical Tikhonov regularization method. The method was tested for a single force moving over a simply supported beam, showing exact results for noiseless simulated strain data and a good accuracy in case of noise-polluted simulated measurements. The dynamic programming for MFI problems has also been previously used by LAW et al. (2001b), in order to avoid the large spurious fluctuations of the results given by classical force identification approaches. A further dynamic programming approach for MFI was developed by GONZÁLES et al. (2008). It is based on strain measurements and uses the Tikhonov regularization in order to suppress the ill-conditioning of the inverse problem. The algorithm – tested with simulated measurements – requires more sensors than the number of forces to be identified, whereas the structure can be modeled with an arbitrary FE software.

WANG et al. (2011) delivered one of the few publications dealing explicitly with MFI on railway bridges. It addresses the force reconstruction problem by using the concept of dynamic displacement influence line, which is defined as the displacement time history of a point under a unit load moving along the bridge. It is based on the fact that the total dynamic displacement of a linear bridge system subjected to the passage of a multi-axle vehicle can be expressed as a superposition of the responses generated by each moving force. This approach, which implies a FE model of a continuous steel truss girder bridge, ensures an increased computational efficiency of the direct analysis. The inverse problem is represented by the minimization of the error between the measured and the computed displacements. It is solved using a combined formulation based on the simulated annealing strategy and a genetic algorithm. QU et al. (2011) presents a similar approach, which in fact preceded the work of WANG et al. (2011). The main differences are that QU et al. (2011) defines the structural system using the analytical expressions of an Euler-Bernoulli beam and solves the equation of motion by means of classical methods, without making use of the dynamic displacement influence line. Both approaches were only tested with simulated measurements and can only handle constant forces. Nevertheless, a high accuracy was reported, also in case of a high noise level included in the simulated data.

Many of the previously mentioned MFI methods are based on the modal superposition principle. However, the modal truncation represents an additional error source, since the number of modes in the measurement data is usually higher than the number of modes considered in the identification (ZHU et al., 2002a). Therefore, LAW et al. (2004) tackled the identification of axle loads based on the finite element method (FEM) and a reduced system condensation technique. The measured displacements are expressed using the shape functions of the finite elements, without performing the modal transformation, whereas the main problem is formulated as a regularized least-squares minimization. The investigations were subsequently extended to MFI based on stochastic finite element (FE) models with Gaussian distributions of both the system parameters and the random external forces (WU et al., 2010; WU et al., 2012).

While the previously discussed MFI approaches generally seek to reconstruct the force in every time instant, there are also some attempts to address the reconstruction process by modeling the force using a set of basis functions, such as Legendre polynomials or Fourier series (YU et al., 2008a; CHEN et al., 2021). This technique implies that the coefficients of the basis functions are identified instead of the force values at each time instant, reducing thus the number of design variables in comparison to other methods. However, it is concluded that the efficiency of the computation strongly depends on the pattern and order of the basis functions. These have to be properly selected, in order to achieve satisfactory results. In general, the representation of the basis functions by means of Fourier series provides better results, whereas higher orders of the basis functions lead to improved accuracy. Using appropriate basis functions, the method of YU et al. (2008a) is proved to be more effective than the classical TDM method. The study is complemented by an experimental validation in laboratory conditions (YU et al., 2008b). A similar approach is used by LAW et al. (2010), which addresses the load identification in the context of condition assessment of a bridge structure. It describes the force time history using the Chebyshev polynomials as basis functions. In addition, CHEN et al. (2021) deals with the identification of pedestrian induced loads, whereas the force time histories are represented by Fourier series.

Besides the time domain methods, which clearly dominate the field of MFI, some attempts to include frequency domain computations in the identification process have also been made. LAW et al. (1999) proposed a combined frequency and time domain method (FTDM), which implies the modal decomposition of the differential equations of motion, followed by their Fourier transformation. Subsequently, the axle loads are identified in the frequency domain by solving a least-square problem, which is regularized by means of the singular value decomposition (SVD). The time history of the axle forces can be determined by inverse Fourier transformation. It is concluded that the FTDM method is very sensitive to sensor locations, whereas its computational efficiency is lower than the one of the TDM (LAW et al., 1999). YU et al. (2003) took up on the investigation of LAW et al. (1999) and compared the solution of FTDM identification obtained using the direct pseudoinverse and the SVD-based pseudoinverse (cf. Section 4.3.3). It is concluded that the latter approach delivered more accurate and robust results.

CHAN et al. (2001a) and CHAN et al. (2001b) present an extensive comparative study of the IMI, IMII, TDM and FTDM. It is concluded that IMII and TDM are not sensitive to the vehicle speed, while IMI presents a high sensitivity to noisy signals and to the ratio between the axle spacing and the bridge span. TDM delivered good results for the investigated cases, however at high computational cost. The study recommends that the number of sensors used is equal to the number of axles to be identified plus one for the time domain approach (TDM) and plus two for inverse computation in frequency-time domain (FTDM). Furthermore, the results indicate that – for a fixed number of sensors – the use of only acceleration sensors leads to the best results. Finally, TDM proved itself to be the least sensitive method to the location of the sensors. LAW et al. (2000) also performed a comparative study of MFI on different beam models, i.e. Timoshenko and Euler-Bernoulli. The study found that the Timoshenko formulation provides better results and recommends the number of sensors to be at least equal to the number of modes considered in the analysis. Moreover, it is stated that the sampling frequency should be at least two times higher than the highest frequency of interest (either natural frequency of the bridge or excitation frequency component).

In order to overcome the practical problems caused by uncertain models, some attempts for joint estimation of the external forces and the system parameters have been made. In this sense, FENG et al. (2015) developed a powerful method, which allows the simultaneous identification of the bridge parameters and the axle loads from a limited set of acceleration measurements. The basic idea is to minimize the sum of squared differences between the measured and computed accelerations, whereas the main optimization variables are the bridge parameters. The identification of the moving loads is achieved under consideration of a Bayesian inference regularization and is treated as a subproblem, which is incorporated within the primary optimization process of system identification. This leads to an iterative process, which terminates when the convergence criteria for both the force and system identification are satisfied. The method was only tested with simulated measurements for a single-span simply supported bridge and a three-span continuous structure under action of two moving forces. The numerical investigations generally revealed a good quality of the results, except for the regions in the proximity of the supports, which are typically characterized by low signal levels. Besides the work of FENG et al. (2015), there are some further approaches dealing with simultaneous identification of loads and structural parameters. ZHU et al. (2007) proposed a damage detection method, which implies the iterative determination of moving force time histories and structural parameters using dynamic response measurements. A considerable number of sensors is necessary in order to achieve satisfactory accuracy. A similar method was employed by LAW et al. (2010), which deals with the condition assessment of a three-span continous box-girder concrete bridge. The investigation used the response-sensitivity method to iteratively identify both the system parameters and the coefficients of the basis functions describing the force time histories. The dynamic response sensitivity approach was also employed by LU et al. (2011), which tackled the identification of both structural damage and multi-body vehicle parameters, based on dynamic measurements of bridge responses. A simultaneous determination of moving loads and prestressing forces from dynamic measurements is addressed in LAW et al. (2008).

A common drawback of all previously discussed approaches is that they completely lack on validations by means of full scale experiments under operational conditions. However, it is to be mentioned, that some laboratory tests have been performed in order to verify the reliability of the developed methods (LAW et al., 1997; LAW et al., 1999; LAW et al., 2000; ZHU et al., 2001; LAW et al., 2001a; LAW et al., 2001b; LAW et al., 2004; PINKAEW, 2006; YU et al., 2008b; ASNACHINDA et al., 2008; PAN et al., 2017). The experimental set-up mostly consisted of a beam with uniform cross-section instrumented with different sensors. The excitation was ensured by a model car with known static axle loads, which is pulled along the beam. Since usually no dynamic force measurements during the model car movement over the beam are possible, the quality of the results is mostly evaluated with respect to the quality of the match between the re-built[5] and measured responses and by comparing the static axle loads, which can be easily approximated as average values of the identified force time histories. ZHU et al. (2002b) gives a comparative study on practical aspects for different MFI methods (ZHU et al., 2000; ZHU et al., 2001), based on experimental investigations on a laboratory beam structure passed by a model-car. It concludes that the sampling frequency used in the identification should cover the frequency content of the responses, while the frequency of the highest mode considered in the identification should exceed the frequency range of the forces to be identified. In addition, it specifies that the number of measurement locations should be larger than the number of modes used in the identification. Finally, it was found that the investigated methods are sensitive to the surface roughness and to a large variation in the vehicle speed. Besides the beam-like laboratory structures, some experimental investigations on an experimental orthotropic bridge deck were performed by ZHU et al. (2006b), in order to validate its previously mentioned state-space MFI method.

The moving force identification problem for more than two axles is addressed only in few cases. YU et al. (2004) treated a three axle model vehicle using the same laboratory structure as in the original work of LAW et al. (1997). ASNACHINDA et al. (2008) used a more complex experimental set-up, which consists of a continuous bridge model and two double-axled vehicle models. The vehicles include also spring suspensions and – in contrast to the previously mentioned experiments – are instrumented with load cells above the wheel set level, in order to measure the dynamic forces during the passage. Moreover, they can drive on three different paths, allowing also investigations on overtaking events. The experimental set-up of ASNACHINDA et al. (2008) represents an improved version of the experiment presented in PINKAEW et al. (2007), which deals with a similar study on a simply supported bridge.

Perhaps one of the first full-scale experiments for moving force reconstruction on bridges was performed by CHAN et al. (2000a), which used the method of LAW et al. (1997) to identify the forces induced by a two-axle heavy vehicle driving over a prestressed

[5] It is to be noted that the responses generated by direct analysis with the identified forces are denoted as *re-built* responses.

concrete bridge. This investigation was a a continuation of the theoretical study presented by CHAN et al. (2000b). The results show that the gross weight of the vehicle could be retrieved with acceptable accuracy, while the dynamic forces can be considered as valid, based on comparison of the measured responses with the re-built responses. DENG et al. (2011) presented an extensive full-scale experimental study for dynamic axle load reconstruction in case of a truck passing over a bridge consisting of three simply supported spans. The theoretical background of the used MFI method was formulated in the companion paper (DENG et al., 2010). It is based on the superposition principle and the influence surface concept, whereas it allows the consideration of detailed FE models generated with specialized software instead of using very simplified beam models. The method of DENG et al. (2010) and DENG et al. (2011) is one of the first approaches, which considers displacement measurements for the MFI. However, it also provides a solution for strain measurements. The method presents a good accuracy, excepting the time instants when the axle enters and leaves the bridge. This effect is explained by the low values of the influence surface in this regions. Moreover, it is concluded – as in most of the papers dealing with MFI – that a well calibrated model is required, in order to achieve satisfactory results.

2.3.2 Identification of non-moving forces

This section addresses a few state of the art aspects and some recent developments on dynamic identification of non-moving forces. The reason for this is, on the one side, the fact that several concepts and methods of the non-moving force identification are similar to those used for the identification of moving forces. On the other side, the very broad and extensively investigated field of non-moving dynamic force identification may reveal possible trends for future developments in the field of MFI.

The Tikhonov regularization (TIKHONOV et al., 1995) is used by the majority of force identification methods, in order to tackle the ill-conditioning of the inverse problems (cf. Section 4.3.2). In this respect, CHOI et al. (2007) compares different approaches for determining the optimal regularization parameter for applications in the force reconstruction. It is stated that the L-curve is probably the most appropriate method for many practical situations. Furthermore, JACQUELIN et al. (2003) gives comprehensive insights into the general numerical treatment of the force identification problem formulated using the deconvolution approach. Besides the Tikhonov approach, the truncated singular value decomposition (TSVD) is introduced as an alternative regularization method in the force identification. The paper emphasizes that the Tikhonov regularization is a powerful method, mostly leading to robust results. However, it presents – as already mentioned in the previous sections – an important drawback with respect to the efficiency of finding the optimal regularization parameter, especially when addressing large problems (FENG et al., 2015). Therefore, AUCEJO et al. (2017) proposed a regularization strategy based on a multiplicative iterative approach, instead of the popular Tikhonov-like additive methods.

Its main advantage is that no regularization parameter needs to be determined in advance. Hence, important savings of computational time can be achieved. Without going into detail, it is mentioned that the regularization of dynamic force identification problems has been lately also addressed by means of Bayesian theory (ZHANG et al., 2012; FAURE et al., 2017; LI et al., 2019). These approaches seek to include known a priori information regarding the distribution of unknown variables into the solution process. Moreover, they can account to some extent for the influence of measurement and modeling errors on the uncertainty of the reconstructed forces (ZHANG et al., 2012).

Another disadvantage of the available force identification methods in time domain is represented by the expensive matrix computations, which makes solving large problems (e.g. large number of forces and time steps) very difficult or even impossible (LAI et al., 2016; ZHU et al., 2016). BERNAL et al. (2015) made a first step towards overcoming this issue by proposing a method for identification of non-moving forces based on sequential deconvolution. It implies a window of finite length, which moves along the temporal axis with a certain rate, whereas the forces are determined separately in each window position. By defining a reasonable number of time steps for the moving window, the dimension of the operation matrix can be controlled in such a way, that the computational effort lies in a reasonable range. Hence, the method could represent a great advantage with respect to a possible on-line applicability of force identification methods. LAI et al. (2016) has taken up the developments of BERNAL et al. (2015) and performed several parametric studies with respect to the window parameters (size and moving rate), noise level and sensor configuration. It is concluded that all three parameters have significant influence on the accuracy of the results. Moreover, a large window size and a small sliding rate were found to have positive influence on the solution accuracy. LAI et al. (2016) also mentions an important constraint with respect to the number of sensors, which should exceed the number of forces to be identified. However, the optimal position of the sensors for an efficient force identification is mentioned as a still open question among the researchers on this field. LIU et al. (2014a) proposed an additional method for a sequential solution of the problem, which implies the discretization of the total time window in a set of local intervals, in which the force is approximated as a linear combination of some high order continuous basis functions, whereas the shape functions are reconstructed using the least squares fitting and the regularization is performed by means of TSVD.

HUANG et al. (2009) formulated an inverse problem for identification of spatial and temporal-dependent forces applied on cutting tools. The approach is based on the formulation of an error function as time-integral of the least-squared errors between measured and calculated displacements, whereas the structure (cutting tool) is defined by means of the analytical expressions for an Euler-Bernoulli beam. The problem is solved by the iterative conjugate gradient (CG) method. It is shown that the force applied on the cutting tool could be very accurately reconstructed, even though both the temporal and spatial distribution of the force were assumed as unknown. However, the procedure was only validated based on exact and noisy simulated measurements. Moreover, it considers a very dense layout of displacement measurement points, which is rather difficult to implement in practical applications. The same author addresses in another work the non-linear force reconstruction of a system with displacement dependent system parameters (HUANG, 2001a). The problem was studied on a single degree of freedom (SDOF) system and tested with exact and noisy simulated displacement measurement data. The solution method is similar to the one used in HUANG et al. (2009). Since the system consists of only one degree of freedom (DOF) and no system uncertainties are considered, the force reconstruction in the noiseless case delivered – as expected – (nearly) exact results. Even with noise polluted data, the reconstructed forces were accurate. The method of HUANG (2001a) was later extended to nonlinear multiple degree of freedom (MDOF) systems under simultaneous action of several forces (HUANG, 2005). Using similar formulations, HUANG (2001b) solved a force identification problem for a linear SDOF system with time-dependent system parameters. The force reconstruction for time-variant systems is also discussed by NORDSTRÖM (2006), which solves the problem by means of dynamic programming. The dynamic programming method is also addressed in more recent publications (ZHU et al., 2014), which employed it to successfully reconstruct trigonometrical or stochastic forces acting on a system with four degrees of freedom (DOFs). Here, the system is formulated in the state-space, while the inverse problem is expressed as a least-square problem with an additional (Tikhonov) regularization term.

While many of the dynamic load identification methods are based on the deconvolution approach, DING et al. (2013) proposed an approach based on the average acceleration time integration method of Newmark. The problem is formulated in state space, while the ill-posedness is tackled by the Tikhonov regularization. The method requires that the number of sensors is higher than half of the number of external excitations and presents limited performance with noisy measurements. LIU et al. (2014b) used the explicit form of the implicit time-integration method of Newmark, in order to allow a formulation in the classical form of an inverse problem ($\boldsymbol{Ax} = \boldsymbol{b}$). This leads to a reduction of the computational effort in comparison to approaches based on the convolution integral in matrix form, as used, for example, by LAW et al. (1997). The method of LIU et al. (2014b) has only been formulated for non-moving forces and was only tested with simulated measurement data, revealing a better accuracy of the results compared to the state-space approach.

LU et al. (2006) presented a time-domain approach for reconstructing non-moving forces. It is based on dynamic response sensitivities to the parameters of the external periodic forces (frequency and amplitude). The sensitivities are used within an iterative process to determine the optimal parameters, which minimize the difference between the measured and computed accelerations. The method, which also included a Tikhonov regularization term, delivered good results with simulated and real noisy measurements. Moreover, it seems to operate even with only one acceleration sensor (in case of a system with 22 degrees of freedom). A similar formulation has already been adapted for identification of moving forces (YU et al., 2008a; YU et al., 2008b; LAW et al., 2010). Further approaches, which assume a certain profile of the force solution are given by GUNAWAN et al. (2006) for a B-spline profile, GUNAWAN et al. (2008a) and QIAO et al. (2015) for a quadratic spline and by GUNAWAN et al. (2008b) for the approximation of the unknown load time history using a set of scaled harmonic sine functions. In a similar context, LIU et al. (2016) proposed a time domain Galerkin method for dynamic force identification.

The dynamic load identification was tackled by GUNAWAN also by means of iterative optimization algorithms. In this sense, GUNAWAN (2012) proposed an extended Levenberg-Marquard method (LEVENBERG, 1944; MARQUARDT, 1963) to solve the ill-posed impact force identification problem in conjunction with a trust region strategy. The iterative regularization in the impact force reconstruction is discussed also by GUNAWAN et al. (2004), which adopted the CG method for solving the inverse problem. Further methods for localizing and identifying impact loads are given by LI et al. (2016) and LI et al. (2017). LI et al. (2016) addressed the problem by applying the positivity constraint and making use of the Tikhonov regularization method. LI et al. (2017) adopted a two step iterative approach, which implies the localization of the force prior to the identification of its magnitude in each iteration step, whereas the Tikhonov method is used for regularization. Considering the sparse characteristic of the impact forces, QIAO et al. (2017) introduces the concept of sparse deconvolution based on the l_1 norm regularization into the field of force identification. Due to the iterative solution based on interior point optimization algorithms, the method allows the solution of large-scale ill-posed problems of impact load identification. QIAO et al. (2016) proposes a further variation of the sparsity-based impact force reconstruction from highly incomplete and inaccurate measurements.

While most of the publications discussed in this section require a well calibrated structural model, several publications address the simultaneous identification of dynamic loads and system parameters. The work of FENG et al. (2015) has already been discussed in the context of moving force identification. The same author adapted the method for a simultaneous system and non-moving force identification, which relies on non-contact vision-based displacement measurements (FENG et al., 2017). The non-contact character of the measurements is related to a set of advantages, especially with regard to the analysis of real structures in operation. The investigation, supported also by a laboratory experiment, showed promising results. The use of non-contact displacement measurements for dynamic force identification is adopted also by DJAMAA et al. (2007), which deals with the reconstruction of distributed forces acting on a cylindrical thin shell. In order to solve the noise related issues of the inverse problem, a regularization implying windowing and spatial filtering in the wave number domain was adopted.

Without providing a detailed discussion, it should also be mentioned that a series of works for joint input-state estimation have been recently published (GILLIJNS et al., 2007; LOURENS et al., 2012a; AZAM et al., 2015; MAES et al., 2016b). They deal with the simultaneous estimation of the forces acting on the structure and of the response at unmeasured locations. The algorithm of GILLIJNS et al. (2007) implies performing of the input estimation before the state estimation step. In a subsequent work, LOURENS et al. (2012a) adapted the algorithm of GILLIJNS et al. (2007) for applications in structural dynamics, also by extending its applicability to reduced order models. MAES et al. (2016b) has undertaken a further development, which takes into account that the instrumented structure is also subjected to unknown stochastic excitations. Alternative formulations based on the classical Kalman filtering were developed for the joint estimation of forces and system states (LOURENS et al., 2012b; AZAM et al., 2015). In the same context, MAES et al. (2015) derived invertibility conditions for instantaneous input estimation or joint input-state estimation. They are related to the identifiability, stability and uniqueness of the loads and the system states. One of the few full-scale experimental verifications of joint input-state estimation methods is performed by MAES et al. (2016c), which considers the algorithm developed in MAES et al. (2016b). It is tested for the identification of impact, harmonic and swept sine excitation acting on a pedestrian bridge deck.

In contrast to the approaches of MFI, which are mainly formulated in the time domain, several works on non-moving load identification in the frequency domain have been carried out. The classical solution method in the frequency domain implies computing the pseudoinverse of the system model (USLU et al., 2007). However, the outcome of this approach is not always satisfactory (REZAYAT et al., 2016), which motivated the need for subsequent developments on this field. It is to be mentioned that also in case of frequency domain computations, the addition of a Tikhonov term (TIKHONOV et al., 1995) to the original loss function is the obvious regularization choice. However, this approach presents several limitations in case only few point forces, i.e. a sparse force vector, have to be reconstructed. REZAYAT et al. (2016) addresses this issue by proposing a novel method for force reconstruction, denoted as G-FISTA. It is based on an improved penalty function, which combines the advantages of the l_1 and l_2 norms, together with a modified iterative optimization technique. It allows the reconstruction of both amplitude and location of a single force acting on the structure. WAMBACQ et al. (2019) also deals with the identification of the location and magnitude of the forces in the frequency domain by formulating the minimization problem as a second order cone program, whereas the solution is achieved using an interior point method. JIA et al. (2015) introduced a method for force identification in frequency domain, which uses a weighted regularization approach based on the proper orthogonal decomposition. The regularization parameter is selected using the generalized cross-validation (GCV) method (HANSEN, 1998).

Another class of methods addressing the dynamic force identification is based on stochastic analysis. Unlike the previously discussed deterministic methods, the stochastic approaches take into consideration the uncertainties related to the structural modeling and measurement errors (LIU et al., 2011). HAN et al. (2012) proposed a stochastic method for reconstruction of the impact load acting on a SDOF. However, it requires a priori knowledge of the statistical distribution of the data, which is in many cases rather difficult to formulate. In order to overcome this issue, some methods based on interval analysis have been developed (JIANG et al., 2007; LIU et al., 2011; AHMARI et al., 2013). These only require the lower and upper bounds of the uncertain parameter and lead to lower and upper bounds of the desired quantities, i.e. the unknown loads, without giving any information about their distributions.

Besides the deterministic methods, which represent the vast majority of force identification approaches, and some stochastic methods, the force reconstruction problems can be also addressed by means of artificial intelligence (AI) algorithms. The AI based procedures for force reconstruction require a learning process to establish a relationship between input and output. This can be achieved by employing artificial neural-networks (GHAJARI et al., 2013), fuzzy algorithms (GÓRAL et al., 2002) or evolutionary algorithms (OBATA et al., 2006; CHEN et al., 2021).

2.4 Conclusion

In this chapter, the main features of different direct and indirect force identification techniques available in the literature are discussed. With respect to the direct methods for measurement of dynamic axle forces, it is concluded that they are not feasible for a continuous application implying an extensive number of vehicles containing a large number of axles. The reason for this is that the currently available methods require the instrumentation of each axle, in order to get a complete load pattern of the entire vehicle. The Weigh-In-Motion (WIM) and Bridge Weigh-In-Motion (B-WIM) are two methods that overcome the issue of instrumenting every axle by either using sensors installed at track level (WIM) or using a bridge structure as a scale for the passing vehicle axles (B-WIM). However, both methods are so far only able to retrieve static axle loads.

The moving force identification (MFI) is a concept that involves reconstructing the dynamic axle forces passing a bridge structure by means of dynamic measurements of structural responses. Even though the MFI is a rather new concern in the research field of bridge dynamics, several deterministic (PINKAEW, 2006; LAW et al., 2011; FENG et al., 2015; PAN et al., 2017), stochastic (WU et al., 2010; WU et al., 2012) and AI based formulations (JIANG et al., 2003; OBATA et al., 2006; WANG et al., 2011; CHEN et al., 2021) are already available in the literature. However, it can be clearly concluded that the deterministic approaches represent the vast majority of the currently existing MFI methods. They can be formulated either in the time domain (PINKAEW, 2006; LAW et al., 2011; FENG et al., 2015; PAN et al., 2017) or in the combined time and frequency domain (LAW et al., 1999; YU et al., 2003) and require to establish a linear dependency between the input and output of a system. Accelerations and strains are typical measurement quantities used in the identification process (LAW et al., 2011; ZHU et al., 2016; YU et al., 2016). Nevertheless, initial attempts were also made to use displacement measurements (DENG et al., 2010; DENG et al., 2011). The ill-posedness and ill-conditioning related to the inverse character of the MFI problems are usually suppressed using the Tikhonov regularization in various formulations (LAW et al., 2001b; ZHU et al., 2002a; PAN et al., 2017).

Despite the extensive research in the field of MFI over the last two decades, there is surprisingly little work addressing the MFI in the context of railway bridges (WANG et al., 2011; QU et al., 2011). One possible reason for this could be the fact that this type of application is usually characterized by a large number of axle loads to be reconstructed. This eventually leads to large problem dimensions that can easily become very computationally expensive, especially when the solution implies a classical Tikhonov regularization (PINKAEW, 2006; FENG et al., 2015). Furthermore, it can be concluded that – regardless of the MFI apporach – a very accurate structural model is required, in order to achieve a good accuracy of the results. However, most of the currently available MFI methods are using very simplified bridge models, such as 1D beam models or 2D orthotropic plate models (LAW et al., 2011; ZHU et al., 2016; PAN et al., 2017). These are expected to prove unsuitable for applications to certain real structures, clearly highlighting the need for new approaches that allow force reconstruction based on arbitrarily detailed 3D models. A further conclusion regarding the MFI is that a systematic experimental validation on real structures in operation is generally missing. This also implies a total lack of discussion regarding other difficulties that may arise when dealing with real practical applications.

In addition to discussing direct and indirect methods for MFI, this chapter also briefly reviews some recent developments in the field of the identification of dynamic non-moving forces. The reason for this is that this area of research has been more comprehensively studied and may reveal possible trends for future developments in relation to the MFI. For example, it is noticeable that a more comprehensive discussion regarding the regularization strategies is provided. They are either concerned with different variations of the Tikhonov regularization (JACQUELIN et al., 2003; CHOI et al., 2007; AUCEJO et al., 2017) or imply the Bayesian theory (ZHANG et al., 2012; FAURE et al., 2017; LI et al., 2019). Furthermore, the idea of a sequential solution of large problems is firstly introduced in context of non-moving forces (LIU et al., 2014a; BERNAL et al., 2015; LAI et al., 2016). In addition, several publications are concerned with the solution of the force identification problem by means of optimization algorithms instead of using the normal equations (GUNAWAN et al., 2004; HUANG et al., 2009; GUNAWAN, 2012; LI et al., 2016). This fact is closely related to the concept of dynamic response sensitivity, which was already used in context of non-moving forces (HUANG, 2001a; LU et al., 2006; HUANG et al., 2009). Another innovative approach applied for the identification of non-moving forces is the use of non-contact displacement measurements (DJAMAA et al., 2007; FENG et al., 2017). This is expected to present a great potential also for MFI applications on bridge structures. Finally, it should be mentioned that the simultaneous estimation of forces and responses at unmeasured locations has already been explored in the context of non-moving forces (GILLIJNS et al., 2007; LOURENS et al., 2012a; AZAM et al., 2015; MAES et al., 2016b). It also seems plausible that this concept will soon find its way to MFI on bridge structures.

3 Fundamentals of bridge dynamics under moving loads

3.1 Generals

There are mainly two methods for dealing with the structural analysis of bridge structures under moving loads. On the one hand, an analytical solution is provided for beam systems with uniformly distributed mass and elasticity (FRÝBA, 1972; CLOUGH et al., 1995; CHOPRA, 2014; PETERSEN et al., 2017). On the other hand, an efficient solution can be achieved using the FEM (CLOUGH et al., 1995; PETERSEN et al., 2017; WERKLE, 2021). The latter idealizes the system by a set of elements interconnected at nodal points, which are used to form a coupled system of differential equations of motion (BATHE, 2014; WERKLE, 2021). This chapter briefly introduces some relevant aspects of structural dynamics and presents typical methods of dynamic analysis of bridge structures under action of moving loads. These are necessary for the understanding of the theoretical derivations and experimental investigations presented in the further course of this thesis. The main sources for this chapter are the reference books of FRÝBA (1972), CLOUGH et al. (1995), CHOPRA (2014), PETERSEN et al. (2017) and WERKLE (2021), which contain more detailed descriptions of the topics covered here.

3.2 Analytical solution for a simply supported beam with uniformly distributed mass

3.2.1 Undamped free vibrations of a beam

The undamped free vibration of an Euler-Bernoulli beam with uniformly distributed mass $\bar{m}$ and bending stiffness EI can be described by the following partial differential equation of motion[1]:

$$EI \cdot \frac{\partial^4 u(x,t)}{\partial x^4} + \bar{m} \cdot \frac{\partial^2 u(x,t)}{\partial t^2} = 0 \quad \text{or} \quad EI \cdot u''''(x,t) + \bar{m} \cdot \ddot{u}(x,t) = 0 \qquad (3.1)$$

[1] It is to be noted that Eq. 3.1 represents an approximate formulation, since the rotational inertia and the shear deformation have been neglected. This assumption is fully justified for typical slenderness grades of beams prone to vibrations (PETERSEN et al., 2017). A more elaborate solution, including rotational inertia, shear deformation effects but also the influence of axial forces in the beam is given in PETERSEN et al. (2017).

A. Firus, *A Contribution to Moving Force Identification in Bridge Dynamics*, Mechanik, Werkstoffe und Konstruktion im Bauwesen 65, https://doi.org/10.1007/978-3-658-39838-5_3

where the prime $(\ )'$ and dot $(\dot{\ })$ notations represent the first derivative of $u(x,t)$ with respect to the length coordinate x and the time t, respectively. Considering that EI and $\bar{m}$ are constants, the separation of the variables using:

$$u(x,t) = \phi(x) \cdot Y(t) \tag{3.2}$$

is an intuitive approach for the solution of the Eq. 3.1. It indicates that the motion of the undamped beam can can be described by a specific shape $\phi(x)$ vibrating with a time dependent amplitude $Y(t)$. The substitution of Eq. 3.2 into Eq. 3.1 and dividing by $EI \cdot \phi(x) \cdot Y(t)$ yields:

$$\frac{\phi''''(x)}{\phi(x)} = -\frac{\bar{m}}{EI} \cdot \frac{\ddot{Y}(t)}{Y(t)} \tag{3.3}$$

Since the term on the left hand side only depends on x and the expression on the right hand side is a function depending solely on t, Eq. 3.3 can only be fulfilled if both sides are constant, i.e. equal with a constant value a^4 (the exponent of a^4 was chosen for later mathematical convenience). This allows the separation of the partial differential Eq. 3.3 into two ordinary differential equations of fourth and second order, respectively:

$$\phi''''(x) - a^4 \cdot \phi(x) = 0 \tag{3.4a}$$
$$\ddot{Y}(t) + \omega^2 \cdot Y(t) = 0 \tag{3.4b}$$

where

$$a^4 = \frac{\omega^2 \cdot \bar{m}}{EI} \quad \Longrightarrow \quad \omega^2 = \frac{a^4 \cdot EI}{\bar{m}} \tag{3.5}$$

The time component $Y(t)$ of Eq. 3.2 can be obtained from Eq. 3.4b using the exponential solution approach $Y(t) = \mathrm{e}^{\lambda \cdot t}$. Under consideration of the Euler relation[2], the solution can be expressed in terms of trigonometrical functions[3]:

$$Y(t) = A \cdot \cos(\omega \cdot t) + B \cdot \sin(\omega \cdot t) \tag{3.6}$$

The constants A and B have to be determined from the initial conditions $Y(t=0)$ and $\dot{Y}(t=0)$. This results in:

$$Y(t) = Y(0) \cdot \cos(\omega \cdot t) + \frac{\dot{Y}(0)}{\omega} \cdot \sin(\omega \cdot t) \tag{3.7}$$

[2] The Euler formula establishes the relation between the trigonometrical functions and the complex exponential function: $\mathrm{e}^{\mathrm{i} \cdot y} = \cos(y) + \mathrm{i} \cdot \sin(y)$.

[3] Eq. 3.4b looks similar to the equation of motion describing the free vibration of an undamped SDOF system. The detailed derivation of its solution is given in CLOUGH et al. (1995), CHOPRA (2014) and PETERSEN et al. (2017).

Eq. 3.4a can be solved as well using an exponential solution approach: $\phi(x) = e^{\lambda \cdot x}$. Considering the trigonometric and hyperbolic equivalents of the exponential functions, this leads to the following general solution (CLOUGH et al., 1995; PETERSEN et al., 2017):

$$\phi(x) = A_1 \cdot \cos(ax) + A_2 \cdot \sin(ax) + A_3 \cdot \cosh(ax) + A_4 \cdot \sinh(ax) \tag{3.8}$$

While the constants A and B for the solution of the time dependent differential Eq. 3.6 are determined from the initial conditions, the real constants A_1 to A_4 in Eq. 3.8 have to be determined by considering the boundary conditions (displacement, slope, moment or shear) at the two ends of the beam. This allows the expression of any three constants in terms of the fourth, which defines the arbitrary amplitude of the shape function $\phi(x)$. This leads to the so called frequency equation, which provides a solution for a and thus – under consideration of Eq. 3.5 – the angular frequency ω.

Considering the present case of an uniform simply supported beam of length l, the boundary conditions can be expressed as:

$$\phi(0) = 0; \quad M(0) = -EI \cdot \phi''(0) = 0; \quad \phi(l) = 0; \quad M(l) = -EI \cdot \phi''(l) = 0; \tag{3.9}$$

By introducing them into Eq. 3.8 and its derivatives and performing a set of mathematical operations, the following frequency equation is obtained:

$$A_2 \cdot \sin(a \cdot l) = 0 \tag{3.10}$$

The nontrivial solution of Eq. 3.10 is given by

$$a = \frac{j \cdot \pi}{l} \qquad j = 1, 2, 3, \ldots \tag{3.11}$$

The angular frequency can be obtained by substituting Eq. 3.11 into Eq. 3.5, yielding:

$$\omega_j = \frac{j^2 \cdot \pi^2}{l^2} \sqrt{\frac{EI}{\bar{m}}} \tag{3.12}$$

By dividing this equation by 2π, the natural frequency f_j of the beam is obtained:

$$f_j = \frac{\omega_j}{2 \cdot \pi} = \frac{j^2 \cdot \pi}{2 \cdot l^2} \sqrt{\frac{EI}{\bar{m}}} \tag{3.13}$$

Finally, the natural mode of vibration $\phi_j(x)$, corresponding to the j-th angular frequency ω_j, is obtained by substituting Eq. 3.11 into Eq. 3.8 with $A_1 = A_3 = A_4 = 0$:

$$\phi_j(x) = A_2 \cdot \sin\left(\frac{j \cdot \pi \cdot x}{l}\right) \tag{3.14}$$

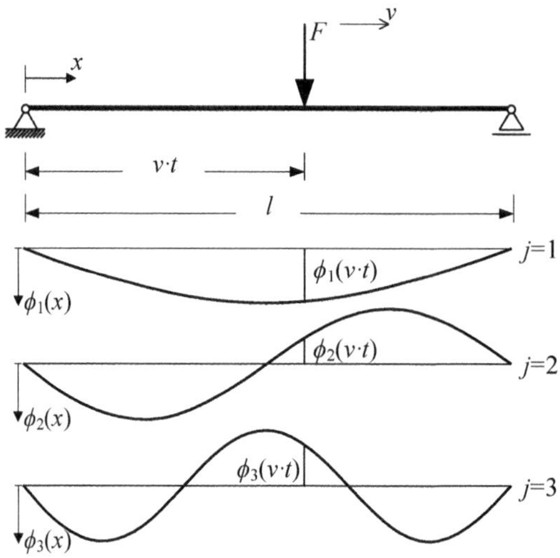

Figure 3.1 Mode shapes of a simply supported beam under action of a single moving load

The constant A_2 may take arbitrary values[4]. For example, $A_2 = 1$ ensures that the maximum of each mode shape $\phi_j(x)$ is equal to unity. It is obvious that for the simply supported beam, an infinite series of mode shapes – each with its own vibration frequency – can be determined. The first three modes are illustrated in Fig. 3.1.

3.2.2 Damped vibrations of a beam subjected to moving loads

In order to address the more general case of a damped system subjected to an external time and location dependent load, Eq. 3.1 is extended by a velocity proportional (viscous) damping term and a force term $q(x,t)$:

$$EI \cdot \frac{\partial^4 u(x,t)}{\partial x^4} + \bar{m} \cdot \frac{\partial^2 u(x,t)}{\partial t^2} + c \cdot \frac{\partial u(x,t)}{\partial t} = q(x,t) \tag{3.15}$$

While the mass and stiffness of a structure can be evaluated with reasonable effort based on certain physical considerations, the accurate determination of the damping coefficient is only possible in rather isolated cases. The reason for this is that the energy-loss mechanisms of most practical systems are not yet fully understood (CLOUGH et al., 1995). Even though they are likely to be more complicated that the simple assumption of a velocity proportional damping, this approach is usually adopted because it allows a convenient formulation of

[4] It is to be mentioned that the initial conditions $Y(0)$ and $\dot{Y}(0)$ in Eq. 3.7 have to be consistent with the arbitrary chosen amplitude A_2 of the shape function $\phi(x)$ in Eq. 3.14. This is achieved by ensuring that Eq. 3.2 is satisfied also in the initial point, i.e. $Y(0) = u(x,0)/\phi(x)$ and $\dot{Y}(0) = \dot{u}(x,0)/\phi(x)$.

the equation of motion. In this sense, the effects of all damping mechanisms acting within a system are comprised under an equivalent viscous-damping, which can be experimentally determined with reasonable effort (CLOUGH et al., 1995; PETERSEN et al., 2017). It is usually expressed using the damping coefficient c or the damping ratio ξ, which is defined as the ratio between the given damping and the critical value[5]:

$$\xi = \frac{c}{c_{\text{cr}}} = \frac{c}{2 \cdot \bar{m} \cdot \omega} \quad \Longrightarrow \quad c = 2 \cdot \xi \cdot \bar{m} \cdot \omega \tag{3.16}$$

The equation of motion for a single constant force F moving over the bridge structure with a constant velocity v (Fig. 3.1) reads:

$$EI \cdot \frac{\partial^4 u(x,t)}{\partial x^4} + \bar{m} \cdot \frac{\partial^2 u(x,t)}{\partial t^2} + c \cdot \frac{\partial u(x,t)}{\partial t} = \begin{cases} \delta(x - v \cdot t) \cdot F & \text{for } 0 \leq t \leq T_{\text{u}} \\ 0 & \text{for } t \geq T_{\text{u}} \end{cases} \tag{3.17}$$

where $\delta(x - v \cdot t)$ is the Dirac delta function centered at $x = v \cdot t$. It represents a mathematical description of the moving concentrated force. $T_{\text{u}} = l/v$ is the time needed by the force to cross the bridge.

The solution $u(x,t)$ of Eq. 3.17 can be determined with reasonable effort based on the modal superposition principle, which implies that the total response of a system can be expressed as a linear combination of all its modal components[6]:

$$u(x,t) = \sum_{j=1}^{\infty} \alpha_j(t) \cdot \phi_j(x) \tag{3.18}$$

here $\alpha_j(t)$ represents the time history of the modal coordinate for the j-th mode. Without presenting a detailed derivation, the expression of $u_{\text{F}}(x,t)$ during the passage of the constant force ($0 \leq t \leq T_{\text{u}}$) is defined as (FRÝBA, 1972; PETERSEN et al., 2017):

$$\begin{aligned} u_{\text{F}}(x,t) = \sum_{j=1}^{\infty} & \frac{(2 \cdot F)/(\bar{m} \cdot l)}{\left(\omega_j^2 - \Omega_j^2\right)^2 + (2 \cdot \xi_j \cdot \omega_j \cdot \Omega_j)^2} \cdot \Big\{ \left(\omega_j^2 - \Omega_j^2\right) \cdot \sin(\Omega_j \cdot t) - \\ & 2 \cdot \xi_j \cdot \omega_j \cdot \Omega_j \cdot \cos(\Omega_j \cdot t) + e^{-\xi_j \cdot \omega_j \cdot t} \cdot \Big[\left(2 \cdot \xi_j^2 \cdot \omega_j^2 - (\omega_j^2 - \Omega_j^2)\right) \cdot \\ & \frac{\Omega_j}{\omega_{\text{d},j}} \cdot \sin(\omega_{\text{d},j} \cdot t) + 2 \cdot \xi_j \cdot \omega_j \cdot \Omega_j \cdot \cos(\omega_{\text{d},j} \cdot t) \Big] \Big\} \cdot \sin\left(j \cdot \pi \cdot \frac{x}{l}\right) \end{aligned} \tag{3.19}$$

[5] The critical damping is defined as the smallest amount of damping at which no oscillation about the zero deflection position occurs in the free-vibration response (CLOUGH et al., 1995). For a system with distributed mass, it can be expressed as $c_{\text{cr}} = 2 \cdot \bar{m} \cdot \omega$.

[6] The general principle of modal superposition will be presented in detail in Section 3.3.

where $\Omega_j = j \cdot \pi \cdot v/l$ and $\omega_{\mathrm{d},j} = \omega_j \cdot \sqrt{1-\xi_j^2}$, the latter representing the j-th damped angular frequency of the bridge. A comparison between Eq. 3.19 and Eq. 3.18 reveals that the last sine term of Eq. 3.19 corresponds to the term $\phi_j(x)$ in Eq. 3.18, while the remaining terms from the summation expression in Eq. 3.19 depict the time history of the modal coordinate $\alpha_j(t)$.

After the force leaves the bridge, the system undergoes free vibration. The corresponding response $u_{\mathrm{D}}(x,t)$, assigned to the decay phase ($t \geq T_{\mathrm{u}}$), can be calculated by superposing the modal components of the damped free vibration:

$$u_{\mathrm{D}}(x,t) = \sum_{j=1}^{\infty} \mathrm{e}^{-\xi_j \cdot \omega_j \cdot (t-T_{\mathrm{u}})} \cdot \Big[A_j \cdot \cos\left(\omega_{\mathrm{d},j} \cdot (t-T_{\mathrm{u}})\right) + B_j \cdot \sin\left(\omega_{\mathrm{d},j} \cdot (t-T_{\mathrm{u}})\right)\Big] \cdot \sin\left(j \cdot \pi \cdot \frac{x}{l}\right) \tag{3.20}$$

where the real constants A_j and B_j for each mode j are determined from the initial conditions resulting from the evaluation of Eq. 3.19 at $t = T_{\mathrm{u}}$. In terms of total displacements this means that $u_{\mathrm{D}}(x,T_{\mathrm{u}}) = u_{\mathrm{F}}(x,T_{\mathrm{u}})$ and $\dot{u}_{\mathrm{D}}(x,T_{\mathrm{u}}) = \dot{u}_{\mathrm{F}}(x,T_{\mathrm{u}})$. The initial conditions for the modal displacements and velocities can be determined in a similar way: $\alpha_{\mathrm{D},j}(T_{\mathrm{u}}) = \alpha_{\mathrm{F},j}(T_{\mathrm{u}})$ and $\dot{\alpha}_{\mathrm{D},j}(T_{\mathrm{u}}) = \dot{\alpha}_{\mathrm{F},j}(T_{\mathrm{u}})$. The total expression for the displacement during the decay phase is:

$$u_{\mathrm{D}}(x,t) = \sum_{j=1}^{\infty} \mathrm{e}^{-\xi_j \cdot \omega_j \cdot (t-T_{\mathrm{u}})} \cdot \Big[\alpha_{\mathrm{F},j}(T_{\mathrm{u}}) \cdot \cos\left(\omega_{\mathrm{d},j} \cdot (t-T_{\mathrm{u}})\right) + \frac{\xi_j \cdot \omega_j \cdot \alpha_{\mathrm{F},j}(T_{\mathrm{u}}) + \dot{\alpha}_{\mathrm{F},j}(T_{\mathrm{u}})}{\omega_{\mathrm{d},j}} \cdot \sin\left(\omega_{\mathrm{d},j} \cdot (t-T_{\mathrm{u}})\right)\Big] \cdot \sin\left(j \cdot \pi \cdot \frac{x}{l}\right) \tag{3.21}$$

The total solution $u(x,t)$ of Eq. 3.17 can be summarized as follows:

$$u(x,t) = \begin{cases} u_{\mathrm{F}}(x,t) & \text{acc. to Eq. 3.19 for } 0 \leq t \leq T_{\mathrm{u}} \\ u_{\mathrm{D}}(x,t) & \text{acc. to Eq. 3.21 for } t \geq T_{\mathrm{u}} \end{cases} \tag{3.22}$$

The solution for a set of n_{F} successive forces F_i – as a typical load case for a railway vehicle (Fig. 3.2) – can be obtained in different ways. FRÝBA (2001) gives a reformulation of Eq. 3.17, where the load term on the right hand side is extended for n_{F} forces using the Dirac and Heaviside functions. The corresponding closed analytical solution is determined based on the Fourier and Laplace transformations. However, a more intuitive way for determining the structural response for a set of constant forces passing a linear system implies the evaluation of Eq. 3.22 for a unit force (i.e. $F = 1$ in Eq. 3.19). By scaling it according to each load value F_i and superposing the results of all forces in time with a delay corresponding to the time t_i needed by the force F_i to pass the distance d_i between the forces F_1 and F_i ($t_i = d_i/v$), the exact solution is obtained (PIRCHER et al., 2009).

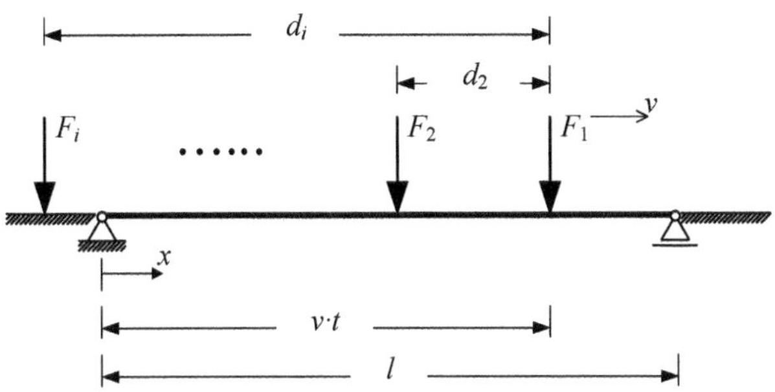

Figure 3.2 Simply supported beam under action of a set of moving loads

3.3 Finite Element solution for dynamic analysis

A powerful alternative to the analytical solutions presented in the previous section is the modeling of the structure using the finite element method (FEM). In the context of bridges under moving loads, it implies the modeling of the structure as a discrete system with a finite number of DOFs. This system type is denoted in the following as MDOF system. In order to facilitate the comparison to the analytical solutions given in Section 3.2, the analysis of MDOF systems will be introduced using the example of a simply supported beam. However, it is to be mentioned that all the following equations are valid without restrictions for any type of structure.

3.3.1 Equations of motion

The first step of a finite element analysis (FEA) consists in the discretization of the system into an appropriate number of elements, which are connected to each other in nodes (Fig. 3.3). By evaluating the properties of each individual element separately and assembling them in an appropriate way, the properties of the complete system can be determined. This discretization transforms the system with continuously distributed mass and an infinite number of degrees of freedom (described by Eq. 3.15) into a MDOF system with a finite number n_{dof} of DOFs. Its linear dynamic response is governed by a system of n_{dof} linear differential equations (of motion) of second order:

$$\boldsymbol{K} \cdot \boldsymbol{u}(t) + \boldsymbol{C} \cdot \dot{\boldsymbol{u}}(t) + \boldsymbol{M} \cdot \ddot{\boldsymbol{u}}(t) = \boldsymbol{P}(t) \tag{3.23}$$

where $\boldsymbol{K}^{n_{\text{dof}} \times n_{\text{dof}}}$, $\boldsymbol{C}^{n_{\text{dof}} \times n_{\text{dof}}}$ and $\boldsymbol{M}^{n_{\text{dof}} \times n_{\text{dof}}}$ are the stiffness, damping and mass matrices, respectively; $\boldsymbol{P}(t)^{n_{\text{dof}} \times 1}$ is the time dependent vector of external loads and $\boldsymbol{u}(t)^{n_{\text{dof}} \times 1}$, $\dot{\boldsymbol{u}}(t)^{n_{\text{dof}} \times 1}$ and $\ddot{\boldsymbol{u}}(t)^{n_{\text{dof}} \times 1}$ are the time dependent vectors of nodal displacement, velocity and acceleration of the FE system.

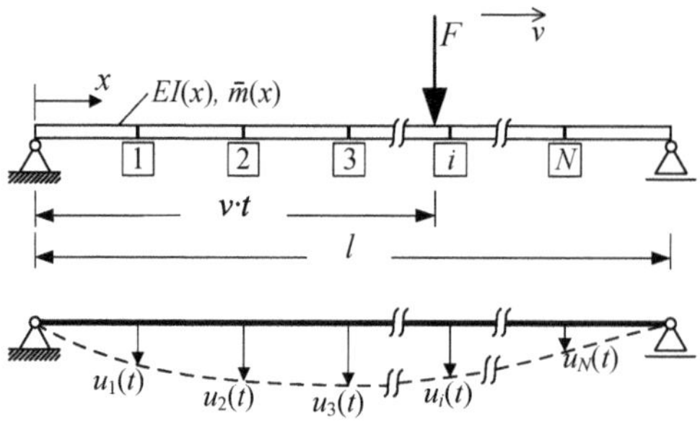

Figure 3.3 Discretization of a beam structure under action of a moving load

Stiffness matrix Without going into into details, it is mentioned that the system stiffness matrix $\boldsymbol{K}$ is determined by assembling all element stiffness matrices $\boldsymbol{K}_{\mathrm{e}}^{(j)}$ (with j being the element index). These are generally computed based on variational principles, using a set of shape functions that approximate the displacements within the element (ZIENKIEWICZ, 2005; BATHE, 2014; WERKLE, 2021). It is to be noted that the analytical solutions of the equilibrium equations formulated for a Bernoulli beam can be used to determine the exact element stiffness matrices of beam structures[7]. In this case, the FE approach is in fact a matrix form of the displacement method of analysis of beam and trusses, also known as the direct stiffness method. For this reason, the term FEA was originally used only in relation with plate and shell structures (BATHE, 2014; WERKLE, 2021). Unlike the analysis of beam structures, the solution of the FEA implying plate and shell elements has only an approximate character, since the derivation of the elements is based solely on shape functions, which can only approximate the exact solution. A detailed discussion on the derivation of different element types would go far beyond the scope of this thesis. In this sense, the author refers to the reference books of ZIENKIEWICZ (2005), BATHE (2014) and WERKLE (2021).

Mass matrix A rigorous derivation of the mass matrix $\boldsymbol{M}$ implies a similar procedure as used in the derivation of the stiffness matrix. In order to calculate the element mass matrix, the acceleration within an element is approximated using the same shape functions considered for the derivation of element stiffness matrices (BATHE, 2014; PETERSEN et al., 2017; WERKLE, 2021). The system mass matrix can be subsequently assembled from all the element mass matrices $\boldsymbol{M}_{\mathrm{e}}^{(j)}$. This type of formulation leads to the so called *consistent* mass matrix, which is – like the stiffness matrix – symmetric. However, a much simpler form of the mass matrix, denoted as the *lumped* mass matrix, is used in the most practical

[7] It is to be mentioned, that the general variational approach can be also used to calculate the stiffness matrices of beam structures. This approach leads to exact solution, only if the exact internal displacements are adopted as shape functions (BATHE, 2014).

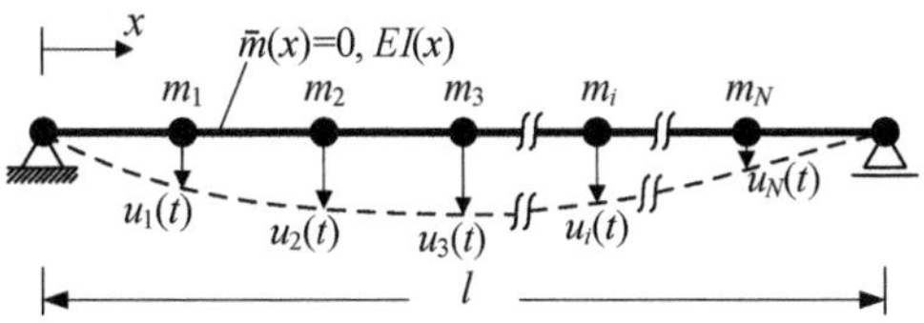

Figure 3.4 Assumption of mass concentration for a diagonal mass matrix

applications. It implies the concentration of the masses at the nodes of the discretized structure (Fig. 3.4). The mass at each node is the sum of all contributions from the elements connected to that node. For example, in case of two neighboring beam elements of equal length l_e, connected at a node i and having an uniformly distributed mass per unit length $\bar{m}$, the nodal mass is $m_i = \bar{m} \cdot l_e$. In a more general case of an arbitrarily distributed mass, the corresponding nodal masses can be computed by means of static resultants (CLOUGH et al., 1995; CHOPRA, 2014).

The lumped mass matrix of a system is diagonal, since the mass components are uncoupled, i.e. an acceleration of any mass point only produces an acceleration at that point. If a node possesses more than one translational DOF, the same nodal mass will be assigned to all DOFs. The mass associated with the rotational DOFs is usually considered to be zero, since the basic assumption is that the mass is lumped in points, which possess no rotational inertia (CLOUGH et al., 1995). The main advantage of the lumped mass matrix approach in comparison to the consistent formulation is the significantly lower computational effort (CLOUGH et al., 1995). On the one hand, this occurs due to the vanishing of the mass coupling (off-diagonal zero terms) and on the other side due to the possibility of eliminating the degrees of freedom with zero mass by static condensation (explained later). For example, by condensing out the horizontal and rotational DOFs of the simply supported beam in Fig. 3.4, the following diagonal lumped mass matrix, consisting of only translational DOFs (i.e. $N = n_{\text{dof}}$), is obtained:

$$M = \begin{pmatrix} m_1 & 0 & 0 & \cdots & 0 & \cdots & 0 \\ 0 & m_2 & 0 & \cdots & 0 & \cdots & 0 \\ 0 & 0 & m_3 & \cdots & 0 & \cdots & 0 \\ \vdots & \vdots & \vdots & \ddots & \ddots & \cdots & 0 \\ 0 & 0 & 0 & \cdots & m_i & \cdots & 0 \\ \vdots & \vdots & \vdots & \ddots & \ddots & \ddots & \vdots \\ 0 & 0 & 0 & \cdots & 0 & \cdots & m_N \end{pmatrix} \tag{3.24}$$

Static condensation It has already been mentioned that the lumped mass matrix usually only includes non-zero entries for the translational DOFs. However, all the DOFs are taken into account in the definition of the global stiffness matrix. In order to overcome this incompatibility and to reduce the size of the system of equations, the static condensation of the stiffness matrix can be applied. It represents a procedure to eliminate "unwanted" degrees of freedom from the equilibrium equations. In case of dynamic analysis, the static condensation is usually adopted to eliminate the DOFs, to which a zero mass was assigned. These are mainly the rotational DOFs, which are expected to have a much lower influence in the dynamic analysis of typical (slender) beam structures than the translational components, especially in the low vibration modes (CLOUGH et al., 1995; CHOPRA, 2014; PETERSEN et al., 2017; WERKLE, 2021). In the special case of the beam structure illustrated in Fig. 3.4, also the horizontal DOFs (in x-direction) may be eliminated. The reason for this is that a vertical load moving over the structure does not generate inertial forces in x-direction, justifying thus the assumption made in Eq. 3.24.

The static condensation implies in a first step the rearrangement of Eq. 3.23 – neglecting the damping term – in the following form:

$$\begin{pmatrix} \boldsymbol{K}_{\text{tt}} & \boldsymbol{K}_{\text{t0}} \\ \boldsymbol{K}_{\text{0t}} & \boldsymbol{K}_{\text{00}} \end{pmatrix} \cdot \begin{pmatrix} \boldsymbol{u}_{\text{t}} \\ \boldsymbol{u}_{0} \end{pmatrix} + \begin{pmatrix} \boldsymbol{M}_{\text{tt}} & \boldsymbol{0} \\ \boldsymbol{0} & \boldsymbol{0} \end{pmatrix} \cdot \begin{pmatrix} \ddot{\boldsymbol{u}}_{\text{t}} \\ \ddot{\boldsymbol{u}}_{0} \end{pmatrix} = \begin{pmatrix} \boldsymbol{P}_{\text{t}}(t) \\ \boldsymbol{0} \end{pmatrix} \tag{3.25}$$

which presents a subdivided displacement $\boldsymbol{u}$: $\boldsymbol{u}_{\text{t}}$ depicts the DOFs which have mass[8] and $\boldsymbol{u}_0$ the DOFs with zero mass, i.e. the "unwanted" DOFs. The mass and stiffness matrices and the force vector were subdivided in a similar way (index "t" or "tt" for DOFs with mass and "0" or "00" for the DOFs without mass). The stiffness matrix includes also the coupling terms $\boldsymbol{K}_{\text{t0}}$ and $\boldsymbol{K}_{\text{0t}}$ with $\boldsymbol{K}_{\text{t0}}^{\top} = \boldsymbol{K}_{\text{0t}}$. Subsequently, $\boldsymbol{u}_0$ can be expressed in terms of $\boldsymbol{u}_{\text{t}}$ from a static consideration, since the DOFs associated with $\boldsymbol{u}_0$ present no inertial forces and are not subjected to external forces:

$$\boldsymbol{u}_0 = \boldsymbol{K}_{00}^{-1} \cdot \boldsymbol{K}_{0\text{t}} \cdot \boldsymbol{u}_{\text{t}} \tag{3.26}$$

By substituting Eq. 3.26 in the first matrix equation in Eq. 3.25, the following equation – expressed only in terms of $\boldsymbol{u}_{\text{t}}$ – can be obtained:

$$\hat{\boldsymbol{K}}_{\text{tt}} \cdot \boldsymbol{u}_{\text{t}} + \boldsymbol{M}_{\text{tt}} \cdot \ddot{\boldsymbol{u}}_{\text{t}} = \boldsymbol{P}_{\text{t}}(t) \tag{3.27}$$

with the condensed stiffness matrix $\hat{\boldsymbol{K}}_{\text{tt}}$ being expressed as:

$$\hat{\boldsymbol{K}}_{\text{tt}} = \boldsymbol{K}_{\text{tt}} - \boldsymbol{K}_{0\text{t}}^{\top} \cdot \boldsymbol{K}_{00}^{-1} \cdot \boldsymbol{K}_{0\text{t}} \tag{3.28}$$

Based on the relation between the flexibility and the stiffness matrix, a more practical method for performing the static condensation can be used. In the first step, it implies the definition of separate static load cases of unit forces acting at the DOFs, for which the

[8] In some literature sources, the DOFs to which a nonzero mass was assigned, are denoted as dynamic DOFs.

stiffness matrix has to be determined. In the dynamic analysis, these are usually the DOFs occupied by mass, comprised in the vector $\boldsymbol{u}_{\mathrm{t}}$. In a second step, the resulting displacements are computed. They represent the flexibility values, which can be merged column-wise into the flexibility matrix $\boldsymbol{F}_{\mathrm{flex}}$. By inversion of the flexibility matrix, the condensed stiffness matrix can be obtained.

Force vector Eq. 3.23 represents the kinetic equilibrium of the system at each time instant. While the system matrices $\boldsymbol{K}$, $\boldsymbol{C}$ and $\boldsymbol{M}$ are assumed – due to the linearity of the system – to remain constant during the computation, the load vector $\boldsymbol{P}(t)$ has to be updated at each time instant t_i. Since the size of the finite elements is usually much higher than the distance passed by the force within a time step, special attention has to be drawn to the distribution of the moving load F acting between the FE nodes. In case of a structure idealized using beam elements, the moving force acting within one element is distributed to the two connecting nodes j and k. This results in two nodal forces $F_j(t_i)$ at node j and $F_k(t_i)$ at node k. Based on the principle of static resultants, they can be computed as:

$$F_j(t_i) = \frac{d_{\mathrm{F},k}}{d_{\mathrm{F},j} + d_{\mathrm{F},k}} \cdot F(t_i) \quad \text{and} \quad F_k(t_i) = \frac{d_{\mathrm{F},j}}{d_{\mathrm{F},j} + d_{\mathrm{F},k}} \cdot F(t_i) \tag{3.29}$$

where $d_{\mathrm{F},j}$ and $d_{\mathrm{F},k}$ are the distances between the force position and the neighboring nodes j and k. $F_j(t_i)$ and $F_k(t_i)$ have to be integrated in the load vector $\boldsymbol{P}(t_i)$, in the entries corresponding to the nodes j and k. It is evident that this approach only considers nodal forces in the translational DOFs. However, the errors should be small in case of a fine discretization.

The nodal forces for arbitrary load positions within a finite element can be determined also by means of the FE concept. In a similar way to the derivation of the stiffness and consistent mass matrices, the principle of virtual displacements can be used for the determination of the *consistent* nodal loads. Since their presentation is not particulary relevant for the scope of this thesis, the author only refers to the reference works of CLOUGH et al. (1995), BATHE (2014) and WERKLE (2021).

Damping matrix As mentioned above, the rigorous formulation of the mass and stiffness matrices is based on variational principles using a set of shape functions. They describe the variation of the displacement (for the stiffness matrix) and acceleration (for the mass matrix) within an element. Assuming that the damping properties of a structure could be determined accurately, the same FE concept could be used for the definition of the damping matrix $\boldsymbol{C}$. This would imply the formulation of the element damping matrix and the subsequent assemblage to the system damping matrix. However, while the constants needed for the definition of the stiffness and mass matrices (Young's modulus E and the mass distribution $\bar{m}$) can be determined with sufficient accuracy, the explicit evaluation of damping properties (damping coefficient c) is not possible in most practical applications (CLOUGH et al., 1995;

WERKLE, 2021). Therefore, the damping of MDOF systems is usually also expressed in terms of damping ratios, which account for all energy-dissipating mechanisms and can be determined through experiments on similar structures (CLOUGH et al., 1995; PETERSEN et al., 2017). This concept was already briefly introduced in Section 3.2.2 for a beam structure with distributed mass.

The solution of Eq. 3.23 can be generally addressed by two approaches: the modal superposition and the direct integration. They will be briefly presented in Sections 3.3.3 and 3.3.4. Due to the particularities of each method with respect to the formulation of the damping term $\boldsymbol{C} \cdot \dot{\boldsymbol{u}}(t)$, this aspect will be separately addressed for each method.

3.3.2 Natural frequencies and mode shapes

The free undamped vibration of a MDOF system is governed by the following system of equations, which is obtained by omitting the damping and force terms in Eq. 3.23:

$$\boldsymbol{K} \cdot \boldsymbol{u}(t) + \boldsymbol{M} \cdot \ddot{\boldsymbol{u}}(t) = \boldsymbol{0} \tag{3.30}$$

Under assumption of a harmonic free-vibration motion, a sinusoidal time history is adopted as solution for the equation system given in Eq. 3.30:

$$\boldsymbol{u}(t) = \boldsymbol{\phi} \cdot \sin(\omega \cdot t + \varphi) \tag{3.31}$$

where φ is a phase angle and $\boldsymbol{\phi}$ represents the shape of the system, which remains constant over time. Only the amplitude will vary. The acceleration can be expressed as second derivative with respect to time of Eq. 3.31:

$$\ddot{\boldsymbol{u}}(t) = -\omega^2 \cdot \boldsymbol{\phi} \cdot \sin(\omega \cdot t + \varphi) = -\omega^2 \cdot \boldsymbol{u}(t) \tag{3.32}$$

Substituting Eq. 3.31 and Eq. 3.32 into Eq. 3.30 yields:

$$\left(\boldsymbol{K} - \omega^2 \cdot \boldsymbol{M}\right) \cdot \boldsymbol{\phi} = \boldsymbol{0} \tag{3.33}$$

Eq. 3.33 represents the *generalized eigenproblem* for the eigenvalues ω^2 and the eigenvectors $\boldsymbol{\phi}$. In the physical sense, the eigenvalues represent the squared angular free-vibration frequencies and the eigenvectors describe the mode shapes of the system. Since $\boldsymbol{M}$ and $\boldsymbol{K}$ are real, positive definite and symmetric matrices, the eigenvalues will adopt real and positive values (CLOUGH et al., 1995; BATHE, 2014; WERKLE, 2021).

Obviously, Eq. 3.33 always presents a trivial solution $\boldsymbol{\phi} = \boldsymbol{0}$. However, this is not of interest, since it implies no motion of the system. Nontrivial solutions are only given for:

$$\det\left[\boldsymbol{K} - \omega^2 \cdot \boldsymbol{M}\right] = 0 \tag{3.34}$$

By expanding Eq. 3.34, a polynomial of order n_{dof} is obtained, widely denoted as the *characteristic* polynomial of the eigenvalue problem. Solving the eigenproblem in Eq. 3.33 is equivalent to finding the roots ω^2 of the characteristic polynomial. This is usually done by means of iterative numerical methods. Overviews of typical algorithms for solving eigenproblems and discussions on their performance can be found in CLOUGH et al. (1995), BATHE (2014), CHOPRA (2014), PETERSEN et al. (2017) and WERKLE (2021).

The solution of the eigenproblem delivers a total number of natural frequencies n_{mod}, which is equal to the number of DOFs occupied by mass. If the DOFs without mass are eliminated prior to the solution of the eigenproblem (e.g. using static condensation), the number of natural frequencies equals the number of DOFs of the system matrices: $n_{\text{mod}} = n_{\text{dof}}$. The n_{mod} angular frequencies ω_j $(j = 1, 2, ..., n_{\text{mod}})$ are arranged in ascending order: $\omega_1 < \omega_2 < ... < \omega_j < ... < \omega_{n_{\text{mod}}}$. The corresponding natural frequencies f_j can be computed through division of the angular frequencies by 2π (cf. Eq. 3.13). Each natural frequency f_j has a corresponding mode shape $\boldsymbol{\phi}_j$. This can be obtained as solution of the system of equations that results by introducing the computed ω_j values into Eq. 3.33:

$$\left(\boldsymbol{K} - \omega_j^2 \cdot \boldsymbol{M}\right) \cdot \boldsymbol{\phi}_j = \boldsymbol{0} \tag{3.35}$$

where the mode shape $\boldsymbol{\phi}_j$ consists of n_{dof} entries $\phi_{i,j}$, representing the mode shape component of the j-th mode for the i-th degree of freedom:

$$\boldsymbol{\phi}_j = \begin{pmatrix} \phi_{1,j} \\ \phi_{2,j} \\ \vdots \\ \phi_{i,j} \\ \vdots \\ \phi_{n_{\text{dof}},j} \end{pmatrix} \tag{3.36}$$

By imposing the condition of a zero determinant to the coefficient matrix of the homogeneous system of linear equations in Eq. 3.33, no unique solution can be achieved[9]. This means that the amplitude of the mode shape $\boldsymbol{\phi}_j$ cannot be obtained explicitly. However, its shape can be determined by expressing all components $\phi_{i,j}$ in terms of a freely chosen one, which can be assumed to take an arbitrary value. This leads practically to an infinite number of possible solutions, only defined up to a scalar multiplication factor.

Without presenting a detailed derivation, it is mentioned that the mode shapes fulfill the orthogonality conditions with respect to the mass and the stiffness matrix[10]. This is an essential property used in the modal superposition procedure (Section 3.3.3):

[9] A homogeneous system of linear equations may possess either the trivial solution or a solution set with an infinite number of solutions. The latter case occurs for a singular coefficient matrix, i.e. a zero determinant.

[10] The orthogonality of the eigenvectors with respect to the mass and stiffness matrix is widely denoted as $\boldsymbol{M}$-orthogonality and $\boldsymbol{K}$-orthogonality, respectively.

$$\begin{aligned}
&\phi_i^\top \cdot \boldsymbol{K} \cdot \phi_j = 0 \quad \text{and} \quad \phi_i^\top \cdot \boldsymbol{M} \cdot \phi_j = 0 \quad \text{for} \quad i \neq j, \quad i,j = 1,2,...,n_{\text{mod}} \\
&\phi_i^\top \cdot \boldsymbol{K} \cdot \phi_j \neq 0 \quad \text{and} \quad \phi_i^\top \cdot \boldsymbol{M} \cdot \phi_j \neq 0 \quad \text{for} \quad i = j, \quad i,j = 1,2,...,n_{\text{mod}}
\end{aligned} \tag{3.37}$$

Since the mode shapes are defined up to a constant multiplier and describe thus only the shape of the vibration and not its amplitude, they can be normalized in different ways. One possibility is to set the absolute maximum value of each mode shape to 1. A further, widely used normalization approach is formulated under consideration of the orthogonality conditions (Eq. 3.37) as:

$$\phi_j^\top \cdot \boldsymbol{M} \cdot \phi_j = 1 \qquad \Longrightarrow \qquad \phi_j^\top \cdot \boldsymbol{K} \cdot \phi_j = \omega_j^2 \tag{3.38}$$

This normalization presents certain computational advantages within the analysis of the vibration response using the modal superposition method.

3.3.3 Solution using mode superposition

The modal superposition (also denoted as modal analysis) is a powerful method for computing the vibration response of a MDOF system subjected to time dependent loads. It is based on the fact that the displaced position of the system – defined by the n_{dof} entries of the vector $\boldsymbol{u}(t)$ – can be expressed as a linear combination of the n_{mod} mode shapes (with $n_{\text{mod}} \leq n_{\text{dof}}$), i.e. as a sum (superposition) of all modal contributions:

$$\boldsymbol{u}(t) = \sum_{i=1}^{n_{\text{mod}}} \alpha_i(t) \cdot \phi_i \tag{3.39}$$

where $\alpha_i(t)$ represents the i-th modal (or generalized) coordinate ($i = 1,2,...,n_{\text{mod}}$). From the mathematical point of view, Eq. 3.39 can be understood as a change of basis from the basis of geometric coordinates of the FE model to the basis consisting of the mode shapes. It is obvious that the modal superposition principle implies known natural frequencies and mode shapes (cf. Section 3.3.2).

Substituting Eq. 3.39 and its first and second time derivatives into Eq. 3.23 gives:

$$\sum_{i=1}^{n_{\text{mod}}} \boldsymbol{K} \cdot \phi_i \cdot \alpha_i(t) + \sum_{i=1}^{n_{\text{mod}}} \boldsymbol{C} \cdot \phi_i \cdot \dot{\alpha}_i(t) + \sum_{i=1}^{n_{\text{mod}}} \boldsymbol{M} \cdot \phi_i \cdot \ddot{\alpha}_i(t) = \boldsymbol{P}(t) \tag{3.40}$$

Premultiplying each component of Eq. 3.40 with a transposed mode shape $\phi_j^\top$ yields:

$$\sum_{i=1}^{n_{\text{mod}}} \phi_j^\top \cdot \boldsymbol{K} \cdot \phi_i \cdot \alpha_i(t) + \sum_{i=1}^{n_{\text{mod}}} \phi_j^\top \cdot \boldsymbol{C} \cdot \phi_i \cdot \dot{\alpha}_i(t) + \sum_{i=1}^{n_{\text{mod}}} \phi_j^\top \cdot \boldsymbol{M} \cdot \phi_i \cdot \ddot{\alpha}_i(t) = \phi_j^\top \cdot \boldsymbol{P}(t) \tag{3.41}$$

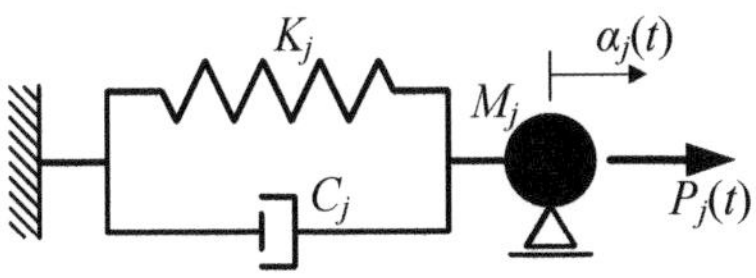

Figure 3.5 Equivalent (generalized) SDOF system for the j-th vibration mode

Under consideration of the orthogonality conditions given in Eq. 3.37, Eq. 3.41 becomes:

$$\phi_j^\top \cdot \boldsymbol{K} \cdot \phi_j \cdot \alpha_j(t) + \sum_{i=1}^{n_{\text{mod}}} \phi_j^\top \cdot \boldsymbol{C} \cdot \phi_i \cdot \dot{\alpha}_i(t) + \phi_j^\top \cdot \boldsymbol{M} \cdot \phi_j \cdot \ddot{\alpha}_j(t) = \phi_j^\top \cdot \boldsymbol{P}(t) \quad (3.42)$$

Assuming that the system is classically (proportionally) damped[11], the orthogonality conditions are valid also with respect to the damping matrix:

$$\begin{aligned} \phi_i^\top \cdot \boldsymbol{C} \cdot \phi_j &= 0 \quad \text{for} \quad i \neq j, \quad i,j = 1,2,...,n_{\text{mod}} \\ \phi_i^\top \cdot \boldsymbol{C} \cdot \phi_j &\neq 0 \quad \text{for} \quad i = j, \quad i,j = 1,2,...,n_{\text{mod}} \end{aligned} \quad (3.43)$$

This leads to the following equation with the modal coordinate $\alpha_j(t)$ as unknown:

$$K_j \cdot \alpha_j(t) + C_j \cdot \dot{\alpha}_j(t) + M_j \cdot \ddot{\alpha}_j(t) = P_j(t) \quad (3.44)$$

where

$$\begin{aligned} K_j &= \phi_j^\top \cdot \boldsymbol{K} \cdot \phi_j && \text{is the modal (generalized) stiffness} \\ C_j &= \phi_j^\top \cdot \boldsymbol{C} \cdot \phi_j && \text{is the modal (generalized) damping} \\ M_j &= \phi_j^\top \cdot \boldsymbol{M} \cdot \phi_j && \text{is the modal (generalized) mass} \\ P_j(t) &= \phi_j^\top \cdot \boldsymbol{P}(t) && \text{is the modal (generalized) load} \end{aligned} \quad (3.45)$$

Eq. 3.44 exists for each of the n_{mod} modes and governs solely the j-th modal coordinate $\alpha_j(t)$, whereas the parameters K_j, C_j, M_j and P_j only depend on the j-th mode. This means that the response of the MDOF system described by the system of differential equations given in Eq. 3.23 can be obtained by solving n_{mod} decoupled (modal) equations – one for each mode – and superposing their effects according to Eq. 3.39. In fact, Eq. 3.44 is equivalent to the equation of motion of a SDOF system with the stiffness constant K_j, the mass M_j and the viscous damping coefficient C_j, subjected to a time dependent load $P_j(t)$ (Fig. 3.5), whereas the displacement of the equivalent SDOF system is $\alpha_j(t)$. The following

[11] A system is considered to be classically or proportionally damped if the damping matrix can be expressed as a linear combination of the mass and stiffness matrix, e.g. as in case of the *Rayleigh* damping, which will be presented in detail in Section 3.3.4.

relations of the equivalent SDOF hold:

$$\omega_j = \sqrt{\frac{K_j}{M_j}} \quad \text{and} \quad C_j = 2 \cdot \xi_j \cdot \omega_j \cdot M_j \tag{3.46}$$

The damping approach given in Eq. 3.46 implies the assignment of a damping ratio ξ_j to each mode, i.e. to each of the equivalent SDOF systems. A great advantage of this formulation is that the damping ratios can be usually determined from experimental investigations performed on similar existing structures (CLOUGH et al., 1995; CHOPRA, 2014; PETERSEN et al., 2017), avoiding thus a damping definition based on the explicit damping matrix (Eq. 3.45). Dividing Eq. 3.44 by M_j and considering the relations given in Eq. 3.46, the following equation is obtained[12]:

$$\omega_j^2 \cdot \alpha_j(t) + 2 \cdot \xi_j \cdot \omega_j \cdot \dot{\alpha}_j(t) + \ddot{\alpha}_j(t) = \frac{P_j(t)}{M_j} \tag{3.47}$$

The equation of motion of a SDOF system (Eqs. 3.44 and 3.47) can be solved in time domain using, for example, the *Duhamel* or convolution integral:

$$\begin{aligned} \alpha_j(t) = & \frac{1}{M_j \cdot \omega_{\mathrm{D},j}} \cdot \int_0^t P_j(\tau) \cdot \mathrm{e}^{-\xi_j \cdot \omega_j \cdot (t-\tau)} \cdot \sin\left(\omega_{\mathrm{D},j} \cdot (t-\tau)\right) \mathrm{d}\tau + \\ & \mathrm{e}^{-\xi_j \cdot \omega_j \cdot t} \cdot \left(\frac{\xi_j \cdot \omega_j \cdot \alpha_j(0) + \dot{\alpha}_j(0)}{\omega_{\mathrm{d},j}} \cdot \sin(\omega_{\mathrm{D},j} \cdot t) + \alpha_j(0) \cdot \cos(\omega_{\mathrm{D},j} \cdot t) \right) \end{aligned} \tag{3.48}$$

where $\omega_{\mathrm{D},j} = \omega_j \cdot \sqrt{1 - \xi_j^2}$ represents the damped angular frequency of the j-th mode and the initial conditions $\alpha_j(0)$ and $\dot{\alpha}_j(0)$ are defined as:

$$\alpha_j(0) = \frac{\boldsymbol{\phi}_j^\top \cdot \boldsymbol{M} \cdot \boldsymbol{u}(0)}{M_j} \quad \text{and} \quad \dot{\alpha}_j(0) = \frac{\boldsymbol{\phi}_j^\top \cdot \boldsymbol{M} \cdot \dot{\boldsymbol{u}}(0)}{M_j} \tag{3.49}$$

The Duhamel integral (Eq. 3.48) considers the loading term $P_j(t)$ as a succession of impulses of very short duration. The structural response of each impulse represents a contribution to the total response, which is obtained by time integration (summation) over all impulse responses[13]. Eq. 3.48 can be evaluated analytically or numerically (PETERSEN et al., 2017), whereas the analytical solution is rather suitable for force functions, which can be easily analytically integrated. However, in most practical applications with arbitrary

[12] It is obvious, that a normalization of the mode shapes according to Eq. 3.38, i. e. $M_j = 1$ simplifies the computation procedure in Eq. 3.47.

[13] Since the Duhamel integral implies the superposition of separate responses resulting from the successive impulse loads, it can only be used with linear systems.

force functions, the modal equations of motion (Eq. 3.44) are usually solved using numerical time-stepping integration methods, such as the *Newmark* procedure (CLOUGH et al., 1995; PETERSEN et al., 2017; WERKLE, 2021). The investigations performed within the present thesis are mainly based on the latter approach, which will be presented in Section 3.3.4 in the context of a direct integration solution for MDOF systems. Since a SDOF system represents in fact a particular case of a MDOF system with only one DOF, the equations developed for the MDOF system are applicable without any restrictions also for the solution of the modal equations.

The great advantage of the modal superposition method is the considerable reduction of the computational time compared to the direct integration method (cf. Section 3.3.4). The reason for this is that – in most practical applications – only a few modes are necessary to achieve an accurate approximation of the vibration behavior of the system. This holds even in case of systems containing several hundreds or thousands of DOFs with mass. Therefore, only few modal equations have to be solved. Considering a computation under consideration of only n out of n_{mod} modes, Eq. 3.39 and its second derivative become:

$$\boldsymbol{u}(t) = \sum_{i=1}^{n} \alpha_i(t) \cdot \boldsymbol{\phi}_i \quad \text{and} \quad \ddot{\boldsymbol{u}}(t) = \sum_{i=1}^{n} \ddot{\alpha}_i(t) \cdot \boldsymbol{\phi}_i \tag{3.50}$$

Exact solutions for Eq. 3.50 are only obtained when all mode shapes are taken into consideration, i.e. $n = n_{\text{mod}}$.

A limitation of the modal analysis method is its applicability to linear systems only, which is related to the use of the superposition principle. Moreover, the modal superposition can only be used for classically damped systems, as the equations of motion can not be decoupled in case of a system with unproportional damping (cf. Eq. 3.42). As opposed to this, the time step direct integration methods – discussed in the following section – are generally applicable, including both nonlinear and nonclasically damped systems.

3.3.4 Solution using direct numerical integration

The structural response of a MDOF system based on direct integration is obtained by integrating the coupled system of equations given in Eq. 3.23 using step-by-step integration methods, without any prior transformation of the equation system into another form. Many step-by-step (also denoted as *time-stepping*) integration methods divide the loading and the response into a set of time steps of length Δt. This means that they do not seek to satisfy the dynamic equilibrium equations (Eq. 3.23) at any time t, but to fulfill them only at discrete time points t_i, lying Δt apart from each other (BATHE, 2014), with:

$$\Delta t = t_{i+1} - t_i \tag{3.51}$$

The *numerical* step-by-step integration methods, which are relevant for the scope of this thesis, are approximation methods, since they only approximately fulfill the equilibrium equations at the discrete time points t_i. The total response within one step is generally calculated based on the initial conditions (displacement and velocities) at the beginning of the step and on load variation within the step. Hence, the response in each step can be obtained from an independent analysis, with no need of superposing different contributions. This also allows the dynamic analysis of structures with nonlinear behavior, under the assumption that the system matrices remain constant during the time step (CLOUGH et al., 1995).

Furthermore, it is to be mentioned that the size of Δt has a significant influence on the accuracy of the solution. It has to be selected in such a way that the dynamic excitation and the vibration response characteristics of the structure can be properly represented. However, the stability of some time-stepping integration procedures also depends on the size of the time step (BATHE, 2014; WERKLE, 2021). In this sense, one can differentiate between *conditionally stable* and *unconditionally stable* time-stepping integration methods. Unconditionally stable procedures lead to bounded solutions for any size of the time step Δt and arbitrary initial conditions, while conditionally stable procedures deliver bounded solutions only for time steps smaller than a certain critical time step Δt_{cr} (BATHE, 2014; WERKLE, 2021).

The general concept for the numerical integration of the equations of motion is essentially based on the well known integral relationships between acceleration, velocity and displacement, expressed for a time step of length Δt:

$$\dot{\boldsymbol{u}}_{i+1} = \dot{\boldsymbol{u}}_i + \int_0^{\Delta t} \ddot{\boldsymbol{u}}(\tau)\mathrm{d}\tau \quad \text{and} \quad \boldsymbol{u}_{i+1} = \boldsymbol{u}_i + \int_0^{\Delta t} \dot{\boldsymbol{u}}(\tau)\mathrm{d}\tau \tag{3.52}$$

It is evident that the velocity $\dot{\boldsymbol{u}}_{i+1}$ at the end of a time step can be expressed as a summation of the initial velocity $\dot{\boldsymbol{u}}_i$ and the integral of the acceleration history over the time step. Similarly, the displacement $\boldsymbol{u}_{i+1}$ is computed from the initial displacement $\boldsymbol{u}_i$ plus the integrated velocity. Obviously, a formulation according to Eq. 3.52 is only possible if an assumption regarding the variation of the acceleration within the time step is made. In this way, the velocity variation and subsequently the displacement history over the time step can be determined. For instance, considering a constant acceleration, Eq. 3.52 becomes:

$$\dot{\boldsymbol{u}}_{i+1} = \dot{\boldsymbol{u}}_i + \ddot{\boldsymbol{u}} \cdot \Delta t \quad \text{and} \quad \boldsymbol{u}_{i+1} = \boldsymbol{u}_i + \dot{\boldsymbol{u}}_i \cdot \Delta t + \frac{1}{2} \cdot \ddot{\boldsymbol{u}} \cdot \Delta t^2 \tag{3.53}$$

Newmark method One of the most frequently used time-stepping integration methods in structural dynamics is the *Newmark* procedure (NEWMARK, 1959; BATHE, 2014). It was adopted for the dynamic analyses performed within this thesis and will be briefly presented in the following. It is to be mentioned on beforehand that the equations are generally formulated for MDOF systems, i.e. for FE models. However, they are also valid for SDOF

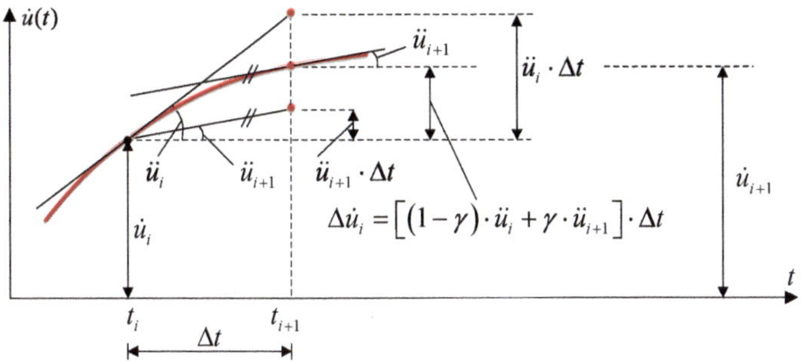

Figure 3.6 The Newmark method (WERKLE, 2021)

systems, if the system matrices describing the MDOF system and its response vectors are replaced with the corresponding scalar quantities of the SDOF system. In this way, the modal equations expressed in Eq. 3.44 can be easily solved for arbitrary force time-histories. In this case, the same time step Δt has to be selected for all modes.

The integration method of NEWMARK (1959) is based on the following equations:

$$\dot{\boldsymbol{u}}_{i+1} = \dot{\boldsymbol{u}}_i + [(1-\gamma)\cdot\ddot{\boldsymbol{u}}_i + \gamma\cdot\ddot{\boldsymbol{u}}_{i+1}]\cdot\Delta t \quad (3.54a)$$

$$\boldsymbol{u}_{i+1} = \boldsymbol{u}_i + \Delta t\cdot\dot{\boldsymbol{u}}_i + [(0.5-\beta)\cdot\ddot{\boldsymbol{u}}_i + \beta\cdot\ddot{\boldsymbol{u}}_{i+1}]\cdot\Delta t^2 \quad (3.54b)$$

It can be seen that a constant acceleration is assumed over the time step. It is formulated as a weighted average of the acceleration at $\ddot{\boldsymbol{u}}_i$ at the beginning and $\ddot{\boldsymbol{u}}_{i+1}$ at the end of the time step. γ and β are integration parameters used for weighting the accelerations $\ddot{\boldsymbol{u}}_i$ and $\ddot{\boldsymbol{u}}_{i+1}$ in the velocity and displacement expressions, respectively. They were introduced into the procedure to allow the control of the accuracy and stability of the integration (CHOPRA, 2014; BATHE, 2014). The understanding of Eq. 3.54a can be facilitated by Fig. 3.6. It illustrates the determination of the velocity $\dot{\boldsymbol{u}}_{i+1}$ as a sum of the velocity $\dot{\boldsymbol{u}}_i$ at the beginning of the time step and a constant weighted average acceleration (weighting factors $1-\gamma$ and γ) multiplied by the time step Δt.

Generally, the stability of the Newmark procedure is determined by the integration parameters γ and β. In this sense, unconditional stability can be achieved if:

$$\gamma \geq \frac{1}{2} \quad \text{and} \quad \beta \geq \frac{1}{4}\cdot\left(\frac{1}{2}+\gamma\right)^2 \quad (3.55)$$

For a parameter set γ and β not fulfilling Eq. 3.55, the procedure becomes conditionally stable for:

$$\frac{\Delta t}{T_{\min}} \leq \frac{1}{2\cdot\pi}\cdot\frac{1}{\sqrt{\gamma/2-\beta}} \quad (3.56)$$

where $T_{\min}$ represents the shortest vibration period to be considered in analysis.

The reference work of BATHE (2014) recommends the parameter set $\gamma = 1/2$ and $\beta = 1/4$, which delivers satisfactory results from both the accuracy and the stability point of view. This assumption leads – according to Eq. 3.54 – to:

$$\dot{\boldsymbol{u}}_{i+1} = \dot{\boldsymbol{u}}_i + \frac{\Delta t}{2}\left(\ddot{\boldsymbol{u}}_i + \ddot{\boldsymbol{u}}_{i+1}\right) \tag{3.57a}$$

$$\boldsymbol{u}_{i+1} = \boldsymbol{u}_i + \Delta t \cdot \dot{\boldsymbol{u}}_i + \left[\frac{1}{4} \cdot \ddot{\boldsymbol{u}}_i + \frac{1}{4} \cdot \ddot{\boldsymbol{u}}_{i+1}\right] \cdot \Delta t^2 \tag{3.57b}$$

This approach was used for all investigations performed within the present work. It is denoted as the *constant average acceleration method*, since the average $(\ddot{\boldsymbol{u}}_i + \ddot{\boldsymbol{u}}_{i+1})/2$ is assumed for both the displacement and velocity expressions in Eq. 3.54.

The kinetic equation of equilibrium at the time point t_{i+1} is expressed as:

$$\boldsymbol{K} \cdot \boldsymbol{u}_{i+1} + \boldsymbol{C} \cdot \dot{\boldsymbol{u}}_{i+1} + \boldsymbol{M} \cdot \ddot{\boldsymbol{u}}_{i+1} = \boldsymbol{P}_{i+1} \tag{3.58}$$

Its solution according to Eq. 3.54 requires iteration, since the unknown term $\ddot{\boldsymbol{u}}_{i+1}$ occurs on the right side. This implies the reevaluation of the acceleration $\ddot{\boldsymbol{u}}_{i+1}$ in each iteration step, after determining $\boldsymbol{u}_{i+1}$ and $\dot{\boldsymbol{u}}_{i+1}$ according to Eq. 3.54. $\ddot{\boldsymbol{u}}_{i+1}$ can be evaluated at the end of each iteration as follows (cf. Eq. 3.58):

$$\ddot{\boldsymbol{u}}_{i+1} = \boldsymbol{M}^{-1} \cdot \left(\boldsymbol{P}_{i+1} - \boldsymbol{C} \cdot \dot{\boldsymbol{u}}_{i+1} - \boldsymbol{K} \cdot \boldsymbol{u}_{i+1}\right) \tag{3.59}$$

The result of Eq. 3.59 has to substituted in Eq. 3.54a and 3.54b in each iteration step. However, since $\ddot{\boldsymbol{u}}_{i+1}$ is unknown at the beginning of the first iteration, an initial value has to be assumed, e.g. (PETERSEN et al., 2017):

$$\ddot{\boldsymbol{u}}_{i+1} = \ddot{\boldsymbol{u}}_i \qquad \text{or} \qquad \ddot{\boldsymbol{u}}_{i+1} = \ddot{\boldsymbol{u}}_i + \dot{\boldsymbol{u}}_i \cdot \Delta t \tag{3.60}$$

Eq. 3.59 is also used for determination of the initial acceleration $\ddot{\boldsymbol{u}}_0$, based on the initial conditions $\boldsymbol{u}_0$ and $\dot{\boldsymbol{u}}_0$:

$$\ddot{\boldsymbol{u}}_0 = \boldsymbol{M}^{-1} \cdot \left(\boldsymbol{P}_0 - \boldsymbol{C} \cdot \dot{\boldsymbol{u}}_0 - \boldsymbol{K} \cdot \boldsymbol{u}_0\right) \tag{3.61}$$

The iteration can be stopped, when a proper termination criterion is fulfilled, e.g. a certain relative change of the acceleration between two iteration steps.

In case of dynamic analysis of linear systems, the Newmark method can be reformulated to achieve a solution without iterations. In a first step, the acceleration $\ddot{\boldsymbol{u}}_{i+1}$ has to be expressed in terms of $\boldsymbol{u}_{i+1}$, using the relation given in Eq. 3.54b:

$$\ddot{\boldsymbol{u}}_{i+1} = \frac{1}{\beta \cdot \Delta t^2} \cdot \left(\boldsymbol{u}_{i+1} - \boldsymbol{u}_i\right) - \frac{1}{\beta \cdot \Delta t} \cdot \dot{\boldsymbol{u}}_i - \left(\frac{1}{2 \cdot \beta} - 1\right) \cdot \ddot{\boldsymbol{u}}_i \tag{3.62}$$

Substituting Eq. 3.62 in Eq. 3.54a yields:

$$\dot{\boldsymbol{u}}_{i+1} = \frac{\gamma}{\beta \cdot \Delta t} \cdot (\boldsymbol{u}_{i+1} - \boldsymbol{u}_i) + \left(1 - \frac{\gamma}{\beta}\right) \cdot \dot{\boldsymbol{u}}_i + \Delta t \cdot \left(1 - \frac{\gamma}{2 \cdot \beta}\right) \cdot \ddot{\boldsymbol{u}}_i \tag{3.63}$$

The next step implies the substitution of Eq. 3.62 and 3.63 into Eq. 3.58 and an appropriate gathering of the terms, which results in:

$$\hat{\boldsymbol{K}} \cdot \boldsymbol{u}_{i+1} = \hat{\boldsymbol{P}}_{i+1} \tag{3.64}$$

where

$$\hat{\boldsymbol{K}} = \boldsymbol{K} + \frac{\gamma}{\beta \cdot \Delta t} \cdot \boldsymbol{C} + \frac{1}{\beta \cdot \Delta t^2} \cdot \boldsymbol{M} \tag{3.65}$$

and

$$\begin{aligned} \hat{\boldsymbol{P}}_{i+1} = & \boldsymbol{P}_{i+1} + \left[\frac{1}{\beta \cdot \Delta t^2} \cdot \boldsymbol{M} + \frac{\gamma}{\beta \cdot \Delta t} \cdot \boldsymbol{C}\right] \cdot \boldsymbol{u}_i + \left[\frac{1}{\beta \cdot \Delta t} \cdot \boldsymbol{M} + \left(\frac{\gamma}{\beta} - 1\right) \cdot \boldsymbol{C}\right] \cdot \dot{\boldsymbol{u}}_i \\ & + \left[\left(\frac{1}{2 \cdot \beta} - 1\right) \cdot \boldsymbol{M} + \Delta t \cdot \left(\frac{\gamma}{2 \cdot \beta} - 1\right) \cdot \boldsymbol{C}\right] \cdot \ddot{\boldsymbol{u}}_i \end{aligned} \tag{3.66}$$

$\hat{\boldsymbol{K}}$ and $\hat{\boldsymbol{P}}_{i+1}$ represent the effective stiffness and the effective load of the equivalent static equilibrium equation given in Eq. 3.64. They only depend on the motion quantities from the previous time step $\boldsymbol{u}_i$, $\dot{\boldsymbol{u}}_i$ and $\ddot{\boldsymbol{u}}_i$, the system matrices $\boldsymbol{K}$, $\boldsymbol{C}$ and $\boldsymbol{M}$, the integration parameters γ and β and the time step Δt. Hence, the displacement $\boldsymbol{u}_{i+1}$ at the time point t_{i+1} can be computed by solving the linear equation system expressed in Eq. 3.64:

$$\boldsymbol{u}_{i+1} = \hat{\boldsymbol{K}}^{-1} \cdot \hat{\boldsymbol{P}}_{i+1} \tag{3.67}$$

Subsequently, $\ddot{\boldsymbol{u}}_{i+1}$ and $\dot{\boldsymbol{u}}_{i+1}$ can be directly obtained from Eq. 3.62 and Eq. 3.63, whereas the initial acceleration is determined according to Eq. 3.61. The investigations presented in this thesis are based on this noniterative formulation of the Newmark method.

The Newmark method is denoted as an *implicit* method, since the displacements are computed by solving a linear equation system in each time step (cf. Eq. 3.67). On the contrary, the solution of *explicit* integration procedures, e.g. the central difference method, can be computed without requiring the solution of an equation system. They present certain advantages with respect to the computational effort, but usually have important stability issues (WERKLE, 2021). For example, the central difference method requires:

$$\Delta t \leq \frac{T_{\min}}{\pi} \tag{3.68}$$

This criterion depends on the smallest period of vibration $T_{\min}$ of the FE model and implicitly on the fineness of the model discretization. Small element sizes require very small time steps. Therefore, explicit integration schemes are usually only used in exceptional cases, e.g. for studying impact problems (WERKLE, 2021).

The accuracy of the time-stepping integration methods can be analyzed in terms of amplitude decays of an undamped system, i.e. numerical (artifical) damping[14], and frequency shift during its free vibration phase, i.e. period elongation or shortening. Extensive studies on the accuracy of the Newmark method are presented in BATHE (2014), CHOPRA (2014) and PETERSEN et al. (2017). Without going into a detailed discussion, it is mentioned that the selection of integration parameters according to Eq. 3.55 does not cause artificial damping, while the frequency drift is dependent on the ratio between the time step Δt and the shortest vibration period $T_{\min}$. BATHE (2014) and CHOPRA (2014) indicate that reasonably accurate results are obtained for:

$$\Delta t \leq \frac{T_{\min}}{10} \quad \text{or even better for} \quad \Delta t \leq \frac{T_{\min}}{20} \tag{3.69}$$

Eq. 3.69 represents in fact a typical selection of the time step for dynamic analysis of MDOF-systems, which ensures a proper representation of the vibration for the highest mode to be considered. In the vast majority of the practical cases of bridge structures subjected to moving loads, the rate of variation of the applied loading $\boldsymbol{P}(t)$ is also properly represented using the selection given in Eq. 3.69.

Besides the Newmark procedure, various time-stepping integration methods can be used for the numerical solution of a MDOF system. The Houbolt method, the Wilson-θ method, the Park method and the HHT method are only a few of them, which were specially developed for solving differential equations of motion. In addition, solutions based on well known methods from the field of numerical mathematical integration, as the central difference method and the procedures of Euler-Cauchy, Heun or Runge-Kutta are also possible. Their reproduction, even in a brief manner, would go far beyond the scope of this thesis. Therefore, the author refers to the relevant literature (CLOUGH et al., 1995; CHOPRA, 2014; BATHE, 2014; PETERSEN et al., 2017) and the further sources mentioned there.

Damping approach The direct integration of Eq. 3.23 implies that the damping has to be formulated in matrix form. However, an explicit determination of the damping matrix $\boldsymbol{C}$ in the engineering practice is only possible in exceptional cases. Therefore, it is usually defined as a linear combination of the mass and the stiffness matrix, which should account for all energy dissipating mechanisms:

[14] In general, it is sought to avoid the numerical damping in the dynamic analysis. However, due to its property of a higher damping for shorter vibration periods (i.e. higher damping for higher modes), it can be used in certain cases for the damping of unwanted contributions of higher modes (WERKLE, 2021).

$$\boldsymbol{C} = a_0 \cdot \boldsymbol{M} + a_1 \cdot \boldsymbol{K} \tag{3.70}$$

This damping formulation is widely denoted as *Rayleigh* damping and the systems for which it is adopted are denoted as proportionally or classically damped systems[15]. a_0 and a_1 in Eq. 3.70 represent the weighting factors of the Rayleigh damping. They have the units $[\mathrm{s}]^{-1}$ and $[\mathrm{s}]$, respectively. The damping ratio of the j-th mode of vibration is determined from Eq. 3.47 under consideration of Eq. 3.44, 3.46 and 3.70:

$$\frac{C_j}{M_j} = a_0 + a_1 \cdot \omega_j^2 = 2 \cdot \xi_j \cdot \omega_j \tag{3.71}$$

as

$$\xi_j = \frac{1}{2} \cdot \left(\frac{a_0}{\omega_j} + a_1 \cdot \omega_j \right) \tag{3.72}$$

The weighting factors a_0 and a_1 can be determined by imposing two known damping ratios ξ_i and ξ_k associated with two frequencies (modes) ω_i and ω_k. Expressing Eq. 3.72 for the two modes i and k and solving the resulting equation system for a_0 and a_1 yields:

$$a_0 = 2 \cdot \omega_i \cdot \omega_k \cdot \frac{\xi_i \cdot \omega_k - \xi_k \cdot \omega_i}{\omega_k^2 - \omega_i^2} \tag{3.73a}$$

$$a_1 = 2 \cdot \frac{\xi_k \cdot \omega_k - \xi_i \cdot \omega_i}{\omega_k^2 - \omega_i^2} \tag{3.73b}$$

In general, ω_i is assumed to be the first (fundamental) natural frequency, while ω_k is the frequency of a higher mode with significant contribution to the dynamic response. Fig. 3.7 shows the variation of the damping ratio with the natural frequency. It can be seen that the derivation presented above ensures that the specified damping ratios ξ_i and ξ_k are obtained for the corresponding modes i and k. All modes with frequencies between ω_i and ω_k will have slightly lower damping ratios than ξ_i and ξ_k, while the damping ratios of the modes with frequencies higher than ω_k are monotonically increasing with the frequency. The higher damping is likely to eliminate the corresponding modal contributions from the computed dynamic responses.

In addition, it is to be mentioned that different approaches for proportional damping, which allow the formulation of the damping matrix $\boldsymbol{C}$ based on an arbitrary number of

[15] It is to be noted that classical damping can be assumed if similar damping mechanisms are distributed throughout the structure, i.e. it is not appropriate if significantly different levels of damping are associated to different parts of a structure (Chopra, 2014). Typical examples of nonclassically or nonproportional damping are the soil-structure interaction problem (the soil presents a much higher damping than the structure) or a structure instrumented with special energy-dissipating devices (e.g. a tuned mass damper, which presents as well a significantly higher damping than the structure). Formulation approaches of nonproportional damping are given in Clough et al. (1995), Chopra (2014) and Petersen et al. (2017).

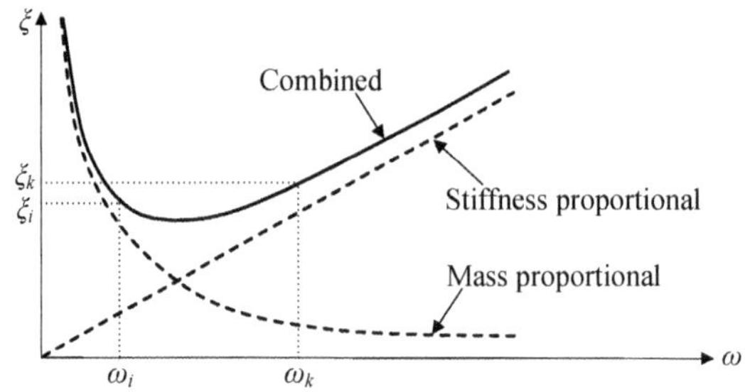

Figure 3.7 Rayleigh damping: variation of modal damping ratios with natural frequency

given damping ratios ξ_j, are available in the literature (CLOUGH et al., 1995; BATHE, 2014; CHOPRA, 2014).

There is also an alternative method for determination of a classical damping matrix from modal damping ratios. Its expression is given in the following, without providing its derivation, which can be found in CLOUGH et al. (1995) and CHOPRA (2014):

$$\boldsymbol{C} = \boldsymbol{M} \cdot \left(\sum_{j=1}^{n_{\text{mod}}} \frac{2 \cdot \xi_j \cdot \omega_j}{M_j} \cdot \boldsymbol{\phi}_j \cdot \boldsymbol{\phi}_j^{\top} \right) \cdot \boldsymbol{M} \tag{3.74}$$

The contribution of the j-th mode to the total damping matrix $\boldsymbol{C}$ – expressed by the j-th term in Eq. 3.74 – is proportional to the corresponding damping ratio ξ_j. This means that all modes not included in the summation are undamped (CLOUGH et al., 1995).

3.4 Conclusion

In this chapter, the fundamentals of dynamic analysis of bridge structures subjected to the action of moving loads are briefly recapitulated. The analytical solution of a beam system with uniformly distributed mass and stiffness may indeed facilitate the understanding of the dynamic behavior of a bridge excited by a passing vehicle. However, the solution of most practical bridge dynamics problems is nowadays usually addressed by means of numerical methods. This concerns both the definition of the structural models, e.g. based on the finite element method (FEM), and the numerical solution of the equations of motion. Regarding the latter, the widely used Newmark method is presented in detail. It can be used in the context of both direct integration and modal superposition, the latter being the rather typical approach for the analysis of bridge structures under moving loads, whose structural responses are usually dominated by a few lower vibration modes. Therefore, the developments of this thesis, which are presented in the following chapters, are based on the modal superposition approach in conjunction with the Newmark method.

4 Introduction to inverse problems

4.1 Generals

In most cases of dynamic analysis of bridges under moving loads, civil engineers have to deal with the *forward* problem formulation (FRÝBA, 1972). This implies a realistic structural model and known input forces, while the system behavior is sought to be determined. In contrast, the *inverse* problems in bridge dynamics, which have become increasingly important in recent years, can be divided in two main categories:

- System identification problem: it implies the identification of the system model parameters based on known forces and measured structural responses. This type of problem formulation is used, for example, in model updating.
- Input identification problem: it implies the reconstruction of the excitation under assumption of known structural responses and system model parameters. The MFI, which is the core aspect of this thesis, falls within this category.

In this context, it should be noted that there is in fact no formal mathematical definition of the term *inverse problem* (RICHTER, 2015). Nevertheless, a rather intuitive definition can be formulated, also taking into account the previously mentioned applications in bridge dynamics. It states that an inverse problem is concerned with determining the *causes* knowing the *effects*. Thus, this problem type is the opposite of the *forward* or *direct* problem, which consists of determining the effects on the basis of known causes (KERN, 2016). The following sections will provide a brief insight into the general topic of inverse problems from mathematical point of view and present the main solution approaches.

4.2 Inverse and ill-posed problems

Inverse problems are usually expressed by a linear system of equations of the form:

$$\boldsymbol{A}\boldsymbol{x} = \boldsymbol{b} \quad \text{with } \boldsymbol{x} \in \mathbb{R}^n,\ \boldsymbol{b} \in \mathbb{R}^m,\ \boldsymbol{A} \in \mathbb{R}^{m,n} \tag{4.1}$$

where $\boldsymbol{x}$ is the vector of the unknowns (*causes*), $\boldsymbol{b}$ is the right side of the equation system (*effects*, usually *observed* from experiments) and $\boldsymbol{A}$ depicts the coefficient matrix of the system, which describes the causal relationship between $\boldsymbol{x}$ and $\boldsymbol{b}$. To find the unknown vector $\boldsymbol{x}$ that fits the measured data, one might be able to invert the system matrix $\boldsymbol{A}$:

A. Firus, *A Contribution to Moving Force Identification in Bridge Dynamics*, Mechanik, Werkstoffe und Konstruktion im Bauwesen 65, https://doi.org/10.1007/978-3-658-39838-5_4

$$x = A^{-1}b \tag{4.2}$$

However, the direct solution according to Eq. 4.2 is only possible if A is a full-rank square matrix[1]. It is to be mentioned that even square matrices ($m = n$) might not be invertible. This occurs when A is rank deficient, leading thus to a non-unique solution. In this case, both the matrix A and the system are referred to be singular.

In most practical cases, the number of measurement data and the number of unknowns differ, so that the matrix A is not square. If $m > n$, the system consists of more equations than unknowns in the vector x, i.e. it is overdetermined. Slight inaccuracies in the elements of A and b usually lead in this case to contradictions in the individual equations, so that there is no x that exactly satisfies Eq. 4.1 (RICHTER, 2015). The case $m < n$ implies more unknowns than equations, i.e. the system is underdetermined. This situation results in an infinite number of solutions.

Basically, all problems of type $Ax = b$ described using rectangular matrices ($m < n$ or $m > n$) or by systems with a singular square coefficient matrix A, are denoted as ill-posed problems in the sense of Hadamard's criteria (HADAMARD, 1923; TIKHONOV et al., 1995). This means that they usually violate at least one of the following conditions, which characterize a well-posed problem:

- a solution exists (existence criterion),
- the solution is unique (uniqueness criterion),
- the solution depends continuously on the data (stability criterion).

All three conditions of well-posedness are only fulfilled for a full-rank square matrix A. In this case, the equation system given in Eq. 4.1 has a unique solution for any right side b. However, even if the problem is formally well posed and the stability condition is fulfilled ($\|A^{-1}\|_2 < \infty$)[2], the system can be practically unstable (KABANIKHIN, 2011). This situation occurs in case of an *ill-conditioned* system matrix A.

A matrix is said to be ill-conditioned if arbitrarily small perturbations of b (e.g. due to measurement noise) result in considerable changes in the solution. The criterion for classifying a matrix as ill-conditioned is the *condition number*. Is is defined as the product between the norm of the system matrix A and the norm of its inverse:

$$\mu(A) = \|A\|_2 \|A^{-1}\|_2 \tag{4.3}$$

In the following, different approaches for the solution of ill-posed inverse problems will be discussed. Basically, a distinction can be made between *direct methods* (presented in

[1] A square matrix has a full rank if its determinant is nonzero, i.e. the matrix is nonsingular. The opposite of the full rank square matrix is the rank deficient square matrix, which has a zero determinant and thus zero eigenvalues, i.e. the matrix is also singular.

[2] The 2-norm of a matrix A is equal to the highest singular value of the matrix, $\|A\|_2 = \sigma_{\max}(A)$ (KABANIKHIN, 2011).

Section 4.3), which imply that the solution is achieved in a predetermined number of operations, and methods based on successive approximations using *mathematical optimization algorithms* (discussed in Section 4.4).

4.3 Direct methods for solving inverse problems

4.3.1 Normal equations

Since singular, under- and overdetermined systems do not allow an exact solution, an alternative approach has to be used. It attempts to find the vector $\boldsymbol{x}$ so that $\boldsymbol{Ax}$ is the best approximation to $\boldsymbol{b}$. An intuitive approach to achieve this is to make the residual vector $\boldsymbol{r}(\boldsymbol{x}) = \boldsymbol{Ax} - \boldsymbol{b}$ small (BJÖRCK, 2015). This can be expressed mathematically as[3] (RICHTER, 2015):

$$\text{find } \hat{\boldsymbol{x}} \text{ so that } \|\boldsymbol{r}(\hat{\boldsymbol{x}})\|_2 \leq \|\boldsymbol{r}(\boldsymbol{x})\|_2 \text{ for all } \boldsymbol{x} \in \mathbb{R}^n \tag{4.4}$$

It is to be mentioned that the following derivations are based – unless otherwise specified – on overdetermined systems ($m \geq n$), which are typical for inverse problems of MFI (YU et al., 2016). The minimization problem in Eq. 4.4 is equivalent to:

$$\min_{\boldsymbol{x}} \mathcal{F}(\boldsymbol{x}) = \|\boldsymbol{Ax} - \boldsymbol{b}\|_2^2 = \|\boldsymbol{r}(\boldsymbol{x})\|_2^2 \tag{4.5}$$

It becomes clear that Eq. 4.5 denotes the minimization of the sum of squares of the elements of $\boldsymbol{Ax} - \boldsymbol{b}$. The corresponding solution is called – for obvious reasons – a *least squares* solution, whereas problems implying the finding of a least squares solution to $\boldsymbol{Ax} = \boldsymbol{b}$ are generally denoted as *least squares problems* (FORD, 2015).

The minimizer of Eq. 4.5 is found as a solution of the normal equations:

$$\boldsymbol{A}^\top \boldsymbol{Ax} = \boldsymbol{A}^\top \boldsymbol{b} \qquad \Longrightarrow \qquad \boldsymbol{x} = \underbrace{\left(\boldsymbol{A}^\top \boldsymbol{A}\right)^{-1} \boldsymbol{A}^\top}_{\boldsymbol{A}^\dagger} \boldsymbol{b} = \boldsymbol{A}^\dagger \boldsymbol{b} \tag{4.6}$$

where $\boldsymbol{A}^\dagger = (\boldsymbol{A}^\top \boldsymbol{A})^{-1} \boldsymbol{A}^\top$ is called the pseudoinverse or the Moore-Penrose generalized inverse of $\boldsymbol{A}$. A short derivation of the normal equations is provided in Appendix A.1.

However, even if Eq. 4.6 provides a unique solution of the inverse problem, the solution might still be unevenly magnified by small perturbations in the vector $\boldsymbol{b}$. In order to cope with this issue, regularization methods can be used. The Tikhonov[4] regularization

[3] $\|\cdot\|_2$ denotes here and below the Euclidean norm of a vector. Recall that it is defined as the square root of the sum of squares of all vector entries, i.e. it represents the vector length in the n dimensional space.

[4] The *Tikhonov regularization* is one of the most popular methods for solving ill-posed problems, originally published by TIKHONOV (1963). However, the same regularization method was independently proposed at about the same time by PHILLIPS (1962).

(TIKHONOV et al., 1995) and the truncated singular value decomposition (TSVD) (HANSEN, 1987; MOTTERSHEAD et al., 1991) are two widely used regularization methods. They are elaborated in Section 4.3.2 and Section 4.3.3. .

4.3.2 Tikhonov regularization

The *Tikhonov regularization* represents a powerful tool for solving ill-posed linear least squares problems (GOLUB et al., 1997). Its basic idea is to impose a priori knowledge on the solution as constraints, by adding a penalty term with an adjustable weighting (regularization) parameter into the objective function (SANTAMARINA et al., 2005):

$$\min_{\boldsymbol{x}} \mathcal{F}(\boldsymbol{x},\alpha) = \|\boldsymbol{A}\boldsymbol{x}-\boldsymbol{b}\|_2^2 + \alpha\,\|\boldsymbol{\Gamma}\boldsymbol{x}\|_2^2 \tag{4.7}$$

where $\alpha \geq 0$ is the regularization parameter. $\boldsymbol{\Gamma}$ denotes the Tikhonov matrix, which describes the a priori information on the solution (HANSEN, 1998). For example, $\boldsymbol{\Gamma} = \boldsymbol{I}$ leads to the penalty term $\alpha\,\|\boldsymbol{I}\boldsymbol{x}\|_2^2$, which prevents the norm of the solution from becoming too large. The corresponding solution is called the *damped least squares solution* (SANTAMARINA et al., 2005). Further formulations of the Tikhonov matrix considering different types of a priori information (e.g. derivative information or expected solution) are provided in DOYLE (2004) and SANTAMARINA et al. (2005).

Similar to the procedure used for obtaining the unregularized solution (Section 4.3.1), a unique solution can be obtained for Eq. 4.7 (cf. also Appendix A.2):

$$\boldsymbol{x} = \left(\boldsymbol{A}^\top\boldsymbol{A} + \alpha\boldsymbol{\Gamma}^\top\boldsymbol{\Gamma}\right)^{-1}\boldsymbol{A}^\top\boldsymbol{b} \tag{4.8}$$

It can be noted that $\alpha = 0$ reduces the problem to the unregularized system of normal equations in Eq. 4.6, which can be solved using the pseudoinverse matrix $\boldsymbol{A}^\dagger$.

An important difficulty of applying the Tikhonov regularization is represented by the proper choice of the regularization parameter α (GOLUB et al., 1979; MOROZOV, 1984; GFRERER, 1987; HANSEN, 1992). The generalized cross-validation (GCV) (GOLUB et al., 1979) and the L-curve criterion (HANSEN, 1992) represent two widely used approaches used in this context. Although the performance of the two methods is still subject to dispute, the latter seems to be more widely used in the field of dynamic force reconstruction so far (BUSBY et al., 1997; HANSEN, 2001; CHOI et al., 2007; LAW et al., 2011). Therefore, only the L-curve method will be briefly discussed in the following. The reader is referred to the original work of GOLUB et al. (1979) and the reference works of HANSEN (1998) and ENGL et al. (2000) for insights in the GCV.

The L-curve represents a double logarithmic plot of the norm $\|\boldsymbol{\Gamma}\boldsymbol{x}\|_2$ of a regularized solution against the corresponding residual norm $\|\boldsymbol{A}\boldsymbol{x}-\boldsymbol{b}\|_2$:

$$\mathcal{L} = \{\log\|\boldsymbol{A}\boldsymbol{x}-\boldsymbol{b}\|_2\,, \log\|\boldsymbol{\Gamma}\boldsymbol{x}\|_2\} \tag{4.9}$$

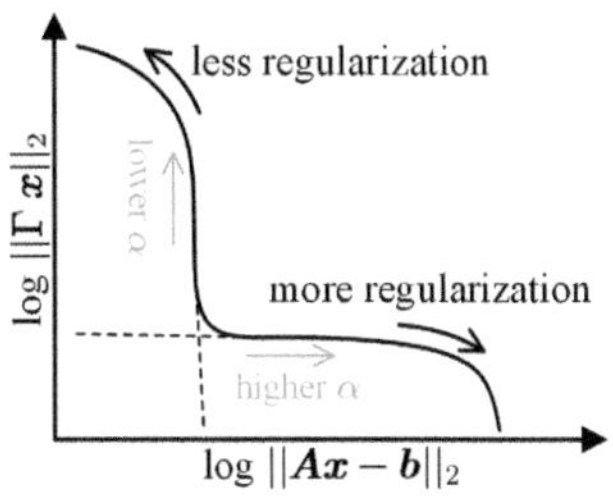

Figure 4.1 General form of the L-curve (HANSEN, 1998)

The curve described by Eq. 4.9 and exemplarily illustrated in Fig. 4.1 often presents an L-shaped appearance (HANSEN, 1998), which provides a good insight into the trade-off between minimizing the solution norm and the residual norm in dependency of α. The regularization parameter close to the "corner" of the curve is considered as optimal. A more formal definition of the optimal α refers to the point of maximal curvature of the L-curve (HANSEN, 1998; HANSEN, 2001; KERN, 2016). However, it should be noted that this method can be computationally demanding in case of large discrete problems, since a very dense set of discrete regularization parameters has to be considered when generating the L-curve, especially in the region of the "corner" (HANSEN, 1998).

4.3.3 Singular value decomposition

The singular value decomposition (SVD) is an alternative, simple and accurate way to compute the pseudoinverse matrix $\boldsymbol{A}^{\dagger}$. Moreover, it can be used to analyze the numerical characteristic of the ill-posedness of an inverse problem by evaluating the decreasing rate of the singular values. High decreasing rates indicate a high grade of instability of the solution with respect to small perturbations in the measurement vector $\boldsymbol{b}$. The SVD of a real matrix $\boldsymbol{A} \in \mathbb{R}^{m,n}$ of rank $r \leq n$ is given by the following factorization of three matrices with special properties:

$$\boldsymbol{A} = \boldsymbol{U}\boldsymbol{\Sigma}\boldsymbol{V}^{\top} \tag{4.10}$$

where $\boldsymbol{\Sigma} \in \mathbb{R}^{m,n}$ is a diagonal matrix[5] representing the singular values σ_i of $\boldsymbol{A}$ in descending order. They are determined as the nonnegative square roots of the eigenvalues of $\boldsymbol{A}\boldsymbol{A}^{\top}$ or $\boldsymbol{A}^{\top}\boldsymbol{A}$. It is to be noted that the first r singular values are greater that zero ($\sigma_1 > \sigma_2 > ... > \sigma_r > 0$), while the following singular values are zero or very close to zero ($\sigma_{r+1} \approx ... \approx \sigma_n \approx 0$). $\boldsymbol{U} \in \mathbb{R}^{m,m}$ is an orthogonal matrix, whose columns represent the eigenvectors of $\boldsymbol{A}\boldsymbol{A}^{\top}$ ordered according to the singular values σ_i in $\boldsymbol{\Sigma}$. The columns of

[5] Recall that the definition of a diagonal matrix is not limited to square matrices. An arbitrary rectangular matrix $\boldsymbol{\Sigma} \in \mathbb{R}^{m,n}$ is denoted as diagonal matrix if all the terms $\Sigma_{i,k} = 0$ for $i \neq k$.

the matrix $V \in \mathbb{R}^{n,n}$ depict the eigenvectors of $A^\top A$, arranged as well according to the singular values in Σ.

The condition number $\mu(A)$ of a matrix $A \in \mathbb{R}^{m,n}$ of rank r can be generally expressed as the ratio between its greatest and lowest singular values[6] (cf. Eq. 4.3):

$$\mu(A) = \frac{\sigma_1}{\sigma_r} = \|A\|_2 \left\|A^\dagger\right\|_2 \tag{4.11}$$

The pseudoinverse $A^\dagger$ can be determined by introducing Eq. 4.10 into Eq. 4.6 and performing some algebraic matrix operations:

$$A^\dagger = V\Sigma^\dagger U^\top \tag{4.12}$$

Using Eq. 4.12, the solution $x = A^\dagger b$ (cf. Eq. 4.6) can be expressed as:

$$x = V\Sigma^\dagger U^\top b \tag{4.13}$$

Since Σ is a rectangular matrix, it can not be inverted directly. Therefore, the term $\Sigma^\dagger$ in Eq. 4.12 and 4.13 denotes the pseudoinverse of Σ (KABANIKHIN, 2011).

In the following, let the i-th columns of U and V be denoted as u_i and v_i, respectively. The SVD from Eq. 4.10 can be expressed in compact form as:

$$A = U\Sigma V^\top = \sum_{i=1}^{r} \sigma_i u_i v_i^\top \tag{4.14}$$

Considering Eq. 4.14 and Eq. 4.13, the explicit form of the solution $x = A^\dagger b$ becomes:

$$x = \sum_{i=1}^{r} \frac{v_i u_i^\top b}{\sigma_i} \tag{4.15}$$

Eq. 4.15 provides valuable information about the sensitivity of the solution x to measurement errors in b. Dividing the errors by small singular values σ_i can magnify the errors to an indefinite level, affecting thus the stability of the solution.

In order to overcome this issue, the truncated singular value decomposition (TSVD) can be applied. Its basic idea is to limit the number of summation terms r in Eq. 4.15 by adopting a new value $p < r$, which ensures that only the largest singular values are taken into consideration:

$$x = \sum_{i=1}^{p} \frac{v_i u_i^\top b}{\sigma_i} \tag{4.16}$$

[6] Eq. 4.3 implied a non-singular matrix A, while Eq. 4.11 is formulated for any matrix $A \in \mathbb{R}^{m,n}$ of rank r. This is why the norm based expression of the condition number includes the pseudoinverse $A^\dagger$ instead of the classical inverse, as in Eq. 4.3.

Eq. 4.16 can be interpreted as the replacement of the system matrix $\boldsymbol{A}$ with an approximated matrix $\boldsymbol{A}_p$ of lower rank and neglecting the components on the right hand side of equation Eq. 4.14 corresponding to all $i > p$ (HANSEN, 1987). Choosing an appropriate value for p implies that the condition number $\mu(\boldsymbol{A}) = \sigma_1/\sigma_p$ will be moderate, ensuring the numerical stability of the inverse problem. A first step in finding the optimal number p is to sort the singular values in order to find jumps (steep decreases) between two subsequent values. If there are no jumps, p has to be manually adjusted, so that the numerical stability of the solution is achieved (SANTAMARINA et al., 2005).

4.4 Solutions based on optimization algorithms

4.4.1 General form of least squares problems

The least squares problem (cf. Eq. 4.5) can be written in a more general form as:

$$\min_{\boldsymbol{x}} \mathcal{F}(\boldsymbol{x}) = \sum_{i=1}^{m} (u_i(\boldsymbol{x}) - \bar{u}_i)^2 = \sum_{i=1}^{m} r_i(\boldsymbol{x})^2 = \|\boldsymbol{r}(\boldsymbol{x})\|_2^2 \tag{4.17}$$

where $u_i(\boldsymbol{x})$ and $\bar{u}_i$ represent the computed and observed (measured) quantities, respectively. $\boldsymbol{r}$ is a residual function from $\mathbb{R}^n$ to $\mathbb{R}^m$ consisting of the m components $r_i(\boldsymbol{x})$. If $\boldsymbol{r}(\boldsymbol{x})$ can be expressed in the form

$$\boldsymbol{r}(\boldsymbol{x}) = \boldsymbol{A}\boldsymbol{x} - \boldsymbol{b}, \tag{4.18}$$

the problem can be solved according to the approaches presented in Section 4.3.1 to 4.3.3. However, it is to be mentioned that these solution methods might be inconvenient in some cases. This holds, for example, when dealing with large-size matrix operations, which can require an extensive amount of computation time or might even become impossible at all. Furthermore, a representation according to Eq. 4.18, which corresponds to an inverse problem of type $\boldsymbol{A}\boldsymbol{x} = \boldsymbol{b}$, is not always convenient or achievable. The inverse problem has to be then formulated and solved in its general form (Eq. 4.17). This situation occurs, for example, when dealing with inverse problems implying nonlinear systems or when the system matrix $\boldsymbol{A}$ of a linear system can not be explicitly formulated. The latter case arises, for example, when the (linear) system behavior is described by a finite element (FE) model, while a set of structural parameters (e.g. stiffness) are to be identified based on a set of measured displacements. The classical representation of the inverse problem according to Eq. 4.1 implies that the term $\boldsymbol{A}\boldsymbol{x}$ describes the computed displacement vector at the measured locations, while $\boldsymbol{x}$ contains the set of parameters to be optimized. However, the direct results of a finite element analysis (FEA) are the structural displacements, i.e. the term $\boldsymbol{A}\boldsymbol{x}$ is directly retrieved. Under consideration of the numerical nature and approximate character of the finite element method (FEM) solutions, it is obvious that an explicit representation of the type $\boldsymbol{Ax} = \boldsymbol{b}$ is usually not possible.

4.4.2 Concept of unconstrained optimization

In order to solve least squares problems formulated in their general form (Eq. 4.17), different optimization algorithms[7] can be employed. The subcategory of algorithms for unconstrained optimization is relevant within the framework of this thesis. In unconstrained optimization, one seeks to minimize an *objective* function $\mathcal{F}(\boldsymbol{x})$, which only depends on real variables, without any restrictions with respect to the values of the variables. The mathematical expression of the minimization problem is:

$$\min_{\boldsymbol{x}} \mathcal{F}(\boldsymbol{x}) \tag{4.19}$$

where $\boldsymbol{x} \in \mathbb{R}^n$ is a real vector consisting of n entries and $\mathcal{F}(\boldsymbol{x})$ is a differentiable function. The necessary condition for the n-dimensional minimization problem in Eq. 4.19 is that the first partial derivatives of the objective function, comprised in gradient vector $\nabla\mathcal{F}(\boldsymbol{x})$, vanish in the optimal point $\boldsymbol{x}^*$ (HARZHEIM, 2019):

$$\nabla\mathcal{F}(\boldsymbol{x}^*) = \begin{pmatrix} \frac{\partial \mathcal{F}}{\partial x_1}(\boldsymbol{x}^*) \\ \frac{\partial \mathcal{F}}{\partial x_2}(\boldsymbol{x}^*) \\ \vdots \\ \frac{\partial \mathcal{F}}{\partial x_n}(\boldsymbol{x}^*) \end{pmatrix} = \begin{pmatrix} 0 \\ 0 \\ \vdots \\ 0 \end{pmatrix} = \mathbf{0} \tag{4.20}$$

If Eq. 4.20 is satisfied, the function has no ascent direction anymore at $\boldsymbol{x}^*$. However, this is a property for all minimum, maximum and saddle points. This means that a further – sufficient – condition is needed, in order to find a minimizer of Eq. 4.19. It states that the Hessian matrix, consisting of the second partial derivatives of the objective function:

$$\boldsymbol{H}(\boldsymbol{x}^*) = \begin{pmatrix} \frac{\partial^2 \mathcal{F}}{\partial x_1 \partial x_1}(\boldsymbol{x}^*) & \cdots & \frac{\partial^2 \mathcal{F}}{\partial x_1 \partial x_n}(\boldsymbol{x}^*) \\ \vdots & \ddots & \vdots \\ \frac{\partial^2 \mathcal{F}}{\partial x_n \partial x_1}(\boldsymbol{x}^*) & \cdots & \frac{\partial^2 \mathcal{F}}{\partial x_n \partial x_n}(\boldsymbol{x}^*) \end{pmatrix} \tag{4.21}$$

[7] From mathematical point of view, any inverse problem can be interpreted as an optimization problem, which aims to find the optimal parameter set $\boldsymbol{x}$ minimizing the objective function $\mathcal{F}(\boldsymbol{x})$. Therefore, classical algorithms for mathematical optimization can be used to solve both linear and nonlinear inverse problems.

has to be positive definite[8].

All algorithms for unconstrained optimization problems require the setting of an initial point $\boldsymbol{x}^{(0)}$. Beginning at the starting point, the algorithm generates a set of iterates $\boldsymbol{x}^{(k)}$. Each new set of variables $\boldsymbol{x}^{(k+1)}$ delivers a lower value of the objective function than in the previous iteration step: $\mathcal{F}(\boldsymbol{x}^{(k+1)}) < \mathcal{F}(\boldsymbol{x}^{(k)})$. The iteration process stops either when no more progress is made or if a sufficient accuracy of the approximated solution is achieved.

Detailed descriptions of different optimization algorithms are given in FLETCHER (1987), DENNIS JR. et al. (1996), NOCEDAL et al. (2006), SCHUMACHER (2013) and HARZHEIM (2019). In general, a distinction can be made between two main optimization strategies: line search and trust region methods. The developments of the present work are based on the latter. Therefore, it will be briefly introduced in the following section.

4.4.3 Trust region strategy

4.4.3.1 Outline of the trust region concept

The *trust region* strategy implies the approximation of the objective function $\mathcal{F}(\boldsymbol{x})$ at the current point $\boldsymbol{x}^{(k)}$ using a simpler function $m^{(k)}$, which has a similar behavior as the original objective function in the neighborhood of $\boldsymbol{x}^{(k)}$. Within this region, the model function is *trusted* to be a good representation of the original objective function. The concept can be explained using Fig. 4.2, which illustrates a comparison between the line search and trust-region strategies for a function with two variables. A line search algorithm attempts to search the minimum of the approximated function along the line search direction, but

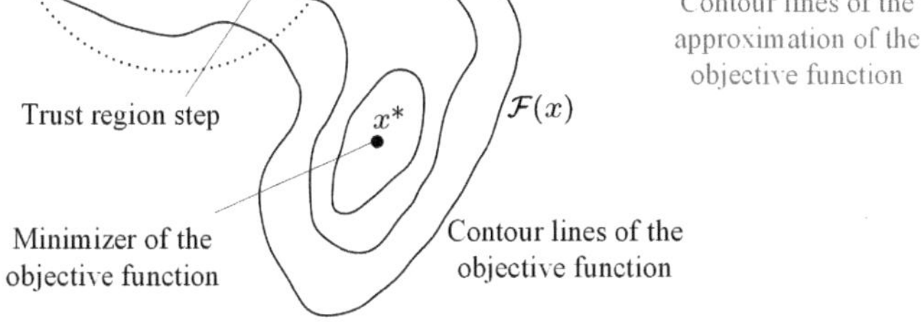

Figure 4.2 Illustration of the line search and trust region steps (NOCEDAL et al., 2006)

[8] A matrix is denoted as positive definite when all its eigenvalues are greater that zero. This also means that the function has positive curvatures in all directions.

this process will reveal a rather small reduction of the objective function in the illustrated case. The trust region method aims to minimize the approximated function $m^{(k)}$ within the trusted region. This leads to a stronger reduction of the objective function and thus to a better progress towards the minimizer $\boldsymbol{x}^*$.

Usually, the model function $m^{(k)}$ is defined as a quadratic approximation of $\mathcal{F}(\boldsymbol{x})$:

$$m^{(k)}(\boldsymbol{p}) = \mathcal{F}(\boldsymbol{x}^{(k)}) + \nabla^\top \mathcal{F}(\boldsymbol{x}^{(k)}) \cdot \boldsymbol{p} + \frac{1}{2} \boldsymbol{p}^\top \tilde{\boldsymbol{H}}^{(k)} \boldsymbol{p} \tag{4.22}$$

where $(\boldsymbol{x}^{(k)} + \boldsymbol{p})$ lies inside the trust region. The matrix $\tilde{\boldsymbol{H}}$ denotes the Hessian $\boldsymbol{H} = \nabla^2 \mathcal{F}(\boldsymbol{x}^{(k)})$ or some approximation of it (NOCEDAL et al., 2006).

An important aspect of the optimization algorithms based on the trust region strategy is the selection of the trust region radius $\Delta^{(k)}$ at each iteration step. It can be chosen based on the agreement between the function $\mathcal{F}(\boldsymbol{x}^{(k)})$ and its approximation $m^{(k)}$:

$$\rho^{(k)} = \frac{\mathcal{F}(\boldsymbol{x}^{(k)}) - \mathcal{F}(\boldsymbol{x}^{(k)} + \boldsymbol{p}^{(k+1)})}{m^{(k)}(\boldsymbol{0}) - m^{(k)}(\boldsymbol{p}^{(k+1)})} \tag{4.23}$$

The denominator of Eq. 4.23 denotes the so called *predicted reduction*, while the numerator describes the *actual reduction* (NOCEDAL et al., 2006). Since $\boldsymbol{p}^{(k+1)}$ represents the step obtained by minimizing the approximation $m^{(k)}$ within the region of radius $\Delta^{(k)}$ (explained later, cf. Eq. 4.24), which also includes $\boldsymbol{p} = \boldsymbol{0}$, the denominator will always provide a positive value. Therefore, a negative $\rho^{(k)}$ can only occur due to an increase of the objective function, i.e. negative numerator in Eq. 4.23. In this case, the step has to be rejected and the region has to be reduced for the next iteration. Moreover, $\Delta^{(k)}$ has to be lowered also when the value of $\rho^{(k)}$ is positive, but very close to zero. On the contrary, if the value of the ratio $\rho^{(k)}$ is close to 1, $m^{(k)}$ is a good approximation of the original function within the region of radius $\Delta^{(k)}$. In this case, the trust region can be expanded for the next iteration step. In general, the region may be left unchanged for positive values of $\rho^{(k)}$, which are significantly lower than 1 but not very close to zero.

After fixing the radius $\Delta^{(k)}$, the trust region method simultaneously selects the step size and the direction $\boldsymbol{p}$, which lead to the strongest reduction of the objective function subjected to the constraint of the maximal distance (NOCEDAL et al., 2006; DENNIS JR. et al., 1996). This is achieved by solving the following subproblem:

$$\min_{\boldsymbol{p}} m^{(k)}(\boldsymbol{p}) = \mathcal{F}(\boldsymbol{x}^{(k)}) + \nabla^\top \mathcal{F}(\boldsymbol{x}^{(k)}) \cdot \boldsymbol{p} + \frac{1}{2} \boldsymbol{p}^\top \tilde{\boldsymbol{H}}^{(k)} \boldsymbol{p} \quad \text{so that } \|\boldsymbol{p}\|_2 \leq \Delta^{(k)} \tag{4.24}$$

If $\tilde{\boldsymbol{H}}^{(k)}$ is a positive definite matrix and $\|-\tilde{\boldsymbol{H}}^{(k)^{-1}} \nabla \mathcal{F}(\boldsymbol{x}^{(k)})\|_2 \leq \Delta^{(k)}$, the solution of Eq. 4.24 can be determined as the unconstrained minimum of the quadratic function $m^{(k)}$:

$$p_{\mathrm{H}}^{*(k+1)} = -\tilde{H}^{(k)^{-1}} \nabla \mathcal{F}(x^{(k)}) \tag{4.25}$$

However, in other cases the solution $p^{*(k+1)}$ of Eq. 4.24 is not straightforward. Nevertheless, an approximate solution that ensures convergence can still be computed with reasonable computational effort (NOCEDAL et al., 2006). In this sense, $p^{*(k+1)}$ is a solution of the subproblem Eq. 4.24 only if it satisfies:

$$(\tilde{H}^{(k)} + \lambda I) \cdot p^{*(k+1)} = -\nabla \mathcal{F}(x^{(k)}) \tag{4.26}$$

and the following conditions are fulfilled:

$$\lambda \cdot (\Delta^{(k)} - \|p^{*(k+1)}\|) = 0 \tag{4.27a}$$

$$\tilde{H}^{(k)} + \lambda I \text{ is positive semidefinite} \tag{4.27b}$$

Eq. 4.27a indicates that when the solution is strictly included in the trust region, λ has to be zero and the solution $p^{*(k+1)}$ is given by Eq. 4.25. In other cases, λ may take positive values, since we have $\Delta^{(k)} = \|p^{*(k+1)}\|$. Eq. 4.26 yields under consideration of $\lambda \neq 0$:

$$\lambda \cdot p^{*(k+1)} = -\tilde{H}^{(k)} \cdot p^{*(k+1)} - \nabla \mathcal{F}(x^{(k)}) = -\nabla m^{(k)}(p^{*(k+1)}) \tag{4.28}$$

which shows that the solution $p^{*(k+1)}$ and the negative gradient of the model function $m^{(k)}$ are collinear. Moreover, $p^{*(k+1)}$ is perpendicular to the contour lines of $m^{(k)}$ (NOCEDAL et al., 2006). In the following, three methods for finding approximate solutions of Eq. 4.24 will be briefly discussed in Section 4.4.3.2 to 4.4.3.4. For the sake of completeness, an additional (nearly) exact iterative solution approach is briefly introduced in Appendix A.3.

4.4.3.2 Cauchy point

The *Cauchy point* method is similar to the basic concept of the steepest descent[9] in the line search strategy (NOCEDAL et al., 2006; HARZHEIM, 2019). The Cauchy point in iteration k is denoted as $p_{\mathrm{C}}^{(k+1)}$. It represents the point lying on the gradient, which minimizes the quadratic function $m^{(k)}$ considering that the step length has to lie within the trust region. It can be expressed by the following closed form relations (NOCEDAL et al., 2006):

$$p_{\mathrm{C}}^{(k+1)} = -\tau^{(k)} \frac{\Delta^{(k)}}{\|\nabla \mathcal{F}(x^{(k)})\|_2} \nabla \mathcal{F}(x^{(k)}) \tag{4.29}$$

where

[9] The steepest descent is the most intuitive method for determining the search direction within a line search algorithm. It makes use of the property of the gradient stating that it is always oriented towards the steepest ascent (HARZHEIM, 2019).

$$\tau^{(k)} = \begin{cases} 1 & \text{if } \nabla^\top \mathcal{F}(\boldsymbol{x}^{(k)})\tilde{\boldsymbol{H}}^{(k)}\nabla\mathcal{F}(\boldsymbol{x}^{(k)}) \leq 0 \\ \min\left(\dfrac{\|\nabla\mathcal{F}(\boldsymbol{x}^{(k)})\|_2^3}{\Delta^{(k)}\nabla^\top\mathcal{F}(\boldsymbol{x}^{(k)})\tilde{\boldsymbol{H}}^{(k)}\nabla\mathcal{F}(\boldsymbol{x}^{(k)})}, 1\right) & \text{otherwise} \end{cases} \tag{4.30}$$

It is obvious that the Cauchy step $\boldsymbol{p}_{\mathrm{C}}^{(k+1)}$ can be computed with reasonable effort, since no matrix factorizations are needed. Moreover, it represents a key criterion in deciding whether the solution of the subproblem is acceptable or not. However, in many cases this algorithm proves inefficient with respect to its convergence behavior, since the Cauchy point is only slightly dependent on the matrix $\tilde{\boldsymbol{H}}^{(k)}$, which is used only for the determination of the step length (Eq. 4.30). A significant improvement of the convergence is achieved, if the matrix $\tilde{\boldsymbol{H}}^{(k)}$ contains curvature information of the original function and contributes to the definition of the direction and length of the step.

In general, it can be stated that the global convergence of any trust region algorithm can be only achieved if its step $\boldsymbol{p}^{(k+1)}$ provides a decrease of $m^{(k)}$, which is at least some fixed positive multiple of the reduction generated by the Cauchy point $\boldsymbol{p}_{\mathrm{C}}^{(k+1)}$. Therefore, many algorithms imply the computation of the Cauchy point and subsequently seek to improve it, whereas the consideration of $\tilde{\boldsymbol{H}}^{(k)}$ as the exact Hessian matrix $\boldsymbol{H}^{(k)}$ results in super-linear convergence (DENNIS JR. et al., 1996; NOCEDAL et al., 2006).

4.4.3.3 Dogleg method

The *dogleg* method, firstly published by POWELL (1970), is a procedure used to approximately solve the subproblem in Eq. 4.24. It can be used for positive definite matrices $\tilde{\boldsymbol{H}}^{(k)}$. The basic idea is to determine a V-shaped trajectory of the iterations. The first line segment results from the movement from the origin along the steepest descent direction:

$$\boldsymbol{p}_{\mathrm{U}}^{(k+1)} = -\frac{\nabla^\top\mathcal{F}(\boldsymbol{x}^{(k)})\cdot\nabla\mathcal{F}(\boldsymbol{x}^{(k)})}{\nabla^\top\mathcal{F}(\boldsymbol{x}^{(k)})\cdot\tilde{\boldsymbol{H}}^{(k)}\cdot\nabla\mathcal{F}(\boldsymbol{x}^{(k)})}\cdot\nabla\mathcal{F}(\boldsymbol{x}^{(k)}). \tag{4.31}$$

The polygonal trajectory, denoted as $\tilde{\boldsymbol{p}}^{(k+1)}$, can be expressed for a factor $\tau^{(k)} \in [0,2]$ as:

$$\tilde{\boldsymbol{p}}^{(k+1)}(\tau^{(k)}) = \begin{cases} \tau^{(k)}\cdot\boldsymbol{p}_{\mathrm{U}}^{(k+1)} & \text{for } 0 \leq \tau^{(k)} \leq 1 \\ \boldsymbol{p}_{\mathrm{U}}^{(k+1)} + (\tau^{(k)}-1)(\boldsymbol{p}_{\mathrm{H}}^{(k+1)} - \boldsymbol{p}_{\mathrm{U}}^{(k+1)}) & \text{for } 1 \leq \tau^{(k)} \leq 2 \end{cases} \tag{4.32}$$

where $\boldsymbol{p}_{\mathrm{H}}^{(k+1)}$ is the optimal solution of the quadratic model $m^{(k)}(\boldsymbol{p})$ (cf. Eq. 4.25). The dogleg method minimizes the model function $m^{(k)}$ along the path $\tilde{\boldsymbol{p}}^{(k+1)}(\tau^{(k)})$. It is to be noted that $\tilde{\boldsymbol{p}}^{(k+1)}(\tau^{(k)})$ intersects the boundary of the trust region at exactly one point if $\boldsymbol{p}_{\mathrm{H}}^{(k+1)} \geq \Delta^{(k)}$ and nowhere for $\boldsymbol{p}_{\mathrm{H}}^{(k+1)} < \Delta^{(k)}$. The latter case implies $\tilde{\boldsymbol{p}}^{(k+1)} = \boldsymbol{p}_{\mathrm{H}}^{(k+1)}$, while for the first case ($p_{\mathrm{H}}^{(k+1)} \geq \Delta^{(k)}$), an appropriate value of $\tau^{(k)}$ has be computed from the following scalar quadratic equation:

$$\left\| \boldsymbol{p}_{\mathrm{U}}^{(k+1)} + (\tau^{(k)} - 1)(\boldsymbol{p}_{\mathrm{H}}^{(k+1)} - \boldsymbol{p}_{\mathrm{U}}^{(k+1)}) \right\|_2^2 = \Delta^{(k)} \tag{4.33}$$

As mentioned above, the dogleg method is suitable for solving the trust region subproblem with a positive definite matrix $\tilde{\boldsymbol{H}}$. The two-dimensional subspace minimization, presented in Section 4.4.3.4, is a method that includes the directions of negative curvature (indefinite $\tilde{\boldsymbol{H}}$) in the space of candidate trust region steps.

4.4.3.4 Two-dimensional subspace minimization

The two-dimensional subspace minimization method (BYRD et al., 1988; SCHULTZ et al., 1985) implies the widening of the solution space of $\boldsymbol{p}$ to the two-dimensional subspace spanned by $\boldsymbol{p}_{\mathrm{U}}$ and $\boldsymbol{p}_{\mathrm{H}}$. Since $\boldsymbol{p}_{\mathrm{U}}$ is oriented in the steepest descent direction (cf. Eq. 4.31) and $\boldsymbol{p}_{\mathrm{H}}$ is oriented along the Newton direction (cf. Eq. 4.25), the subspace span$[\boldsymbol{p}_{\mathrm{U}}, \boldsymbol{p}_{\mathrm{H}}]$ is equivalent to the subspace spanned by $-\nabla\mathcal{F}(\boldsymbol{x}^{(k)})$ and $-\tilde{\boldsymbol{H}}^{(k)^{-1}}\nabla\mathcal{F}(\boldsymbol{x}^{(k)})$. In this sense, the trust region subproblem (Eq. 4.24) is slightly modified to:

$$\begin{aligned} &\min_{\boldsymbol{p}} m^{(k)}(\boldsymbol{p}) = \mathcal{F}(\boldsymbol{x}^{(k)}) + \nabla^\top \mathcal{F}(\boldsymbol{x}^{(k)}) \cdot \boldsymbol{p} + \frac{1}{2}\boldsymbol{p}^\top \tilde{\boldsymbol{H}}^{(k)} \boldsymbol{p} \\ &\text{so that } \|\boldsymbol{p}\|_2 \leq \Delta^{(k)} \text{ and } \boldsymbol{p} \in \text{span}\left[\nabla\mathcal{F}(\boldsymbol{x}^{(k)}),\ \tilde{\boldsymbol{H}}^{(k)^{-1}}\nabla\mathcal{F}(\boldsymbol{x}^{(k)})\right] \end{aligned} \tag{4.34}$$

Since Eq. 4.34 is now a problem in two variables, its solution becomes trivial (under assumption of a given subspace). After some algebraic manipulations, the problem can be reduced to the root-finding of a fourth degree polynomial (NOCEDAL et al., 2006). Therefore, the main task in the two-dimensional subspace approach is not the solution of the minimization problem Eq. 4.34 itself, but the determination of the subspace.

Eq. 4.34 indicates that, in case of positive definite matrix $\tilde{\boldsymbol{H}}^{(k)}$, the two-dimensional subspace minimization is an extension of the dogleg method, since the entire dogleg search trajectory is included in span$[\boldsymbol{p}_{\mathrm{U}}, \boldsymbol{p}_{\mathrm{H}}]$. A further modification allows the treatment of problems, where $\tilde{\boldsymbol{H}}^{(k)}$ possesses negative eigenvalues. This implies the modification of the subspace given in Eq. 4.34 as follows:

$$\text{span}\left[\nabla\mathcal{F}(\boldsymbol{x}^{(k)}), (\tilde{\boldsymbol{H}}^{(k)} + \alpha\boldsymbol{I})^{-1}\nabla\mathcal{F}(\boldsymbol{x}^{(k)})\right] \text{ for some } \alpha \in (-\lambda_1, -2\lambda_1] \tag{4.35}$$

where λ_1 represents the most negative eigenvalue of $\tilde{\boldsymbol{H}}$. SCHULTZ et al. (1985) and BYRD et al. (1988) provide a methodology for determination of α in Eq. 4.35, which in any case ensures the positive definiteness of $\tilde{\boldsymbol{H}}^{(k)} + \alpha\boldsymbol{I}$.

If $\|(\tilde{\boldsymbol{H}}^{(k)}+\alpha\boldsymbol{I})^{-1}\nabla\mathcal{F}(\boldsymbol{x}^{(k)})\|_2 \leq \Delta^{(k)}$, the subspace search according to Eq. 4.34 and 4.35 is discarded. In this case, the step is defined as follows:

$$\boldsymbol{p}^{(k+1)} = -(\tilde{\boldsymbol{H}}^{(k)}+\alpha\boldsymbol{I})^{-1}\nabla\mathcal{F}(\boldsymbol{x}^{(k)})+\boldsymbol{v} \tag{4.36}$$

where $\boldsymbol{v}$ denotes a vector that satisfies $\boldsymbol{v}^\top(\tilde{\boldsymbol{H}}^{(k)}+\alpha\boldsymbol{I})^{-1}\nabla\mathcal{F}(\boldsymbol{x}^{(k)}) \leq 0$, thus accounting for $\|\boldsymbol{p}^{(k+1)}\|_2 \geq \|(\tilde{\boldsymbol{H}}^{(k)}+\alpha\boldsymbol{I})^{-1}\nabla\mathcal{F}(\boldsymbol{x}^{(k)})\|_2$ (SCHULTZ et al., 1985; BYRD et al., 1988). For a matrix $\tilde{\boldsymbol{H}}^{(k)}$ with zero eigenvalues, the step is defined as the Cauchy point: $\boldsymbol{p}^{(k+1)} = \boldsymbol{p}_{\mathrm{C}}^{(k+1)}$ (NOCEDAL et al., 2006).

If the exact Hessian can be provided ($\tilde{\boldsymbol{H}}^{(k)} = \nabla^2\mathcal{F}(\boldsymbol{x}^{(k)})$), then $-\tilde{\boldsymbol{H}}^{(k)^{-1}}\nabla\mathcal{F}(\boldsymbol{x}^{(k)}$ represents the Newton step. Therefore, for points $\boldsymbol{x}^{(k)}$ lying in the vicinity of the optimal point $\boldsymbol{x}^{*(k)}$ (where the Hessian is expected to be positive definite) and for a sufficiently large $\Delta^{(k)}$, the solution of Eq. 4.34 is given by the Newton step.

The two-dimensional subspace minimization is computationally effective, since it requires only one matrix factorization (of $\tilde{\boldsymbol{H}}^{(k)}$ or $\tilde{\boldsymbol{H}}^{(k)}+\alpha\boldsymbol{I}$). Moreover, it achieves a reduction of the quadratic model function m, which is often close to the (nearly) exact – but more computationally demanding – iterative method presented in Appendix A.3.

4.4.4 Sensitivity analysis

All previously discussed unconstrained optimization algorithms require the knowledge of the gradients (sensitivities) of the objective function:

$$\nabla\mathcal{F}(\boldsymbol{x}) = \begin{pmatrix} \dfrac{\partial\mathcal{F}}{\partial x_1}(\boldsymbol{x}) & \dfrac{\partial\mathcal{F}}{\partial x_2}(\boldsymbol{x}) & \cdots & \dfrac{\partial\mathcal{F}}{\partial x_n}(\boldsymbol{x}) \end{pmatrix}^\top \tag{4.37}$$

The determination of the gradients plays a crucial role within a gradient-based optimization process, since it may take up most of the computing time. It is therefore a decisive factor for the effectiveness of a gradient-based optimization method (HARZHEIM, 2019).

Besides the gradient computation, several optimization algorithms require also the knowledge of the Hessian matrix, which contains the second derivatives of the function:

$$\boldsymbol{H}(\boldsymbol{x}) = \begin{pmatrix} \dfrac{\partial^2\mathcal{F}}{\partial x_1\partial x_1}(\boldsymbol{x}) & \cdots & \dfrac{\partial^2\mathcal{F}}{\partial x_1\partial x_n}(\boldsymbol{x}) \\ \vdots & \ddots & \vdots \\ \dfrac{\partial^2\mathcal{F}}{\partial x_n\partial x_1}(\boldsymbol{x}) & \cdots & \dfrac{\partial^2\mathcal{F}}{\partial x_n\partial x_n}(\boldsymbol{x}) \end{pmatrix} \tag{4.38}$$

In some cases, the function allows for easy calculation of its derivatives, e.g. by symbolic differentiation, which eventually provides algebraic expressions for each component of the gradient and the Hessian. However, in many applications, the functions are too complicated to be analytically differentiated or they can not be expressed analytically at all. Therefore, different approaches for approximation of the derivatives are available. Two widely used methods will be discussed in the following sections.

4.4.4.1 Finite difference method

The idea of the finite difference method is to analyze the change of the function values with respect to small perturbations of a variable around a given point. The sensitivity is defined as the ratio between the computed function difference and the variable difference. The *forward difference* method is widely used for determining the partial derivative $\partial\mathcal{F}/\partial x_k$ of a multivariable function $\mathcal{F}(\boldsymbol{x})$ with respect to a certain variable x_k:

$$\frac{\partial\mathcal{F}}{\partial x_k} \approx \frac{\mathcal{F}(\boldsymbol{x}+\Delta x_k\cdot\boldsymbol{e}_k)-\mathcal{F}(\boldsymbol{x})}{\Delta x_k} \tag{4.39}$$

where Δx_k represents a small positive scalar used as perturbation factor of the variable x_k. $\boldsymbol{e}_k$ depicts the unit vector along the direction of x_k, that is the vector whose elements are all 0 except for a 1 in the k-th position. It is obvious that the forward difference approach requires $n+1$ function evaluations at each point. A more accurate finite difference approximation of the gradient is given by the so called *central difference* method:

$$\frac{\partial\mathcal{F}}{\partial x_k} \approx \frac{\mathcal{F}(\boldsymbol{x}+\Delta x_k\cdot\boldsymbol{e}_k)-\mathcal{F}(\boldsymbol{x}-\Delta x_k\cdot\boldsymbol{e}_k)}{2\cdot\Delta x_k} \tag{4.40}$$

Although Eq. 4.40 delivers more precise results than the forward difference approach (Eq. 4.39), it is computationally twice as expensive, requiring $2n+1$ function evaluations at each point.

The selection of the perturbation parameter Δx_k is an important issue of the finite difference method. A natural choice would be to select Δx_k as small as possible[10]. However, when implemented on a computer, very small perturbation values can lead to difficulties with respect to the roundoff errors (NOCEDAL et al., 2006; KEMMLER, 2014). Too large values of the perturbation steps may also lead to erroneous sensitivities, especially in case of functions with high curvatures. In this regard, it is to be mentioned that the perturbation size can be adapted for each variable. This is required, for example, when the function includes variables representing different quantities with different magnitudes. The size of the perturbation parameter is usually adjusted until a convergent behavior is achieved.

In some situations, the gradient $\nabla\mathcal{F}(\boldsymbol{x})$ can be computed analytically, but not the Hessian $\boldsymbol{H}(\boldsymbol{x}) = \nabla^2\mathcal{F}(\boldsymbol{x})$. In this case, the components $\partial^2\mathcal{F}/\partial x_k\partial x_i$ of $\boldsymbol{H}(\boldsymbol{x})$ can be approximated by applying Eq. 4.39 to the function $\mathcal{F}\widehat{=}\nabla\mathcal{F}_k$ (NOCEDAL et al., 2006):

---

[10] This fact is motivated by the formal definition of the derivatives as a measure of the sensitivity of the function to *infinitesimal* changes in the values of the variables (NOCEDAL et al., 2006).

$$\frac{\partial^2 \mathcal{F}}{\partial x_k \partial x_i} \approx \frac{\nabla \mathcal{F}_k(\boldsymbol{x} + \Delta x_i \cdot \boldsymbol{e}_i) - \nabla \mathcal{F}_k(\boldsymbol{x})}{\Delta x_i} \tag{4.41}$$

where $\nabla \mathcal{F}_k$ represent the k-th entry $\partial \mathcal{F} / \partial x_k$ of $\nabla \mathcal{F}$. It is therefore obvious that the Hessian can be interpreted as the Jacobian matrix of the gradient vector. Furthermore, if the gradient $\nabla \mathcal{F}(\boldsymbol{x})$ is not available, $\partial^2 \mathcal{F} / \partial x_i \partial x_k$ can be also approximated only by means of function evaluations (DENNIS JR. et al., 1996; NOCEDAL et al., 2006):

$$\frac{\partial^2 \mathcal{F}}{\partial x_k \partial x_i} \approx \frac{\mathcal{F}(\boldsymbol{x} + \Delta x_k \cdot \boldsymbol{e}_k + \Delta x_i \cdot \boldsymbol{e}_i) - \mathcal{F}(\boldsymbol{x} + \Delta x_k \cdot \boldsymbol{e}_k) - \mathcal{F}(\boldsymbol{x} + \Delta x_i \cdot \boldsymbol{e}_i) + \mathcal{F}(\boldsymbol{x})}{\Delta x_k \cdot \Delta x_i} \tag{4.42}$$

It is to be mentioned that the numerical computation of the Hessian according to Eq. 4.41 or Eq. 4.42 is rather unlikely to be perfectly symmetric. In this case, symmetry is enforced by replacing the numerical results of the Hessian $\boldsymbol{H}$ by:

$$\tilde{\boldsymbol{H}} = \frac{1}{2} \cdot \left(\boldsymbol{H} + \boldsymbol{H}^\top \right) \tag{4.43}$$

The numerical evaluation of the Hessian according to Eq. 4.41 to 4.43 is – from computational point of view – very demanding, since it requires $n(n+1)/2$ evaluations of the objective function in each iteration step. Therefore, it is usually only practicable in isolated cases for functions with a low number of variables. Nevertheless, the numerical determination of the gradients and Hessians is often only used to ensure the correct derivation and implementation of analytical sensitivities, which are discussed in the following sections.

4.4.4.2 Direct sensitivity

The *direct sensitivity* analysis is elaborated using the example of the linear structural analysis based on the FEM. This implies that the structural displacements $\boldsymbol{u}$ are determined from the equilibrium equation:

$$\boldsymbol{K}(\boldsymbol{x}) \cdot \boldsymbol{u}(\boldsymbol{x}) = \boldsymbol{P}(\boldsymbol{x}) \tag{4.44}$$

where $\boldsymbol{K}^{n_{\text{dof}} \times n_{\text{dof}}}$ represents the stiffness matrix depending on the variable set $\boldsymbol{x}$ and $\boldsymbol{P}^{n_{\text{dof}} \times 1}$ denotes the force vector. The displacement-based formulation of the FEM problem (given in Eq. 4.44) implies the following dependency of an optimization criterion $\mathcal{F}$ on the optimization variables[11] x_k, which are comprised in the vector $\boldsymbol{x}$:

$$\mathcal{F} = \mathcal{F}(\boldsymbol{x}, \boldsymbol{u}(\boldsymbol{x})) \tag{4.45}$$

[11] In structural optimization, the unknown optimization variables are widely denoted as *design variables*.

It is clear that both explicit and implicit dependencies of $\mathcal{F}$ with respect to x_k may occur. Therefore, the chain rule has to be used in order to determine the total derivative:

$$\frac{\mathrm{d}\mathcal{F}}{\mathrm{d}x_k} = \underbrace{\sum_{h=1}^{n_{\mathrm{dof}}} \left(\frac{\partial \mathcal{F}}{\partial u_h} \cdot \frac{\partial u_h}{\partial x_k} \right)}_{\text{implicit}} + \underbrace{\frac{\partial \mathcal{F}}{\partial x_k}}_{\text{explicit}} \tag{4.46}$$

Eq. 4.46 can be expressed also in vector notation, which facilitates the implementation[12]:

$$\frac{\mathrm{d}\mathcal{F}}{\mathrm{d}x_k} = \underbrace{\left(\frac{\partial \mathcal{F}}{\partial \boldsymbol{u}} \right)^{\top} \cdot \frac{\partial \boldsymbol{u}}{\partial x_k}}_{\text{implicit}} + \underbrace{\frac{\partial \mathcal{F}}{\partial x_k}}_{\text{explicit}} \tag{4.47}$$

While the partial derivatives $\partial \mathcal{F}/\partial x_k$ and $\partial \mathcal{F}/\partial \boldsymbol{u}$ can be usually determined by a straightforward analytical differentiation of the objective function, the partial derivative of the displacement with respect to the design variables $\partial \boldsymbol{u}/\partial x_k$ represents the actual challenge of the sensitivity problem. The direct sensitivity analysis implies the determination of $\partial \boldsymbol{u}/\partial x_k$ starting from Eq. 4.44, which has to be firstly differentiated on both sides with respect to x_k:

$$\frac{\partial \boldsymbol{K}}{\partial x_k} \cdot \boldsymbol{u} + \boldsymbol{K} \cdot \frac{\partial \boldsymbol{u}}{\partial x_k} = \frac{\partial \boldsymbol{P}}{\partial x_k} \tag{4.48}$$

Rearranging Eq. 4.48 yields:

$$\boldsymbol{K} \cdot \frac{\partial \boldsymbol{u}}{\partial x_k} = \underbrace{\frac{\partial \boldsymbol{P}}{\partial x_k} - \frac{\partial \boldsymbol{K}}{\partial x_k} \cdot \boldsymbol{u}}_{\boldsymbol{P}_{\mathrm{Pseudo}}} \tag{4.49}$$

It is obvious that Eq. 4.49 has a similar form as Eq. 4.44, whereas the unknown displacement vector is replaced by the unknown sensitivity vector $\partial \boldsymbol{u}/\partial x_k$ and the force vector $\boldsymbol{P}$ is replaced by an pseudo force vector $\boldsymbol{P}_{\mathrm{Pseudo}}$. The pseudo force vector contains the explicit differentiation of the load vector and of the stiffness matrix. While the first one is usually zero, since the external forces are mostly independent of the design variables[13], the latter one is usually non-zero and can be determined either by analytical differentiation or by

---

[12] It is to be mentioned that $\partial \mathcal{F}/\partial \boldsymbol{u}$ and $\partial \boldsymbol{u}/\partial x_k$ are, in fact, standing vectors. In order to allow their multiplication, the first one has to be transposed. However, in case of further manipulations – as it will occur for example in case of the second derivative in Eq. 4.51 – the derivation of the chain rule equations under consideration of the dimension compatibility for matrix multiplication is getting difficult or even impossible. Therefore, several sources in the literature define the chain rule without transposing $\partial \mathcal{F}/\partial \boldsymbol{u}$.

[13] In case the force is dependent on the design variables, its explicit expression should be provided, allowing thus an analytical differentiation.

finite differences (HARZHEIM, 2019). Basically, the sensitivities $\partial \boldsymbol{u}/\partial x_k$ are determined by solving an additional pseudo load case of Eq. 4.44 for each variable x_k:

$$\frac{\partial \boldsymbol{u}}{\partial x_k} = \boldsymbol{K}^{-1} \cdot \underbrace{\left[\frac{\partial \boldsymbol{P}}{\partial x_k} - \frac{\partial \boldsymbol{K}}{\partial x_k} \cdot \boldsymbol{u} \right]}_{\boldsymbol{P}_{\text{Pseudo}}} \tag{4.50}$$

The solution of an additional pseudo load case for each design variable requires a significant computational effort, especially for problems with a large numbers of variables. Therefore, the application of the direct senstivity analysis is generally recommended for problems with rather low number of variables. At this point, it should be mentioned that the adjoint sensitivity analysis is another concept, which may reduce computational effort in case of problems with large number of variables, as it only requires the solution of a single pseudo load case. However, the concept of adjoint sensitivity is not applied within the present thesis. Nevertheless, the author refers to KEMMLER (2014) and HARZHEIM (2019) for detailed derivations of the adjoint method.

The entries of the Hessian can be determined by a subsequent total differentiation of Eq. 4.46 with respect to x_i under consideration of the chain rule:

$$\begin{aligned}
\frac{\mathrm{d}^2 \mathcal{F}}{\mathrm{d}x_i \mathrm{d}x_k} = & \sum_{h=1}^{n_{\text{dof}}} \left[\frac{\mathrm{d}}{\mathrm{d}x_i} \left(\frac{\partial \mathcal{F}}{\partial u_h} \cdot \frac{\partial u_h}{\partial x_k} \right) \right] + \frac{\mathrm{d}}{\mathrm{d}x_i} \left(\frac{\partial \mathcal{F}}{\partial x_k} \right) = \\
& \sum_{h=1}^{n_{\text{dof}}} \left[\frac{\mathrm{d}}{\mathrm{d}x_i} \left(\frac{\partial \mathcal{F}}{\partial u_h} \right) \cdot \frac{\partial u_h}{\partial x_k} + \left(\frac{\partial \mathcal{F}}{\partial u_h} \right) \cdot \frac{\mathrm{d}}{\mathrm{d}x_i} \left(\frac{\partial u_h}{\partial x_k} \right) \right] + \frac{\mathrm{d}}{\mathrm{d}x_i} \left(\frac{\partial \mathcal{F}}{\partial x_k} \right) = \\
& \sum_{h=1}^{n_{\text{dof}}} \left\{ \left[\frac{\partial}{\partial u_h} \left(\frac{\partial \mathcal{F}}{\partial u_h} \right) \cdot \frac{\partial u_h}{\partial x_i} + \frac{\partial}{\partial x_i} \left(\frac{\partial \mathcal{F}}{\partial u_h} \right) \right] \cdot \frac{\partial u_h}{\partial x_k} \right\} + \\
& \sum_{h=1}^{n_{\text{dof}}} \left\{ \frac{\partial \mathcal{F}}{\partial u_h} \cdot \left[\frac{\partial}{\partial u_h} \left(\frac{\partial u_h}{\partial x_k} \right) \cdot \frac{\partial u_h}{\partial x_i} + \frac{\partial^2 u_h}{\partial x_i \partial x_k} \right] \right\} + \\
& \frac{\partial}{\partial \boldsymbol{u}} \left(\frac{\partial \mathcal{F}}{\partial x_k} \right) \cdot \frac{\partial \boldsymbol{u}}{\partial x_i} + \frac{\partial^2 \mathcal{F}}{\partial x_i \partial x_k}
\end{aligned} \tag{4.51}$$

The partial derivatives $\partial u_h/\partial x_k$ are available from Eq. 4.50. Excepting $\partial^2 u_h/\partial x_i \partial x_k$, all the other components of Eq. 4.51 can be usually determined by a straightforward analytical differentiation. The computation of $\partial^2 u_h/\partial x_i \partial x_k$ is based on a further differentiation of Eq. 4.48 with respect to an arbitrary variable x_i, which delivers the following expression:

$$\frac{\partial^2 \boldsymbol{K}}{\partial x_i \partial x_k} \cdot \boldsymbol{u} + \frac{\partial \boldsymbol{K}}{\partial x_k} \cdot \frac{\partial \boldsymbol{u}}{\partial x_i} + \frac{\partial \boldsymbol{K}}{\partial x_i} \cdot \frac{\partial \boldsymbol{u}}{\partial x_k} + \boldsymbol{K} \cdot \frac{\partial^2 \boldsymbol{u}}{\partial x_i \partial x_k} = \frac{\partial^2 \boldsymbol{P}}{\partial x_i \partial x_k} \tag{4.52}$$

Eq. 4.52 can be rearranged in a similar way to Eq. 4.49, leading to the following solution:

$$\frac{\partial^2 \boldsymbol{u}}{\partial x_i \partial x_k} = \boldsymbol{K}^{-1} \cdot \left[\frac{\partial^2 \boldsymbol{P}}{\partial x_i \partial x_k} - \frac{\partial^2 \boldsymbol{K}}{\partial x_i \partial x_k} \cdot \boldsymbol{u} - \frac{\partial \boldsymbol{K}}{\partial x_k} \cdot \frac{\partial \boldsymbol{u}}{\partial x_i} - \frac{\partial \boldsymbol{K}}{\partial x_i} \cdot \frac{\partial \boldsymbol{u}}{\partial x_k} \right] \tag{4.53}$$

Eq. 4.53 shows that the direct determination of the second derivatives requires an extremely large number of operations, motivating thus the general use of iteratively approximated Hessian matrices, e.g. by means of the Broyden–Fletcher–Goldfarb–Shanno (BFGS) update (HARZHEIM, 2019).

4.5 Conclusion

In this chapter, the concept of inverse problems in bridge dynamics is introduced and two main methods for their solution are discussed: the direct approach based on the concept of normal equations and the iterative solution based on optimization algorithms. The direct solution method is suitable for solving problems formulated as a linear system of equations of the type $\boldsymbol{Ax} = \boldsymbol{b}$. However, it is well known that inverse problems are typically ill-conditioned. This means, that arbitrarily small perturbations in the vector $\boldsymbol{b}$ can lead to an uneven magnification of the solution. To suppress this effect, regularization methods have to be employed. In this respect, two widely used approaches, the Tikhonov regularization and the truncated singular value decomposition (TSVD), are presented.

Another method for solving inverse problems is based on mathematical optimization algorithms. They are usually employed if the inverse problem can not be directly expressed in the form of a system of equations $\boldsymbol{Ax} = \boldsymbol{b}$ or if its direct solution is not convenient, e.g. due to a high computational cost. The developments presented in this thesis are mainly based on the trust region optimization strategy. In this context, three different algorithms are discussed. The Cauchy point and the dogleg methods are two basic approaches, which are presented to facilitate the understanding of the general trust region concept and of the more complex two-dimensional subspace minimization strategy, which was used in this work. An important feature of the trust region methods is that they can achieve super-linear convergence if the exact gradient and Hessian of the objective function can be provided. In this respect, this chapter also explicitly discusses the principle of direct sensitivity analysis. This is used as the basis for computing the sensitivities of the MFI problem discussed in Chap. 5. In addition, the determination of the sensitivity information using the finite difference method is also presented. However, it is usually very demanding from computational point of view and is therefore used only in isolated practical case. Nevertheless, the finite difference method represents a valuable tool for verifying the derivation and implementation of the analytical direct sensitivities.

5 Formulation of an inverse problem for moving force identification

The essential objective of this chapter is to develop the mathematical framework for the new MFI procedure. On the one side, it should allow the robust solution of large problems based on measurement data from a limited number of sensors and, on the other side, it should be applicable for arbitrarily detailed FE systems.

5.1 Objective function

The basic idea of the proposed MFI problem is to find the optimal time dependent moving forces acting on a bridge structure, which generated a set of acceleration and displacement time histories measured at discrete locations over a finite number of time samples. For this, an optimization problem is formulated, which seeks – under consideration of a calibrated linear structural model and known force locations at each time step – to minimize the difference between the measured and computed displacements and accelerations.

5.1.1 Vector of unknown variables

Before proceeding to the actual definition of the objective function, the formulation of the vector of unknowns is firstly discussed. It is denoted as $\boldsymbol{x} \in \mathbb{R}^{n_{\mathrm{var}}}$ and comprises the n_{var} unknown force values x_k. Each x_k represents the value of one out of n_{F} moving forces in a certain time step. The structure of the vector of unknown variables for an arbitrarily chosen computation time window is illustrated in Fig. 5.1.

Considering that each moving force $\boldsymbol{x}_l^{n_{\mathrm{tp},l} \times 1}$ is applied in $n_{\mathrm{tp},l}$ time steps on the bridge, the vector $\boldsymbol{x}$ can be defined by stacking all n_{F} individual force vectors $\boldsymbol{x}_l$:

$$\boldsymbol{x}^{n_{\mathrm{var}} \times 1} = \begin{pmatrix} \boldsymbol{x}_1^{n_{\mathrm{tp},1} \times 1} & \boldsymbol{x}_2^{n_{\mathrm{tp},2} \times 1} & \cdots & \boldsymbol{x}_l^{n_{\mathrm{tp},l} \times 1} & \cdots & \boldsymbol{x}_{n_{\mathrm{F}}}^{n_{\mathrm{tp},n_{\mathrm{F}}} \times 1} \end{pmatrix}^{\top} \tag{5.1}$$

$\boldsymbol{x}_l$ represents the vector of the l-th force. At this point, it is worth mentioning that the computation time window may be shorter than the time required for crossing the structure (cf. Fig. 5.1). This implies different dimensions of the force vectors $\boldsymbol{x}_l$ and motivates thus the formulation based on individual number of time steps $n_{\mathrm{tp},l}$ for each force. Each entry

A. Firus, *A Contribution to Moving Force Identification in Bridge Dynamics*, Mechanik, Werkstoffe und Konstruktion im Bauwesen 65, https://doi.org/10.1007/978-3-658-39838-5_5

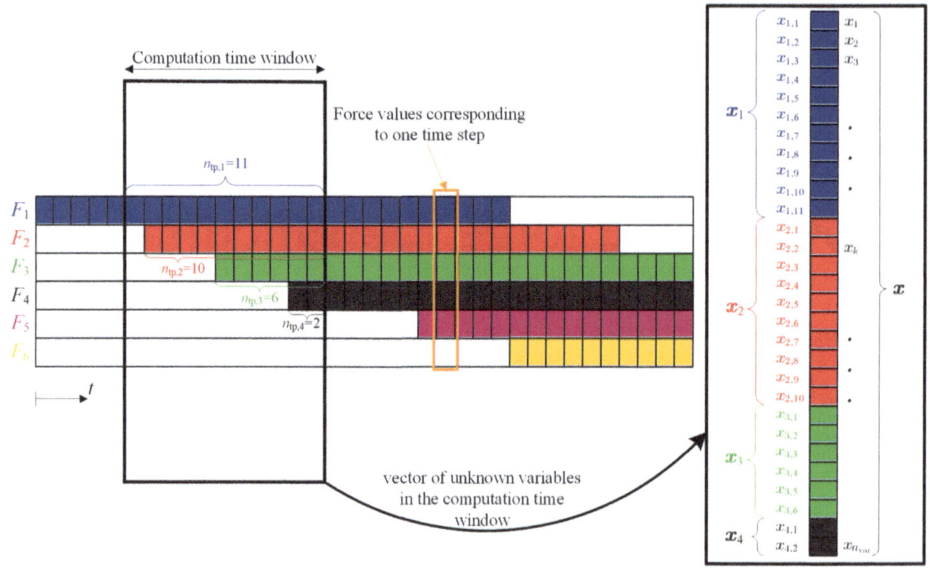

Figure 5.1 Structure of the vector of unknown variables

$x_{l,i}$ in Eq. 5.1 depicts the force value in the i-th out of $n_{tp,l}$ time steps, in which the l-th force is present on the structure:

$$\boldsymbol{x}_l = \begin{pmatrix} x_{l,1} & x_{l,2} & \cdots & x_{l,i} & \cdots & x_{l,n_{tp,l}} \end{pmatrix}^\top \tag{5.2}$$

Under consideration of Eq. 5.1 and 5.2, a more detailed expression of the total vector of optimization variables $\boldsymbol{x}$ can be formulated:

$$\boldsymbol{x} = \begin{pmatrix} x_{1,1} & \cdots & x_{1,n_{tp,1}} & x_{2,1} & \cdots & x_{2,n_{tp,2}} & \cdots & x_{n_F,1} & \cdots & x_{n_F,n_{tp,n_F}} \end{pmatrix}^\top \tag{5.3}$$

Eq. 5.3 points out that the number of variables n_{var} can be expressed as summation of all number of time steps $n_{tp,l}$, which are needed by the n_F forces to pass the structure:

$$n_{var} = \sum_{l=1}^{n_F} n_{tp,l} \tag{5.4}$$

5.1.2 Measured and computed data

The formulation of the proposed identification method considers n_{md} displacement and n_{ma} acceleration measured time histories. They are recorded over n_t time samples. The corresponding measurement vectors can be expressed for an arbitrary h-th measurement point (MP) as:

$$\begin{aligned}\bar{\boldsymbol{u}}_h &= \begin{pmatrix}\bar{u}_{h,1} & \bar{u}_{h,2} & \cdots & \bar{u}_{h,i} & \cdots & \bar{u}_{h,n_t}\end{pmatrix}\\ \bar{\ddot{\boldsymbol{u}}}_h &= \begin{pmatrix}\bar{\ddot{u}}_{h,1} & \bar{\ddot{u}}_{h,2} & \cdots & \bar{\ddot{u}}_{h,i} & \cdots & \bar{\ddot{u}}_{h,n_t}\end{pmatrix}\end{aligned} \tag{5.5}$$

where $\bar{u}_{h,i}$ and $\bar{\ddot{u}}_{h,i}$ represent the acceleration and displacement of the h-th MP in the i-th time step, respectively.

The computed vectors of the h-th point are denoted as $\boldsymbol{u}_h$ for the displacements and $\ddot{\boldsymbol{u}}_h$ for the accelerations, respectively. They must have the same length n_t as the measured vectors and are formulated similar to Eq. 5.5:

$$\begin{aligned}\boldsymbol{u}_h &= \begin{pmatrix}u_{h,1} & u_{h,2} & \cdots & u_{h,i} & \cdots & u_{h,n_t}\end{pmatrix}\\ \ddot{\boldsymbol{u}}_h &= \begin{pmatrix}\ddot{u}_{h,1} & \ddot{u}_{h,2} & \cdots & \ddot{u}_{h,i} & \cdots & \ddot{u}_{h,n_t}\end{pmatrix}\end{aligned} \tag{5.6}$$

The computed displacements and accelerations are obtained by a forward dynamic analysis under assumption of a calibrated structural model. For this, the mode superposition approach is used (cf. Section 3.3.3). This fact requires the assumption of a linear sytem, i.e. the system matrices $\boldsymbol{K}$, $\boldsymbol{C}$ and $\boldsymbol{M}$ (cf. Eq. 3.23) are known and assumed as constant over the computation time. Besides the significant reduction of the computational time, the modal superposition method has the great advantage that the modal equations of motion (cf. Eq. 3.44) can be formulated for arbitrarily large and detailed FE models using only a few regular output results of the FE modal analysis. These are the natural frequencies f_j of the modes of interest, the corresponding modal masses M_j and the modes shapes along the excitation path $\phi_{\mathrm{p},j}$ and at the computation points ($\phi_{\mathrm{cd},j}$ for displacements and $\phi_{\mathrm{ca},j}$ for accelerations).

The generalized load can be determined as:

$$P_j(t) = \phi_{\mathrm{p},j}^{\top} \cdot \boldsymbol{P}_{\mathrm{p}}(t) \tag{5.7}$$

where $\boldsymbol{P}_{\mathrm{p}}(t)$ represents the time dependent load vector formulated only for the nodes disposed along the load path. It can be determined according to Eq. 3.29.

The modal stiffness K_j and the modal damping coefficient C_j can be subsequently determined according to the following equations:

$$\begin{aligned}K_j &= (2 \cdot \pi \cdot f_j)^2 \cdot M_j\\ C_j &= 2 \cdot \xi_j \cdot \sqrt{K_j \cdot M_j}\end{aligned} \tag{5.8}$$

where ξ_j denotes the modal damping ratio of the j-th mode. By retrieving all the parameters needed for Eq. 3.44, the modal equations can be formulated independently of the finite element software. Thus, they can be manipulated and solved using any computer algebra system, e.g. Matlab (PETERSEN et al., 2017).

The superposition of the modal contributions (Eq. 3.50) only needs to be expressed for the computation points:

$$\boldsymbol{u}(t) = \sum_{j=1}^{n} \alpha_j(t) \cdot \boldsymbol{\phi}_{\mathrm{cd},j} \quad \text{and} \quad \ddot{\boldsymbol{u}}(t) = \sum_{j=1}^{n} \ddot{\alpha}_j(t) \cdot \boldsymbol{\phi}_{\mathrm{ca},j} \tag{5.9}$$

The number and position of the computation points have to be selected according to the real locations of the n_{md} displacement and n_{ma} acceleration measurements points. This means that $\boldsymbol{u}(t)$ and $\ddot{\boldsymbol{u}}(t)$ are also expressed for n_{md} displacement and n_{ma} acceleration measurement points, respectively.

5.1.3 Main optimization criterion

After introducing the vector of unknown variables and the formulation of the measured and computed data, an initial form of the objective function can be defined as an unconstrained and unbounded optimization problem:

$$\min_{\boldsymbol{x}} \mathcal{F}(\boldsymbol{x}) = \underbrace{\sum_{h=1}^{n_{\mathrm{md}}} \sum_{j=1}^{n_{\mathrm{t}}} w_{\mathrm{d},h} \cdot (u_{h,j} - \bar{u}_{h,j})^2 + \sum_{h=1}^{n_{\mathrm{ma}}} \sum_{j=1}^{n_{\mathrm{t}}} w_{\mathrm{a},h} \cdot (\ddot{u}_{h,j} - \bar{\ddot{u}}_{h,j})^2}_{\text{Main optimization criterion: } \mathcal{F}_{\mathrm{mc}}} \tag{5.10}$$

Eq. 5.10 is denoted as the main optimization criterion $\mathcal{F}_{\mathrm{mc}}$. It obviously seeks to minimize the discrepancy between the computed and measured acceleration and displacement time histories. Since the main optimization criterion consists of two terms, the weighting factors $w_{\mathrm{d},h}$ and $w_{\mathrm{a},h}$ are introduced, in order to account for the importance of the individual acceleration and displacement measurements. It is to be noted that the displacements and accelerations usually present different orders of magnitude. This fact can also be compensated by choosing appropriate values for $w_{\mathrm{d},h}$ and $w_{\mathrm{a},h}$. An alternative solution is to scale the measured and computed vectors so that they have dimensionless units. A natural choice for this is the normalization with respect to the norm of the measured vector at each MP, which implies the following replacement in Eq. 5.15:

$$u_{h,j} - \bar{u}_{h,j} \quad \text{and} \quad \ddot{u}_{h,j} - \bar{\ddot{u}}_{h,j} \quad \text{by} \quad \frac{u_{h,j} - \bar{u}_{h,j}}{\|\bar{\boldsymbol{u}}_h\|_2} \quad \text{and} \quad \frac{\ddot{u}_{h,j} - \bar{\ddot{u}}_{h,j}}{\|\bar{\ddot{\boldsymbol{u}}}_h\|_2} \tag{5.11}$$

where $\bar{\boldsymbol{u}}_h$ and $\bar{\ddot{\boldsymbol{u}}}_h$ represent the displacement and acceleration of the h-th measurement point (cf. Eq. 5.5). Eq. 5.11 is equivalent to the following selection of $w_{\mathrm{d},h}$ and $w_{\mathrm{a},h}$:

$$w_{\mathrm{d},h} = \frac{1}{\|\bar{\boldsymbol{u}}_h\|_2^2} \quad \text{and} \quad w_{\mathrm{a},h} = \frac{1}{\|\bar{\ddot{\boldsymbol{u}}}_h\|_2^2} \tag{5.12}$$

Another option is to normalize the measured and computed vectors with respect to the absolute maximum of the measured vector at each location. This requires the following replacement in Eq. 5.15:

$$u_{h,j} - \bar{u}_{h,j} \quad \text{and} \quad \ddot{u}_{h,j} - \bar{\ddot{u}}_{h,j} \quad \text{by} \quad \frac{u_{h,j} - \bar{u}_{h,j}}{\max|\bar{\boldsymbol{u}}_h|} \quad \text{and} \quad \frac{\ddot{u}_{h,j} - \bar{\ddot{u}}_{h,j}}{\max|\bar{\ddot{\boldsymbol{u}}}_h|} \tag{5.13}$$

Eq. 5.13 can also be achieved by the following definition of the weighting parameters:

$$w_{\mathrm{d},h} = \frac{1}{(\max|\bar{\boldsymbol{u}}_h|)^2} \quad \text{and} \quad w_{\mathrm{a},h} = \frac{1}{(\max|\bar{\ddot{\boldsymbol{u}}}_h|)^2} \tag{5.14}$$

In addition, it is to be mentioned that the weighting parameters $w_{\mathrm{d},h}$ and $w_{\mathrm{a},h}$ could also be used to account for the quality of the measurement data, for example in case of different signal-to-noise ratios of the sensors. It is obvious that measured data of small amplitudes are greatly affected by any systematic errors and noise. In this situation, appropriate weighting parameters can be defined based on the covariance matrix of the measured data (SANTAMARINA et al., 2005; LAW et al., 2011).

As mentioned in the previous sections, inverse problems in general, and implicitly the MFI formulation given in Eq. 5.10 are typically ill-posed and ill-conditioned. Without going into a detailed quantitative discussion of the results at this point, Fig. 5.2 shows two qualitative possible undesirable solution behaviors that arise due to the ill-posedness and ill-conditioning of Eq. 5.10. Fig. 5.2a highlights that a solution with strongly varying quasi-static components of the moving forces can be retrieved. This behavior is not realistic in the context of dynamic traffic forces acting on bridges, which are usually expected to vary around a fixed quasi-static component (e.g. axle weight). Fig. 5.2b shows a typical behavior that occurs due to the noise components of the measured signals. It can be seen, that the solution is partly dominated by very high frequent force components with isolated peaks that can be up to several times higher than the real force magnitude. A more detailed discussion of this issue, including quantitative aspects, is provided in Section 6.5.

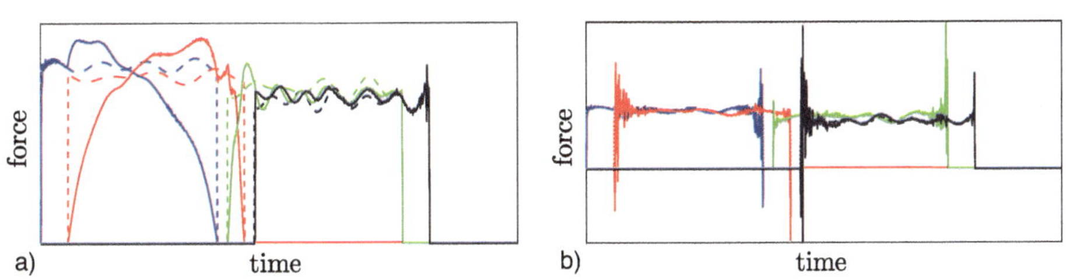

a) b)

Figure 5.2 Possible unwanted behavior of the solution under consideration of the main optimization criterion (solid lines: identified forces, dashed lines: reference forces)

5.1.4 Penalty functions

In order to suppress as much as possible the previously mentioned effects related to the ill-posedness and ill-conditioning of the of the MFI problem (Eq. 5.10), the main optimization criterion is complemented by two penalty (regularization) terms as follows:

$$\min_{\boldsymbol{x}} \mathcal{F}(\boldsymbol{x}) = \underbrace{\sum_{h=1}^{n_{\text{md}}}\sum_{j=1}^{n_{\text{t}}} w_{\text{d},h}\cdot(u_{h,j}-\bar{u}_{h,j})^2 + \sum_{h=1}^{n_{\text{ma}}}\sum_{j=1}^{n_{\text{t}}} w_{\text{a},h}\cdot(\ddot{u}_{h,j}-\bar{\ddot{u}}_{h,j})^2}_{\text{Main optimization criterion: } \mathcal{F}_{\text{mc}}} +$$
$$\underbrace{w_{\text{p1}}\cdot\sum_{l=1}^{n_{\text{F}}}\left\|\boldsymbol{Q}_{\text{loc},l}^{n_{\text{tp},l}\times n_{\text{tp},l}}\cdot\boldsymbol{x}_l^{n_{\text{tp},l}\times 1} - \boldsymbol{Q}_{\text{glo},l}^{n_{\text{tp},l}\times n_{\text{tp},l}}\cdot\boldsymbol{x}_l^{n_{\text{tp},l}\times 1}\right\|_2^2}_{\text{Penalty function 1: } \mathcal{F}_{\text{p1}}} + \tag{5.15}$$
$$\underbrace{w_{\text{p2}}\cdot\sum_{l=1}^{n_{\text{F}}}\left\|\boldsymbol{I}^{n_{\text{tp},l}\times n_{\text{tp},l}}\cdot\boldsymbol{x}_l^{n_{\text{tp},l}\times 1} - \boldsymbol{Q}_{\text{loc},l}^{n_{\text{tp},l}\times n_{\text{tp},l}}\cdot\boldsymbol{x}_l^{n_{\text{tp},l}\times 1}\right\|_2^2}_{\text{Penalty function 2: } \mathcal{F}_{\text{p2}}}$$

Eq. 5.15 is denoted as the total objective function and can be formally interpreted as a regularized inverse problem. In the following, the penalty terms $\mathcal{F}_{\text{p1}}$ and $\mathcal{F}_{\text{p2}}$ will be discussed in detail, in order to facilitate the understanding of the total objective function.

5.1.4.1 Penalty function 1

A particular characteristic of traffic loads on bridges is that their dynamic components generally vary around the corresponding static force components. This applies equally to railway, road and pedestrian bridges (LUDESCHER et al., 2009; RACIC et al., 2009). This signal feature is denoted as discrete finite signal with stable average value (DFS-SAV) and has already been used by PAN et al. (2017) in the context of MFI based on a direct solution method under consideration of a modified Tikhonov regularization approach. The a priori knowledge on the solution related to the DFS-SAV feature is considered in the formally unconstrained and unbounded optimization problem given in Eq. 5.15 through the first penalty function ($\mathcal{F}_{\text{p1}}$). It seeks to minimize the squared differences between the vectors of local and global average values of the n_{F} forces $\boldsymbol{x}_l$. In this way, it is ensured that the dynamic components of the solution will lie not too far from the static component, which is expressed as the global average. This is expected to reduce the solution space of the optimization problem and thus to increase the accuracy of the results.

The local average values of the l-th force are obtained by multiplying a local smoothing matrix $\boldsymbol{Q}_{\text{loc},l}$ by the force vector $\boldsymbol{x}_l$:

$$\boldsymbol{x}_{\text{loc},l} = \boldsymbol{Q}_{\text{loc},l}\cdot\boldsymbol{x}_l \tag{5.16}$$

where the entries of the local smoothing matrix $\boldsymbol{Q}_{\text{loc}}$ are defined in dependency of an averaging radius r_{a} as:

$$\boldsymbol{Q}_{\text{loc},l}^{(i,j)} = \begin{cases} \dfrac{1}{r_{\text{a}}+i} & \text{if} \quad i \leq r_{\text{a}} \text{ and } j \leq r_{\text{a}}+i \\ \dfrac{1}{2 \cdot r_{\text{a}}+1} & \text{if} \quad r_{\text{a}} < i \leq (n_{\text{tp},l}-r_{\text{a}}) \text{ and } |i-j| \leq r_{\text{a}} \\ \dfrac{1}{r_{\text{a}}+n_{\text{tp},l}-i+1} & \text{if} \quad i > n_{\text{tp},l}-r_{\text{a}} \text{ and } j \geq i-r_{\text{a}} \\ 0 & \text{otherwise} \end{cases} \tag{5.17}$$

For example, the evaluation of Eq. 5.17 for $r_{\text{a}} = 2$ results in the following matrix:

$$\left[\boldsymbol{Q}_{\text{loc},l}\right]_{r_{\text{a}}=2} = \begin{pmatrix} \frac{1}{3} & \frac{1}{3} & \frac{1}{3} & 0 & \cdots & 0 & 0 & 0 \\ \frac{1}{4} & \frac{1}{4} & \frac{1}{4} & \frac{1}{4} & 0 & \cdots & 0 & 0 \\ \frac{1}{5} & \frac{1}{5} & \frac{1}{5} & \frac{1}{5} & \frac{1}{5} & 0 & \cdots & 0 \\ 0 & \frac{1}{5} & \frac{1}{5} & \frac{1}{5} & \frac{1}{5} & \frac{1}{5} & 0 & \cdots \\ & & \cdots & \cdots & \cdots & \cdots & \cdots & \\ 0 & \cdots & 0 & \frac{1}{5} & \frac{1}{5} & \frac{1}{5} & \frac{1}{5} & \frac{1}{5} \\ 0 & 0 & \cdots & 0 & \frac{1}{4} & \frac{1}{4} & \frac{1}{4} & \frac{1}{4} \\ 0 & 0 & 0 & \cdots & 0 & \frac{1}{3} & \frac{1}{3} & \frac{1}{3} \end{pmatrix} \tag{5.18}$$

It is obvious that the moving average approach adopted in this work considers a general averaging window of $(2 \cdot r_{\text{a}} + 1)$ samples. Exceptions are only the first and the last r_{a} averaging windows, which contain a lower number of samples.

In a similar way as in Eq. 5.16, the global average is computed by multiplication of a global smoothing matrix $\boldsymbol{Q}_{\text{glo},l}$ by $\boldsymbol{x}_l$:

$$\boldsymbol{x}_{\text{glo},l} = \boldsymbol{Q}_{\text{glo},l} \cdot \boldsymbol{x}_l \tag{5.19}$$

where $\boldsymbol{Q}_{\text{glo},l}$ is defined as follows:

$$\boldsymbol{Q}_{\text{glo},l} = \begin{pmatrix} \frac{1}{n_{\text{tp},l}} & \frac{1}{n_{\text{tp},l}} & \cdots & \frac{1}{n_{\text{tp},l}} & \frac{1}{n_{\text{tp},l}} & \frac{1}{n_{\text{tp},l}} & \frac{1}{n_{\text{tp},l}} & \frac{1}{n_{\text{tp},l}} \\ \frac{1}{n_{\text{tp},l}} & \frac{1}{n_{\text{tp},l}} & \frac{1}{n_{\text{tp},l}} & \cdots & \frac{1}{n_{\text{tp},l}} & \frac{1}{n_{\text{tp},l}} & \frac{1}{n_{\text{tp},l}} & \frac{1}{n_{\text{tp},l}} \\ \vdots & \vdots & \vdots & \vdots & \ddots & \vdots & \vdots & \vdots \\ \frac{1}{n_{\text{tp},l}} & \frac{1}{n_{\text{tp},l}} & \frac{1}{n_{\text{tp},l}} & \frac{1}{n_{\text{tp},l}} & \cdots & \frac{1}{n_{\text{tp},l}} & \frac{1}{n_{\text{tp},l}} & \frac{1}{n_{\text{tp},l}} \end{pmatrix} \tag{5.20}$$

Fig. 5.3 should facilitate the understanding of the functioning principle of the first penalty function. It illustrates the determination of the local and global averages for an arbitrary signal $\boldsymbol{x}_l$ with DFS-SAV characteristics (shown in Fig. 5.3a). As expected, the global average vector $\boldsymbol{x}_{\mathrm{glo},l}$ takes a constant value for all entries, while the appearance of the local average vector $\boldsymbol{x}_{\mathrm{loc},l}$ is highly dependent on the averaging radius r_a (Fig. 5.3b). Increasing values of r_a cause a more pronounced filtering of the signal, eventually leading to a flattening of the curve. The penalty function seeks in fact to minimize the sum of squared differences between the blue line (global average, see Fig. 5.3b) and the other lines, corresponding to local averages for different radii r_a. In this respect, it is clear that the difference between the global and local averages becomes zero if all values of any vector $\boldsymbol{x}_l$ are constant and equal to the global average value $\boldsymbol{x}_{\mathrm{glo},l}$. However, this is an unwanted result, which can be avoided by selecting a proper weighting factor w_{p1}. This defines the contribution of the first regularization term within the optimization problem described by Eq. 5.15. Its determination will be discussed later.

Finally, it is to be mentioned that the formulation of the local smoothing matrix given in Eq. 5.17 is different from the proposal of PAN et al. (2017). Further differences between the two approaches refer to the selection of the weighting parameter (cf. Section 5.1.4.3) and of the averaging radius as well as to the used solution approaches. The present work is based on a gradient-based optimization algorithm, while PAN et al. (2017) relies on the direct solution based on the normal equations and a Tikhonov regularization approach (cf. Section 4.3). Nevertheless, the penalty term $\mathcal{F}_{\mathrm{p1}}$ can also be expressed in a similar form to the Tikhonov term in Eq. 4.7 if the Tikhonov matrix $\boldsymbol{\Gamma}$ is assumed as the difference between the global and the local smoothing matrix: $\boldsymbol{\Gamma} = \boldsymbol{Q}_{\mathrm{p1},l} = \boldsymbol{Q}_{\mathrm{loc},l} - \boldsymbol{Q}_{\mathrm{glo},l}$. However, some adaptations would be required in this case due to the summation over all moving forces, which are likely to contain different numbers of unknown variables (force values).

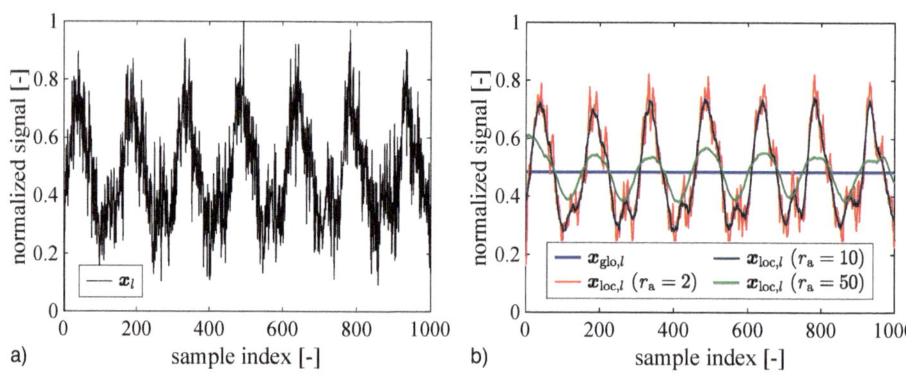

Figure 5.3 Principle of the first penalty function: a) An arbitrary normalized vector and b) the effect of the smoothing matrices $\boldsymbol{Q}_{\mathrm{glo},l}$ and $\boldsymbol{Q}_{\mathrm{loc},l}$ for different averaging radii r_a

5.1.4.2 Penalty function 2

The second penalty function of Eq. 5.15 ($\mathcal{F}_{p2}$) regulates the smoothness of the solution, which can be affected by the noise component of the measurement signals. This is achieved by minimizing the distance between each value in the solution vector $\boldsymbol{x}_l$ and the local average $\boldsymbol{x}_{\text{loc},l}$ for each of the n_F forces. In order to facilitate the mathematical manipulation, the identity matrix $\boldsymbol{I}$ is used in the formulation of the penalty function. Its multiplication with $\boldsymbol{x}_l$ formally delivers the vector:

$$\boldsymbol{x}_{\text{sm},l} = \boldsymbol{I} \cdot \boldsymbol{x}_l \tag{5.21}$$

which represents in fact the unchanged vector $\boldsymbol{x}_l$. This means that:

$$\boldsymbol{x}_{\text{sm},l} \hat{=} \boldsymbol{x}_l \tag{5.22}$$

Fig. 5.4 illustrates the determination of the two components $\boldsymbol{x}_{\text{sm},l}$ and $\boldsymbol{x}_{\text{loc},l}$ of the second penalty function for an averaging radius $r_a = 10$, based on an arbitrary vector $\boldsymbol{x}_l$ (shown in Fig. 5.4a). The term $\mathcal{F}_{p2}$ aims to minimize the sum of squared differences between each value of the blue line and the corresponding value of the red line in Fig. 5.4b. The moving average approach obviously ensures a much smoother curve $\boldsymbol{x}_{\text{loc},l}$ than the original vector $\boldsymbol{x}_l$. Therefore, the minimization of the squared differences between the two components has a similar smoothing effect, which can be also controlled by the weighting factor w_{p2} (discussed later). It is to be mentioned that a smoothing effect can be also obtained by choosing a different definition of the Tikhonov matrix $\boldsymbol{\Gamma}$, e.g. as discrete approximation of the first derivative of the solution instead of the actual approach, which implies: $\boldsymbol{\Gamma} = \boldsymbol{Q}_{\text{p2},l} = \boldsymbol{I} - \boldsymbol{Q}_{\text{loc},l}$ (DOYLE, 2004; SANTAMARINA et al., 2005; BJÖRCK, 2015).

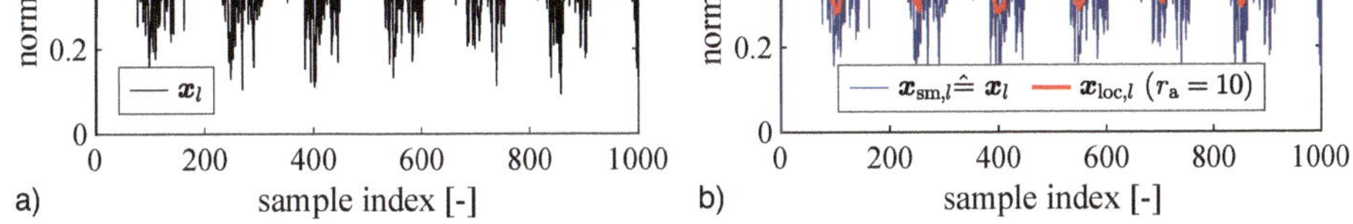

Figure 5.4 Principle of the second penalty function: a) An arbitrary normalized vector and b) the effect of the smoothing matrices $\boldsymbol{Q}_{\text{loc},l}$ and $\boldsymbol{I}$ for an averaging radius $r_a = 10$

The formulation of the penalty functions as summations over the n_F forces is required, on the one hand, due to the fact that the computation window can be shorter than the total time of the vehicle (or pedestrian) passage (cf. Fig. 5.1). This leads to different numbers of excitation time steps $n_{tp,l}$ and thus to different matrix dimensions for each force. On the other hand, the forces of different axles comprised in the same window can clearly differ and averaging must therefore be treated separately.

5.1.4.3 Weighting factors of the penalty functions

As already mentioned above, two factors w_{p1} and w_{p2} define the weights of the penalty functions within the total objective function. They have in fact the same purpose as the Tikhonov parameter α in Eq. 4.7, namely to control the amount of regularization introduced into the problem. However, the formulation proposed in this work includes two regularization terms instead of a single one used by the classical Tikhonov approach. The reason for this is that each regularization term depicts a clear a priori known characteristic of the solution, which is aimed to be individually controlled. Therefore, the classical methods for determining the amount of regularization (cf. Section 4.3.2) are not applicable in this case. Nevertheless, intensive numerical parametric studies performed within this work proved that the optimal values for w_{p1} and w_{p2} may be determined only once and then be kept fixed for a given structure, as long as the order of magnitude of the unknown moving forces does not change significantly. This condition is in fact always fulfilled for any type of bridge subjected to moving loads. Moreover, considering that a possible purpose of the MFI methods is their application within a continuous structural health monitoring, the reasonable effort of determining w_{p1} and w_{p2} at the beginning of the investigation through a comprehensive numerical parametric study is fully justified. This seeks to find the optimal values of the weighting factors, leading to the lowest error between an assumed reference force and the identification results obtained using simulated measurement data (Fig. 5.5):

$$x_{\mathrm{err}} = \frac{\|\boldsymbol{x}_{\mathrm{identified}} - \boldsymbol{x}_{\mathrm{reference}}\|_1}{\|\boldsymbol{x}_{\mathrm{reference}}\|_1} \cdot 100\ \% \tag{5.23}$$

The simulated measurement data is generated by a forward structural analysis of the considered structural model subjected to the action of the assumed reference force (comprised in the vector $\boldsymbol{x}_{\mathrm{reference}}$). The magnitude of the assumed reference force should lie in the expected range of the real forces. The results of the direct analysis are subsequently polluted with artificial noise, in order to account – to a certain extent – for the characteristics of real measurement data. A widely used approach for the noise definition is the so called additive white Gaussian noise (ZHU et al., 2002b; ZHU et al., 2006a; LAW et al., 2011):

$$\begin{aligned} \boldsymbol{u}_{\mathrm{sim}} &= \boldsymbol{u} + e_{\mathrm{p}} \cdot \sigma_{\mathrm{d}} \cdot \boldsymbol{N}_{\mathrm{rand}} \\ \ddot{\boldsymbol{u}}_{\mathrm{sim}} &= \ddot{\boldsymbol{u}} + e_{\mathrm{p}} \cdot \sigma_{\mathrm{a}} \cdot \boldsymbol{N}_{\mathrm{rand}} \end{aligned} \tag{5.24}$$

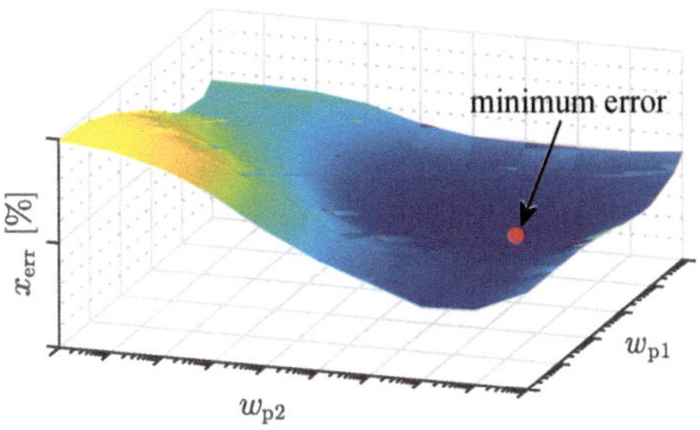

Figure 5.5 Parametric study for determination of the weighing factors $w_{\rm p1}$ and $w_{\rm p2}$

where $\boldsymbol{u}_{\rm sim}$ and $\ddot{\boldsymbol{u}}_{\rm sim}$ are the simulated measurement vectors of displacement and acceleration, respectively; $\boldsymbol{u}$ and $\ddot{\boldsymbol{u}}$ denote the computational results of the forward analysis, i.e. the "measurement" data without noise; $e_{\rm p}$ represents the noise level, $\sigma_{\rm d}$ and $\sigma_{\rm a}$ are the standard deviations of the displacement and acceleration responses without noise, while $\boldsymbol{N}_{\rm rand}$ is a standard normal distribution vector with zero mean and unit standard deviation.

The noise definition according to Eq. 5.24 represents an idealized and convenient noise approach, which should only serve for providing first insights into the performance of the proposed MFI method with noisy signals. The reason for this is that the considered noise does not depend on the state of the system (i.e. no amplitude dependency) and it is time invariant. However, these properties often do not hold for real noise samples. Moreover, it is presumed that the noise samples generated according to Eq. 5.24 present an uniform power across the frequency band (white noise characteristic). This is also a simplifying assumption, as many real sensors are actually characterized by a "colored" noise, which presents variable power in the frequency band (KUTTNER et al., 2019).

5.2 Sensitivity analysis

5.2.1 Gradient of the objective function

It is obvious that the objective function Eq. 5.15 presents both implicit and explicit dependencies with respect to the optimization variables x_k, whereas the implicit dependencies occur both in the acceleration and the displacement terms:

$$\mathcal{F} = \mathcal{F}(\boldsymbol{x}, \boldsymbol{u}(\boldsymbol{x}), \ddot{\boldsymbol{u}}(\boldsymbol{x})) \tag{5.25}$$

According to Section 4.4.4.2, the total derivative of an objective function with respect to a certain variable x_k can be performed analytically, using the chain rule (cf. Eq. 4.46):

$$\frac{\mathrm{d}\mathcal{F}}{\mathrm{d}x_k} = \underbrace{\sum_{h=1}^{n_{\mathrm{md}}}\sum_{j=1}^{n_{\mathrm{t}}}\left(\frac{\partial \mathcal{F}}{\partial u_{h,j}} \cdot \frac{\partial u_{h,j}}{\partial x_k}\right) + \sum_{h=1}^{n_{\mathrm{ma}}}\sum_{j=1}^{n_{\mathrm{t}}}\left(\frac{\partial \mathcal{F}}{\partial \ddot{u}_{h,j}} \cdot \frac{\partial \ddot{u}_{h,j}}{\partial x_k}\right)}_{\text{implicit}} + \underbrace{\frac{\partial \mathcal{F}}{\partial x_k}}_{\text{explicit}} \tag{5.26}$$

The explicit component $\partial \mathcal{F}/\partial x_k$ is solely determined by the differentiation of the penalty functions $\mathcal{F}_{\mathrm{p1}}$ and $\mathcal{F}_{\mathrm{p2}}$ with respect to the load value x_k. On the contrary, the implicit term of the sensitivity is only obtained by differentiating the main optimization criterion $\mathcal{F}_{\mathrm{mc}}$. The implicit differentiation is the most challenging component of the sensitivity analysis. This is due to the determination of $\partial u_{h,j}/\partial x_k$ and $\partial \ddot{u}_{h,j}/\partial x_k$, which describe the sensitivity of the displacement and accelerations of the h-th measured point with respect to the load value x_k. These sensitivity terms will be analytically derived in the following section, based on the direct approach presented in Section 4.4.4.2. However, before proceeding to the next derivations, it should be mentioned that Eq. 5.26 can also be formulated in vector notation, which simplifies the implementation and accelerates the computation:

$$\frac{\mathrm{d}\mathcal{F}}{\mathrm{d}x_k} = \sum_{h=1}^{n_{\mathrm{md}}}\left[\left(\frac{\partial \mathcal{F}}{\partial \boldsymbol{u}_h}\right)^{\top} \cdot \frac{\partial \boldsymbol{u}_h}{\partial x_k}\right] + \sum_{h=1}^{n_{\mathrm{ma}}}\left[\left(\frac{\partial \mathcal{F}}{\partial \ddot{\boldsymbol{u}}_h}\right)^{\top} \cdot \frac{\partial \ddot{\boldsymbol{u}}_h}{\partial x_k}\right] + \frac{\partial \mathcal{F}}{\partial x_k} \tag{5.27}$$

5.2.1.1 Gradient of the main optimization criterion

The derivation of the direct sensitivity analysis given in Section 4.4.4.2 is based on the static FEM equilibrium equation of a linear system (Eq. 4.44). In the following, the direct sensitivity analysis is extended for the dynamic equilibrium expression in Eq. 3.23. For this, Eq. 3.23 has to be differentiated on both sides with respect to a load variable x_k:

$$\begin{aligned}&\cancelto{0}{\frac{\partial \boldsymbol{K}}{\partial x_k}} \cdot \boldsymbol{u}(t) + \boldsymbol{K} \cdot \frac{\partial \boldsymbol{u}(t)}{\partial x_k} + \cancelto{0}{\frac{\partial \boldsymbol{C}}{\partial x_k}} \cdot \dot{\boldsymbol{u}}(t) + \boldsymbol{C} \cdot \frac{\partial \dot{\boldsymbol{u}}(t)}{\partial x_k} + \\ &\cancelto{0}{\frac{\partial \boldsymbol{M}}{\partial x_k}} \cdot \ddot{\boldsymbol{u}}(t) + \boldsymbol{M} \cdot \frac{\partial \ddot{\boldsymbol{u}}(t)}{\partial x_k} = \frac{\partial \boldsymbol{P}(t)}{\partial x_k}\end{aligned} \tag{5.28}$$

$\partial \boldsymbol{u}/\partial x_k$, $\partial \dot{\boldsymbol{u}}/\partial x_k$ and $\partial \ddot{\boldsymbol{u}}/\partial x_k$ represent the sensitivity vectors with respect to x_k for the displacement, velocity and acceleration, respectively. Considering that the system matrices $\boldsymbol{K}$, $\boldsymbol{C}$ and $\boldsymbol{M}$ are not dependent on the external loads, their derivatives with respect to x_k are zero. Hence, Eq. 5.28 can be simplified to:

$$\boldsymbol{K} \cdot \frac{\partial \boldsymbol{u}(t)}{\partial x_k} + \boldsymbol{C} \cdot \frac{\partial \dot{\boldsymbol{u}}(t)}{\partial x_k} + \boldsymbol{M} \cdot \frac{\partial \ddot{\boldsymbol{u}}(t)}{\partial x_k} = \frac{\partial \boldsymbol{P}(t)}{\partial x_k} \tag{5.29}$$

It is obvious that Eq. 5.29 has the same form as Eq. 3.23, whereas the unknown vectors of the motion quantities $\boldsymbol{u}(t)$, $\dot{\boldsymbol{u}}(t)$ and $\ddot{\boldsymbol{u}}(t)$ are replaced by the corresponding unknown sensitivities $\partial \boldsymbol{u}(t)/\partial x_k$, $\partial \dot{\boldsymbol{u}}(t)/\partial x_k$ and $\partial \ddot{\boldsymbol{u}}(t)/\partial x_k$ and the term $\partial \boldsymbol{P}(t)/\partial x_k$ on the right hand side is considered as a pseudo force vector (cf. Section 4.4.4.2). The latter represents the vector of external forces differentiated with respect to a single force in a certain time step. Therefore, $\partial \boldsymbol{P}(t)/\partial x_k$ is determined based on a single force value of 1, corresponding to the entry x_k, while all the other force values are zero.

Considering the previously mentioned similarity between Eq. 3.23 and Eq. 5.29, it can be noted that the modal superposition and the time integration method of Newmark can also be used for the determination the unknown sensitivity values $\partial \boldsymbol{u}(t)/\partial x_k$ and $\partial \ddot{\boldsymbol{u}}(t)/\partial x_k$ as solutions of Eq. 5.29. It is actually strongly recommended to determine the dynamic sensitivity response using exactly the same time integration procedure as used for the forward analysis, in order to avoid additional numerical errors (MEYER, 1998). As Eq. 5.29 is formulated based on the direct sensitivity method (cf. Section 4.4.4.2), it is obvious that an additional load case has to be solved for each optimization variable. This can easily lead to several thousand of evaluations of Eq. 5.29 for large MFI problems.

A similar approach to Eq. 5.29 was also used by LU et al. (2007), LI et al. (2009), LI et al. (2010) and LU et al. (2011) in the context of damage and system identification. In these cases, the differentiation according to Eq. 5.29 was performed with respect to some structural parameters and not to the force values x_k. Furthermore, LU et al. (2006) also used the dynamic response sensitivity in time domain in a force identification context. However, the investigation only dealt with non-moving forces, whereas only the Fourier parameters describing a periodic force were considered as unknown parameters. This implied a differentiation with respect to the amplitude and frequency of the forces instead of each force value x_k, as in the present work.

Finally, the expressions $\partial \mathcal{F}/\partial u_{h,j}$ and $\partial \mathcal{F}/\partial \ddot{u}_{h,j}$ can be determined by a straightforward analytical differentiation of the objective function, yielding:

$$\frac{\partial \mathcal{F}}{\partial u_{h,j}} = 2 \cdot w_{\mathrm{d},h} \cdot (u_{h,j} - \bar{u}_{h,j}) \quad \text{and} \quad \frac{\partial \mathcal{F}}{\partial \ddot{u}_{h,j}} = 2 \cdot w_{\mathrm{a},h} \cdot (\ddot{u}_{h,j} - \bar{\ddot{u}}_{h,j}) \tag{5.30}$$

By solving Eq. 5.29 and taking into account Eq. 5.30, all components of the total derivative of the main optimization criterion are known:

$$\frac{\mathrm{d}\mathcal{F}_{\mathrm{mc}}}{\mathrm{d}x_k} = \sum_{h=1}^{n_{\mathrm{md}}} \sum_{j=1}^{n_{\mathrm{t}}} 2 \cdot w_{\mathrm{d},h} \cdot (u_{h,j} - \bar{u}_{h,j}) \cdot \frac{\partial u_{h,j}}{\partial x_k} + \sum_{h=1}^{n_{\mathrm{ma}}} \sum_{j=1}^{n_{\mathrm{t}}} 2 \cdot w_{\mathrm{a},h} \cdot (\ddot{u}_{h,j} - \bar{\ddot{u}}_{h,j}) \cdot \frac{\partial \ddot{u}_{h,j}}{\partial x_k} \tag{5.31}$$

5.2.1.2 Gradients of the penalty functions

In order to evaluate the total derivative according to Eq. 5.26, the derivatives of the penalty functions still have to be determined. Since the penalty terms only consist of vectors of optimization variables and matrices of constant values, they only account for the explicit term $\partial\mathcal{F}/\partial x_k$ in Eq. 5.26. In order to facilitate the mathematical manipulation within the differentiation process, Eq. 5.15 is rearranged as follows:

$$\begin{aligned}\min_{\boldsymbol{x}} \mathcal{F}(\boldsymbol{x}) = &\underbrace{\sum_{h=1}^{n_{\mathrm{md}}} w_{\mathrm{d},h} \cdot \|\boldsymbol{u}_h - \bar{\boldsymbol{u}}_h\|_2^2 + \sum_{h=1}^{n_{\mathrm{ma}}} w_{\mathrm{a},h} \cdot \|\ddot{\boldsymbol{u}}_h - \bar{\ddot{\boldsymbol{u}}}_h\|_2^2}_{\mathcal{F}_{\mathrm{mc}}} + \\ &\underbrace{w_{\mathrm{p1}} \cdot \sum_{l=1}^{n_{\mathrm{F}}} \|\boldsymbol{Q}_{\mathrm{p1},l} \cdot \boldsymbol{x}_l\|_2^2 + w_{\mathrm{p2}} \cdot \sum_{l=1}^{n_{\mathrm{F}}} \|\boldsymbol{Q}_{\mathrm{p2},l} \cdot \boldsymbol{x}_l\|_2^2}_{\mathcal{F}_{\mathrm{p}}}\end{aligned} \tag{5.32}$$

where

$$\begin{aligned}\boldsymbol{Q}_{\mathrm{p1},l}^{n_{\mathrm{tp},l}\times n_{\mathrm{tp},l}} &= \boldsymbol{Q}_{\mathrm{loc},l}^{n_{\mathrm{tp},l}\times n_{\mathrm{tp},l}} - \boldsymbol{Q}_{\mathrm{glo},l}^{n_{\mathrm{tp},l}\times n_{\mathrm{tp},l}} = \begin{pmatrix}\boldsymbol{q}_{\mathrm{p1},l}^{(1)} & \boldsymbol{q}_{\mathrm{p1},l}^{(2)} & \cdots & \boldsymbol{q}_{\mathrm{p1},l}^{(j)} & \cdots & \boldsymbol{q}_{\mathrm{p1},l}^{(n_{\mathrm{tp},l})}\end{pmatrix}^{\top} \\ \boldsymbol{Q}_{\mathrm{p2},l}^{n_{\mathrm{tp},l}\times n_{\mathrm{tp},l}} &= \boldsymbol{I}^{n_{\mathrm{tp},l}\times n_{\mathrm{tp},l}} - \boldsymbol{Q}_{\mathrm{loc},l}^{n_{\mathrm{tp},l}\times n_{\mathrm{tp},l}} = \begin{pmatrix}\boldsymbol{q}_{\mathrm{p2},l}^{(1)} & \boldsymbol{q}_{\mathrm{p2},l}^{(2)} & \cdots & \boldsymbol{q}_{\mathrm{p2},l}^{(j)} & \cdots & \boldsymbol{q}_{\mathrm{p2},l}^{(n_{\mathrm{tp},l})}\end{pmatrix}^{\top}\end{aligned} \tag{5.33}$$

$\boldsymbol{q}_{\mathrm{p1},l}^{(j)}$ and $\boldsymbol{q}_{\mathrm{p2},l}^{(j)}$ in 5.33 obviously represent the j-th row of the matrices $\boldsymbol{Q}_{\mathrm{p1},l}$ and $\boldsymbol{Q}_{\mathrm{p2},l}$, respectively. In the following, the derivative of the first penalty function with respect to a load variable x_k is addressed separately:

$$\frac{\partial \mathcal{F}_{\mathrm{p1}}}{\partial x_k} = \frac{\partial}{\partial x_k}\left(w_{\mathrm{p1}} \cdot \sum_{l=1}^{n_{\mathrm{F}}} \|\boldsymbol{Q}_{\mathrm{p1},l} \cdot \boldsymbol{x}_l\|_2^2\right) = w_{\mathrm{p1}} \cdot \frac{\partial}{\partial x_k} \sum_{l=1}^{n_{\mathrm{F}}} \sum_{j=1}^{n_{\mathrm{tp},l}} \left(\boldsymbol{q}_{\mathrm{p1},l}^{(j)} \cdot \boldsymbol{x}_l\right)^2 \tag{5.34}$$

The differentiation can be performed within the summation, leading to:

$$w_{\mathrm{p1}} \cdot \sum_{l=1}^{n_{\mathrm{F}}} \sum_{j=1}^{n_{\mathrm{tp},l}} \frac{\partial}{\partial x_k} \left(\boldsymbol{q}_{\mathrm{p1},l}^{(j)} \cdot \boldsymbol{x}_l\right)^2 = w_{\mathrm{p1}} \cdot \sum_{l=1}^{n_{\mathrm{F}}} \sum_{j=1}^{n_{\mathrm{tp},l}} 2 \cdot \left(\boldsymbol{q}_{\mathrm{p1},l}^{(j)} \cdot \boldsymbol{x}_l\right) \cdot \frac{\partial}{\partial x_k}\left(\boldsymbol{q}_{\mathrm{p1},l}^{(j)} \cdot \boldsymbol{x}_l\right) \tag{5.35}$$

The second term on the right hand side of Eq. 5.35 can be elaborated as follows:

$$\frac{\partial}{\partial x_k}\left(\boldsymbol{q}_{\mathrm{p1},l}^{(j)} \cdot \boldsymbol{x}_l\right) = \cancelto{0}{\frac{\partial}{\partial x_k}\left(\boldsymbol{q}_{\mathrm{p1},l}^{(j)}\right)} \cdot \boldsymbol{x}_l + \boldsymbol{q}_{\mathrm{p1},l}^{(j)} \cdot \frac{\partial \boldsymbol{x}_l}{\partial x_k} \tag{5.36}$$

The first term on the right hand side of Eq. 5.36 is zero, since the smoothing matrices defining the term $\boldsymbol{q}_{\mathrm{p1},l}^{(j)}$ are not dependent on the external loads. Inserting Eq. 5.36 into Eq. 5.35 delivers the complete derivative expression of the first penalty function:

$$\frac{\partial \mathcal{F}_{\mathrm{p1}}}{\partial x_k} = w_{\mathrm{p1}} \cdot \sum_{l=1}^{n_{\mathrm{F}}} \sum_{j=1}^{n_{\mathrm{tp},l}} 2 \cdot \left(\boldsymbol{q}_{\mathrm{p1},l}^{(j)} \cdot \boldsymbol{x}_l\right) \cdot \left(\boldsymbol{q}_{\mathrm{p1},l}^{(j)} \cdot \frac{\partial \boldsymbol{x}_l}{\partial x_k}\right) \tag{5.37}$$

where $\partial \boldsymbol{x}_l / \partial x_k$ contains a single value of 1 at the position of x_k, while all other values are zero. The derivative of the second penalty function can be determined in a similar way:

$$\frac{\partial \mathcal{F}_{\mathrm{p2}}}{\partial x_k} = w_{\mathrm{p2}} \cdot \sum_{l=1}^{n_{\mathrm{F}}} \sum_{j=1}^{n_{\mathrm{tp},l}} 2 \cdot \left(\boldsymbol{q}_{\mathrm{p2},l}^{(j)} \cdot \boldsymbol{x}_l \right) \cdot \left(\boldsymbol{q}_{\mathrm{p2},l}^{(j)} \cdot \frac{\partial \boldsymbol{x}_l}{\partial x_k} \right) \tag{5.38}$$

5.2.1.3 Total derivative of the objective function

Under consideration of Eq. 5.31, 5.37 and 5.38, all the components of Eq. 5.26 are known. Hence, the elaborated expression for the sensitivity of the objective function with respect to a load value x_k can be formulated as:

$$\begin{aligned} \frac{\mathrm{d}\mathcal{F}}{\mathrm{d}x_k} = & \sum_{h=1}^{n_{\mathrm{md}}} \sum_{j=1}^{n_{\mathrm{t}}} 2 \cdot w_{\mathrm{d},h} \cdot (u_{h,j} - \bar{u}_{h,j}) \cdot \frac{\partial u_{h,j}}{\partial x_k} + \sum_{h=1}^{n_{\mathrm{ma}}} \sum_{j=1}^{n_{\mathrm{t}}} 2 \cdot w_{\mathrm{a},h} \cdot (\ddot{u}_{h,j} - \bar{\ddot{u}}_{h,j}) \cdot \frac{\partial \ddot{u}_{h,j}}{\partial x_k} + \\ & \sum_{l=1}^{n_{\mathrm{F}}} \sum_{j=1}^{n_{\mathrm{tp},l}} 2 \cdot w_{\mathrm{p1}} \cdot \left(\boldsymbol{q}_{\mathrm{p1},l}^{(j)} \cdot \boldsymbol{x}_l \right) \cdot \left(\boldsymbol{q}_{\mathrm{p1},l}^{(j)} \cdot \frac{\partial \boldsymbol{x}_l}{\partial x_k} \right) + \\ & \sum_{l=1}^{n_{\mathrm{F}}} \sum_{j=1}^{n_{\mathrm{tp},l}} 2 \cdot w_{\mathrm{p2}} \cdot \left(\boldsymbol{q}_{\mathrm{p2},l}^{(j)} \cdot \boldsymbol{x}_l \right) \cdot \left(\boldsymbol{q}_{\mathrm{p2},l}^{(j)} \cdot \frac{\partial \boldsymbol{x}_l}{\partial x_k} \right) \end{aligned} \tag{5.39}$$

Eq. 5.39 can also be expressed in vector notation (cf. Eq. 5.27) as follows:

$$\begin{aligned} \frac{\mathrm{d}\mathcal{F}}{\mathrm{d}x_k} = & \sum_{h=1}^{n_{\mathrm{md}}} 2 \cdot w_{\mathrm{d},h} \cdot (\boldsymbol{u}_h - \bar{\boldsymbol{u}}_h)^{\top} \cdot \frac{\partial \boldsymbol{u}_h}{\partial x_k} + \sum_{h=1}^{n_{\mathrm{ma}}} 2 \cdot w_{\mathrm{a},h} \cdot (\ddot{\boldsymbol{u}}_h - \bar{\ddot{\boldsymbol{u}}}_h)^{\top} \cdot \frac{\partial \ddot{\boldsymbol{u}}_h}{\partial x_k} + \\ & \sum_{l=1}^{n_{\mathrm{F}}} \sum_{j=1}^{n_{\mathrm{tp},l}} 2 \cdot w_{\mathrm{p1}} \cdot \left(\boldsymbol{q}_{\mathrm{p1},l}^{(j)} \cdot \boldsymbol{x}_l \right) \cdot \left(\boldsymbol{q}_{\mathrm{p1},l}^{(j)} \cdot \frac{\partial \boldsymbol{x}_l}{\partial x_k} \right) + \\ & \sum_{l=1}^{n_{\mathrm{F}}} \sum_{j=1}^{n_{\mathrm{tp},l}} 2 \cdot w_{\mathrm{p2}} \cdot \left(\boldsymbol{q}_{\mathrm{p2},l}^{(j)} \cdot \boldsymbol{x}_l \right) \cdot \left(\boldsymbol{q}_{\mathrm{p2},l}^{(j)} \cdot \frac{\partial \boldsymbol{x}_l}{\partial x_k} \right) \end{aligned} \tag{5.40}$$

5.2.2 Hessian of the objective function

The entries of the exact Hessian can be analytically obtained through total differentiation of Eq. 5.26 with respect to another load value x_i:

$$\frac{\mathrm{d}^2\mathcal{F}}{\mathrm{d}x_i \mathrm{d}x_k} = \frac{\mathrm{d}}{\mathrm{d}x_i} \sum_{h=1}^{n_{\mathrm{md}}} \sum_{j=1}^{n_{\mathrm{t}}} \left(\frac{\partial \mathcal{F}}{\partial u_{h,j}} \cdot \frac{\partial u_{h,j}}{\partial x_k} \right) + \frac{\mathrm{d}}{\mathrm{d}x_i} \sum_{h=1}^{n_{\mathrm{md}}} \sum_{j=1}^{n_{\mathrm{t}}} \left(\frac{\partial \mathcal{F}}{\partial \ddot{u}_{h,j}} \cdot \frac{\partial \ddot{u}_{h,j}}{\partial x_k} \right) + \frac{\mathrm{d}}{\mathrm{d}x_i} \left(\frac{\partial \mathcal{F}}{\partial x_k} \right) \tag{5.41}$$

In the following, the Hessian of the main optimization criterion and of the penalty functions are derived separately, in order to enhance the clarity of the explanations.

5.2.2.1 Hessian of the main optimization criterion

For the derivation of the Hessian for the main optimization criterion, the first two terms of Eq. 5.41 are further elaborated (cf. Eq. 4.52), yielding:

$$\begin{aligned}\frac{\mathrm{d}^2\mathcal{F}_{\mathrm{mc}}}{\mathrm{d}x_i\mathrm{d}x_k} = &\sum_{h=1}^{n_{\mathrm{md}}}\sum_{j=1}^{n_{\mathrm{t}}}\left[\frac{\mathrm{d}}{\mathrm{d}x_i}\left(\frac{\partial\mathcal{F}_{\mathrm{mc}}}{\partial u_{h,j}}\right)\cdot\frac{\partial u_{h,j}}{\partial x_k}+\frac{\partial\mathcal{F}_{\mathrm{mc}}}{\partial u_{h,j}}\cdot\frac{\mathrm{d}}{\mathrm{d}x_i}\left(\frac{\partial u_{h,j}}{\partial x_k}\right)\right]+\\ &\sum_{h=1}^{n_{\mathrm{ma}}}\sum_{j=1}^{n_{\mathrm{t}}}\left[\frac{\mathrm{d}}{\mathrm{d}x_i}\left(\frac{\partial\mathcal{F}_{\mathrm{mc}}}{\partial \ddot{u}_{h,j}}\right)\cdot\frac{\partial \ddot{u}_{h,j}}{\partial x_k}+\frac{\partial\mathcal{F}_{\mathrm{mc}}}{\partial \ddot{u}_{h,j}}\cdot\frac{\mathrm{d}}{\mathrm{d}x_i}\left(\frac{\partial \ddot{u}_{h,j}}{\partial x_k}\right)\right]\end{aligned} \tag{5.42}$$

Since the components of Eq. 5.42 also present implicit dependencies with respect to the load values x_i, the chain rule has to be used in the differentiation process. This leads to:

$$\begin{aligned}\frac{\mathrm{d}^2\mathcal{F}_{\mathrm{mc}}}{\mathrm{d}x_i\mathrm{d}x_k} = &\sum_{h=1}^{n_{\mathrm{md}}}\sum_{j=1}^{n_{\mathrm{t}}}\left[\frac{\partial}{\partial u_{h,j}}\left(\frac{\partial\mathcal{F}_{\mathrm{mc}}}{\partial u_{h,j}}\right)\cdot\frac{\partial u_{h,j}}{\partial x_i}+\underbrace{\cancel{\frac{\partial}{\partial x_i}\left(\frac{\partial\mathcal{F}_{\mathrm{mc}}}{\partial u_{h,j}}\right)}}_{*}\right]^{0}\cdot\frac{\partial u_{h,j}}{\partial x_k}+\\ &\sum_{h=1}^{n_{\mathrm{md}}}\sum_{j=1}^{n_{\mathrm{t}}}\frac{\partial\mathcal{F}_{\mathrm{mc}}}{\partial u_{h,j}}\cdot\left[\underbrace{\cancel{\frac{\partial}{\partial u_{h,j}}\left(\frac{\partial u_{h,j}}{\partial x_k}\right)}}_{**}{}^{0}\cdot\frac{\partial u_{h,j}}{\partial x_i}+\underbrace{\cancel{\frac{\partial^2 u_{h,j}}{\partial x_i\partial x_k}}}_{***}{}^{0}\right]+\\ &\sum_{h=1}^{n_{\mathrm{ma}}}\sum_{j=1}^{n_{\mathrm{t}}}\left[\frac{\partial}{\partial \ddot{u}_{h,j}}\left(\frac{\partial\mathcal{F}_{\mathrm{mc}}}{\partial \ddot{u}_{h,j}}\right)\cdot\frac{\partial \ddot{u}_{h,j}}{\partial x_i}+\underbrace{\cancel{\frac{\partial}{\partial x_i}\left(\frac{\partial\mathcal{F}_{\mathrm{mc}}}{\partial \ddot{u}_{h,j}}\right)}}_{*}\right]^{0}\cdot\frac{\partial \ddot{u}_{h,j}}{\partial x_k}+\\ &\sum_{h=1}^{n_{\mathrm{ma}}}\sum_{j=1}^{n_{\mathrm{t}}}\frac{\partial\mathcal{F}_{\mathrm{mc}}}{\partial \ddot{u}_{h,j}}\cdot\left[\underbrace{\cancel{\frac{\partial}{\partial \ddot{u}_{h,j}}\left(\frac{\partial \ddot{u}_{h,j}}{\partial x_k}\right)}}_{**}{}^{0}\cdot\frac{\partial \ddot{u}_{h,j}}{\partial x_i}+\underbrace{\cancel{\frac{\partial^2 \ddot{u}_{h,j}}{\partial x_i\partial x_k}}}_{***}{}^{0}\right]\end{aligned} \tag{5.43}$$

From Eq. 5.30 it can be deducted that the terms $\partial\mathcal{F}_{\mathrm{mc}}/\partial u_{h,i}$ and $\partial\mathcal{F}_{\mathrm{mc}}/\partial \ddot{u}_{h,i}$ only depend implicitly on the load value x_i. This results in the canceling of the corresponding explicit differentiation terms (denoted by $*$ in Eq. 5.42). The terms $**$ are also zero, since the first derivative of the displacements and accelerations with respect to a load value x_k is not directly dependent on the displacements and accelerations, respectively. This is proved by Eq. 5.29, which indicates that the sensitivities are only computed using the system matrices $\boldsymbol{K}$, $\boldsymbol{C}$ and $\boldsymbol{M}$ and the pseudo load vector $\partial\boldsymbol{P}(t)/\partial x_k$. The terms corresponding to the second partial derivatives of the displacements and accelerations, marked as $***$ in Eq. 5.43, are also zero. This can be demonstrated by a further differentiation of Eq. 5.29 with respect to x_i:

$$\underbrace{\frac{\partial \boldsymbol{K}}{\partial x_i}}_{0} \cdot \frac{\partial \boldsymbol{u}(t)}{\partial x_k} + \boldsymbol{K} \cdot \frac{\partial^2 \boldsymbol{u}(t)}{\partial x_i \partial x_k} + \underbrace{\frac{\partial \boldsymbol{C}}{\partial x_i}}_{0} \cdot \frac{\partial \dot{\boldsymbol{u}}(t)}{\partial x_k} + \boldsymbol{C} \cdot \frac{\partial^2 \dot{\boldsymbol{u}}(t)}{\partial x_i \partial x_k} + \underbrace{\frac{\partial \boldsymbol{M}}{\partial x_i}}_{0} \cdot \frac{\partial \ddot{\boldsymbol{u}}(t)}{\partial x_k} + \boldsymbol{M} \cdot \frac{\partial^2 \ddot{\boldsymbol{u}}(t)}{\partial x_i \partial x_k} = \frac{\partial^2 \boldsymbol{P}(t)}{\partial x_i \partial x_k} \tag{5.44}$$

Similar to Eq. 5.28, the fact that the system matrices $\boldsymbol{K}$, $\boldsymbol{C}$ and $\boldsymbol{M}$ are not dependent on the load variable x_i can be used to simplify Eq. 5.44:

$$\boldsymbol{K} \cdot \frac{\partial^2 \boldsymbol{u}(t)}{\partial x_i \partial x_k} + \boldsymbol{C} \cdot \frac{\partial^2 \dot{\boldsymbol{u}}(t)}{\partial x_i \partial x_k} + \boldsymbol{M} \cdot \frac{\partial^2 \ddot{\boldsymbol{u}}(t)}{\partial x_i \partial x_k} = \frac{\partial^2 \boldsymbol{P}(t)}{\partial x_i \partial x_k} \tag{5.45}$$

It can be seen that also the expression of the second derivatives (Eq. 5.45) has the same form as the equation of motion given in Eq. 3.23 (cf. also Eq. 5.28). The pseudo force vector is represented in this case by the second derivative term $\partial^2 \boldsymbol{P}(t)/\partial x_i \partial x_k$. Basically, Eq. 5.45 can be solved – similar to Eq. 5.28 – using the modal superposition principle and the time integration method of Newmark. However, the numerical integration is not necessary at all, since the pseudo load vector $\partial^2 \boldsymbol{P}(t)/\partial x_i \partial x_k$ is always zero. The reason for this is that the term $\partial \boldsymbol{P}/\partial x_k$ is determined based on a single constant force value of 1, corresponding to the entry x_k, while all the other force values are zero (cf. explanations to Eq. 5.28). Its differentiation with respect to the load value x_i obviously delivers the zero vector, since it only implies the differentiation of constant entries. Hence, it is confirmed that:

$$\frac{\partial^2 \boldsymbol{u}(t)}{\partial x_i \partial x_k} = 0 \quad , \quad \frac{\partial^2 \dot{\boldsymbol{u}}(t)}{\partial x_i \partial x_k} = 0 \quad \text{and} \quad \frac{\partial^2 \ddot{\boldsymbol{u}}(t)}{\partial x_i \partial x_k} = 0 \quad \text{for any } i \text{ and } k \tag{5.46}$$

Eq. 5.46 is in fact a consequence of the assumed linearity of the system. This means that the motion quantities depend linearly on the external forces, implying thus a zero second derivative of the motion quantities with respect to the loads.

Introducing Eq. 5.30 in Eq. 5.43 and eliminating the terms canceling to zero yields:

$$\frac{\mathrm{d}^2 \mathcal{F}_{\mathrm{mc}}}{\mathrm{d}x_i \mathrm{d}x_k} = \sum_{h=1}^{n_{\mathrm{md}}} \sum_{j=1}^{n_{\mathrm{t}}} 2 \cdot w_{\mathrm{d},h} \cdot \frac{\partial u_{h,j}}{\partial x_i} \cdot \frac{\partial u_{h,j}}{\partial x_k} + \sum_{h=1}^{n_{\mathrm{ma}}} \sum_{j=1}^{n_{\mathrm{t}}} 2 \cdot w_{\mathrm{a},h} \cdot \frac{\partial \ddot{u}_{h,j}}{\partial x_i} \cdot \frac{\partial \ddot{u}_{h,j}}{\partial x_k} \tag{5.47}$$

Eq. 5.47 can also be expressed in vector notation:

$$\frac{\mathrm{d}^2 \mathcal{F}_{\mathrm{mc}}}{\mathrm{d}x_i \mathrm{d}x_k} = \sum_{h=1}^{n_{\mathrm{md}}} 2 \cdot w_{\mathrm{d},h} \cdot \left(\frac{\partial \boldsymbol{u}_h}{\partial x_i}\right)^{\mathrm{T}} \cdot \frac{\partial \boldsymbol{u}_h}{\partial x_k} + \sum_{h=1}^{n_{\mathrm{ma}}} 2 \cdot w_{\mathrm{a},h} \cdot \left(\frac{\partial \ddot{\boldsymbol{u}}_h}{\partial x_i}\right)^{\mathrm{T}} \cdot \frac{\partial \ddot{\boldsymbol{u}}_h}{\partial x_k} \tag{5.48}$$

5.2.2.2 Hessian of the penalty functions

The Hessian of the penalty functions is described by the last term in Eq. 5.41, which can be formally expressed as follows:

$$\frac{\mathrm{d}}{\mathrm{d}x_i}\left(\frac{\partial \mathcal{F}}{\partial x_k}\right) \widehat{=} \frac{\mathrm{d}}{\mathrm{d}x_i}\left(\frac{\partial \mathcal{F}_\mathrm{p}}{\partial x_k}\right) = \frac{\partial^2 \mathcal{F}_\mathrm{p}}{\partial x_i \partial x_k} \tag{5.49}$$

The penalty functions and their derivatives obviously present only explicit dependencies with respect to the design variables (cf. Eq. 5.15 and 5.40). Therefore, their second partial derivative also consists only of an explicit differentiation term, i.e. $\partial^2 \mathcal{F}_\mathrm{p} / \partial x_i \partial x_k$. It can be elaborated as follows, under consideration of Eq. 5.39:

$$\begin{aligned}\frac{\partial^2 \mathcal{F}_\mathrm{p}}{\partial x_i \partial x_k} =& 2 \cdot w_{\mathrm{p}1} \cdot \frac{\partial}{\partial x_i}\left[\sum_{l=1}^{n_\mathrm{F}} \sum_{j=1}^{n_{\mathrm{tp},l}} \left(\boldsymbol{q}_{\mathrm{p}1,l}^{(j)} \cdot \boldsymbol{x}_l\right) \cdot \left(\boldsymbol{q}_{\mathrm{p}1,l}^{(j)} \cdot \frac{\partial \boldsymbol{x}_l}{\partial x_k}\right)\right] + \\ & 2 \cdot w_{\mathrm{p}2} \cdot \frac{\partial}{\partial x_i}\left[\sum_{l=1}^{n_\mathrm{F}} \sum_{j=1}^{n_{\mathrm{tp},l}} \left(\boldsymbol{q}_{\mathrm{p}2,l}^{(j)} \cdot \boldsymbol{x}_l\right) \cdot \left(\boldsymbol{q}_{\mathrm{p}2,l}^{(j)} \cdot \frac{\partial \boldsymbol{x}_l}{\partial x_k}\right)\right]\end{aligned} \tag{5.50}$$

Introducing the differentiation operators $\partial / \partial x_i$ into the summations and applying the product differentiation rule results in:

$$\begin{aligned}\frac{\partial^2 \mathcal{F}_\mathrm{p}}{\partial x_i \partial x_k} =& 2 \cdot w_{\mathrm{p}1} \cdot \sum_{l=1}^{n_\mathrm{F}} \sum_{j=1}^{n_{\mathrm{tp},l}} \left[\frac{\partial}{\partial x_i}\left(\boldsymbol{q}_{\mathrm{p}1,l}^{(j)} \cdot \boldsymbol{x}_l\right) \cdot \left(\boldsymbol{q}_{\mathrm{p}1,l}^{(j)} \cdot \frac{\partial \boldsymbol{x}_l}{\partial x_k}\right) + \right. \\ & \left. \left(\boldsymbol{q}_{\mathrm{p}1,l}^{(j)} \cdot \boldsymbol{x}_l\right) \cdot \underbrace{\frac{\partial}{\partial x_i}\left(\boldsymbol{q}_{\mathrm{p}1,l}^{(j)} \cdot \frac{\partial \boldsymbol{x}_l}{\partial x_k}\right)}_{*}\right]^{0} + \\ & 2 \cdot w_{\mathrm{p}2} \cdot \sum_{l=1}^{n_\mathrm{F}} \sum_{j=1}^{n_{\mathrm{tp},l}} \left[\frac{\partial}{\partial x_i}\left(\boldsymbol{q}_{\mathrm{p}2,l}^{(j)} \cdot \boldsymbol{x}_l\right) \cdot \left(\boldsymbol{q}_{\mathrm{p}2,l}^{(j)} \cdot \frac{\partial \boldsymbol{x}_l}{\partial x_k}\right) + \right. \\ & \left. \left(\boldsymbol{q}_{\mathrm{p}2,l}^{(j)} \cdot \boldsymbol{x}_l\right) \cdot \underbrace{\frac{\partial}{\partial x_i}\left(\boldsymbol{q}_{\mathrm{p}2,l}^{(j)} \cdot \frac{\partial \boldsymbol{x}_l}{\partial x_k}\right)}_{*}\right]^{0}\end{aligned} \tag{5.51}$$

The terms marked by $*$ in Eq. 5.51 are zero. The reason for this is, one the one hand, that the vectors $\boldsymbol{q}_{\mathrm{p}1,l}^{(j)}$ and $\boldsymbol{q}_{\mathrm{p}2,l}^{(j)}$ only consist of constant values, leading thus to their zero derivatives with respect to a load value x_i. On the other hand, also the second partial derivative component $\partial^2 \boldsymbol{x}_l / \partial x_i \partial x_k$, which occurs when applying the product rule in the terms $*$, is zero, since $\partial \boldsymbol{x}_l / \partial x_k$ only consists of zero entries, excepting a single value of 1 at the location of x_k. This implies its zero derivative with respect to x_i. Eliminating the terms canceling to zero and applying the product differentiation rule for the remaining terms in Eq. 5.51, the total expression of the Hessian for the penalty functions is obtained:

$$\begin{aligned}\frac{\partial^2 \mathcal{F}_\mathrm{p}}{\partial x_i \partial x_k} = &\sum_{l=1}^{n_\mathrm{F}} \sum_{j=1}^{n_{\mathrm{tp},l}} 2 \cdot w_{\mathrm{p}1} \cdot \left(\boldsymbol{q}_{\mathrm{p}1,l}^{(j)} \cdot \frac{\partial \boldsymbol{x}_l}{\partial x_i} \right) \cdot \left(\boldsymbol{q}_{\mathrm{p}1,l}^{(j)} \cdot \frac{\partial \boldsymbol{x}_l}{\partial x_k} \right) + \\ &\sum_{l=1}^{n_\mathrm{F}} \sum_{j=1}^{n_{\mathrm{tp},l}} 2 \cdot w_{\mathrm{p}2} \cdot \left(\boldsymbol{q}_{\mathrm{p}2,l}^{(j)} \cdot \frac{\partial \boldsymbol{x}_l}{\partial x_i} \right) \cdot \left(\boldsymbol{q}_{\mathrm{p}2,l}^{(j)} \cdot \frac{\partial \boldsymbol{x}_l}{\partial x_k} \right)\end{aligned} \tag{5.52}$$

5.2.2.3 Total Hessian of the objective function

The total expression of the Hessian for the objective function is obtained by summation of Eq. 5.48 and 5.52:

$$\begin{aligned}\frac{\mathrm{d}^2 \mathcal{F}}{\mathrm{d}x_i \mathrm{d}x_k} = &\sum_{h=1}^{n_\mathrm{md}} 2 \cdot w_{\mathrm{d},h} \cdot \left(\frac{\partial \boldsymbol{u}_h}{\partial x_i} \right)^\top \cdot \frac{\partial \boldsymbol{u}_h}{\partial x_k} + \sum_{h=1}^{n_\mathrm{ma}} 2 \cdot w_{\mathrm{a},h} \cdot \left(\frac{\partial \ddot{\boldsymbol{u}}_h}{\partial x_i} \right)^\top \cdot \frac{\partial \ddot{\boldsymbol{u}}_h}{\partial x_k} + \\ &\sum_{l=1}^{n_\mathrm{F}} \sum_{j=1}^{n_{\mathrm{tp},l}} 2 \cdot w_{\mathrm{p}1} \cdot \left(\boldsymbol{q}_{\mathrm{p}1,l}^{(j)} \cdot \frac{\partial \boldsymbol{x}_l}{\partial x_i} \right) \cdot \left(\boldsymbol{q}_{\mathrm{p}1,l}^{(j)} \cdot \frac{\partial \boldsymbol{x}_l}{\partial x_k} \right) + \\ &\sum_{l=1}^{n_\mathrm{F}} \sum_{j=1}^{n_{\mathrm{tp},l}} 2 \cdot w_{\mathrm{p}2} \cdot \left(\boldsymbol{q}_{\mathrm{p}2,l}^{(j)} \cdot \frac{\partial \boldsymbol{x}_l}{\partial x_i} \right) \cdot \left(\boldsymbol{q}_{\mathrm{p}2,l}^{(j)} \cdot \frac{\partial \boldsymbol{x}_l}{\partial x_k} \right)\end{aligned} \tag{5.53}$$

The terms $\partial \boldsymbol{u}_h / \partial x_i$, $\partial \boldsymbol{u}_h / \partial x_k$, $\partial \ddot{\boldsymbol{u}}_h / \partial x_i$ and $\partial \ddot{\boldsymbol{u}}_h / \partial x_k$ are available from Eq. 5.29. The components $\partial \boldsymbol{x}_l / \partial x_i$ and $\partial \boldsymbol{x}_l / \partial x_k$ contain a single value of 1 at the location of x_i and x_k, respectively, while all their other values are zero.

5.3 Optimization algorithm

The derivations performed within this thesis are mainly based on the trust region optimization method. A brief insight into the mode of operation of the trust region algorithms was already provided in Section 4.4.3. It should facilitate the general understanding of the solution methods used throughout the work. Without going into further details, the author refers to the reference books of DENNIS JR. et al. (1996) and NOCEDAL et al. (2006) as well as to the works of SORENSEN (1982) and MORÉ et al. (1983). They give more comprehensive overviews of the standard trust region methods and prove – among other things – that they can achieve super-linear convergence when provided with the exact gradient and Hessian of the objective function. However, in addition to the standard formulations in the above-mentioned reference works, various approaches of the trust region methods are available in the literature. They imply a number of adaptations, improvements and enhancements, e.g. various scaling and preconditioning strategies. Their detailed description would go far beyond the scope of this thesis. At this point it should only be mentioned that the solution of the unconstrained optimization problems formulated in the context of this work is based on the two-dimensional subspace minimization strategy, discussed in Section 4.4.3.4 (SCHULTZ et al., 1985; BYRD et al., 1988; BRANCH et al., 1999). It has been enhanced with additional features concerning the scaling of the trust region (COLEMAN et al., 1994; COLEMAN et al., 1996), whereas the two-dimensional subspace is determined using a preconditioned conjugate gradient method (NOCEDAL et al., 2006). All these methods along with the aforementioned additional features are pre-implemented in the default trust region procedure of the computer algebra software MATLAB (THE MATHWORKS INC., 2019), which was adopted for the numerical implementation of the proposed MFI method.

5.4 Moving window approach

5.4.1 Forward sliding window

As already mentioned in Chap. 1, one of the main challenges in dealing with MFI is the solution of problems with a high number of unknowns. They occur, for example, when seeking to identify the loads induced by a vehicle with a large number of axles over the complete time window needed by the vehicle to cross the investigated structure (i.e. computation over a large number of time steps). This situation always occurs in case of railway bridges. The MFI method proposed within the present thesis tackles this issue by means of a forward sliding window approach with a certain overlapping interval, as shown in Fig. 5.6. Its main idea is to formulate and solve an optimization problem according to Eq. 5.15 in each of the n_{w} time windows. In this way, the number of unknown load values n_{var} in the vector of optimization variables $\boldsymbol{x}$ can be controlled. Within the forward optimization, both the unknown variables and the measured time histories are included in the respective time window.

By selecting a reasonable window length, i.e. a limited number of unknown entries in $\boldsymbol{x}$, a substantial increase of the computational efficiency can be achieved in comparison to the direct solution over the whole time window. However, dividing the total time window into a finite number of shorter computational windows requires additional care with respect to kinetic and kinematic compatibility between two successive windows. This is achieved by adopting the initial conditions for each time window from the re-built responses of the previous window. These are the displacement and velocities determined at the beginning of the overlapping interval using the optimal force vector obtained as a solution of Eq. 5.15 in the previous window. In this respect, an assumption regarding the initial conditions has to be made only for the first window, where the consideration of at-rest initial conditions is usually justified. Moreover, the overlapping interval can be used for formulating a plausible initial guess in the subsequent window. The last window is usually longer, since it also comprises the whole decay phase, where no load value is present. However, if the

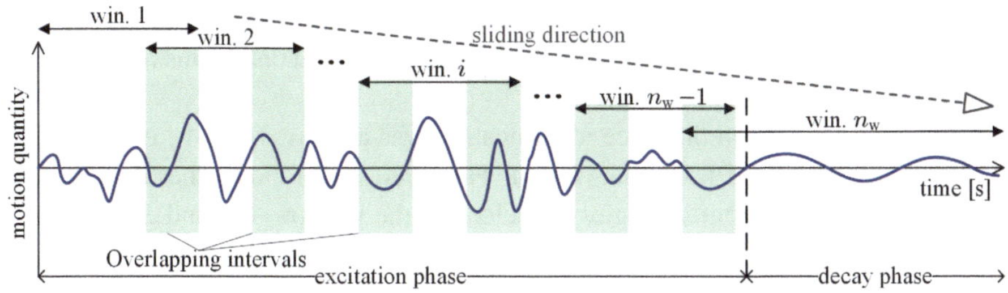

Figure 5.6 Forward sliding window approach

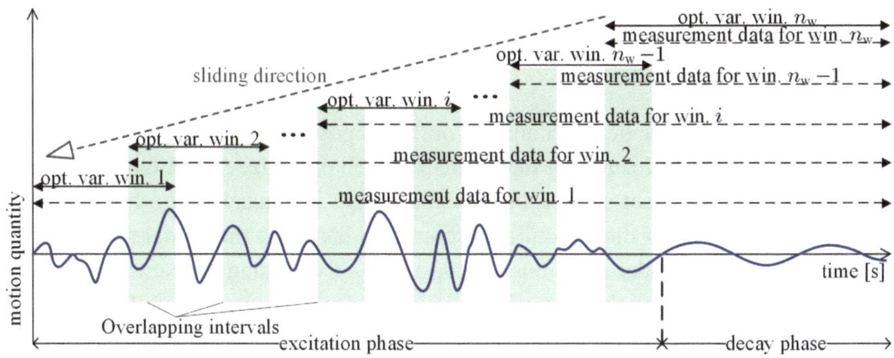

Figure 5.7 Backward sliding window approach

computation time is assumed to be shorter than the measured time histories, the length of the last window may also be lower than the length of the preceding windows.

5.4.2 Backward sliding window

The solution of the MFI problem by means of a forward sliding window, as shown in Fig. 5.6, considers the optimization of the load values from the window i only considering measurement data from the same window. However, the signals assigned to the decay phase (after the last load has left the bridge) may contain important information, which could lead to improved optimization results for all previous windows. Moreover, also the signals of the excitation phase might be influenced by forces acting in the preceding windows. Therefore, a subsequent backward optimization problem according to Fig. 5.7 is enabled after terminating the forward optimization procedure according to Fig. 5.6. It can be seen that this approach allows the optimization of the load values in each window under consideration of a more extensive set of measurement data, which stretches from the starting point of the time window to the end of the decay phase. Furthermore, also in this case, a set of initial conditions has to be provided for each window, in order to allow the direct computation required in the optimization procedure. Adopting the initial conditions from the re-built responses obtained during the forward optimization seems to be a natural choice in this case.

A sequential concept in the force identification was also used by Liu et al. (2014a), Bernal et al. (2015) and Lai et al. (2016). However, they follow other definition approaches and apply different strategies for selecting the window size and eventual overlapping intervals than adopted for the present work. Moreover, they all deal with the reconstruction of non-moving forces and only consider the forward sliding direction of the moving window. A comprehensive discussion on the window length and selection of the overlapping interval for the proposed method is provided in Chap. 6.

5.5 Conclusion

In this chapter, the mathematical formulation of the proposed MFI method is presented. Its main idea is to find the optimal set of force values, which minimizes the difference between a set of computed and a set of measured displacement and acceleration time histories. The direct computation of the structural responses is based on the modal superposition approach, whereas the modal equations of motion are solved using the Newmark method. This approach allows the application of the method for any arbitrarily detailed FE model. The inverse problem is regularized using two weighted penalty terms, which define physical a priori information on the solution. The first one ensures the discrete finite signal with stable average value (DFS-SAV) property of traffic forces on bridges, while the second one seeks to suppress the noise influence on the solution. The weighting parameters of the penalty functions are determined by means of a numerical parametric study, based on simulated noisy measurement data generated by a set of assumed reference forces. The minimum of the objective function is found by means of the trust region optimization strategy. This presents an optimal convergence behavior if it is supplied with exact gradient and Hessian of the objective function. These are determined by analytical differentiation, whereas the Newmark method is used to solve a set of differential equations occurring in the implicit differentiation process. In addition, a forward-backward sliding window approach is formulated, in order to allow the solution of large problems. It implies the division of the total computational time window into a set of separate windows of given length with a certain overlap. Firstly, an optimization problem is solved in each window under consideration of the forward sliding direction. Subsequently, a second optimization for a backward sliding window is performed. In this way, a more extensive set of information, which includes the decay phase of the measured signals, is considered. This allows to improve the accuracy of the identified forces.

6 Numerical validation with simulated measurement data

In this chapter, the MFI method introduced in Chap. 5 is analyzed using a 1D Bernoulli beam model of a simply supported single-tracked bridge. The investigations include a wide set of parameter studies, which give a good insight into the applicability and stability of the method in the context of simulated measurement data.

6.1 System description

The numerical investigations presented in this chapter are based on the structural model of an existing simply supported beam railway bridge with a span of 16.4 m, which is situated on the German railway network. The corresponding structural parameters are provided in PETERSEN et al. (2017). The structure is discretized in 60 Bernoulli beam elements with 61 equally spaced nodes. Each node contains a rotational and a translational (vertical) DOF. The consideration of the boundary conditions in the system of equations and the static condensation of the rotational DOFs (cf. Section 3.3.1) lead to reduced system matrices $\boldsymbol{K}$, $\boldsymbol{C}$ and $\boldsymbol{M}$ in Eq. 3.23, which only contain $n_{\text{dof}} = 59$ translational DOFs (Fig. 6.1). The bending stiffness and the mass distribution of the Bernoulli beam model are $EI = 6.048 \cdot 10^6$ kNm^2 and $\bar{m} = 5063$ kg/m, respectively. The damping ratio is assumed to $\xi = 0.0257$ for all modes.

The first five natural frequencies of the MDOF system are obtained from an eigenvalue analysis (cf. Section 3.3.2) as:

$$\begin{aligned} &f_1 = 6.38 \text{ Hz}; \quad &&f_2 = 25.53 \text{ Hz}; \quad &&f_3 = 57.45 \text{ Hz}; \\ &f_4 = 102.13 \text{ Hz}; \quad &&f_5 = 159.58 \text{ Hz} \end{aligned} \tag{6.1}$$

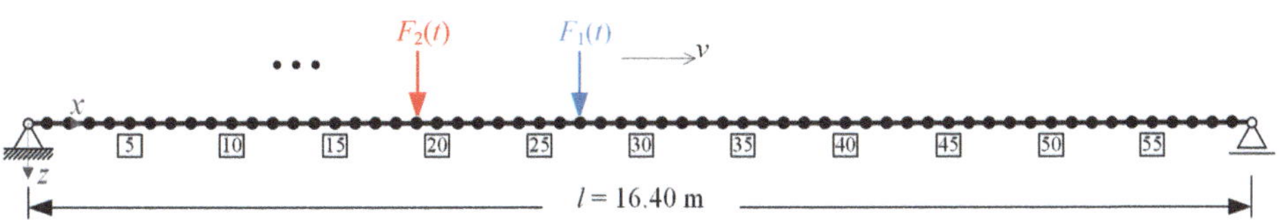

Figure 6.1 Simply supported beam model of a railway bridge under moving loads

A. Firus, *A Contribution to Moving Force Identification in Bridge Dynamics*, Mechanik, Werkstoffe und Konstruktion im Bauwesen 65, https://doi.org/10.1007/978-3-658-39838-5_6

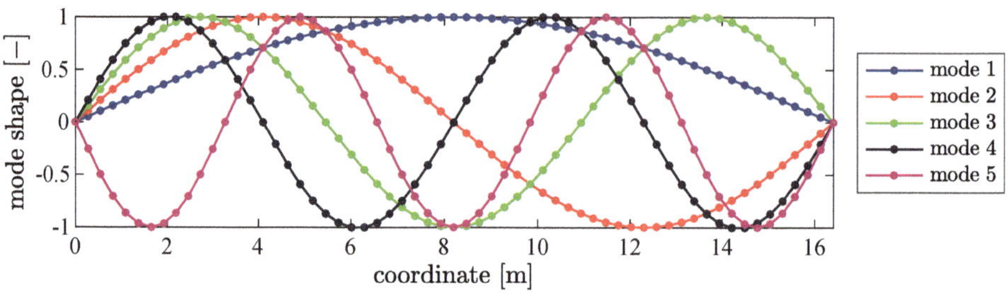

Figure 6.2 Mode shapes of the simply supported Bernoulli beam model

The values given in Eq. 6.1 are practically in perfect agreement with the analytical solutions determined according to Eq. 3.13. The corresponding mode shapes, normalized to the unity maximal value, are illustrated in Fig. 6.2.

6.2 Simulated measurements

The numerical studies presented in the following are based on simulated measurements. They are generated by a forward analysis (cf. Section 3.3.3) under consideration of predefined reference force time histories (Fig. 6.3). The data obtained from the forward analysis is subsequently polluted with artificial noise according to Eq. 5.24, in order to obtain a more realistic data set.

The investigation will firstly focus on a set of four forces, which should facilitate a detailed analysis and interpretation of the results (reference load scenario 1). In addition, a configuration including four forces with impulse-like components is also included (reference load scenario 2). The latter will be used to analyze the performance of the proposed MFI method in the worst case of an excitation including track irregularities or out-of-round wheels. Subsequently, the method will be applied to an extensive set of 32 force time histories (reference load scenario 3), in order to demonstrate its performance in large-size problems typical for railway bridge investigations.

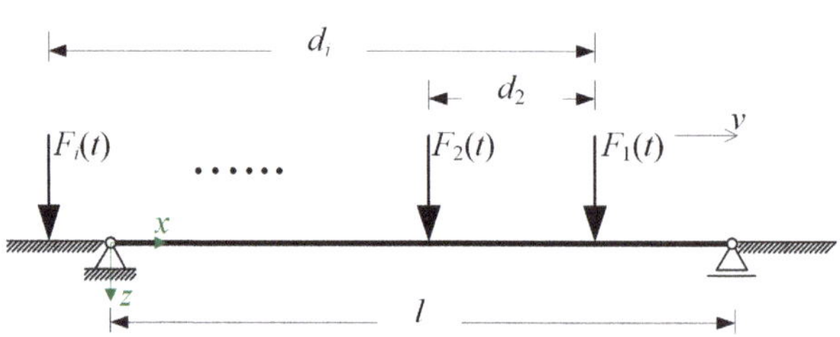

Figure 6.3 Bridge under action of a set of moving loads

6.2.1 Reference load scenario 1

The first reference load scenario implies four dynamic forces, which are defined as a sum of a constant and two harmonic force components:

$$F_1(t) = \begin{cases} 160 + 5 \cdot \sin(2 \cdot \pi \cdot 10 \cdot t) + 2 \cdot \sin(2 \cdot \pi \cdot 7 \cdot t) \quad [\text{kN}] & \text{for } t_{1\text{s}} \leq t < t_{1\text{e}} \\ 0 \quad [\text{kN}] \qquad \text{otherwise} \end{cases}$$
$$F_2(t) = \begin{cases} 150 + 4 \cdot \sin(2 \cdot \pi \cdot 6 \cdot t) + 3.5 \cdot \sin(2 \cdot \pi \cdot 9 \cdot t) \quad [\text{kN}] & \text{for } t_{2\text{s}} \leq t < t_{2\text{e}} \\ 0 \quad [\text{kN}] \qquad \text{otherwise} \end{cases}$$
$$F_3(t) = \begin{cases} 140 + 6 \cdot \sin(2 \cdot \pi \cdot 9 \cdot t) + 5 \cdot \sin(2 \cdot \pi \cdot 12 \cdot t) \quad [\text{kN}] & \text{for } t_{3\text{s}} \leq t < t_{3\text{e}} \\ 0 \quad [\text{kN}] \qquad \text{otherwise} \end{cases}$$
$$F_4(t) = \begin{cases} 130 + 7 \cdot \sin(2 \cdot \pi \cdot 15 \cdot t) + 4 \cdot \sin(2 \cdot \pi \cdot 10 \cdot t) \quad [\text{kN}] & \text{for } t_{4\text{s}} \leq t < t_{4\text{e}} \\ 0 \quad [\text{kN}] \qquad \text{otherwise} \end{cases} \tag{6.2}$$

where t_{is} and t_{ie} represent the time points at which the i-th force enters and leaves the bridge, respectively. They depend on the force velocity and axle coordinates, which are defined as follows (coordinate origin of the vehicle is set at first axle):

$$d_1 = 0\ [\text{m}]; \qquad d_2 = 2.5\ [\text{m}]; \qquad d_3 = 17.375\ [\text{m}]; \qquad d_4 = 19.875\ [\text{m}] \tag{6.3}$$

Considering the natural frequencies of the bridge (cf. Eq. 6.1), it can be seen that the first and the second force in Eq. 6.2 each contain a near-resonant component. The corresponding ratios between the excitation frequencies and the first natural frequency are $6/6.38 = 0.94$ for the first force and $7/6.38 = 1.1$ for the second one. All other excitation frequencies lie further away from the natural frequencies of the structure. Fig. 6.4a shows for example the force-time histories of the load scenario 1 evaluated for the simple supported bridge model described in Section 6.1 and a velocity $v = 160$ km/h.

6.2.2 Reference load scenario 2

The reference force configuration with impulse-like components is achieved by defining four single half-sine shaped impulses, which are added to the dynamic forces from Eq. 6.2. The parameters of the four impulses are given in the following:

$$\begin{aligned} &\text{Amplitude [kN]}: && A_{\text{I}1} = 50,\ A_{\text{I}2} = 60,\ A_{\text{I}3} = 40,\ A_{\text{I}4} = 70 \\ &\text{Impulse duration [ms]}: && t_{\text{I}1} = 10,\ t_{\text{I}2} = 15,\ t_{\text{I}3} = 20,\ t_{\text{I}4} = 13 \\ &\text{Impulse begin [s]}: && t_{\text{aI}1} = 0.15 \cdot t_{\text{out}},\ t_{\text{aI}2} = 0.35 \cdot t_{\text{out}}, \\ & && t_{\text{aI}3} = 0.60 \cdot t_{\text{out}},\ t_{\text{aI}4} = 0.80 \cdot t_{\text{out}} \end{aligned} \tag{6.4}$$

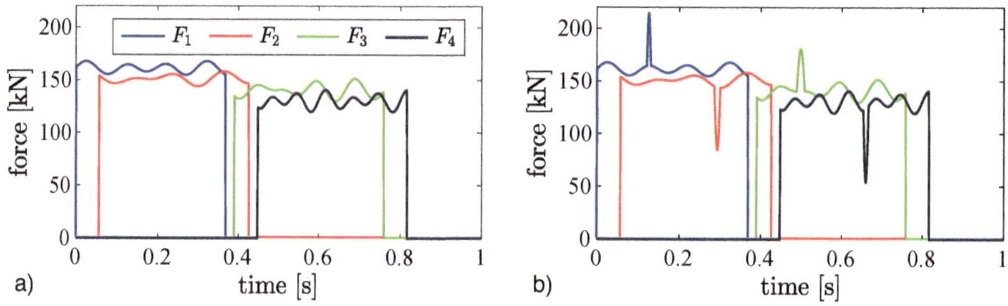

Figure 6.4 Reference load configuration with four moving forces applied to a simply supported bridge model for a vehicle velocity of 160 km/h: a) load scenario 1; b) load scenario 2

where t_{out} denotes the time needed by the load series to cross the structure. For the structure presented above, it can be calculated as follows:

$$t_{\text{out}} = \frac{l + d_4}{v} = \frac{16.4\,[\text{m}] + 19.875\,[\text{m}]}{160\,[\text{km/h}]} = 0.816\,[\text{s}] \tag{6.5}$$

The ratios between the duration of the impulses and the vibration periods of the structure range between 0.06 and 0.13 for the first, 0.26 and 0.51 for the second, 0.57 and 1.5 for the third, 1.02 and 2.04 for the fourth and between 1.59 and 3.17 for the fifth vibration period. It can noted that the impulses have a short relative duration only with respect to the first vibration period. Fig. 6.4b illustrates for example the force time histories including impulse-like components for the considered structure and a vehicle speed of 160 km/h.

6.2.3 Reference load scenario 3

The reference load time histories including 32 moving forces are defined as follows:

$$F_i(t) = \begin{cases} F_{\text{st},i} + \dfrac{F_{\text{st},i}}{28} \cdot \sin(2 \cdot \pi \cdot 5 \cdot t) + \dfrac{F_{\text{st},i}}{20} \cdot \sin(2 \cdot \pi \cdot 8 \cdot t) & \text{for odd } i \\ & \text{and } t_{is} \leq t < t_{ie} \\ F_{\text{st},i} + \dfrac{F_{\text{st},i}}{18} \cdot \sin(2 \cdot \pi \cdot 6 \cdot t) + \dfrac{F_{\text{st},i}}{36} \cdot \sin(2 \cdot \pi \cdot 3 \cdot t) & \text{for even } i \\ & \text{and } t_{is} \leq t < t_{ie} \\ 0 \quad \text{otherwise} & \end{cases} \tag{6.6}$$

where $F_{\text{st},i}$ represents the static component of the i-th moving force. The static loads and the axle distances correspond to a real ICE 3 vehicle[1] (Intercity Express) of the German

[1] The load model of the ICE 3 is given in RIL 804 as a pair of two mirror-symmetrical vehicles. This is indeed a typical operating configuration for the ICE 3, but individual vehicles are also frequently encountered in traffic. The consideration of a single vehicle is sufficient for the purpose of this study.

i	d_i [m]	$F_{st,i}$ [kN]	i	d_i [m]	$F_{st,i}$ [kN]
1	0.000	149.40	17	99.100	148.50
2	2.500	149.40	18	101.600	148.50
3	17.375	148.40	19	116.475	142.50
4	19.875	148.40	20	118.975	142.50
5	24.775	166.30	21	123.875	161.80
6	27.275	166.30	22	126.375	161.80
7	42.150	163.30	23	141.250	159.00
8	44.650	163.30	24	143.750	159.00
9	49.550	154.70	25	148.650	168.00
10	52.050	154.70	26	151.150	168.00
11	66.925	157.60	27	166.025	170.90
12	69.425	157.60	28	168.525	170.90
13	74.325	143.10	29	173.425	151.70
14	76.825	143.10	30	175.925	151.70
15	91.700	144.10	31	190.800	152.30
16	94.200	144.10	32	193.300	152.30

Table 6.1 Axle loads and axle distances of the ICE 3 vehicle (RIL 804)

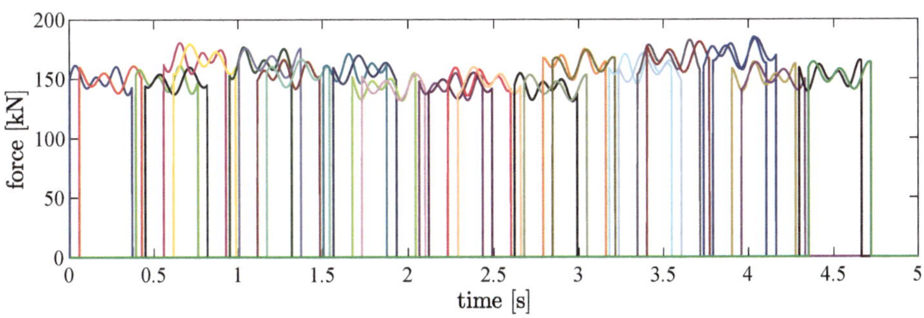

Figure 6.5 Reference load configuration with 32 moving forces (load scenarion 3) applied to a simply supported bridge model for a vehicle velocity of 160 km/h

Railways, as presented in Tab. 6.1 (RIL 804). The large reference force set – determined according to Eq. 6.6 – is illustrated in Fig. 6.5 for the previously presented simply supported beam and a vehicle speed of 160 km/h.

6.2.4 Generation of the simulated measurements

Fig. 6.6 illustrates typical time histories of the simulated acceleration data at some selected (simulated) measurements points (MP). They were obtained by applying the force time histories of the load scenarios 1 and 2 (Fig. 6.4) to the bridge model described in Section 6.1.

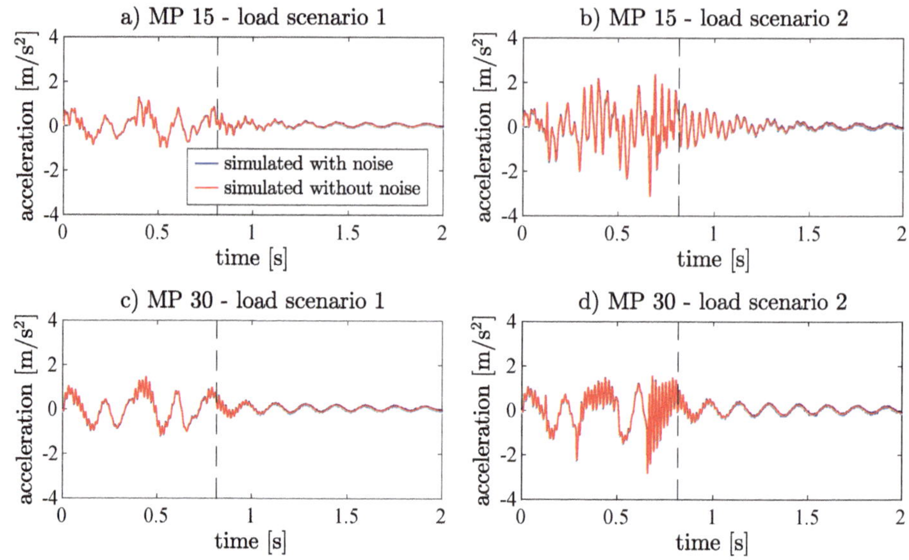

Figure 6.6 Typical time histories of the simulated acceleration measurements generated by load scenarios 1 and 2 using a noise level $e_{\mathrm{p}} = 5\ \%$ (the vertical dashed lines indicate the time points of transition from forced to free vibration)

Corresponding simulated displacement time series are given in Fig. 6.7. As expected, the high-frequency impulse-like force components have a significant influence on the accelerations, whereas the displacement time histories of load scenario 1 and 2 only present slight differences. In addition, Fig. 6.8 shows typical time histories of the displacements and accelerations obtained using the extended set of reference forces (load scenario 3, Fig. 6.5). The noisy signals in Fig. 6.6 to 6.8 were defined under consideration of a noise level $e_{\mathrm{p}} = 5\ \%$, which is considered to be a rather conservative assumption with respect to typical noise levels of common acceleration and displacement sensors. Unless otherwise specified, this value will be used as default noise level for the further investigations.

6.3 Computational parameters

6.3.1 Basic computational parameters

The basic computational configuration used in the following implies a window length of 450 samples and an overlapping interval of one third of the window length. Moreover, the measurement of five accelerations (MP 10, 20, 30, 40, 50) and one displacement (MP 25) was considered as default measurement configuration (Fig. 6.9). The averaging radius of the local smoothing matrix (Eq. 5.17) was firstly assumed to $r_{\mathrm{a}} = 2$. This is – as shown

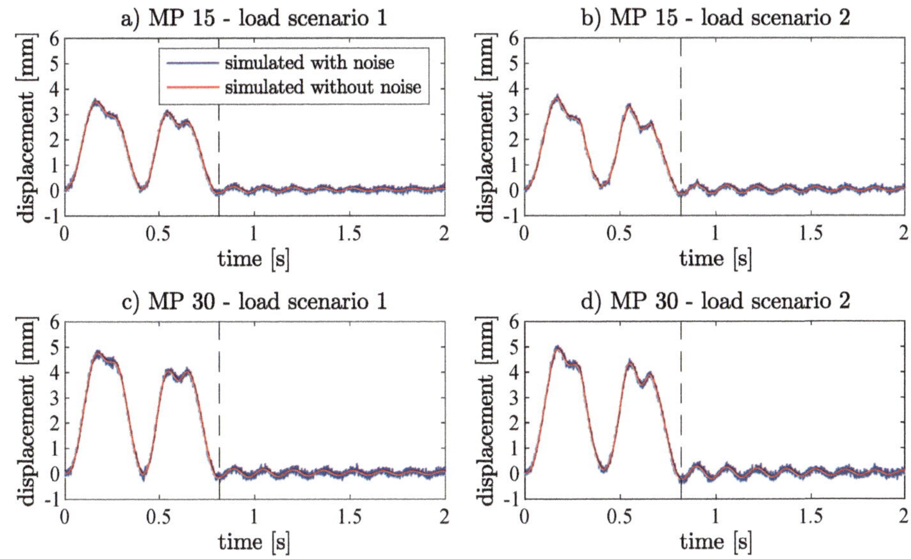

Figure 6.7 Typical time histories of the simulated displacement measurements generated by load scenarios 1 and 2 using a noise level $e_p = 5\ \%$ (the vertical dashed lines indicate the time points of transition from forced to free vibration)

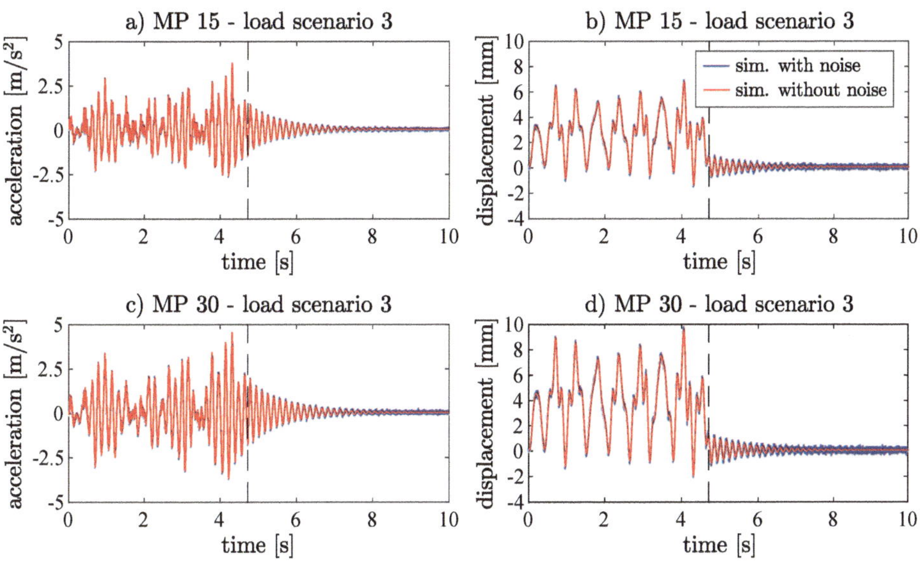

Figure 6.8 Typical time histories of the simulated acceleration and displacement measurements generated by load scenario 3 using a noise level $e_p = 5\ \%$ (the vertical dashed lines indicate the time points of transition from forced to free vibration)

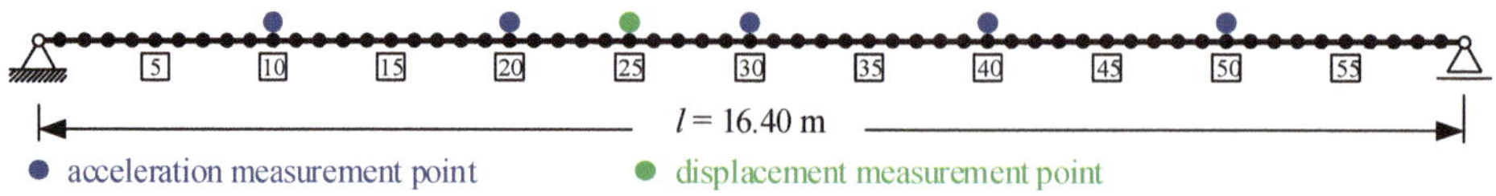

Figure 6.9 Default assumption of the sensor configuration for the beam model

later in Section 6.7.2.2 – a reasonable value, which is usually expected to ensure that the relevant frequency components are not filtered out by the smoothing operation.

In addition, it is to be mentioned that both the computed and the measured vectors of displacement and accelerations (Eq. 5.15) will be normalized with respect to the absolute maximum value of the measured vectors at each location. In this way, the different orders of magnitude of the displacement and acceleration terms are compensated. The normalization is achieved by selecting the weighting parameters $w_{\mathrm{d},h}$ and $w_{\mathrm{a},h}$ according to Eq. 5.14. The numerator 1 in both components of Eq. 5.14 indicates that the same importance is assigned to the displacement and acceleration terms.

Furthermore, the length of the computational time window is fixed to 2 seconds for load scenario 1 and 2 and to 10 seconds for load scenario 3, independent on the investigated velocity. According to RIL 804, the first three modes will be considered in the modal superposition approach[2]. In this sense, a computational time step $\Delta t = 0.001$ s is selected. This allows a sufficiently fine time discretization of the vibration in the third mode with a frequency of 57.45 Hz.

The computation parameters defined above will be considered as default for the following computations, unless separate specifications will be defined in the text. More computational configurations, implying – among others – various window lengths, smoothing radii, overlapping length and various combinations of the measurements points will be investigated in Section 6.7.

6.3.2 Initial values of the optimization process

The assumption used in this work with respect to the initial values of the optimization procedure is given in the following:

- The vector of initial values of the first window for the forward optimization is assumed as the zero vector:

$$\boldsymbol{x}_0^{(1,\mathrm{f})} = \mathbf{0} \tag{6.7}$$

  where $\boldsymbol{x}_0^{(i,\mathrm{f})}$ generally denotes the vector of initial values of the i-th window for the forward optimization.

[2] RIL 804 recommends to choose the upper frequency limit of the considered modes as the maximum between 30 Hz, $1.5 \cdot f_1$ (first bending mode) and f_3 (third bending mode).

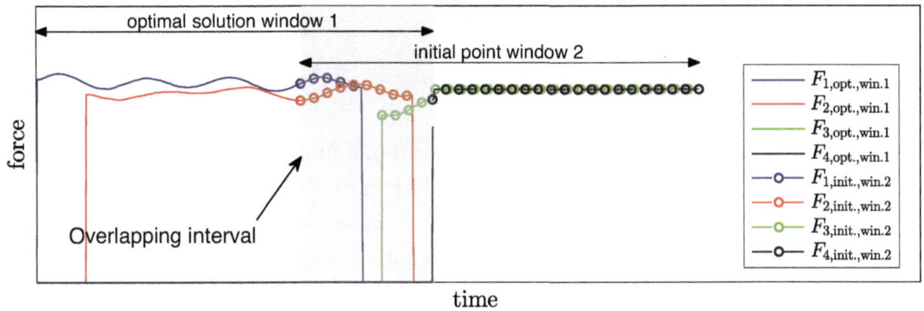

Figure 6.10 Typical assumption of the initial values for the forward optimization

- The vectors of initial values $\boldsymbol{x}_0^{(i,\mathrm{f})}$ of all subsequent windows ($i > 1$) for the forward optimization are defined as follows (cf. also Fig. 6.10):
  - Since an optimal solution $\boldsymbol{x}^{*,(i-1,\mathrm{f})}$ for the previous window ($i-1$) already exists, the entries of the window i located within the overlapping interval are adopted from the solution of the preceding window $\boldsymbol{x}^{*,(i-1,\mathrm{f})}$.
  - The entries outside the overlapping interval are defined as the average value of all optimal solutions obtained for the previous windows. This means that the initial force time history outside the overlapping interval is assumed to be constant. This assumption is well justified, since the axle loads assigned to a specific vehicle or the forces induced by different pedestrians usually present similar orders of magnitude. In this way, it is ensured that the initial values lie in a plausible range, expectedly leading to faster convergence.
- The vector of initial values $\boldsymbol{x}_0^{(i,\mathrm{b})}$ used in the backward optimization for window i is assumed as the optimal point $\boldsymbol{x}^{*,(i,\mathrm{f})}$ resulting from the forward optimization for the same window:

$$\boldsymbol{x}_0^{(i,\mathrm{b})} = \boldsymbol{x}^{*,(i,\mathrm{f})} \tag{6.8}$$

  In this way, the iteration starts with a plausible initial force vector, eventually leading to a faster convergence.

6.3.3 Termination criteria of the optimization

All investigations for MFI presented in the next sections take into account the following convergence criteria of the optimization procedure:

- Relative change of the objective function:

$$\frac{|\mathcal{F}(\boldsymbol{x}^{k+1}) - \mathcal{F}(\boldsymbol{x}^{k})|}{|\mathcal{F}(\boldsymbol{x}^{k+1})|} \leq \varepsilon_{\text{rel}} \quad \text{where} \quad \varepsilon_{\text{rel}} = 10^{-8} \tag{6.9}$$

  This criterion is one of the most common termination criteria used in optimization, since the limit ε_{rel} does not have to be adapted for each objective function. HARZHEIM (2019) states that a typical value of ε_{rel} is 10^{-3}. However, the present investigations require – as it will be explained later – significantly lower values of ε_{rel}.

- Absolute value of the first-order optimality:

$$\left\| \nabla \mathcal{F}(\boldsymbol{x}^{(k+1)}) \right\|_{\infty} < \varepsilon_{\text{lim}} \quad \text{where} \quad \varepsilon_{\text{lim}} = 10^{-8} \tag{6.10}$$

  The first-order optimality represents the infinity norm of the gradient. This is, in fact, the absolute maximum gradient entry. It is expected that this value cancels to zero at the optimal point.

6.4 Verification of the analytical sensitivities

6.4.1 Gradient check

6.4.1.1 Gradient of the main optimization criterion

Before proceeding to the results of the inverse computations, the analytical sensitivity of the main optimization criterion (Eq. 5.31) is verified for the first window. This is done by means of a comparison with the numerical sensitivity, evaluated according to Eq. 4.39 under consideration of $w_{\text{p1}} = w_{\text{p2}} = 0$. The sensitivities are determined at an arbitrary point $\boldsymbol{x} = 2 \cdot \boldsymbol{x}_{\text{reference}}$, whereas $\boldsymbol{x}_{\text{reference}}$ represents the vector comprising the reference forces of load scenario 1 (Fig. 6.4a). The result obtained under consideration of an absolute perturbation factor[3] $\Delta x_k = 1$ [N] is shown in Fig. 6.11. It reveals a nearly exact match between the analytical and the numerical sensitivities of the main optimization criterion. The corresponding relative error computed according to Eq. 5.23 (replacing $\boldsymbol{x}_{\text{reference}}$ by the analytical sensitivity and $\boldsymbol{x}_{\text{identified}}$ by the numerical sensitivity) is 0.003 %. Hence, the correct derivation and implementation of the gradient expression given in Eq. 5.31 are confirmed.

Fig. 6.11 also illustrates which entries of the gradient correspond to the individual forces comprised in the investigated window. In this respect, it is to be mentioned that the gradient vector has the same structure as the vector of unknown variables (cf. Fig. 5.1). It can be

[3] It should be mentioned that the perturbation factors of the numerical gradients (Section 6.4.1) and Hessians (Section 6.4.2) were determined by parametric studies. Their aim is to find the appropriate perturbation factors that provide the best agreement between the numerical and analytical derivatives.

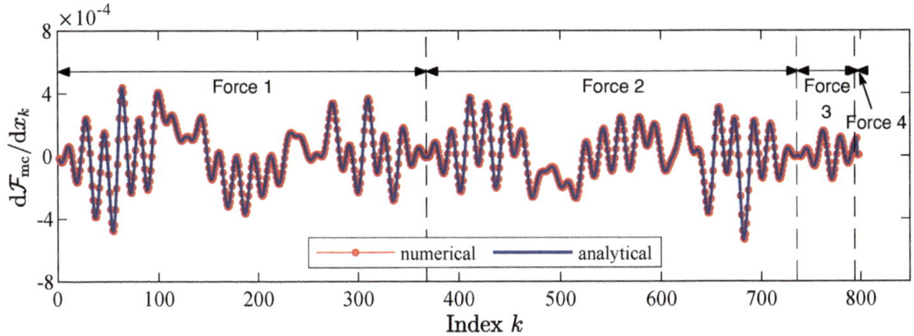

Figure 6.11 Comparison of the numerical and analytical sensitivities of the main optimization criterion for the first window

clearly seen that the sensitivities of the unregularized objective function with respect to the forces located in the proximity of the supports are very low. This explains the low identification accuracy for the forces around the supports, which is often reported in the context of MFI (ZHU et al., 2002a; FENG et al., 2015; PAN et al., 2017).

6.4.1.2 Gradient of the total objective function

Since the previous discussion was only intended to focus on the sensitivity of the main optimization criterion, the weighting factors w_{p1} and w_{p2} of the penalty functions were assumed to be zero. In the following, the derivation and implementation of the sensitivity expressions for the penalty functions (Eq. 5.37 and 5.38) are also verified. This is achieved by firstly computing the total derivative expression for the objective function (Eq. 5.40) with arbitrary selected nonzero weighting parameters $w_{p1} = w_{p2} = 5 \cdot 10^{-8}$. The result of Eq. 5.40 is subsequently compared to the numerical sensitivities determined according to Eq. 4.39. The result obtained using the same perturbation factor $\Delta x_k = 1$ [N] is illustrated in Fig. 6.12. Due to the effect of the regularization terms, the sensitivity values for the forces located around the supports are now not necessarily close to zero. The evaluation of Eq. 5.23 delivers also in this case an insignificant error of 0.008 %. It can be thus concluded that the derivation and implementation of the sensitivities for both the main optimization criterion (Eq. 5.31) and the penalty functions (Eq. 5.37 and 5.38) are correct.

6.4.2 Hessian check

6.4.2.1 Hessian of the main optimization criterion

The analytically determined components of the Hessian matrix of the main optimization criterion (Eq. 5.48) have also been subjected to a verification procedure based on a comparison with the corresponding numerical values. The latter were computed using Eq. 4.42 for

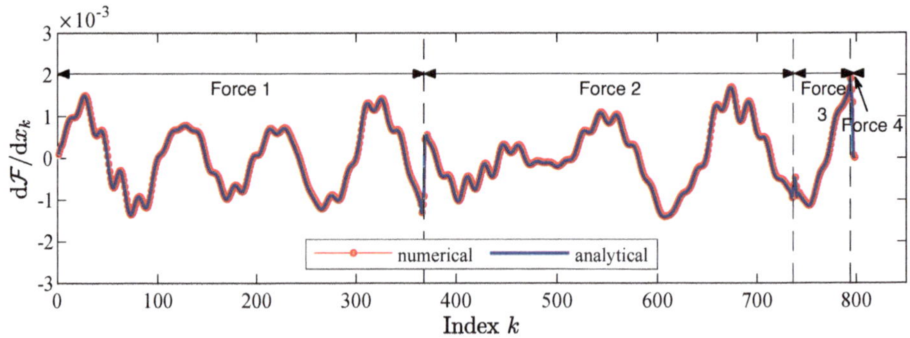

Figure 6.12 Comparison of the numerical and analytical sensitivities of the total objective function for the first window

$w_{p1} = w_{p2} = 0$ and $\Delta x_k = \Delta x_i = 100$ N. Fig. 6.13 shows the graphical comparison of the results for a window size of 50 samples. It only contains force values from the first moving force, since the second force reaches the bridge after 58 samples. The reduced window size was selected for this comparison in order to enhance the clarity of the visual comparison and to limit the computational effort required by the evaluation of Eq. 4.42. It is obvious that the numerical and analytical Hessian entries are in excellent agreement. Concatenating the Hessian columns to a single vector, Eq. 5.23 can be evaluated, in order to provide a rough quantitative comparison between the numerical and analytical approach. An insignificant error of $1 \cdot 10^{-4}$ % is obtained, confirming thus the correct derivation and implementation of the analytical expression of the Hessian for the main optimization criterion (Eq. 5.48).

6.4.2.2 Hessian of the total objective function

Similar to the previously discussed analysis of the gradient, the Hessian of the penalty terms also needs to be investigated. For this, the Hessian of the total objective function is evaluated analytically (according to Eq. 5.53) and numerically (according to Eq. 4.42) for the same weighting factors ($w_{p1} = w_{p2} = 5 \cdot 10^{-8}$). A visual comparison of the results is allowed by Fig. 6.14. A very good agreement can be observed. This can be also quantified by evaluating Eq. 5.23, which delivers an error of $2 \cdot 10^{-5}$ %. It can be thus concluded that the derivation and implementation of analytical Hessian matrices for the main optimization criterion and the penalty functions are correct[4].

[4] It is to be mentioned that the implementation of the sensitivity and Hessian expressions is not dependent on the structure. Therefore, it can be considered that the gradient and Hessian checks are passed for any structural and excitation parameters used within the analysis. In this sense, the verification of the sensitivity information will not be explicitly addressed for the other structures investigated in the further course of this thesis.

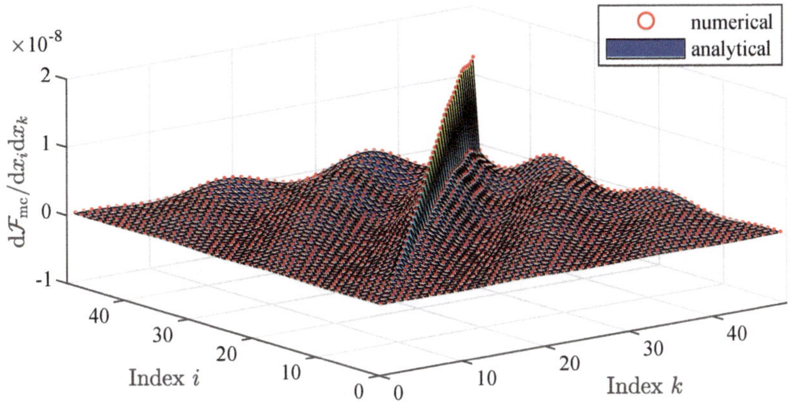

Figure 6.13 Graphical representation of the analytically and numerically determined entries of the Hessian matrix of the main optimization criterion for a first window of 50 samples

6.5 Optimization results without penalty functions

The first example focuses on minimizing an objective function which only consists of the main optimization criterion $\mathcal{F}_{\mathrm{mc}}$. This means that the penalty functions, comprised in the term $\mathcal{F}_{\mathrm{p}}$ (cf. Eq. 5.32), are canceled to zero by setting $w_{\mathrm{p}1} = w_{\mathrm{p}2} = 0$.

6.5.1 Analysis with noisy data

6.5.1.1 Results of the load scenario 1 without penalty functions

The identification results obtained under consideration of noisy data from load scenario 1 are illustrated in Fig. 6.15 for both the forward and backward optimization procedures. In order to allow a visual comparison of the results, the reference forces of the load scenario 1 are as well included in Fig. 6.15. Overall, a quite satisfactory agreement between the identified and the reference forces can be visually observed. However, it is obvious that a lower accuracy is generally obtained when the forces are located in the proximity of the supports. This effect occurs most likely – as already mentioned – due to the low sensitivity of the objective function with respect to forces acting at the ends of the bridge (cf. Fig. 6.11). The error between the reference (Fig. 6.4a) and the identified force can be evaluated according to Eq. 5.23. A value $x_{\mathrm{err,f}} = 1.91\ \%$ is revealed for the forward approach. Even though it is rather difficult to detect a significant visual difference between the results shown in Fig. 6.15a and b, a slightly lower error of $x_{\mathrm{err,b}} = 1.75\ \%$ is obtained from the optimization based on the backward sliding window. This corresponds to the expected tendency of lower errors at the end of the backward optimization.

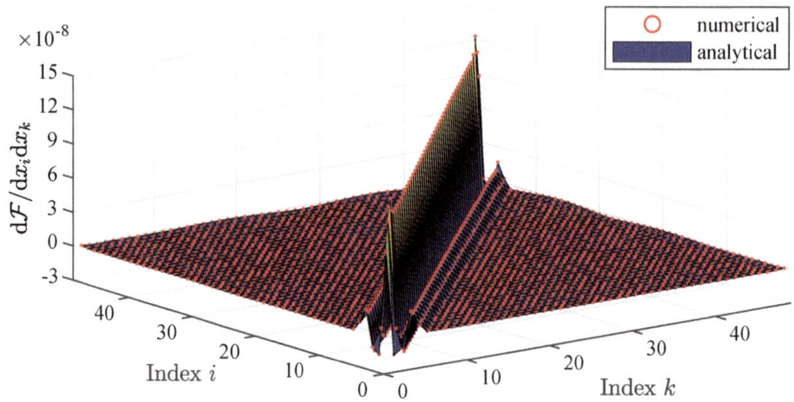

Figure 6.14 Graphical representation of the analytically and numerically determined entries of the Hessian matrix of the total objective function for a first window of 50 samples

The iteration progress of the optimization is shown in Fig. 6.16 for all three windows during the forward and backward optimization procedures. Obviously, the first iteration steps in each window present a very low decay behavior. The reason for this is most likely the program-internal scaling of the trust region radius at the beginning of each iteration. This explains the need of a very strict termination criterion with respect to the relative change of the objective function (cf. Eq. 6.9). The increasing of $\varepsilon_{\mathrm{rel}}$ (e.g. to 10^{-4} or 10^{-5}) eventually leads to premature termination after only a few iterations, where the scaling process is not yet finished. This motivates the introduction of the second convergence criterion given in Eq. 6.10. This was relevant for the termination of all iteration processes shown in Fig. 6.16. In fact, the first-order optimality criterion (Eq. 6.10) is the significant criterion for the vast majority of iteration processes presented in the following sections.

Fig. 6.16a indicates that the convergence in the first window of the forward optimization procedure is achieved after 29 iterations, whereas the iterations of the second and third window are terminated after 23 and 22 steps, respectively. The reason for the higher number of iterations in the first window is the assumption of the vector of initial values as a zero vector (Eq. 6.7). Selecting the initial values of the second and third window according to Fig. 6.10 confirms that a considerably better initial guess is provided. This eventually leads to lower initial values of the objective function[5] and thus to a lower number of iteration steps.

With respect to the backward optimization, it can be seen that the iteration of the third window terminates after only one iteration (Fig. 6.16b). This can be explained by the assumption of the initial point for the backward optimization according to Eq. 6.8. It states

[5] Please note the secondary y axis of Fig. 6.16a.

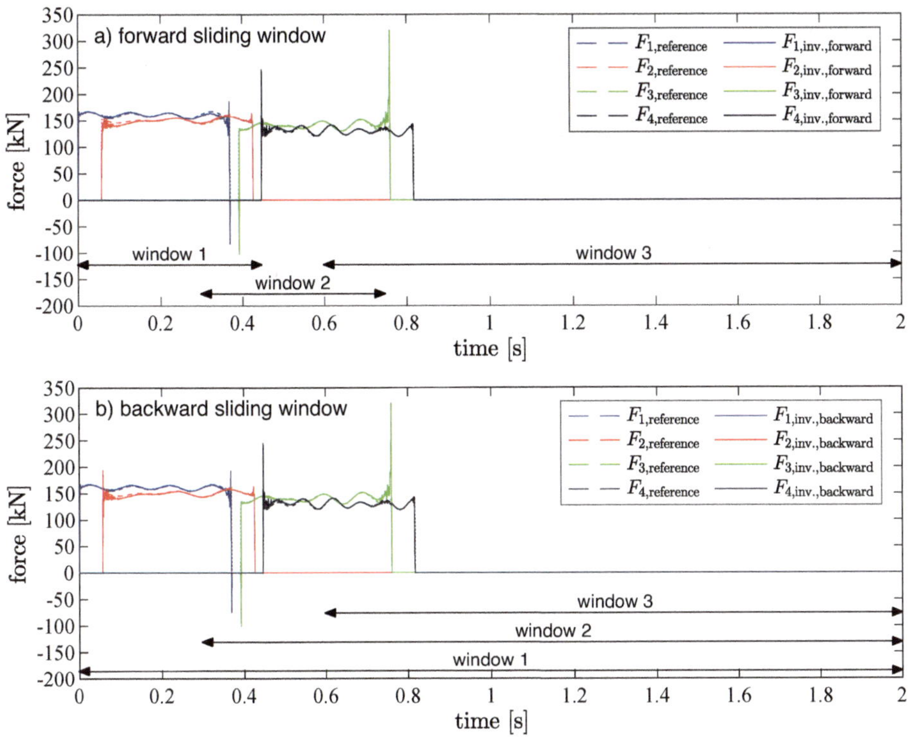

Figure 6.15 Reference and identified force time histories of load scenario 1 without penalty functions (solid lines: identified forces, dashed lines: reference forces): a) results of the forward optimization ($x_{\text{err,f}} = 1.91$ %); b) results of the backward optimization ($x_{\text{err,b}} = 1.75$ %)

that the backward optimization in a certain window is initiated at the optimal point obtained from the forward optimization in the same window. Since the last window is similar for both sliding directions (cf. Fig. 6.15), it is obvious that the convergence of window 3 is achieved at the initial point. Moreover, it can be seen that the assumption of the initial point according to Eq. 6.8 only generates a very slight reduction of the objective function in the backward optimization (Fig. 6.16b and c).

6.5.1.2 Results of the load scenario 2 without penalty functions

The effect of a low accuracy in the proximity of the supports can be also observed in the identification results of the reference load scenario 2 (Fig. 6.17). Nevertheless, the impulse components seem to be identified with a satisfactory accuracy. A total error $x_{\text{err,b}} = 4.14$ % is obtained at the end of the backward optimization procedure. This value is considerably

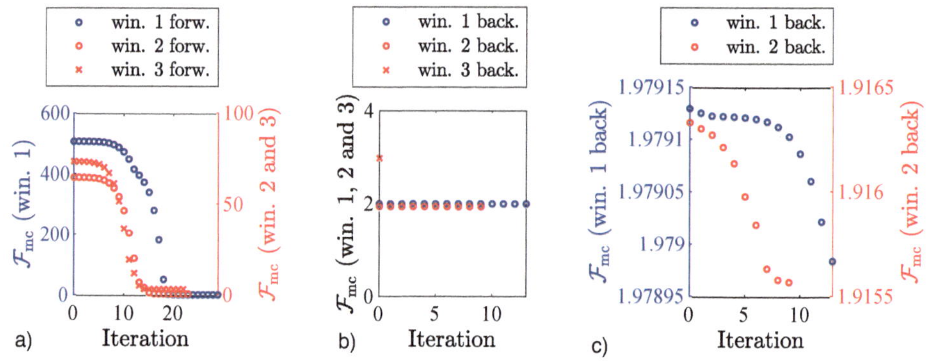

Figure 6.16 Convergence behavior of the MFI problem without penalty functions under consideration load scenario 1: a) results forward optimization; b) results backward optimization; c) detailed view of the iteration behavior for the first two windows using the backward sliding approach

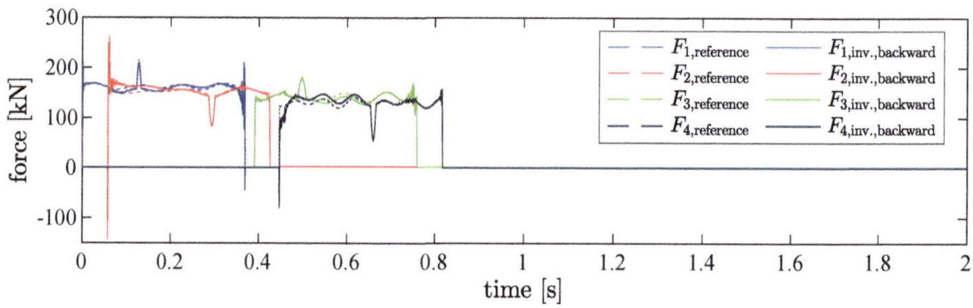

Figure 6.17 Reference and identified force time histories of load scenario 2 without penalty functions – results of the backward optimization ($x_{err,b} = 4.14$ %)

higher than the corresponding results obtained for load scenario 1 (load without impulse components, cf. Fig. 6.15). This effect will be further explored in Section 6.6.

6.5.1.3 Results of the load scenario 3 without penalty functions

The largest relative error between the reference and the identified forces $x_{err,b} = 4.89$ % occurs in relation to load scenario 3. However, this represents in fact a realistic scenario for practical applications in the field of railway bridges. The results are shown in Fig. 6.18 together with the window division for the forward computation. At first sight, a rather unsatisfactory agreement can be observed. It can be clearly seen that the effect of low identification accuracy at force locations close to the supports is very pronounced in this case, leading to isolated force peaks of up to almost 1000 kN and thus about 5 to 6 times higher than the average of the reference forces. However, these high force peaks have a

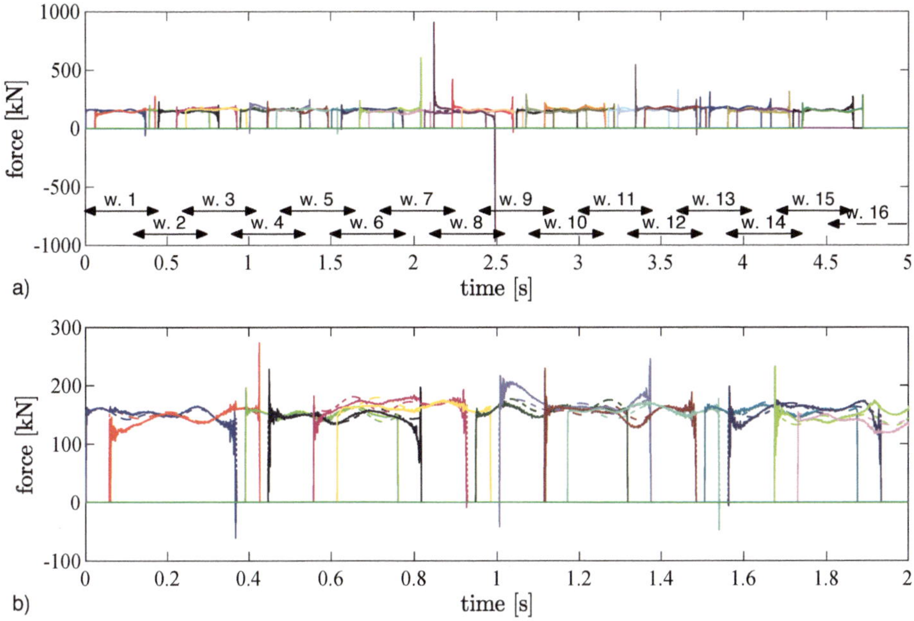

Figure 6.18 Reference and identified force time histories of load scenario 3 without penalty functions (solid lines: identified forces, dashed lines: reference forces): a) results of the backward optimization ($x_{\mathrm{err,b}} = 4.89\ \%$); b) detailed representation of the results for the time interval $0-2$ s

rather insignificant influence on the computed displacement and accelerations due to their low sensitivity with respect to forces acting in the proximity of the supports. This fact can be clearly observed in Fig. 6.19, which illustrates the comparison between the simulated measurement data and the corresponding re-built results. The latter were obtained by a direct analysis (cf. Section 5.1.2) using the identified forces shown in Fig. 6.18. Similar to the evaluation of the relative force error according to Eq. 5.23, the relative errors of the motion quantities can be determined by replacing the force vectors in Eq. 5.23 by the corresponding vectors of the motion quantities:

$$u_{\mathrm{err}} = \frac{\|\boldsymbol{u}_{\mathrm{re-built}} - \boldsymbol{u}_{\mathrm{measured}}\|_1}{\|\boldsymbol{u}_{\mathrm{measured}}\|_1} \cdot 100\ \% \text{ and } \ddot{u}_{\mathrm{err}} = \frac{\|\ddot{\boldsymbol{u}}_{\mathrm{re-built}} - \ddot{\boldsymbol{u}}_{\mathrm{measured}}\|_1}{\|\ddot{\boldsymbol{u}}_{\mathrm{measured}}\|_1} \cdot 100\ \% \tag{6.11}$$

The results obtained in the present case are $\ddot{u}_{\mathrm{err,b}} = 5.54\ \%$ for the accelerations and $u_{\mathrm{err,b}} = 5.18\ \%$ for the displacements. Despite the very good visual agreement between the two signals, the errors seem to reach rather high values. The reason for this is the noise component of the signals, which ensures a high value of the numerators in Eq. 6.11.

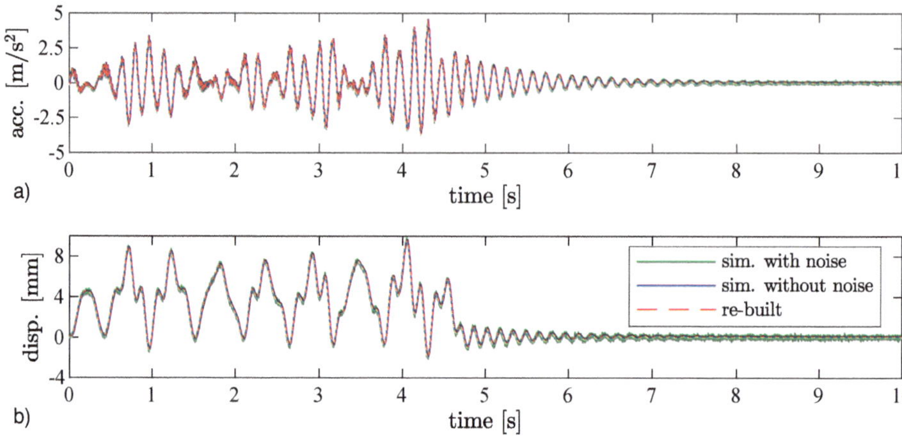

Figure 6.19 Simulated measurements and re-built responses at MP 30 based on the load scenario 3: a) acceleration time histories; b) displacement time histories

Considering the noise-free simulated measurement data as reference values in Eq. 6.11 delivers significantly lower errors: $\ddot{u}_{\text{err,b}} = 1.87$ % and $u_{\text{err,b}} = 0.19$ %. It is thus confirmed that the method successfully reconstructs the initial "measurement" data, even though several force peaks with high amplitudes are present in the identified results.

6.5.2 Analysis with noise-free data

Even though noise-free data has no direct relevance for real practical applications, the performance of the proposed MFI method with noise-free data will be shortly discussed before proceeding to the inverse computations with noisy data and nonzero penalty functions. For this, the time histories without noise obtained from the direct dynamic analysis under consideration of the reference load scenarios (e.g. Fig. 6.6 to 6.8) are used as "measurement" data within the identification problem. Since the data is not polluted with noise, it is expected that the computation retrieves the exact reference forces. On the one side, this analysis is of great importance for the verification of the formulation and implementation of the inverse problem itself. On the other side, it will be used to formulate a set of solvability conditions, which define the relationship between the number of modes and sensors considered in the analysis and the number forces to be identified.

It is to be mentioned that the numerator in both components of Eq. 5.14 was increased from 1 to 10000 for all analyses performed with noise-free data. This leads to the following weighting parameters of the main optimization criterion:

$$w_{d,h} = \frac{10000}{(\max|\bar{\boldsymbol{u}}_h|)^2} \quad \text{and} \quad w_{a,h} = \frac{10000}{(\max|\bar{\ddot{\boldsymbol{u}}}_h|)^2} \tag{6.12}$$

The reason for the increased values of $w_{d,h}$ and $w_{a,h}$ is that the objective function will take extremely low values in case of retrieving the exact solution, eventually leading to a premature termination of the iteration process. In this respect, it is to be mentioned that increasing both parameters by the same value has no influence at all on the optimal solution of Eq. 5.15. An alternative approach to this issue would be the formulation of a more stringent limit ε_{lim} for the convergence criterion given in Eq. 6.10. However, a further reduction of ε_{lim} (actual value: 10^{-8}) implies very low values of the sensitivity and Hessian entries, which are getting close to the computer accuracy. This would eventually lead to increased numerical errors.

6.5.2.1 Solvability condition 1

Fig. 6.20 shows the results of the optimization problem with noise-free "measurement" data for the load scenario 1, obtained using different numbers of modes within the analysis[6]. It is obvious that the consideration of three and two modes delivers practically exact results. This fact can be quantified by the corresponding error evaluations according to Eq. 5.23: 0.008 % for the analysis using three modes and 0.04 % for the analysis based on two modes. On the contrary, a significantly worse result ($x_{\text{err}} = 3.38$ %) is obtained when the modal superposition considers only one mode. A close visual analysis of Fig. 6.20c reveals that the higher deviation is in fact only generated during the time, in which two forces are

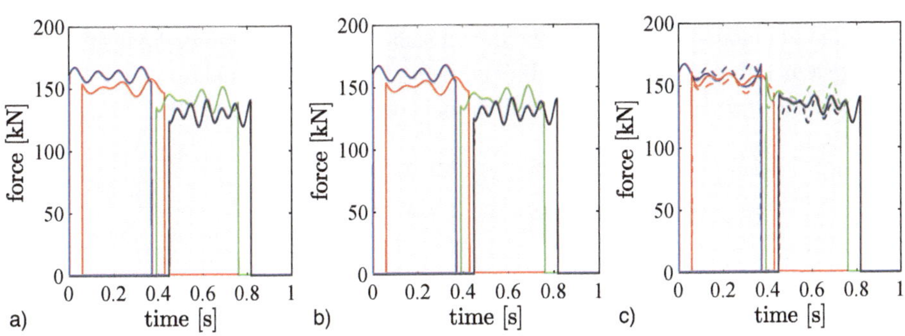

Figure 6.20 Reference and identified force time histories of load scenario 1 for noise-free data and without penalty functions (solid lines: identified forces, dashed lines: reference forces): a) Results computed using three modes ($x_{\text{err,f}} = 0.008$ %); b) Results computed using two modes ($x_{\text{err,f}} = 0.04$ %); c) Results computed using one mode ($x_{\text{err,f}} = 3.38$ %)

[6] The present investigations imply the same number of modes considered for both forward analysis (generation of the simulated measurement data) and the inverse analysis (force identification). The performance of the method with different frequency contents in the forward and inverse analysis will be discussed in Section 6.7.

present on the bridge (grey areas in Fig. 6.20c). This fact indicates that *the number of modes considered in the computation has to be at least equal to the maximum number of forces present on the structure at a given time*, in order to allow an exact force identification[7]. This will be referred to as the first solvability condition. Its proof its given in the following.

The modal forces in a certain time step $z+1$ can be expressed as (cf. Eq. 3.45):

$$\begin{aligned} P_1^{z+1} &= \phi_{\mathrm{F},1,1}^{z+1} \cdot F_1^{z+1} + \phi_{\mathrm{F},1,2}^{z+1} \cdot F_2^{z+1} + \ldots + \phi_{\mathrm{F},1,n_\mathrm{F}}^{z+1} \cdot F_{n_\mathrm{F}}^{z+1} \\ P_2^{z+1} &= \phi_{\mathrm{F},2,1}^{z+1} \cdot F_1^{z+1} + \phi_{\mathrm{F},2,2}^{z+1} \cdot F_2^{z+1} + \ldots + \phi_{\mathrm{F},2,n_\mathrm{F}}^{z+1} \cdot F_{n_\mathrm{F}}^{z+1} \\ &\vdots \\ P_n^{z+1} &= \phi_{\mathrm{F},n,1}^{z+1} \cdot F_1^{z+1} + \phi_{\mathrm{F},n,2}^{z+1} \cdot F_2^{z+1} + \ldots + \phi_{\mathrm{F},n,n_\mathrm{F}}^{z+1} \cdot F_{n_\mathrm{F}}^{z+1} \end{aligned} \tag{6.13}$$

where $\phi_{\mathrm{F},i,j}^{z+1}$ denotes the value of the i-th mode shape at the location of the j-th force F_j in the time step $z+1$. Using Eq. 6.13 and Eq. 3.44, the modal equations can be expressed for a discrete time step $z+1$ as:

$$\begin{aligned} K_1^{z+1}\alpha_1^{z+1} + C_1\dot{\alpha}_1^{z+1} + M_1\ddot{\alpha}_1^{z+1} &= \phi_{\mathrm{F},1,1}^{z+1} \cdot F_1^{z+1} + \phi_{\mathrm{F},1,2}^{z+1} \cdot F_2^{z+1} + \ldots + \phi_{\mathrm{F},1,n_\mathrm{F}}^{z+1} \cdot F_{n_\mathrm{F}}^{z+1} \\ K_2^{z+1}\alpha_2^{z+1} + C_2\dot{\alpha}_2^{z+1} + M_2\ddot{\alpha}_2^{z+1} &= \phi_{\mathrm{F},2,1}^{z+1} \cdot F_1^{z+1} + \phi_{\mathrm{F},2,2}^{z+1} \cdot F_2^{z+1} + \ldots + \phi_{\mathrm{F},2,n_\mathrm{F}}^{z+1} \cdot F_{n_\mathrm{F}}^{z+1} \\ &\vdots \\ K_n^{z+1}\alpha_n^{z+1} + C_n\dot{\alpha}_n^{z+1} + M_n\ddot{\alpha}_n^{z+1} &= \phi_{\mathrm{F},n,1}^{z+1} \cdot F_1^{z+1} + \phi_{\mathrm{F},n,2}^{z+1} \cdot F_2^{z+1} + \ldots + \phi_{\mathrm{F},n,n_\mathrm{F}}^{z+1} \cdot F_{n_\mathrm{F}}^{z+1} \end{aligned} \tag{6.14}$$

where K_i, C_i and M_i represent the i-th modal stiffness, damping and mass, respectively. α_i^{z+1} is the i-th modal coordinate in the time instant $z+1$. Furthermore, the displacement of the system at the nodes corresponding to the n_md displacement measurement locations can be expressed as a superposition of all modal contributions[8] (Eq. 3.39):

$$\begin{aligned} u_1^{z+1} &= \phi_{\mathrm{u},1,1} \cdot \alpha_1^{z+1} + \phi_{\mathrm{u},2,1} \cdot \alpha_2^{z+1} + \ldots + \phi_{\mathrm{u},n,1} \cdot \alpha_n^{z+1} \\ u_2^{z+1} &= \phi_{\mathrm{u},1,2} \cdot \alpha_1^{z+1} + \phi_{\mathrm{u},2,2} \cdot \alpha_2^{z+1} + \ldots + \phi_{\mathrm{u},n,2} \cdot \alpha_n^{z+1} \\ &\vdots \\ u_{n_\mathrm{md}}^{z+1} &= \phi_{\mathrm{u},1,n_\mathrm{md}} \cdot \alpha_1^{z+1} + \phi_{\mathrm{u},2,n_\mathrm{md}} \cdot \alpha_2^{z+1} + \ldots + \phi_{\mathrm{u},n,n_\mathrm{md}} \cdot \alpha_n^{z+1} \end{aligned} \tag{6.15}$$

[7] This finding is also mentioned by MAES et al. (2015) in the context of the instantaneous inversion of modally reduced order models.

[8] The acceleration expression can be expressed in a similar way, by replacing the displacement and the modal coordinates in Eq. 6.15 with their second derivatives (cf. Eq. 5.9).

where $\phi_{u,i,j}$ represents the value of the i-th mode shape at the location of the j-th computation point. u_j^{z+1} denotes the displacement of the j-th measurement point in the time step $z+1$. Eq. 6.15 can be expressed in a more simplified form as:

$$\begin{pmatrix} u_1^{z+1} \\ u_2^{z+1} \\ \vdots \\ u_{n_{\mathrm{md}}}^{z+1} \end{pmatrix} = \begin{pmatrix} \phi_{u,1,1} & \phi_{u,2,1} & \cdots & \phi_{u,n,1} \\ \phi_{u,1,2} & \phi_{u,2,2} & \cdots & \phi_{u,n,2} \\ \vdots & \vdots & \ddots & \vdots \\ \phi_{u,1,n_{\mathrm{md}}} & \phi_{u,2,n_{\mathrm{md}}} & \cdots & \phi_{u,n,n_{\mathrm{md}}} \end{pmatrix} \begin{pmatrix} \alpha_1^{z+1} \\ \alpha_2^{z+1} \\ \vdots \\ \alpha_n^{z+1} \end{pmatrix} \Longrightarrow \underbrace{\boldsymbol{u}^{z+1}}_{n_{\mathrm{md}}\times 1} = \underbrace{\boldsymbol{\phi}_{\mathrm{u}}}_{n_{\mathrm{md}}\times n} \cdot \underbrace{\boldsymbol{\alpha}^{z+1}}_{n\times 1} \tag{6.16}$$

According to the time integration procedure of Newmark (Section 3.3.4) applied for the solution of the modal equations of motion (Eq. 6.14), the i-th modal coordinate α_i^{z+1} can be expressed as (cf. Eq. 3.67):

$$\alpha_i^{z+1} = \hat{K}_i^{-1} \cdot \hat{P}_i^{z+1} = \frac{\hat{P}_i^{z+1}}{\hat{K}_i} \tag{6.17}$$

where

$$\hat{K}_i = K_i + \frac{\gamma}{\beta \cdot \Delta t} \cdot C_i + \frac{1}{\beta \cdot \Delta t^2} \cdot M_i \tag{6.18}$$

and

$$\begin{aligned} \hat{P}_i^{z+1} = & P_i^{z+1} + \left[\frac{1}{\beta \cdot \Delta t^2} \cdot M_i + \frac{\gamma}{\beta \cdot \Delta t} \cdot C_i\right] \cdot \alpha_i^z + \left[\frac{1}{\beta \cdot \Delta t} \cdot M_i + \left(\frac{\gamma}{\beta} - 1\right) \cdot C_i\right] \cdot \dot{\alpha}_i^z \\ & + \left[\left(\frac{1}{2 \cdot \beta} - 1\right) \cdot M_i + \Delta t \cdot \left(\frac{\gamma}{2 \cdot \beta} - 1\right) \cdot C_i\right] \cdot \ddot{\alpha}_i^z \end{aligned} \tag{6.19}$$

$\hat{P}_i^{z+1}$ and $\hat{K}_i$ are the effective modal force and stiffness, respectively (cf. Eq. 3.65 and Eq. 3.66). The last three terms on the right side of Eq. 6.19 only depend on the system parameters and on the modal coordinates in the preceding time step z. They can be comprised to a scalar term $s_{d,i}^z$. Under consideration of Eq. 6.13, Eq. 6.17 becomes:

$$\alpha_i^{z+1} = \frac{\left(\phi_{F,i,1}^{z+1} \cdot F_1^{z+1} + \phi_{F,i,2}^{z+1} \cdot F_2^{z+1} + \ldots + \phi_{F,i,n_F}^{z+1} \cdot F_{n_F}^{z+1}\right) + s_{d,i}^z}{\hat{K}_i} \tag{6.20}$$

By writing Eq. 6.20 in matrix form, the vector of generalized coordinates can be expressed in the time step $z+1$ as:

$$\begin{pmatrix} \alpha_1^{z+1} \\ \alpha_2^{z+1} \\ \vdots \\ \alpha_n^{z+1} \end{pmatrix} = \begin{pmatrix} \frac{\phi_{\mathrm{F},1,1}^{z+1}}{\hat{K}_1} & \frac{\phi_{\mathrm{F},1,2}^{z+1}}{\hat{K}_1} & \cdots & \frac{\phi_{\mathrm{F},1,n_\mathrm{F}}^{z+1}}{\hat{K}_1} \\ \frac{\phi_{\mathrm{F},2,1}^{z+1}}{\hat{K}_2} & \frac{\phi_{\mathrm{F},2,2}^{z+1}}{\hat{K}_2} & \cdots & \frac{\phi_{\mathrm{F},2,n_\mathrm{F}}^{z+1}}{\hat{K}_2} \\ \vdots & \vdots & \ddots & \vdots \\ \frac{\phi_{\mathrm{F},n,1}^{z+1}}{\hat{K}_n} & \frac{\phi_{\mathrm{F},n,2}^{z+1}}{\hat{K}_n} & \cdots & \frac{\phi_{\mathrm{F},n,n_\mathrm{F}}^{z+1}}{\hat{K}_n} \end{pmatrix} \cdot \begin{pmatrix} F_1^{z+1} \\ F_2^{z+1} \\ \vdots \\ F_{n_\mathrm{F}}^{z+1} \end{pmatrix} + \begin{pmatrix} \frac{s_{\mathrm{d},1}^z}{\hat{K}_1} \\ \frac{s_{\mathrm{d},2}^z}{\hat{K}_2} \\ \vdots \\ \frac{s_{\mathrm{d},n}^z}{\hat{K}_n} \end{pmatrix} \tag{6.21}$$

Eq. 6.21 can be written in a compact form as:

$$\underbrace{\boldsymbol{\alpha}^{z+1}}_{n\times 1} = \underbrace{\phi_{\mathrm{F,d}}^{z+1}}_{n\times n_\mathrm{F}} \cdot \underbrace{\boldsymbol{F}^{z+1}}_{n_\mathrm{F}\times 1} + \underbrace{\boldsymbol{s}_\mathrm{d}^z}_{n\times 1} \tag{6.22}$$

Introducing Eq. 6.22 in Eq. 6.16 and transferring a term on the left side, yields:

$$\underbrace{\boldsymbol{u}^{z+1}}_{n_\mathrm{md}\times 1} - \underbrace{\phi_\mathrm{u}}_{n_\mathrm{md}\times n} \cdot \underbrace{\boldsymbol{s}_\mathrm{d}^z}_{n\times 1} = \underbrace{\phi_\mathrm{u}}_{n_\mathrm{md}\times n} \cdot \underbrace{\phi_{\mathrm{F,d}}^{z+1}}_{n\times n_\mathrm{F}} \cdot \underbrace{\boldsymbol{F}^{z+1}}_{n_\mathrm{F}\times 1} \tag{6.23}$$

Eq. 6.23 has a unique solution only if $(\phi_\mathrm{u} \cdot \phi_{\mathrm{F,d}}^{z+1})$ is of rank n_F. However, it is known that the rank of the product of two matrices $\boldsymbol{A}_1 \in \mathbb{R}^{n_1 \times n_2}$ and $\boldsymbol{A}_2 \in \mathbb{R}^{n_2 \times n_3}$ satisfies the following inequality (BRONSTEIN et al., 2012):

$$\mathrm{rank}(\boldsymbol{A}_1 \cdot \boldsymbol{A}_2) \leq \min\left(\mathrm{rank}(\mathrm{A}_1), \mathrm{rank}(\mathrm{A}_2)\right) \tag{6.24}$$

and

$$\begin{aligned} \mathrm{rank}(\boldsymbol{A}_1) &\leq \min(n_1, n_2) \\ \mathrm{rank}(\boldsymbol{A}_2) &\leq \min(n_2, n_3) \end{aligned} \tag{6.25}$$

By replacing $\boldsymbol{A}_1 = \phi_\mathrm{u}$ and $\boldsymbol{A}_2 = \phi_{\mathrm{F,d}}^{z+1}$ in Eq. 6.24 and 6.25 follows immediately that $(\phi_\mathrm{u} \cdot \phi_{\mathrm{F,d}}^{z+1})$ is of rank n_F only if $\mathrm{rank}(\phi_{\mathrm{F,d}}^{z+1}) = n_\mathrm{F}$. This implies that $n_\mathrm{F} \leq n$ and confirms thus the previously formulated solvability condition. Further requirements related to the matrix $\boldsymbol{A}_1$ (ϕ_u in Eq. 6.23) are discussed in Section 6.5.2.2 in the context of the second solvability condition.

Without providing the detailed derivation, it is to be mentioned that Eq. 6.23 can be also expressed for n_ma acceleration measurement points as:

$$\underbrace{\ddot{\boldsymbol{u}}^{z+1}}_{n_\mathrm{ma}\times 1} - \underbrace{\phi_\mathrm{a}}_{n_\mathrm{ma}\times n} \cdot \underbrace{\boldsymbol{s}_\mathrm{a}^z}_{n\times 1} = \underbrace{\phi_\mathrm{a}}_{n_\mathrm{ma}\times n} \cdot \underbrace{\phi_{\mathrm{F,a}}^{z+1}}_{n\times n_\mathrm{F}} \cdot \underbrace{\boldsymbol{F}^{z+1}}_{n_\mathrm{F}\times 1} \tag{6.26}$$

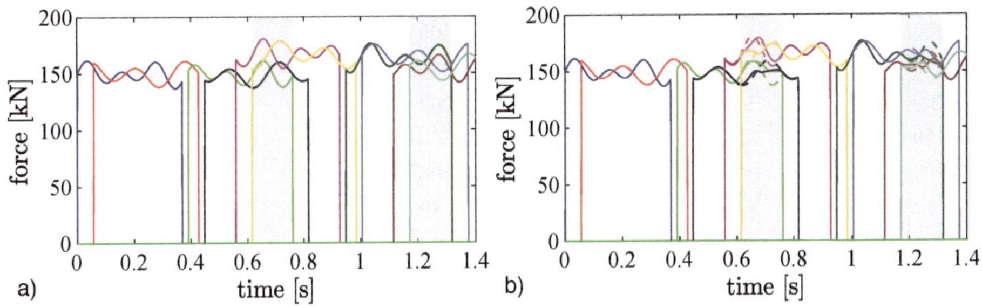

Figure 6.21 Reference and identified force time histories of load scenario 3 for noise-free data and without penalty functions (solid lines: identified forces, dashed lines: reference forces): a) Detailed view of the results computed using four modes ($x_{\mathrm{err,f}} = 0.04\ \%$); b) Detailed view of the results computed using three modes ($x_{\mathrm{err,f}} = 0.72\ \%$)

where $\phi_{\mathrm{F,a}}^{z+1}$ and s_{a}^{z} are designed under consideration of Eq. 3.62 and 6.20. In this case, $(\phi_{\mathrm{F,a}}^{z+1})$ has to be of rank n_{F}, implying as well that the number of modes has to be greater than the number of unknown forces.

The assumption regarding the required number of modes was also verified by analyzing the corresponding results of load scenario 3, which are illustrated in Fig. 6.21. In this case, up to four forces – corresponding to the rear and front bogies of two adjacent carriages – are present on the bridge at a time. In this sense, it is expected that an exact force reconstruction can be achieved when the computation is performed using four modes, while the consideration of only three modes should deliver inexact results. This fact is confirmed by the force time histories represented in Fig. 6.21. The higher error of 0.72 % obtained for the computation based on three modes (compared to 0.04 % when using four modes) is obviously only caused by the deviations occurring during the presence of four forces on the structure (see grey areas in Fig. 6.21b).

6.5.2.2 Solvability condition 2

Based on the requirements $\mathrm{rank}(\phi_{\mathrm{u}} \cdot \phi_{\mathrm{F,d}}^{z+1}) = n_{\mathrm{F}}$ (cf. Eq. 6.23) or $\mathrm{rank}(\phi_{\mathrm{a}} \cdot \phi_{\mathrm{F,a}}^{z+1}) = n_{\mathrm{F}}$ (cf. Eq. 6.26), a second solvability condition can be formulated. Since the dimension of the two matrices is $n_{\mathrm{md}} \times n_{\mathrm{F}}$ in case of displacement or $n_{\mathrm{ma}} \times n_{\mathrm{F}}$ in case of acceleration sensors, it is obvious that the rank n_{F} can only be fulfilled if $n_{\mathrm{F}} \leq n_{\mathrm{md}}$ or $n_{\mathrm{F}} \leq n_{\mathrm{ma}}$, respectively. This means *the number of acceleration or displacement sensors included in the measurement data set has to be at least equal to the number of forces present on the structure in a certain time instant*, in order to allow the exact identification of the moving forces from noise-free data[9]. This condition is demonstrated in Fig. 6.22 using the example of an inverse analysis of the load scenario 1 under consideration of the first two modes. The

[9] A similar requirement has been also formulated by CHAN et al. (2001b), GONZÁLES et al. (2008), MAES et al. (2015) and LAI et al. (2016).

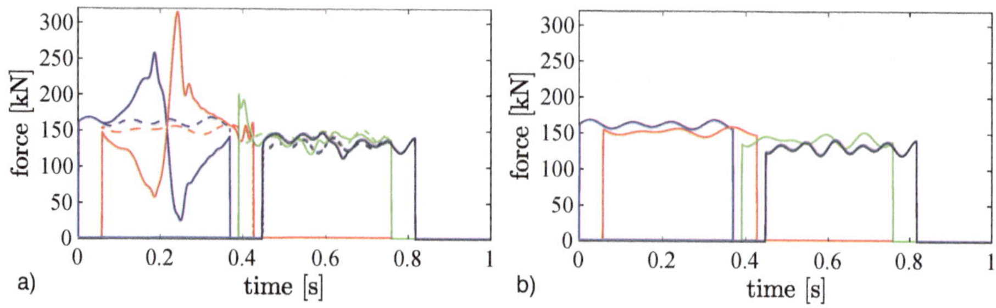

Figure 6.22 Requirement with respect to the number of sensors needed for achieving an exact solution for a maximum of two forces acting on the structure (solid lines: identified forces, dashed lines: reference forces): a) result for one acceleration measurement point (MP 15), $x_{err,f} = 17.41\ \%$; b) results for two acceleration measurement points (MP 15 and 30), $x_{err,f} = 0.1\ \%$.

result shown in Fig. 6.22a is based on one sensor (MP 15), whereas a maximum number of two forces acting simultaneously on the bridge has to be identified. The relative error $x_{err,f} = 17.41\ \%$ lies far outside the acceptable range. In this respect, it can be observed that the discrepancies between the identified and reference force time histories mainly occur in the time intervals, in which two forces are acting simultaneously on the bridge (grey areas in Fig. 6.22a). On the contrary, a more or less exact solution could be retrieved for the time intervals, where only one force excites the structure. This effect is in agreement with the above-mentioned solvability condition. This can be also confirmed by a further computation, which includes an additional measurement point (Fig. 6.22b). It can be seen, that the identification procedure of two forces based on two acceleration sensors (MP 15 and 30) leads to a nearly exact solution with a corresponding error of $x_{err,f} = 0.1\ \%$.

6.5.2.3 Additional remarks

Relation between the number of sensors and number of modes It is to be mentioned, that the comprehensive set of investigations performed within this work might indicate in some situations a further "apparent" solvability condition, which has already been postulated in the literature (LAW et al., 2011). It states that the number of sensors has to be at least equal to the number of mode shapes considered in the analysis, in order to allow an exact reconstruction of the moving forces from noise-free measurement data. This statement is exemplified by the results shown in Fig. 6.23. They are based on the reference load scenario 1 and consider three modes in the dynamic analysis. It can be clearly seen that the solution based on only two acceleration measurement points (MP 15 and 30) is partly in a rather unsatisfactory agreement with the set of reference forces (Fig. 6.23a). The corresponding error is $x_{err} = 2.03\ \%$. On the contrary, the consideration of an additional acceleration MP at node 20 leads to an exact solution with a neglectable error $x_{err} = 0.024\ \%$ (Fig. 6.23b). However, it is to be mentioned that this solvability condition does not entirely hold from mathematical point of view (cf. Eq. 6.23 and Eq. 6.26). In fact, it only occurs because the matrix $\phi_{\mathrm{a}} \cdot \phi_{\mathrm{F,a}}^{z+1}$ is close to rank deficiency in some time steps of the computation with only two acceleration sensors. This can be clearly seen in Fig. 6.24, which illustrates the condition number of $\phi_{\mathrm{a}} \cdot \phi_{\mathrm{F,a}}^{z+1}$ over the first 390 time steps for the two cases presented in Fig. 6.23. The condition number determined for the computation including two sensors is partly several orders of magnitude higher than the condition number obtained under consideration of three sensors. Moreover, the time interval of a high condition number coincides with the interval of an unsatisfactory accuracy of the identified forces in Fig. 6.23a. This effect occurs most likely due to the fact that the forces might pass very close to the mode shape nodes, canceling thus the corresponding modal contributions in certain time steps. This eventually leads to an indirect violation of the first solvability condition and thus to an inexact result.

Requirement of displacement sensors MAES et al. (2015) defines a uniqueness condition in the context of instantaneous inversion of modally reduced order models. It concerns the identification of non-moving forces in the noise-free case and states that their static component can not be correctly retrieved if only acceleration sensors are used. The reason for this is that the acceleration measurements are insensitive to an excitation, which is constant in time. However, this criterion seems to be looser in case of MFI. It could be seen in Fig. 6.22b and 6.23b that the exact forces, including their static components, could be retrieved using only noise-free acceleration data. In order to emphasize this effect more clearly, an additional case is investigated. This uses a set of four constant forces moving over the structure instead of the dynamic load time histories of the reference scenario 1, whereas only the five acceleration measurement points shown in Fig. 6.9 are considered as "simulated measurement data". The identification result can be seen in Fig. 6.25. It shows

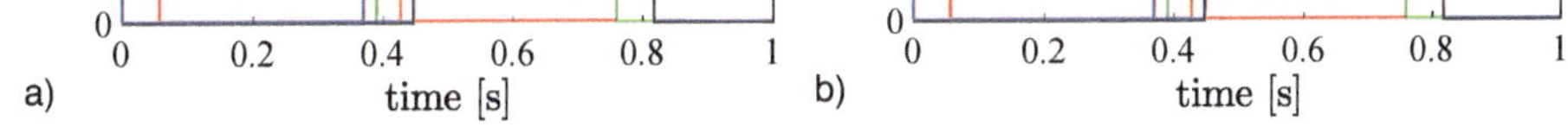

Figure 6.23 "Apparent" requirement with respect to the relation between the number of sensors and number of modes needed for achieving an exact solution (solid lines: identified forces, dashed lines: reference forces): a) results for two acceleration measurement points (MP 15 and 30) and three modes, $x_{\mathrm{err,f}} = 2.03$ %; b) results for three acceleration measurement points (MP 15, 20 and 30) and three modes, $x_{\mathrm{err,f}} = 0.024$ %

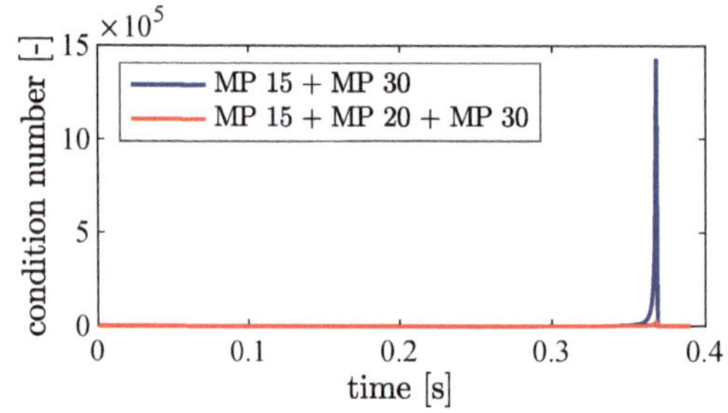

Figure 6.24 Condition number evaluation for the "apparent" solvability condition

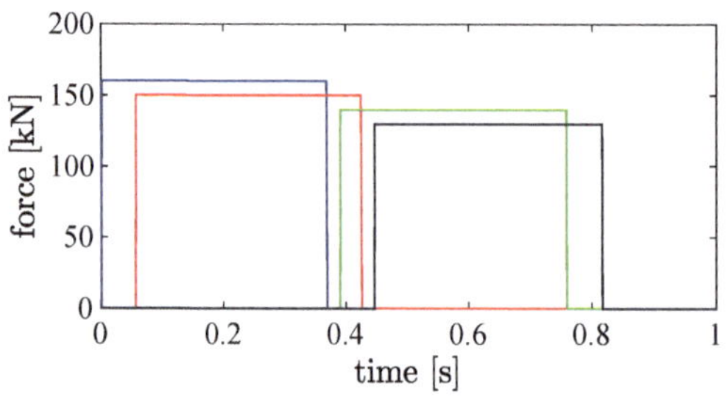

Figure 6.25 Identification results for a set of constant forces based on only simulated accelerations and without penalty terms (solid lines: identified forces, dashed lines: reference forces), $x_{\mathrm{err,f}} = 0.021$ %

that an exact solution is retrieved also in this case. The reason for this can be explained using the concept of modal forces. For a non-moving constant excitation, the modal forces computed according to Eq. 3.45 would be constant as well, since the value of the mode shape at the excitation point $\phi_j^\top$ would be the same for all time steps. On the contrary, the forward locomotion of the forces implies that the value of the mode shape $\phi_j^\top$ changes in every time step. This leads to time variant modal loads, even though the magnitude of the moving forces is constant in time (cf. Fig. 6.26). Hence, the accelerations at the measurement points are expected to be more sensitive with respect to constant moving loads than in case of a non-moving constant excitation. In this respect, Fig. 6.27 illustrates the sensitivity of the acceleration at MP 20 with respect to the constant force magnitudes $F_{\mathrm{const.},i}$. They were obtained by summation of the individual sensitivities $\partial \ddot{u}(t)/\partial x_k$ determined according to Eq. 5.29 for all $n_{\mathrm{tp},i}$ force values x_k assigned to the same force $F_{\mathrm{const.},i}$:

$$\frac{\partial \ddot{\boldsymbol{u}}_h(t)}{\partial F_{\mathrm{const.},i}} = \sum_{k=1}^{n_{\mathrm{tp},i}} \frac{\partial \ddot{\boldsymbol{u}}_h(t)}{\partial x_k} \tag{6.27}$$

The sensitivity time histories show reasonable values, which allow the identification of the constant force components $F_{\mathrm{const.},i}$ from only acceleration measurements.

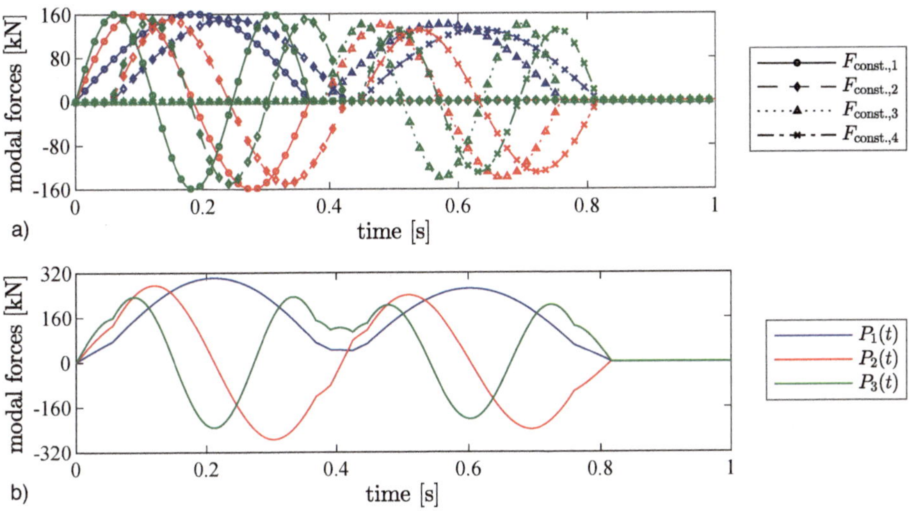

Figure 6.26 Modal forces for a set of constant forces passing the structure: a) modal forces of the individual forces (1st mode: half sine curves, 2nd mode: complete sine curves, 3rd mode: 1.5 sine curves); b) total modal forces as linear addition of the components from all four forces shown in a).

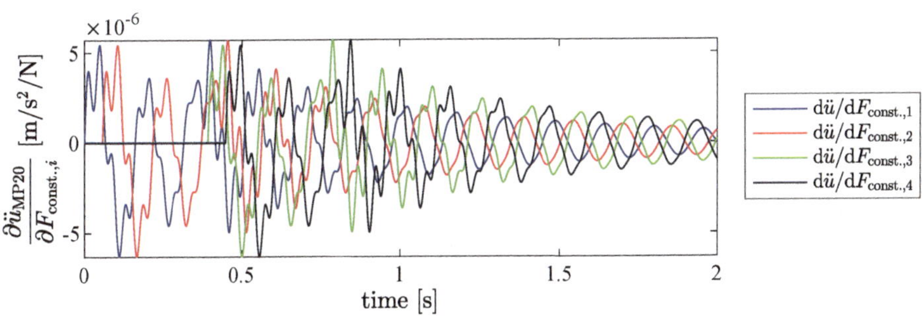

Figure 6.27 Sensitivity of the acceleration at MP 20 with respect to the magnitude of the constant moving forces

6.6 Optimization results including penalty functions

The investigations without penalty functions have shown that the noise component of the measurement signals can have a significant influence on the results, especially for the intervals in which the forces act near the supports. Moreover, a set of computations with noise-free data was performed. Indeed, this case is rather unrealistic in the context of practical applications, but it proved that the formulation of the inverse problem and the implementation of the solution algorithm are correct. In addition, a good insight into the solvability conditions of the mathematical inverse problem was gained. It could be shown that if any of the solvability conditions is harmed, no exact solution can be retrieved in case of analysis with noise-free signals. This section presents a set of investigations performed under consideration of nonzero penalty functions, which are expected to improve the solution behavior in case of noisy signals and violated solvability conditions.

6.6.1 Weighting factors of the penalty functions

The procedure for the determination of the weighting factors w_{p1} and w_{p2} was introduced in Section 5.1.4. Basically, a parametric study implying the variation of the weighting factors within a certain range has to be performed. Its aim is to determine the parameter set $(w_{p1},\ w_{p2})$ that leads to the lowest error x_{err} determined according to Eq. 5.23. The investigation is based on an assumed reference force vector, which is used to generate the simulated measurement data[10]. Hence, the parameters are determined by means of pure numerical analyses. It is important to note that the selected reference forces have to lie within a realistic range with respect to the expected forces to be identified. Moreover, the structural model has to be properly calibrated, so that it represents the real structural behavior as closely as possible.

6.6.1.1 Weighting factors determination based on load scenario 1

Even though the application of the proposed MFI on a railway bridge implies the reconstruction of a high number of forces, the determination of the weighting factors w_{p1} and w_{p2} based on the load scenario 1 (only four forces) is firstly studied in detail. The reason for this is that the – rather computationally inexpensive – analysis based on a reduced number of forces is expected to reveal certain insights into the procedure, which are generally valid, i.e. also in case of a large number of forces. In this sense, Fig. 6.28a shows the results of the parametric study obtained using the forward optimization approach and the load scenario 1 as reference force set (for $v = 80$ km/h). The computational parameters are adopted as specified in Section 6.3. The minimum error of 0.83 % is obtained for $w_{p1} = 10^{-11}$ and $w_{p2} = 10^{-7}$. However, it is to be noted that a partly nearly flat part of the curve (in logarithmic representation!) can be observed around the minimum point. This indicates that similar results can be obtained for a wider parameter range, which obviously stretches longer along the w_{p2} axis. This means that the problem has rather low sensitivity to noise. On the other side, the error function presents rather steeper gradients in the direction of w_{p1}. However, considering the logarithmic representation of the function, it is obvious that even w_{p1} can take different values, while the error is kept below reasonable limits (e.g. from $x_{err,f} \approx 1.7$ % for $w_{p1} = 10^{-10}$ to $x_{err,f} \approx 1.1$ % for $w_{p1} = 10^{-12}$).

Moreover, Fig. 6.28 also includes the corresponding results obtained using the backward moving window (Fig. 6.28b). No significant differences between the general trend of the

[10] It is to be mentioned that the same set of noisy simulated measured data has to be used for all parameter combinations investigated in the parametric study.

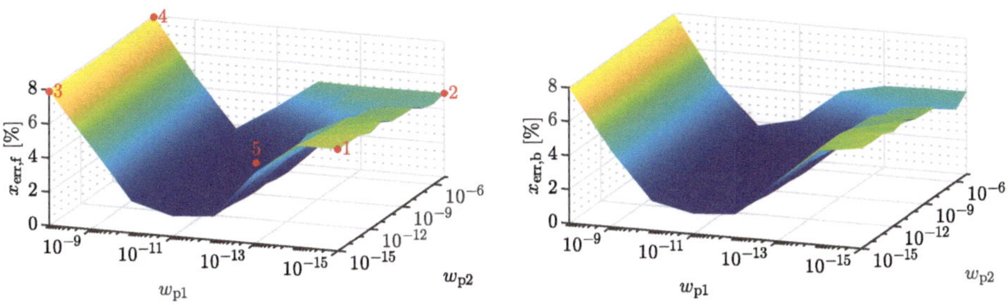

a) $v = 80$ km/h – forward optimization   a) $v = 80$ km/h – backward optimization

Figure 6.28 Parametric study for determination of the weighing factors for w_{p1} and w_{p2} for a velocity of 80 km/h under consideration of the load scenario 1: a) results obtained using only the forward analysis approach; b) results obtained using the backward analysis approach

two diagrams can be observed. The minimum error from the backward computation was retrieved at $w_{\mathrm{p1}} = 10^{-11}$ and $w_{\mathrm{p2}} = 10^{-8}$. It lies thus in the immediate proximity of the minimum resulting from the forward analysis, whereas it has to be noted that the gradient of the error function along the w_{p2} axis is very low. It can therefore be stated that the weighting parameters of the penalty functions can be reliably determined based only on the forward sliding window approach. This expectedly reduces the computation time of the parametric study. In this context, it is noted that the weighting parameters were varied as follows:

$$\begin{aligned} w_{\mathrm{p1}} &= 10^{-15}, 10^{-14}, \ldots, 10^{-8} \\ w_{\mathrm{p2}} &= 10^{-15}, 10^{-14}, \ldots, 10^{-5} \end{aligned} \tag{6.28}$$

This leads to a total of 88 analyses. The parameter range from Eq. 6.28 was selected according to the experience of the author based on empirical findings from a large number of investigations. However, even without prior knowledge, the parameter range can be easily defined and adapted for any computational configuration. For this, the parameter range given in Eq. 6.28 can be used as a starting point. If the results reveal a minimum error lying on one of the boundaries of the considered intervals, the respective range has to be incrementally extended by one entry, until a clear minimum can be identified.

Fig. 6.29 shows the identified and the reference force time histories for the five points marked in Fig. 6.28a, in order to emphasize the role of the penalty functions within the optimization problem. It can be seen that the identified forces corresponding to point 1 (Fig. 6.29a) present major discrepancies with respect to the force amplitudes, especially for the first two forces. Besides that, the results seem to be significantly influenced by the noise components of the measurement signals, which lead to high fluctuations in the

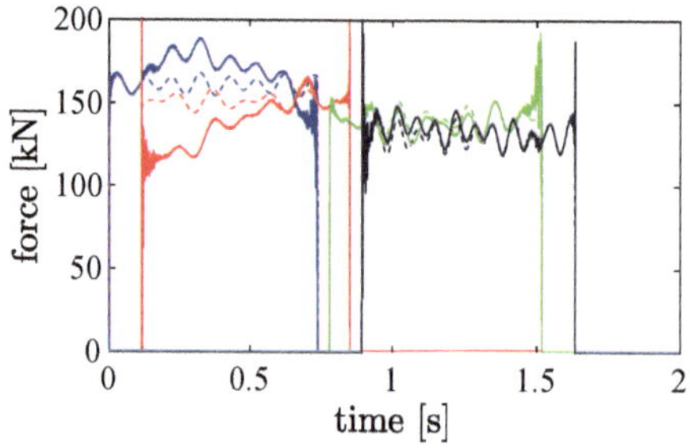

a) Point 1 : $w_{\mathrm{p1}} = 10^{-15}$, $w_{\mathrm{p2}} = 10^{-15}$, $x_{\mathrm{err,f}} = 5.99$ %

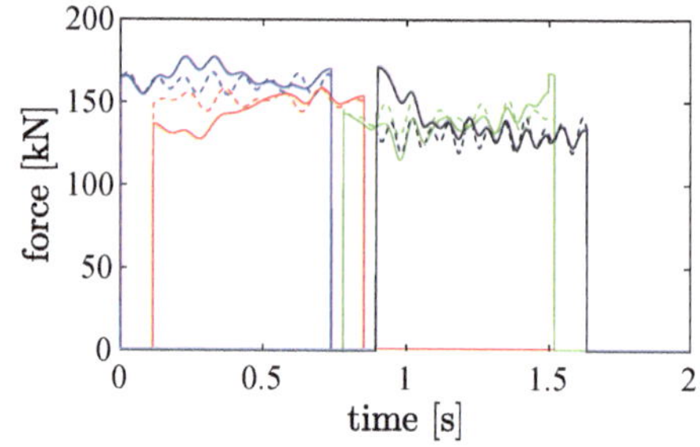

b) Point 2 : $w_{\mathrm{p1}} = 10^{-15}$, $w_{\mathrm{p2}} = 10^{-5}$, $x_{\mathrm{err,f}} = 4.92$ %

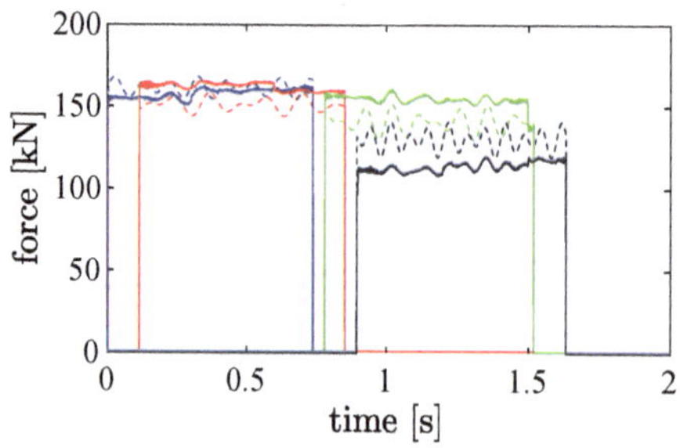

c) Point 3 : $w_{\mathrm{p1}} = 10^{-8}$, $w_{\mathrm{p2}} = 10^{-15}$, $x_{\mathrm{err,f}} = 7.89$ %

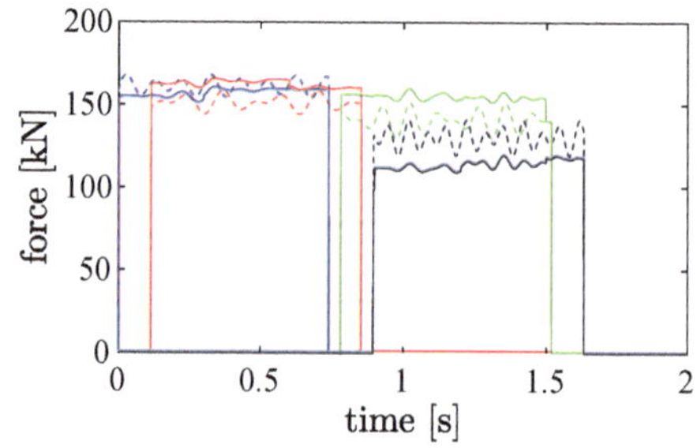

d) Point 4 : $w_{\mathrm{p1}} = 10^{-8}$, $w_{\mathrm{p2}} = 10^{-5}$, $x_{\mathrm{err,f}} = 7.92$ %

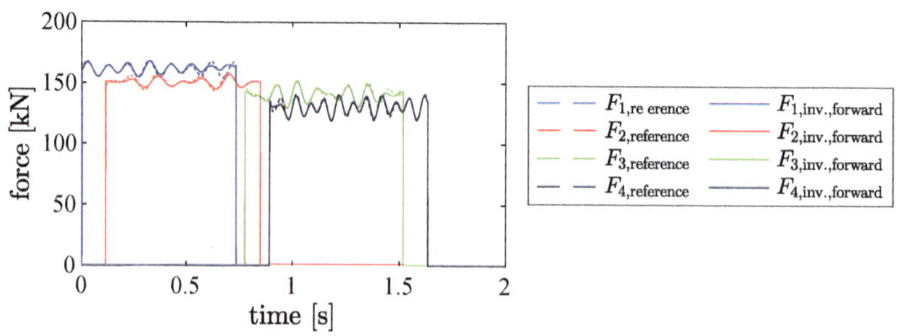

e) Point 5 : $w_{\mathrm{p1}} = 10^{-11}$, $w_{\mathrm{p2}} = 10^{-7}$, $x_{\mathrm{err,f}} = 0.83$ %

Figure 6.29 Influence of the penalty functions on the identification results for a moving force velocity $v = 80$ km/h. The graphs a) to e) are correlated to the points 1-5 marked in Fig. 6.28a

identified forces[11]. In this case, very low values were assigned to both weighting factors ($w_{\mathrm{p1}} = w_{\mathrm{p2}} = 10^{-15}$), leading to similar results as obtained for the computation without penalty functions ($w_{\mathrm{p1}} = w_{\mathrm{p2}} = 0$). The latter lead to an error $x_{\mathrm{err,f}} = 5.89$ % and a force time history which is optically indistinguishable from Fig. 6.29a. It can therefore be stated

[11] It is to be noted that some high frequent, very pronounced isolated peaks occurring for forces acting in the proximity of the supports are not entirely displayed in Fig. 6.29a. They were truncated, so that the limits of the vertical axis are similar to those of all other diagrams in Fig. 6.29.

that the penalty functions are rather inactive for $w_{p1} = w_{p2} = 10^{-15}$. By increasing the second weighting factor to its maximum limit according to Eq. 6.28 ($w_{p2} = 10^{-5}$, point 2 in Fig. 6.28a), a considerable improvement with respect to the high noise-like fluctuations of the identified forces could be achieved (Fig. 6.29b). However, the discrepancies of the force amplitudes are only slightly reduced, since the first penalty function is still (almost) inactive.

Adopting the upper limit of w_{p1} (points 3 and 4 in Fig. 6.28a) obviously leads to even higher errors than in the case without penalty functions (Fig. 6.29c and d). The reason for this is that for high values of w_{p1} the objective function is dominated by the first penalty function, which tends to "flatten" the identified force time histories. It can be seen that the results obtained with the lower (Fig. 6.29c) and upper (Fig. 6.29d) limits of w_{p2} are quite similar, whereas the lower value of $w_{p2} = 10^{-15}$ causes slight noise-like fluctuations in the identified signals. This indicates that the first penalty function also fulfills to a certain extent a signal smoothing role. The minimum error of 0.83 % is obtained at point 5 for $w_{p1} = 10^{-11}$ and $w_{p2} = 10^{-7}$. The corresponding result is plotted in Fig. 6.29e. The combined efficiency of both penalty functions can be clearly observed. On the one side, no noise-like fluctuations can be observed at all. On the other side, the ensuring of the DFS-SAV property by minimizing the distance between the local and global average values (cf. Section 5.1.4) allows a very accurate reconstruction of the moving forces.

The proposed approach for determining the weighting factors w_{p1} and w_{p2} postulates that they can be determined at the beginning of the investigation and be kept constant for a given structure. A long term application obviously requires the force identification for a variety of different vehicles. In this context, it should be mentioned that the structural model and the computational parameters (e.g. window length, smoothing radii, number of measurement locations, time step, etc.) are to be carefully designed at the beginning of the investigation. It makes sense to keep them unchanged during the whole investigation, in order to facilitate the interpretation of the results and to simplify the implementation process. With respect to the structural model, it should be mentioned on beforehand that modeling errors are expected to have a significant influence on the force reconstruction. Considering that the weighting parameters are to be determined only on the basis of a pure numerical study, it is strongly required that the structural model is accurately calibrated with respect to the real structural behavior. A more detailed discussion regarding the effect of modeling errors on the identification accuracy is provided in Sections 6.7.6 and 7.6. Furthermore, it should be noted that even the noise level required for simulating the measurement data can be – more or less – considered as a known parameter[12], since the sensors are not expected to be changed in the middle of the monitoring process. This allows to perform a noise analysis for each sensor before starting the monitoring task, in order to define the actual

[12]This assumption is only valid in case of homoscedastic noise behavior, meaning that the noise components have the same variance, independent on the level of the measured signal. This noise property is assumed for all noisy signals used in the course of this thesis.

Numbering in Fig. 6.30	Velocity [km/h]	$x_{err,f,min}$ [%]	w_{p1}	w_{p2}
a	80	0.83	10^{-11}	10^{-7}
b	100	1.05	10^{-11}	10^{-8}
c	120	1.00	10^{-11}	10^{-8}
d	140	0.92	10^{-13}	10^{-7}
e	160	0.65	10^{-13}	10^{-9}
f	180	0.81	10^{-13}	10^{-9}
g	200	1.48	10^{-12}	10^{-9}
h	220	1.08	10^{-12}	10^{-8}

Table 6.2 Optimal weighting parameters of the penalty functions for different velocities obtained using the load scenario 1

noise parameters to be used in the parametric study. It is therefore clear that the only variable quantities are the amplitudes of the forces and their velocity.

Hence, in order to justify the assumption of constant weighting factors for a given structure, solely the variation of the optimal weighting parameters with respect to different vehicle velocities and changing force levels is to be analyzed. In this sense, Fig. 6.30 illustrates the error functions obtained for moving force velocities between 80 km/h and 220 km/h. It is obvious that all diagrams present a similar general trend, which always includes a clear minimum. The minimum obtained for each velocity is given in Tab. 6.2 together with its location with respect to the w_{p1} and w_{p2} axes. It can be observed that the optimal values of w_{p1} vary between 10^{-13} and 10^{-11} while w_{p2} ranges between 10^{-9} and 10^{-7}. However, as it can be seen in the diagrams shown in Fig. 6.30 (in logarithmic representation!), all error functions are partly flat around the minimum, whereas the "flattened" component basically occurs within the ranges mentioned above. The error differences obtained at two neighbored computation points of the error functions often only differ in the second decimal place. This means that a pure comparison based on the absolute values given in Tab. 6.2 is not always representative. It is more important to consider the flattened components of the functions, which present reasonable errors. In this sense, the selection of the weighting factors as average values of the results obtained for all investigated velocities seems to be a natural choice in this case. This leads to the optimal parameter set obtained based on load scenario 1:

$$(w^*_{p1,4}, w^*_{p2,4}) = (4 \cdot 10^{-12}, 2.9 \cdot 10^{-8}) \tag{6.29}$$

Fig. 6.31 shows the contour plots of the error functions represented in Fig. 6.30. It also highlights both the absolute minimum point $x_{err,f,min}$ according to Tab. 6.2 and the error $x^*_{err,f,4}$ obtained at the average point from Eq. 6.29. It is obvious that the use of fixed values $w^*_{p1,4}$ and $w^*_{p2,4}$ generally delivers only slightly higher errors than the absolute minimum

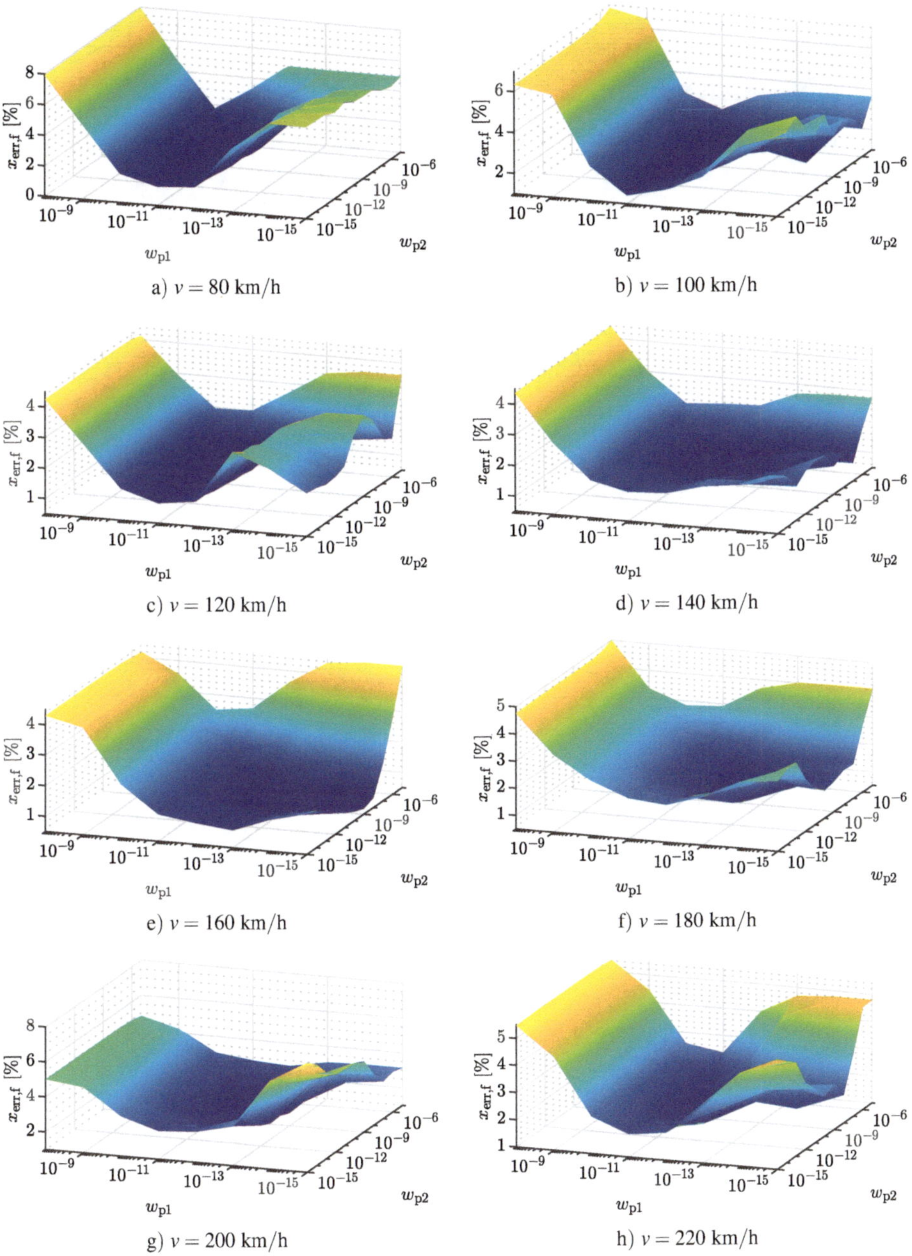

Figure 6.30 Parametric study for determination of the weighing factors w_{p1} and w_{p2} for different moving force velocities based on the load scenario 1

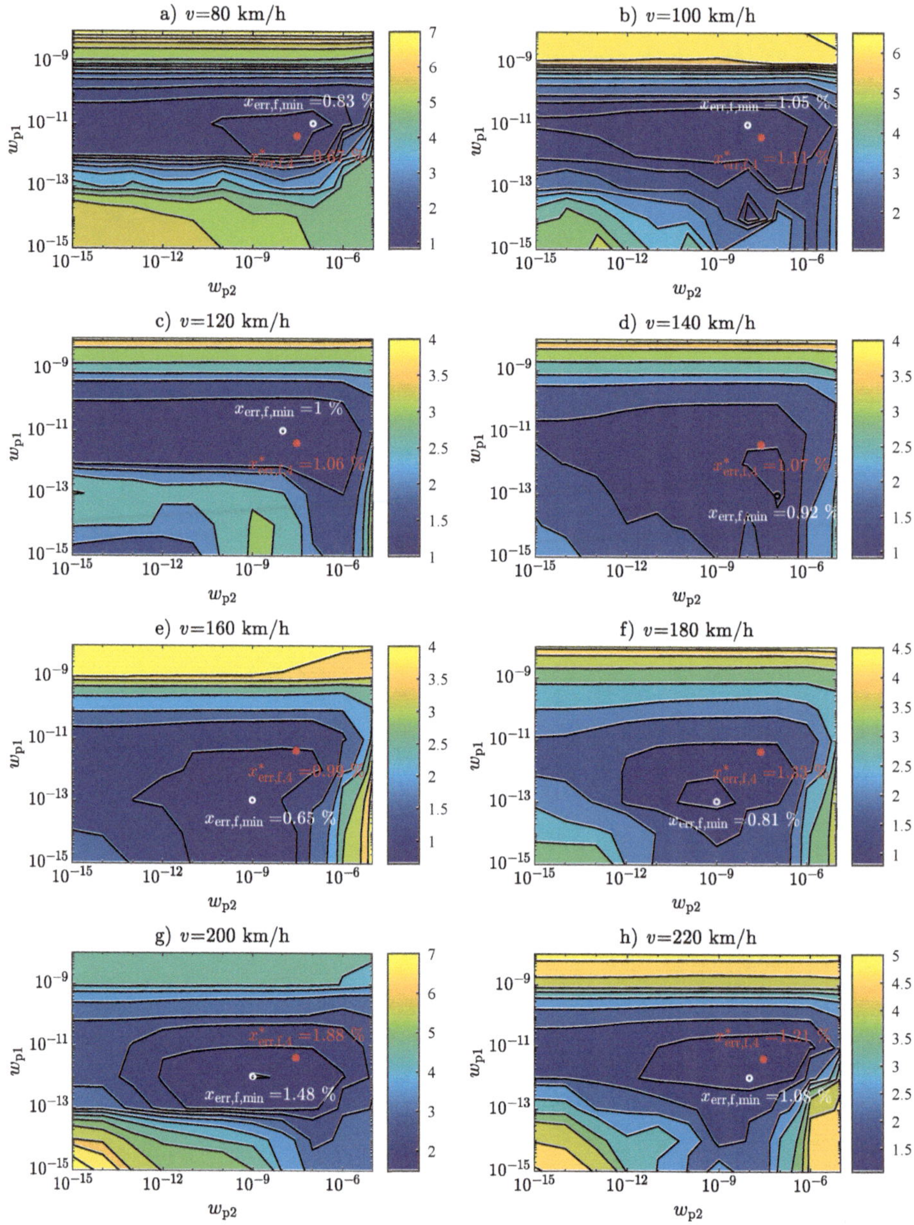

Figure 6.31 Contour plots of the error functions of the parametric study for different moving force velocities – computation based on the load scenario 1

errors given in Tab. 6.2[13]. The reason for this is that the optimal average point of the weighting factors lies for all functions – without exception – within the flattened region of the error surfaces. This fact seems to justify at this point the use of $(w^*_{p1,4}, w^*_{p2,4})$ for the further computations. In fact, under consideration of the rather flat regions in Fig. 6.31a-h (and their imaginary superposition) it can even be assumed that the determination of the weighting factors from a single parametric study, e.g. at 160 km/h, would still deliver sufficiently accurate results for all other velocities.

However, before proceeding to the analysis based on a higher number of forces (e.g. load scenario 3), it has to be shown that the variation of the force amplitudes has also an insignificant influence on the optimal values of w_{p1} and w_{p2}. In this sense, Fig. 6.32a and b illustrate the error surface and the corresponding contour plot obtained for a force set reduced by 50 % with respect to Fig. 6.4a (for $v = 160$ km/h). Fig. 6.32c and d show the corresponding results for increased reference forces by 50 %. The trends of both

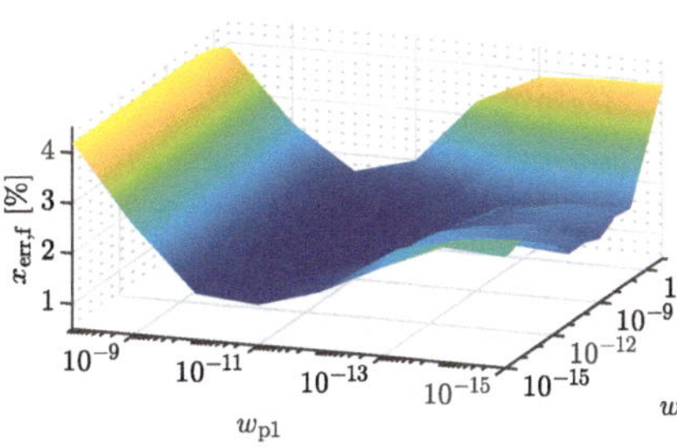

a) 50 % force reduction

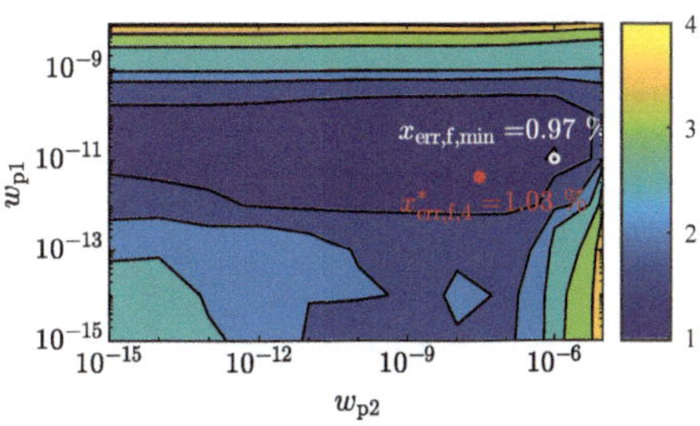

b) 50 % force reduction – contour plot

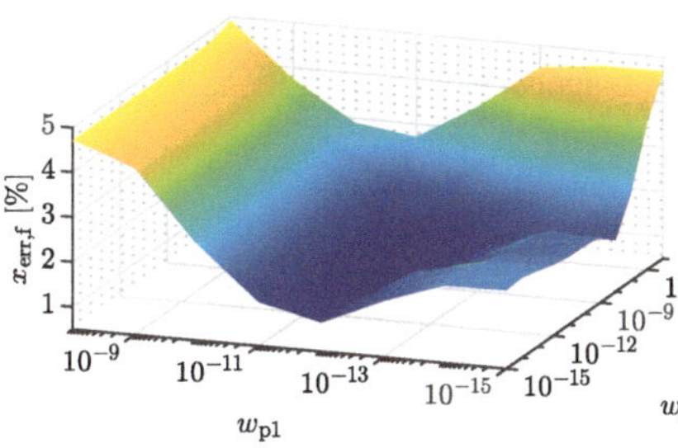

c) 50 % force increasing

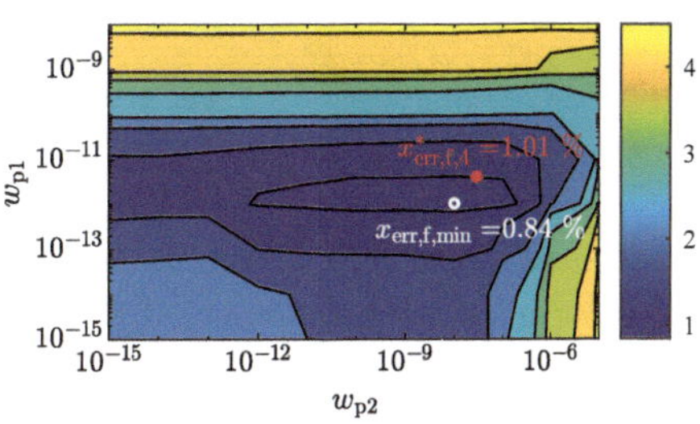

d) 50 % force increasing – contour plot

Figure 6.32 Parametric study for determination of the weighing factors w_{p1} and w_{p2} for different force amplitudes based on load scenario 1. The considered moving force velocity is $v = 160$ km/h

[13] Fig. 6.31a indicates that the error $x^*_{err,f,4}$ obtained at $(w^*_{p1,4}, w^*_{p2,4})$ might be even (slightly) lower than the absolute minimum error $x_{err,f,min}$ given in Tab. 6.2. The reason for this is the fact that the error functions were determined for a discrete grid of values according to Eq. 6.28. The point $(w^*_{p1,4}, w^*_{p2,4})$ obviously falls between the computation points of the error functions. This means that a lower value at this point can not be excluded. However, the raster specified in Eq. 6.28 is expected to be fine enough for sufficiently accurate results.

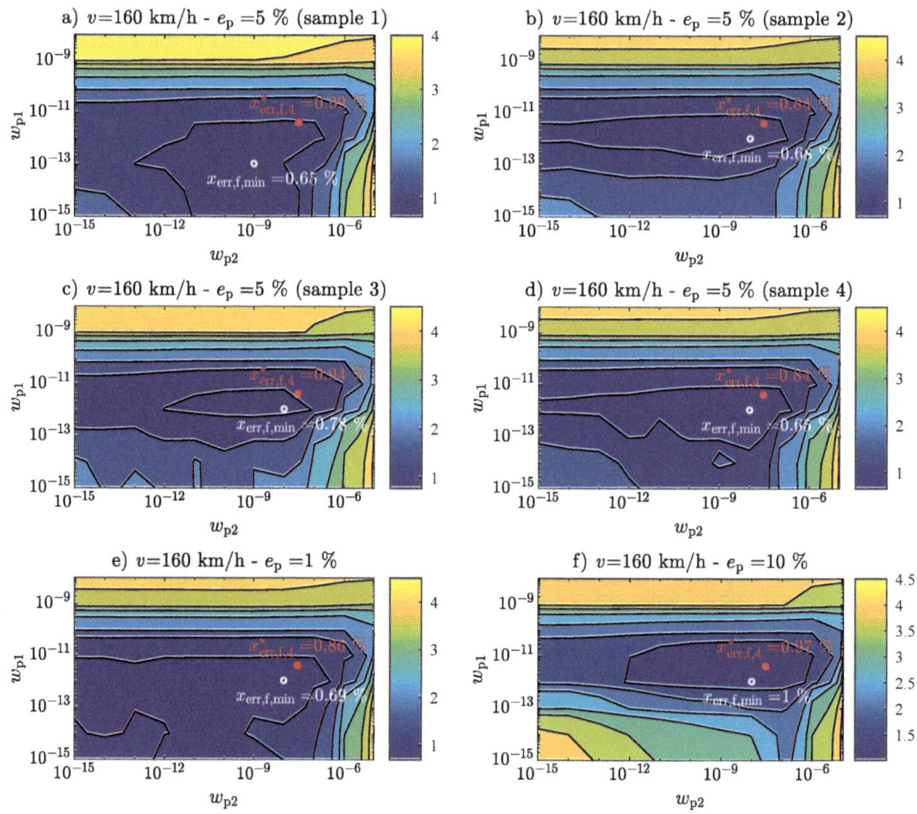

Figure 6.33 Error functions in dependency of the noise components for load scenario 1

error functions are very similar to the results obtained with the initial reference force set (Fig. 6.30e and 6.31e). Moreover, the error values obtained by applying the average optimal weighting factors $(w^*_{p1,4}, w^*_{p2,4})$ are only slightly different from the absolute minimum value of each error function. The average optimal point $(w^*_{p1,4}, w^*_{p2,4})$ obviously lies in both cases within the flattened area of the error functions. This means that even larger deviations of the w_{p1} and w_{p2} would not significantly affect the computation accuracy.

Another short study is concerned with the noise dependency of the optimal weighting parameters w_{p1} and w_{p2}. The results of each diagram shown in Fig. 6.30 to 6.32 were determined for a single sample of noisy measurement data, which was kept unchanged for all parameter combinations of each parametric study. However, it is well known that the noise component changes for each measurement signal, even though the statistical parameters of the noise might remain more or less unchanged. In order to check the sensitivity of w_{p1} and w_{p2} with respect to different noise samples, a set of three additional analyses are performed similar to Fig. 6.30e, i.e. for a velocity of 160 km/h and a noise level $e_\mathrm{p} = 5$ %. The results are shown in Fig. 6.33b-d. In order to facilitate the visual comparison, the initial result given in Fig. 6.30e is also shown in Fig. 6.33a. It can be observed that the latter three computations for $e_\mathrm{p} = 5$ % (Fig. 6.33b-d) reveal the exact same values for the optimal weighting parameters, whereas the point $(w^*_{\mathrm{p}1,4}, w^*_{\mathrm{p}2,4})$ lies in their immediate proximity. It delivers thus error values, which are only slightly higher than the absolute minima of the error functions. The results of the first computation (Fig. 6.33a) obviously present a wider flattened area, reaching the lower boundary of the diagram. This means that even a value of $w_{\mathrm{p}2} = 10^{-15}$ still reveals satisfactory results. This clearly indicates that the "measurement" signals used in this analysis included – by chance – a more convenient noise component. However, in all cases, the optimal point $(w^*_{\mathrm{p}1,4}, w^*_{\mathrm{p}2,4})$ revealed very satisfactory results with respect to the minimum error, justifying thus once again its use for the further computations.

The latter two plots shown in Fig. 6.33 (e and f) investigate the dependency of $w_{\mathrm{p}1}$ and $w_{\mathrm{p}2}$ with respect to different noise levels. In this sense, Fig. 6.33e shows the error function obtained for $e_\mathrm{p} = 1$ % and Fig. 6.33f the corresponding results for $e_\mathrm{p} = 10$ %. The absolute minimum error resulting from the computation with $e_\mathrm{p} = 10$ % is indeed slightly higher than the values obtained for $e_\mathrm{p} = 1$ % or $e_\mathrm{p} = 5$ %, but its location is retrieved at a similar point as in Fig. 6.33b-e. Moreover, a reasonable flattened component of the error function could also be identified in this case. However, it is to be noted that the error function seems to have steeper gradients towards the boundaries of the investigated parameter range. On the contrary, the flattened area obtained for $e_\mathrm{p} = 1$ % covers a much wider range of parameters. It can therefore be assumed that the noise level might influence the steepness of the error functions. However, even for a – with respect to practical applications – rather unrealistic noise level $e_\mathrm{p} = 10$ %, the error function presents a reasonable flattened area in the range, in which the most of the parameter were previously identified (cf. Tab. 6.2).

6.6.1.2 Weighting factors determination based on load scenario 2

The investigations from the previous section have shown that the values of the weighting parameters of the penalty functions are rather robust with respect to the velocity, noise level and force amplitudes. This can be explained by the (partly) very flat gradients of the error functions around the minimum points (cf. Fig. 6.31 to 6.33). Therefore, one can assume, that it is actually sufficient to perform the parametric study for a single set of velocity and force levels. The corresponding minimum of the error function is expected to lie within the low gradient ("flat") regions of the error functions one would obtain for other velocities and force amplitudes.

However, the reference load case 2 defined in Section 6.2.2 presents some particularities due to its pronounced impulse-like components. The aim of the following study is to check whether the impulse like-components have a major influence on the determination of the weighting factors $w_{\mathrm{p}1}$ and $w_{\mathrm{p}2}$. In this sense, a similar parametric study as presented in

Section 6.2.2 was performed for the load scenario 2, whereas the considered moving force velocity was 160 km/h and the noise level $e_{\mathrm{p}} = 5\ \%$. The result is shown in Fig. 6.34. The minimum of the error function could be identified at the point:

$$\left(w^{*}_{\mathrm{p1,4,imp}},\ w^{*}_{\mathrm{p2,4,imp}}\right) = \left(1 \cdot 10^{-12},\ 1 \cdot 10^{-9}\right) \tag{6.30}$$

Indeed the point $(w^{*}_{\mathrm{p1,4,imp}}, w^{*}_{\mathrm{p2,4,imp}})$ does not lie in the very proximity of $(w^{*}_{\mathrm{p1,4}}, w^{*}_{\mathrm{p2,4}})$, which was determined based on load scenario 1. Nevertheless, it can be clearly observed that the gradients of the error function illustrated in Fig. 6.34 are quite low in both directions around the minimum point. The error computed for the load scenario 2 using the values $w^{*}_{\mathrm{p1,4}}$ and $w^{*}_{\mathrm{p2,4}}$ from Eq. 6.29 is $x_{\mathrm{err,f}} = 2.14\ \%$, which represents in any case a very satisfactory result. Following the statement formulated in Section 6.6.1.1 with respect to the independency of the weighting factors on the velocity and noise level, $(w^{*}_{\mathrm{p1,4,imp}}, w^{*}_{\mathrm{p2,4,imp}})$ will be assumed in the following as the optimal weighting parameter set configuration for the load scenario 2.

6.6.1.3 Weighting factors determination based on load scenario 3

The previously presented investigations with respect to the determination of the optimal weighting parameters w_{p1} and w_{p2} were based either on load scenario 1 (Section 6.6.1.1) or on load scenario 2 (Section 6.6.1.2), i.e. on a set of four forces. Indeed the scenario of four moving forces delivers great insights in the performance of the method and facilitates the interpretation of the results. It is, however, a rather uncommon scenario for the real application on a railway bridge, which usually implies the passage of vehicles with several dozens of axles. Therefore, a further study for analyzing the sensitivity of the optimal w_{p1} and w_{p2} with respect to the number of forces is performed in the following. Its purpose is to find out whether the parameters determined for the reduced set of forces can be also kept for identifying the axle loads of a complete railway vehicle. In this sense, a similar

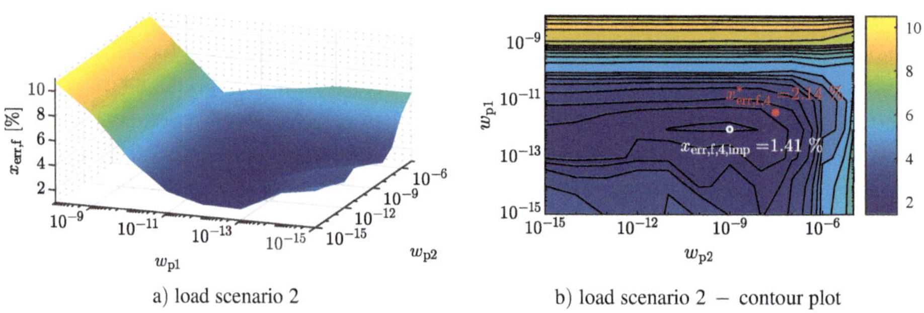

Figure 6.34 Parametric study for determination of the weighing factors w_{p1} and w_{p2} under consideration of load scenario 2. The considered moving force velocity is $v = 160$ km/h

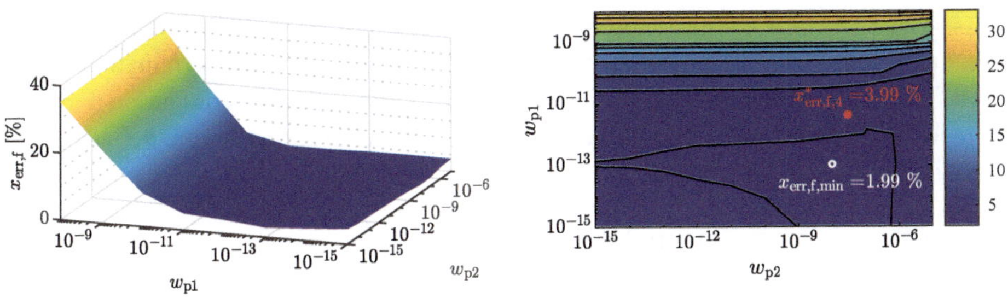

a) load scenario 3 b) load scenario 3 – contour plot

Figure 6.35 Parametric study for determination of the weighing factors w_{p1} and w_{p2} under consideration of load scenario 3. The considered moving force velocity is $v = 160$ km/h

parametric study as presented above is carried out under consideration of load scenario 3 (i.e. 32 forces, cf. Fig. 6.5) as reference force. This scenario is obviously computationally more demanding[14] than the consideration of only four forces (ca. factor 7). The results are given in Fig. 6.35. At first glance, it can be observed that the error function presents very steep gradients occurring for increasing values of w_{p1}. This leads to errors of up to 36 % at the upper w_{p1}-boundary of the investigated interval and indicates that the error function obtained for the extended set of reference forces is more sensitive to changes of w_{p1}. The absolute value of the minimum error $x_{\mathrm{err,f,min}} = 2.00$ % is obviously higher than in the previously investigated cases. This might be partly explained by the violation of the first solvability condition[15] and partly by the significantly higher number of forces determining the error value. The error obtained under consideration of $(w^*_{\mathrm{p1,4}}, w^*_{\mathrm{p2,4}})$ is $x^*_{\mathrm{min,f,4}} = 3.99$ %, being thus about two times higher than the absolute minimal error of 2.00 %. It can be seen that the curve in this case also presents a rather flat region (Fig. 6.35b), which also includes the point $(w^*_{\mathrm{p1,4}}, w^*_{\mathrm{p2,4}})$. However, the steeper gradients have a significant influence on the error obtained at two computation points located in the immediate proximity of each other[16].

Even though the result obtained for load scenario 3 under consideration of $(w^*_{\mathrm{p1,4}}, w^*_{\mathrm{p2,4}})$ is not completely unsatisfactory, a set of optimal weighting parameters $(w^*_{\mathrm{p1,32}}, w^*_{\mathrm{p2,32}})$ will be defined in the following, in order to achieve the best possible identification results for

[14]The author implemented the MFI problem and its solution using the computer algebra system MATLAB (The MathWorks Inc., 2019). He focused mainly on the development of the mechanic-mathematical based method and does not claim having implemented the problem in the most optimal way from the programming point of view. As a rough orientation value, the current implementation needs up to 20 minutes computational time for the worst case scenario studied in this section, i.e. a vehicle including 32 axles traveling at 80 km/h. However, this value has already been significantly reduced within a follow-up project at the Institute of Structural Mechanics and Design by using more appropriate programming languages and strategies.

[15]The results shown in Fig. 6.35 are based on analysis with three modes, while the maximum number of forces present on the bridge at a time is four.

[16]The flatness of the curve shown in Fig. 6.35b for the lower error range can be not clearly visually recognized, due to high upper limit of the third axis.

the relevant case of a complete vehicle driving over a bridge. Taking into account the assumption formulated in the previous section, which states that the determination of w_{p1} and w_{p2} can be performed with sufficient accuracy from a single parametric study, the optimal weighting parameters for the extended set of 32 reference forces are defined as the location of the minimum error shown in Fig. 6.35:

$$\left(w^*_{\text{p1},32},\, w^*_{\text{p2},32}\right) = \left(1 \cdot 10^{-13},\ 1 \cdot 10^{-8}\right) \tag{6.31}$$

This approach seems even more plausible, when considering the previously discussed findings with respect to lower gradients (i.e. flatter curves) occurring for lower noise levels, since the noise level $e_{\text{p}} = 5\ \%$ considered in the most analyses performed up to this point is a very conservative approach. Therefore, the flat region of the error curve obtained under consideration of real noise parameter is expected to be wider.

6.6.1.4 Trade-off selection of the weighting parameters for the penalty functions

So far, three sets of "optimal" weighting parameters for the penalty functions were defined, since the results obtained for each load scenario did not match perfectly. However, it can be generally stated that the three optimal points lie approximately in the same range, where all error functions present a nearly flat region. Considering that this work intends to provide a method, which is – to the greatest possible extent – feasible for the practical applications, the definition of a single set of weighting parameters is aimed. This would facilitate and considerably simplify the practical implementation of the proposed method and the analysis of the results. In this respect, the natural choice would be to determine the weighting parameters using a typical, realistic set of reference forces, containing a large number of axles, e.g. as discussed in Section 6.6.1.3 for the load scenario 3. Indeed this approach implies an increased computation time, but under consideration of the findings from Section 6.6.1.1, it is usually sufficient to perform a single parametric study for a freely selected velocity and force amplitude level.

Following this assumption, the weighting parameters from Eq. 6.31 retrieved for load scenario 3 are used to perform a further set of computations. The purpose is to confirm that they deliver reasonable results for different load configurations and vehicle velocities. Besides the three load scenarios defined in Section 6.2, four additional load cases were considered in order to check the performance of the weighting parameters with different number of moving forces. The additional load cases were defined on the basis of the load scenario 3 by considering only the first 4, 8, 16 or 24 out of the total of 32 forces. The results $x_{\text{err,b}}$ obtained at the end of the backward optimization are given in Tab. 6.3. It can be seen that all error values are below 3 %. They can be therefore considered as very satisfactory. A tendency of slightly increasing errors with increasing number of moving forces can be observed. On the one side, this effect is thought to be partly attributed to the

Load scenario	Velocity [km/h]							
	80	100	120	140	160	180	200	220
1	3.00	1.75	2.38	1.11	0.57	1.29	1.82	1.76
2	2.08	1.60	1.57	1.39	1.49	1.64	1.68	1.89
3 (complete)	2.42	2.36	2.46	2.53	2.13	2.36	2.63	2.01
3 (24 forces)	2.75	2.17	2.21	2.12	2.04	2.18	2.16	1.88
3 (16 forces)	2.50	1.71	1.99	1.72	1.81	2.05	2.06	1.83
3 (8 forces)	1.74	1.54	1.44	1.41	1.68	1.73	1.83	1.57
3 (4 forces)	1.32	0.91	1.21	1.08	1.06	0.98	1.67	1.90

Table 6.3 Error values $x_{\text{err,b}}$ in % for different load cases obtained using the optimal weighting parameters for load case 3 ($w^*_{\text{p1,32}} = 1 \cdot 10^{-13}$ and $w^*_{\text{p2,32}} = 1 \cdot 10^{-8}$)

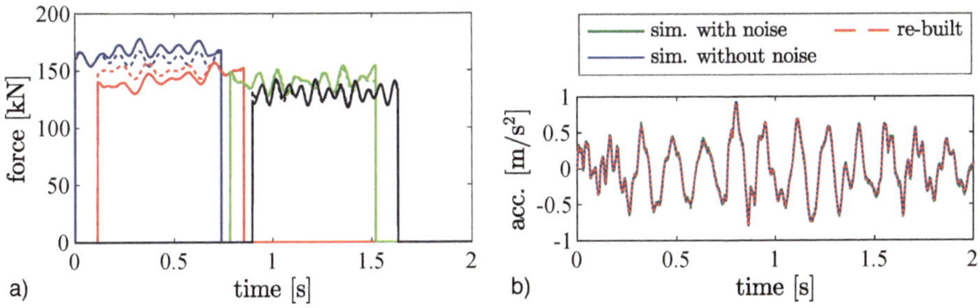

Figure 6.36 Identification result for load scenario 1 and a vehicle velocity of 80 km/h: a) reference (dashed lines) and identified forces (solid lines) at the end of the backward identification ; b) re-built signals at the unmeasured location node 15

violation of the solvability condition 1, as the analyses were performed under consideration of only three modes, whereas the maximum number of forces present on the bridge at a time is four. Increasing the number of wagons obviously implies an increased number of intervals with four forces acting simultaneously, which are characterized by higher inaccuracies of the force identification (cf. Fig. 6.21). On the other side, each result from Tab. 6.3 is based on a single sample of noisy measurement data, which does not allow to exclude any statistical influence on the observed result tendency.

The highest error of about 3.00 % is obtained for the load scenario 1 and a velocity of 80 km/h. The corresponding result is illustrated in Fig. 6.36. It can be seen that – despite some inaccuracies occurring in the reconstruction of the first and second force (Fig. 6.36a) – the re-built acceleration at the unmeasured location of node 15 (cf. Fig. 6.9) presents a (visually) nearly perfect match with respect to the reference results (Fig. 6.36b). The corresponding errors obtained according to Eq. 6.11 are $\ddot{u}_{\text{err}} = 4.83$ % when considering the noisy measurement data as reference value and $\ddot{u}_{\text{err}} = 1.10$ % for the evaluation with respect to the simulated data without noise. This confirms also from quantitative point of view the successful prediction of responses at unmeasured locations, even for a slightly inaccurate force identification. Hence, it can be stated that the weighting parameter combination $(w^*_{p1,32}, w^*_{p2,32})$ delivers sufficiently accurate results for all investigated cases, without needing any case-specific modifications (e.g. the normalization with respect to the number of forces). They will be used for all analyses performed in the succeeding part of this thesis.

6.6.2 Effect of the backward sliding window approach

The investigations without penalty functions (Section 6.5.1.1) already provided first insights into the differences between the forward and backward computation (cf. Fig. 6.15). As expected, lower errors were revealed at the end of the backward computation, even though the visual comparison of Fig. 6.15a and b does not allow to recognize clear differences. This indicates that the forward optimization would have been sufficient in that case. In fact – according to the experience of the author after several thousands of computation runs – the role of the backward identification is not significant for most computations with simulated measurement data[17]. However, there are certain cases, in which the backward computation plays a major role with respect to the identification accuracy. A relevant example is illustrated in Fig. 6.37. It represents a detailed view of the identification results[18] obtained using load scenario 3 considering a velocity of 140 km/h and weighting parameters according to Eq. 6.31. All other computational parameters were adopted from Section 6.3. The forward optimization revealed an error $x_{\mathrm{err,f}} = 2.78$ % for the total computation time of 10 s, while the corresponding error of the backward computation is $x_{\mathrm{err,b}} = 2.53$ %. The two errors $x_{\mathrm{err,f}}$ and $x_{\mathrm{err,b}}$ obviously lie not too far from each other, which could erroneously

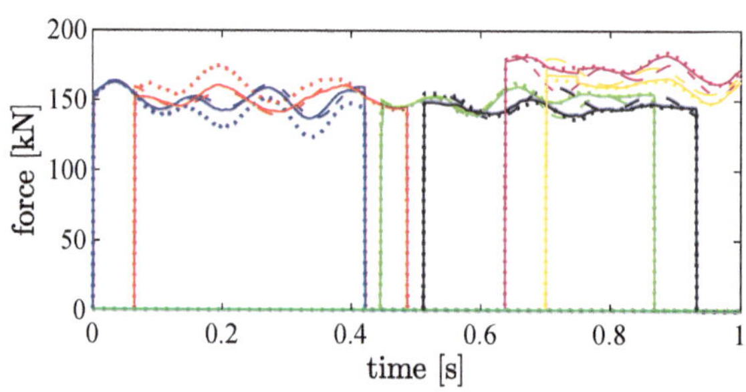

Figure 6.37 Effect of the backward sliding window approach within the reconstruction of the load scenario 3 for a vehicle velocity of 140 km/h (dashed lines: reference forces; dotted lines: results based on the forward optimization; solid lines: results at the end of the backward optimization)

[17]It will be shown in Chap. 7 that the backward approach plays a very important role in every computation when dealing with real measurement data.

[18]Fig. 6.37 illustrates the results for a reduced time interval, in order to allow a visual comparison of the forward and backward computation approach. The computation was performed for a total analysis time of 10 s.

lead to the conclusion that the forward computation delivers sufficiently accurate results. However, when looking at Fig. 6.37, it is obvious that the forward optimization identifies the first and the second force with rather poor errors of $x^{(1)}_{\text{err,f}} = 5.53$ % and $x^{(2)}_{\text{err,f}} = 5.56$ %, respectively. It is clearly visible that the accuracy of the identification is significantly improved by the subsequent backward optimization. This fact is confirmed also by the error values, which dropped to $x^{(1)}_{\text{err,b}} = 1.33$ % and $x^{(2)}_{\text{err,b}} = 1.28$ %, respectively. Hence, it can be stated that high errors obtained for certain forces seem to be averaged out when evaluating the results by means of the total error (for all forces at once). This is a consequence of using the l_1 norm in the error evaluation (Eq. 5.23), which is less sensitive to a few outliers (SANTAMARINA et al., 2005).

In the present numerical case based on known reference forces, it is easy to (visually) identify such an effect and to decide if the backward computation is really needed. However, no reference forces are available in real applications. Therefore, it makes sense and it is strongly recommended to always perform the additional backward computation, even though it increases the computational time and only reveals significant differences with respect to the forward results in a limited number of cases. In this sense, unless specified otherwise, only results obtained at the end of the backward approach will be discussed in the following sections. In this context, it should also be noted that the backward computation might even reveal slightly higher errors than the forward optimization. This effect can only be analyzed if the reference forces are available. It occurs due to the fact that the optimization problem defined in Eq. 5.15 actually tries to minimize the distance between the measured and computed measured quantities. Hence, even if the error evaluated with respect to the reference forces (Eq. 5.23) at the end of the backward optimization ($x_{\text{err,b}}$) is higher than the corresponding error resulting from the forward computation ($x_{\text{err,f}}$), the errors evaluated at the level of the measured motion quantities (Eq. 6.11) are generally expected to drop after the additional backward computation.

6.6.3 Selected results of the optimization with noisy data

To facilitate the understanding of the effect of the penalty functions, the cases presented in Section 6.5.1 in the context of an objective function without penalty functions (Fig. 6.15, 6.17 and 6.18) will be reanalyzed under consideration of the optimal weighting factors defined in Section 6.6.1.4.

6.6.3.1 Results of load scenario 1

The solution of the inverse problem for load scenario 1 under consideration of the weighting factors according to Eq. 6.31 is shown in Fig. 6.38 together with the reference force time histories defined in Fig. 6.4a. It can be visually observed that the identified and reference force time histories are in excellent agreement. This can be quantified by means of the total error evaluated according to Eq. 5.23. The error at the end of the backward computation is $x_{\mathrm{err,b}} = 0.57$ %, while the forward computation revealed $x_{\mathrm{err,f}} = 1.00$ %. This proves that the inaccuracies occurred within the computation without penalty functions, especially with respect to the force values in the proximity of the supports (cf. Fig. 6.15), could be successfully corrected by the activation of the penalty functions.

The iteration progress of the total objective function is illustrated in Fig. 6.39a for the three moving windows, whereas the results are obtained using the forward optimization approach. The convergence of the first window is achieved after 22 iterations, whereas the computations of the second and third window are terminated after 22 and 20 iterations, respectively. Hence, a considerable reduction of the iteration number for the first window was achieved by activating the penalty functions (cf. Fig. 6.16). This effect is thought to occur due to the fact that the smoothing capacities of the penalty functions do not allow the fitting of some high frequent isolated peaks, as they occurred in Fig. 6.15.

Moreover, it can be seen that a low decay behavior during the first iteration steps is also present in this case, indicating a similar general behavior to the corresponding computation without penalty functions (Fig. 6.16). Since the initial point of the first window is defined as $\boldsymbol{x}_0^{(1,1)} = \mathbf{0}$ (Eq. 6.7), the initial value of the objective function of the first window in Fig. 6.39a is identical to the corresponding value from the computation without penalty functions (Fig. 6.16a). This effect is, however, not seen for the second and third window, since the penalty functions are active as soon as the iteration process retrieves nonzero force vectors. This can be clearly seen when separately analyzing the iteration histories for the main optimization criterion $\mathcal{F}_{\mathrm{mc}}$ and for the two penalty functions $\mathcal{F}_{\mathrm{p1}}$ and $\mathcal{F}_{\mathrm{p2}}$. They are represented in Fig. 6.39b, c and d, respectively. According to Eq. 5.15, the total objective function is expressed as a sum of the main optimization criterion and the two penalty functions. This means that the diagram shown in Fig. 6.39a can be obtained as a sum of the corresponding curves from Fig. 6.39b-d. It can be clearly seen that the main optimization criterion (Fig. 6.39b) is by far the dominant component of the total objective function. The penalty functions (Fig. 6.39c and d) only take rather low values in comparison

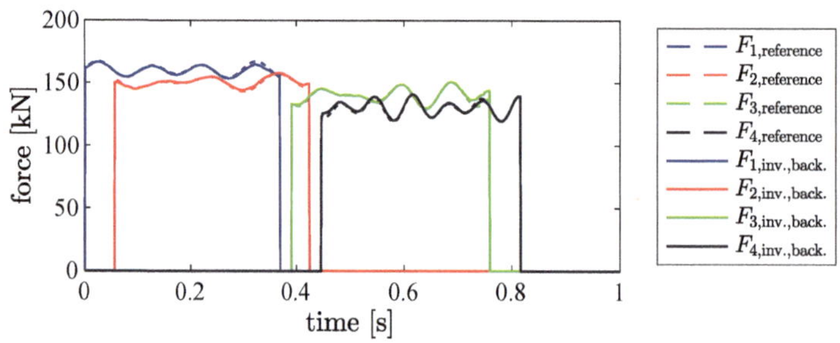

Figure 6.38 Identified and reference force time histories of load scenario 1 for a vehicle velocity of 160 km/h and optimal weighting parameters of the penalty functions: $w^*_{\mathrm{p1},32} = 1 \cdot 10^{-13}$ and $w^*_{\mathrm{p2},32} = 1 \cdot 10^{-8}$

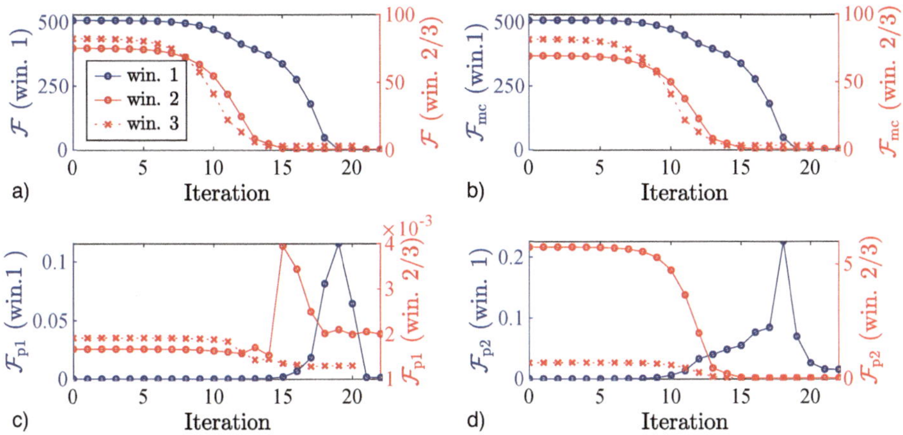

Figure 6.39 Iteration history of the optimization problem obtained using the forward computation approach under consideration of the load scenario 1: a) total objective function; b) main optimization criterion; c) first penalty function; d) second penalty function

to the main optimization criterion. However, they are sufficient to ensure the correction of the force time histories as soon as either the DFS-SAV property is harmed and/or the noise-like components of the solution are growing. Such effects occur for example in the 19-th iteration of the first window for the first penalty function and in the 18-th iteration of the first window for the second penalty function. This is visible by the increased peaks of the blue curve in Fig. 6.39c and d. It can thus be deducted, that the isolated components of the total objective function ($\mathcal{F}_{\text{mc}}$, $\mathcal{F}_{\text{p1}}$, $\mathcal{F}_{\text{p2}}$) do not necessarily have to monotonically decrease. It is only their sum, i.e. the total objective function, which must always present a monotonically decreasing behavior.

The statement mentioned above can be also confirmed by the iteration progress of the first two windows obtained for the backward computation (Fig. 6.40). The main optimization criterion is almost monotonically decreasing in the first window (with one exception in iteration 5), while an increasing behavior can be noticed for the second window (Fig. 6.40b). At the same time, the penalty functions (Fig. 6.40c and d) present a mixed behavior. However, the total objective function is – as it is supposed to – strictly monotonically decreasing (Fig. 6.40a). It is to be mentioned that Fig. 6.40 does not include the results of the third moving window. The reason for this is, as already mentioned in Section 6.5.1.1, the formulation of the problem, which implies an identical last window for the forward and backward optimizations. Considering the selection of the initial values according to Eq. 6.8, the convergence of the backward computation is in fact achieved at the initial point. Moreover, the choice of the initial values for the backward computation as the optimal solutions of the forward optimization (Eq. 6.8) causes also in this case only very slight changes in the values of the total objective function and its separate components.

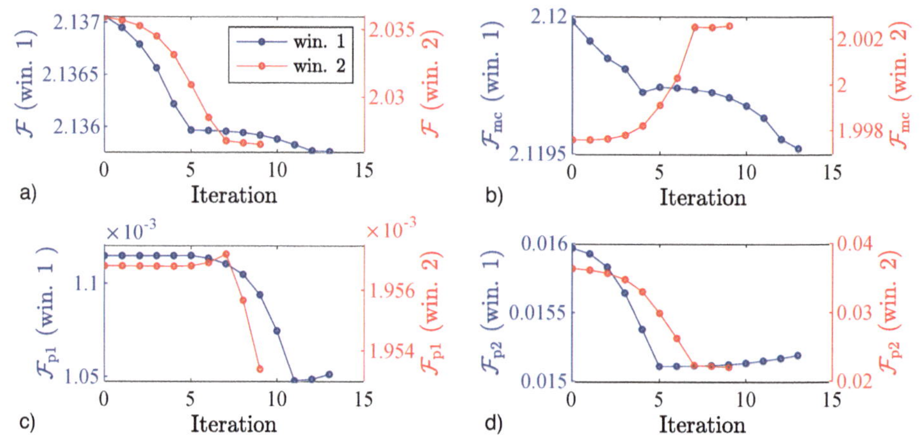

Figure 6.40 Iteration history of the optimization problem obtained using the backward computation approach under consideration of the load scenario 1: a) total objective function; b) main optimization criterion; c) first penalty function; d) second penalty function

In addition, it is worth mentioning that the termination criterion is also fulfilled after less iterations, 13 for the first window and 9 for the second window. Here, it is also to be noted that the convergence of all iteration processes shown in Fig. 6.39 and Fig. 6.40 was achieved according to first-order optimality criterion specified in Eq. 6.10.

6.6.3.2 Results of load scenario 2

The second load scenario was defined in order to analyze the performance of the proposed method with respect to the identification of force time histories including impulse-like components. The result of the optimization procedure without penalty functions was presented in Fig. 6.17 and delivered a total error of $x_{\mathrm{err,b}} = 4.14\ \%$, whereas considerable inaccuracies could be observed especially for the force values corresponding to locations in the proximity of the supports. The solution obtained under consideration of the weighting parameters according to Eq. 6.31 is illustrated in Fig. 6.41. The corresponding error is $x_{\mathrm{err,b}} = 1.49\ \%$. Hence, a considerable accuracy improvement could be achieved in comparison to the result without penalty functions. Moreover, it can be clearly seen that the location and duration of the impulse-like components could be successfully retrieved. However, the corresponding peaks of the impulse components were identified with rather moderate accuracy, showing errors ΔI_i of up to ca. 13.2 % for the impulse component of the fourth moving force (Tab. 6.4), whereas the error ΔI_i is defined as:

$$\Delta I_i = \frac{I_i - I_{\mathrm{ref},i}}{I_{\mathrm{ref},i}} \cdot 100\ \% \tag{6.32}$$

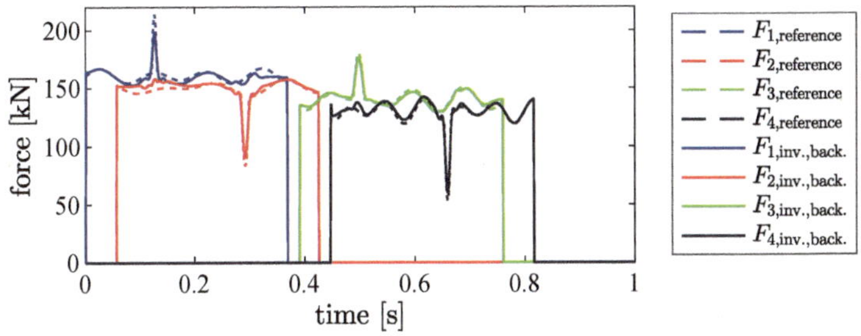

Figure 6.41 Identified and reference force time histories for the load scenario 2 and a vehicle velocity of 160 km/h under consideration of the weighting parameters $w^*_{p1,32} = 1 \cdot 10^{-13}$ and $w^*_{p2,32} = 1 \cdot 10^{-8}$

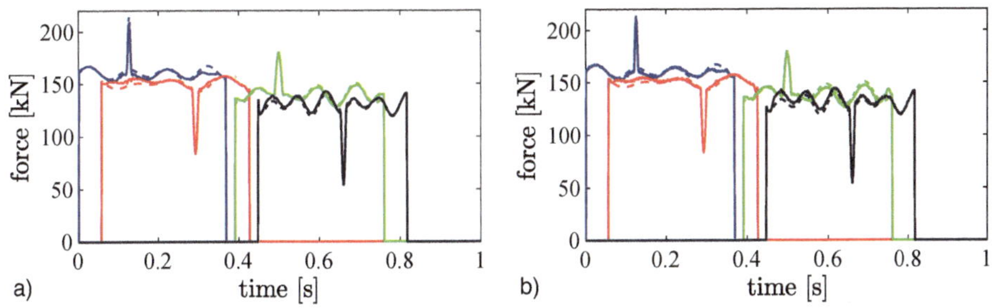

Figure 6.42 Identified and reference force time histories for the load scenario 2 and a vehicle velocity of 160 km/h under consideration of a lower parameter for the second penalty function: a) results for $w_{p1} = 1 \cdot 10^{-13}$ and $w_{p2} = 1 \cdot 10^{-9}$; b) results for $w_{p1} = 1 \cdot 10^{-13}$ and $w_{p2} = 1 \cdot 10^{-10}$ (dashed lines: reference forces; solid lines: identified forces)

where I_i represents the amplitude of the identified impulse assigned to the i-th moving force and $I_{\text{ref},i}$ is the corresponding value in the time histories of the reference forces. The reason for the inaccuracies in the identification of the impulse amplitudes is the smoothing effect of the second penalty function. In this sense, it should be noted that all results discussed so far are obtained for a smoothing radius $r_a = 2$ (cf. Eq. 5.18). This means that the local moving averages are computed using five consecutive values (except for the first two smoothing windows). Considering the impulse durations between 10 ms and 20 ms (cf. Eq. 6.4) and the computation time step $\Delta t = 0.001$ s, it is obvious that the impulses are described by only 10 to 20 samples. This obviously leads to the smoothing of the peaks by the moving local averaging. It can therefore be presumed that reducing the weight of the second penalty function could improve the identification of the impulse amplitudes. Since the majority of the error functions determined in Section 6.6.1 present rather low gradients in the direction of w_{p2}, it is expected that the overall error will not increase significantly. In this sense, Fig. 6.42 illustrates the results for two further sets of weighting

i	$I_{i,\text{ref}}$	$w_{\text{p1}} = 10^{-13}$ $w_{\text{p2}} = 10^{-8}$ I_i [kN]	ΔI_i [%]	$w_{\text{p1}} = 10^{-13}$ $w_{\text{p2}} = 10^{-9}$ I_i [kN]	ΔI_i [%]	$w_{\text{p1}} = 10^{-13}$ $w_{\text{p2}} = 10^{-10}$ I_i [kN]	ΔI_i [%]
1	213.4	197.0	-7.68	207.9	-2.58	210.7	-1.27
2	83.6	90.8	+8.61	85.0	+1.67	83.8	+0.24
3	179.8	175.9	-2.17	179.0	-0.44	178.2	-0.89
4	54.0	61.2	+13.33	55.9	+3.52	56.6	+4.81

Table 6.4 Comparison of the reference and identified impulse components of the axle loads for the load scenario 2

parameters: $(w_{\text{p1}}, w_{\text{p2}}) = (1 \cdot 10^{-13},\ 1 \cdot 10^{-9})$ and $(w_{\text{p1}}, w_{\text{p2}}) = (1 \cdot 10^{-13},\ 1 \cdot 10^{-10})$. The corresponding errors are $x_{\text{err,b}} = 1.33\ \%$ for the reduction of w_{p2} to 10^{-9} (Fig. 6.42a) and $x_{\text{err,b}} = 1.65\ \%$ for $w_{\text{p2}} = 10^{-10}$ (Fig. 6.42b). Both total errors are comparable to the error retrieved for the results shown in Fig. 6.41. However, as it can be seen in Tab. 6.4, the impulse amplitudes could be retrieved in both cases with a much higher accuracy. This fact confirms the presumption of a better accuracy of the identified impulse-like components for a lower weighting factor of the second penalty function.

Considering the findings presented above, it can be concluded that a reduction of the weighting parameter w_{p2} has to be considered, if the user is interested in the exact value of the impulse-like components of the moving forces. With respect to the practical applications, it is recommended to perform a second computation with reduced w_{p2} only for the passages, for which the computation with the regular weighting parameters reveals any kind of impulse components. However, it is to be noted that the purpose of a real application with respect to the impulse-like force components could be, for example, the identification of the axles presenting out-of-round wheels. In this case it is rather important to only identify the presence of the impulses, while their amplitude expectedly plays only a minor role.

6.6.3.3 Results of load scenario 3

While the solution of the optimization problem for load scenario 3 without penalty functions delivered an error 4.89 % (Fig. 6.18), the activation of the penalty functions by using the weighting factors defined in Eq. 6.31 reveals a significantly improved accuracy of the identified force time histories. The result is illustrated in Fig. 6.43 together with the reference forces. The corresponding error evaluated according to Eq. 5.23 is $x_{\text{err,b}} = 2.13\ \%$, which can be considered as a very satisfactory result. The errors $x_{\text{err,b},i}$ evaluated separately for each of the 32 forces are given in Tab. 6.5. It can be seen that the lowest errors occur for the first two and the last two axles. The reason for this is that these are the only axles which never experience the simultaneous presence of more than two forces on the bridge. The

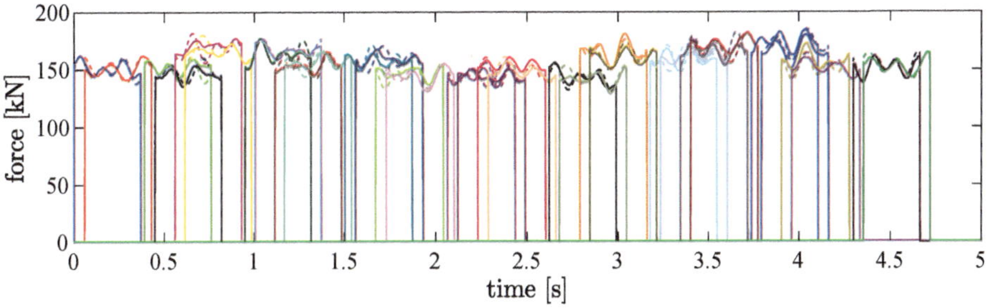

Figure 6.43 Identified and reference force time histories for the load scenario 3 and a vehicle velocity of 160 km/h under consideration of the weighting parameters $w^*_{\mathrm{p1},32} = 1 \cdot 10^{-13}$ and $w^*_{\mathrm{p2},32} = 1 \cdot 10^{-8}$ (dashed lines: reference forces; solid lines: identified forces)

i	$F_{\mathrm{st,id},i}$ [kN]	$F_{\mathrm{st},i}$ [kN]	$\Delta F_{\mathrm{st},i}$ [%]	$x_{\mathrm{err,b},i}$ [%]	i	$F_{\mathrm{st,id},i}$ [kN]	$F_{\mathrm{st,ref},i}$ [kN]	$\Delta F_{\mathrm{st},i}$ [%]	$x_{\mathrm{err,b},i}$ [%]
1	150.28	149.40	+0.59	1.13	17	152.51	148.50	+2.70	3.23
2	148.95	149.40	-0.30	1.33	18	146.01	148.50	-1.68	1.94
3	151.53	148.40	+2.11	2.41	19	142.52	142.50	+0.01	1.81
4	145.36	148.40	-2.05	2.83	20	140.75	142.50	-1.23	2.54
5	169.69	166.30	+2.04	3.04	21	164.75	161.80	+1.82	1.83
6	163.47	166.30	-1.70	2.27	22	159.95	161.80	-1.14	1.69
7	162.34	163.30	-0.59	1.33	23	158.28	159.00	-0.45	1.65
8	166.02	163.30	+1.67	2.18	24	160.52	159.00	+0.96	2.32
9	152.49	154.70	-1.43	2.72	25	165.36	168.00	-1.57	2.63
10	155.53	154.70	+0.54	1.99	26	171.27	168.00	+1.95	1.86
11	156.09	157.60	-0.96	1.92	27	173.31	170.90	+1.41	1.30
12	157.83	157.60	+0.15	2.92	28	165.56	170.90	-3.12	2.86
13	145.86	143.10	+1.93	2.94	29	156.35	151.70	+3.07	3.40
14	142.74	143.10	-0.25	1.43	30	150.31	151.70	-0.92	1.43
15	141.79	144.10	-1.60	2.18	31	150.78	152.30	-1.00	1.20
16	143.01	144.10	-0.76	2.73	32	152.54	152.30	+0.16	1.11

Table 6.5 Comparison of the reference and identified static axle loads for the load scenario 3

solvability condition 1 formulated in Section 6.5.2.1 is thus fulfilled only for these forces, since the computation was performed using three modes, while the maximum number of forces present on the bridge at a time is 4. This holds for all axles between $i = 3$ and $i = 30$. Therefore, the slightly higher individual errors of the corresponding identified dynamic axle forces is justified. The average overall error is 2.13 %, while the maximum error $x_{\mathrm{err,b},29} = 3.40$ % was identified for the 29-th axle. This confirms the successful identification of all axles loads with a very satisfactory accuracy.

Besides the errors $x_{\text{err,b},i}$ of the identified dynamic forces, the individual quasi-static component $F_{\text{st,id},i}$ of each identified axle load is another important parameter in practical applications. It can be approximated by a simple averaging of all the values of each moving force for the time instants in which it acts on the structure. The results for the 32 identified forces shown in Fig. 6.43 are given in Tab. 6.5 together with the corresponding error $\Delta F_{\text{st},i}$. The latter is evaluated with respect to the reference static forces $F_{\text{st},i}$ (given in Tab. 6.1):

$$\Delta F_{\text{st},i} = \frac{F_{\text{st,id},i} - F_{\text{st},i}}{F_{\text{st},i}} \cdot 100\ \% \tag{6.33}$$

Considering the good match of the identified dynamic forces, it is expected that their static components are also in a very good agreement with the corresponding values used in the definition of the reference force time histories (cf. Eq. 6.6 and Tab. 6.1). This presumption is clearly confirmed by the error values $\Delta F_{\text{st},i}$ shown in Tab. 6.5. The maximum absolute errors of about 3 % occur for the 28-th and 29-th axles, while the average error value is 1.31 %. Hence, based on the investigations with simulated measurement data, it can be stated that the proposed MFI method shows a clear potential for an accurate identification of both dynamic and static force components for a vehicle with realistic number of axles.

Load scenario 3 with additional impulse components In Section 6.6.3.2 it was indeed shown that the proposed method can successfully identify the impulse components (duration and amplitude) for a set of four moving forces. However, a realistic application of this feature is, for example, the identification of the axles presenting out-of-round wheels for a complete vehicle with a large number of axles. Thus, prior to a real application, the performance of the method has to be tested with simulation data corresponding to a realistic case. In this sense, an additional load case is defined based on the load scenario 3, to which four impulse components are added. The amplitudes and duration of the impulses were adopted from Eq. 6.4, whereas the time points at which the impulses are applied were defined according to Eq. 6.34, so that the impulses are added to the 1st, 9th, 16th and 22nd moving forces:

$$\begin{aligned} \text{Impulse begin [s]: } \quad & t_{\text{aI1}} = 0.04 \cdot t_{\text{out}},\ t_{\text{aI9}} = 0.3 \cdot t_{\text{out}}, \\ & t_{\text{aI16}} = 0.47 \cdot t_{\text{out}},\ t_{\text{aI22}} = 0.64 \cdot t_{\text{out}} \end{aligned} \tag{6.34}$$

where

$$t_{\text{out}} = \frac{l + d_{32}}{v} = \frac{16.4\ [\text{m}] + 193.3\ [\text{m}]}{160\ [\text{km/h}]} = 4.718\ [\text{s}] \tag{6.35}$$

The resulting reference force time history is shown in Fig. 6.44 together with the results of the inverse computation at the end of the backward computation. w_{p1} and w_{p2} were selected according to Eq. 6.31, whereas all other computational parameters were adopted

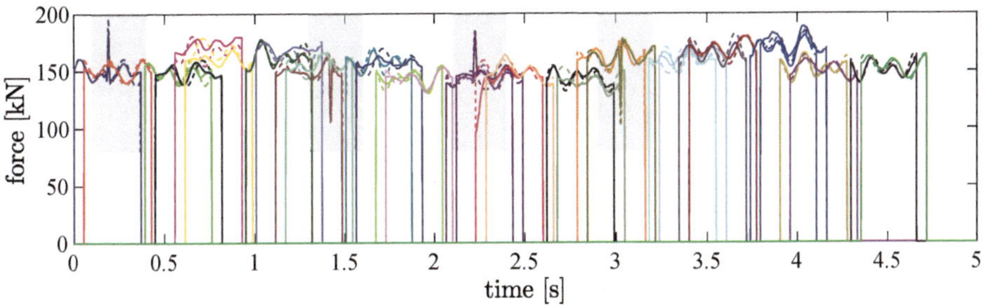

Figure 6.44 Identified and reference force time histories for the load scenario 3 including additional impulse components. The vehicle velocity is 160 km/h and the weighting parameters are $w^*_{\mathrm{p1},32} = 1 \cdot 10^{-13}$ and $w^*_{\mathrm{p2},32} = 1 \cdot 10^{-8}$ (dashed lines: reference forces; solid lines: identified forces)

i	$I_{i,\mathrm{ref}}$	$w_{\mathrm{p1}} = 10^{-13}$ $w_{\mathrm{p2}} = 10^{-8}$ I_i [kN]	ΔI_i [%]	$w_{\mathrm{p1}} = 10^{-13}$ $w_{\mathrm{p2}} = 10^{-9}$ I_i [kN]	ΔI_i [%]	$w_{\mathrm{p1}} = 10^{-13}$ $w_{\mathrm{p2}} = 10^{-10}$ I_i [kN]	ΔI_i [%]
1	196.2	182.6	-6.93	191.4	-2.45	193.6	-1.33
9	103.8	108.0	+4.05	100.3	-3.37	99.5	-4.14
16	186.3	180.0	-3.38	185.0	-0.70	186.1	-0.11
22	101.9	116.8	+14.62	110.4	+8.34	106.5	+4.51

Table 6.6 Comparison of the reference and identified impulse components of the axle loads for the load scenario 3 including additional impulse components

from Section 6.3. The corresponding error is $x_{\mathrm{err,b}} = 2.37$ %, which indicates an excellent general agreement between the reference and the identified force time histories. However, similar to the computations of the load scenario 2 (Section 6.6.3.2), the duration and location of the impulse-like components seem to be correctly identified, while their amplitudes still present certain inaccuracies. The errors evaluated according to Eq. 6.32 range up to about 15 % (Tab. 6.6) for the impulse acting together with the 22nd force.

To facilitate the understanding of the statements made above, Fig. 6.45a illustrates detailed views of the grey areas from Fig. 6.44 including the four impulse components. Since the results obtained with $w^*_{\mathrm{p1},32} = 1 \cdot 10^{-13}$ and $w^*_{\mathrm{p2},32} = 1 \cdot 10^{-8}$ show a similar trend to the corresponding results obtained for the second load scenario (Tab. 6.4), it can be presumed also in this case that a lower weighting factor of the second penalty function, i.e. a reduced smoothing effect, will improve the accuracy of the computation (cf. Section 6.6.3.2). In this sense, Fig. 6.45b and c show the results for the same additional sets of weighting parameters as used in Section 6.6.3.2: $(w_{\mathrm{p1}}, w_{\mathrm{p2}}) = (1 \cdot 10^{-13},\ 1 \cdot 10^{-9})$ and $(w_{\mathrm{p1}}, w_{\mathrm{p2}}) = (1 \cdot 10^{-13},\ 1 \cdot 10^{-10})$. The corresponding errors are $x_{\mathrm{err,b}} = 2.53$ % for the reduction of w_{p2} to 10^{-9} (Fig. 6.45b) and $x_{\mathrm{err,b}} = 2.66$ % for $w_{\mathrm{p2}} = 10^{-10}$ (Fig. 6.45c).

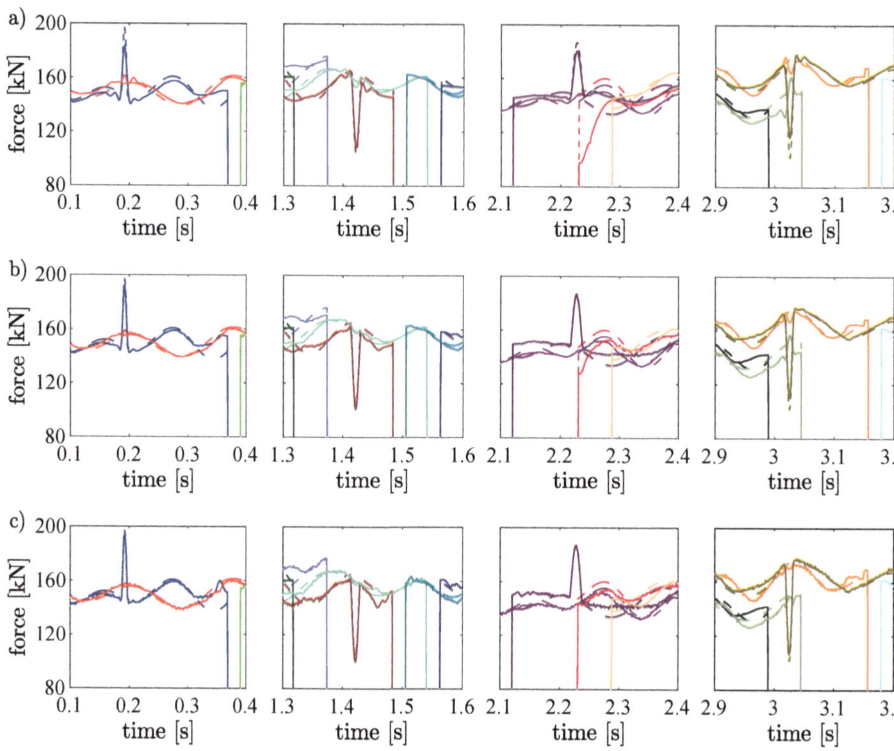

Figure 6.45 Detailed views of the identified and reference force time histories for the load scenario 3 including additional impulse components for different weighting parameters and a vehicle velocity of 160 km/h (dashed lines: reference forces; solid lines: identified forces): a) $w^*_{\mathrm{p1},32} = 1 \cdot 10^{-13}$ and $w^*_{\mathrm{p2},32} = 1 \cdot 10^{-8}$; b) $w_{\mathrm{p1}} = 1 \cdot 10^{-13}$ and $w_{\mathrm{p2}} = 1 \cdot 10^{-9}$; c) $w_{\mathrm{p1}} = 1 \cdot 10^{-13}$ and $w_{\mathrm{p2}} = 1 \cdot 10^{-10}$

These values are similar to the error retrieved for the results shown in Fig. 6.44 with the detailed view in Fig. 6.45a. As expected, the errors with respect to the impulse amplitudes obtained according to Eq. 6.32 were much lower than in the first case (Tab. 6.6). This fact confirms that the findings formulated for the load scenario 2, which state that the accuracy of the identified impulse-like components increases for lower values of w_{p2} (Section 6.6.3.2), are valid also for load configurations with a realistic number of axles.

6.6.4 Solution behavior with violated solvability conditions

The investigations from Section 6.5.2 showed that the violation of the first solvability condition (Section 6.5.2.1) indeed prevents the identification of the exact solution, but the results were stable and showed reasonable errors in the noise free-case (cf. Fig. 6.20 and 6.21). The same robust behavior and sufficiently high accuracy have already been

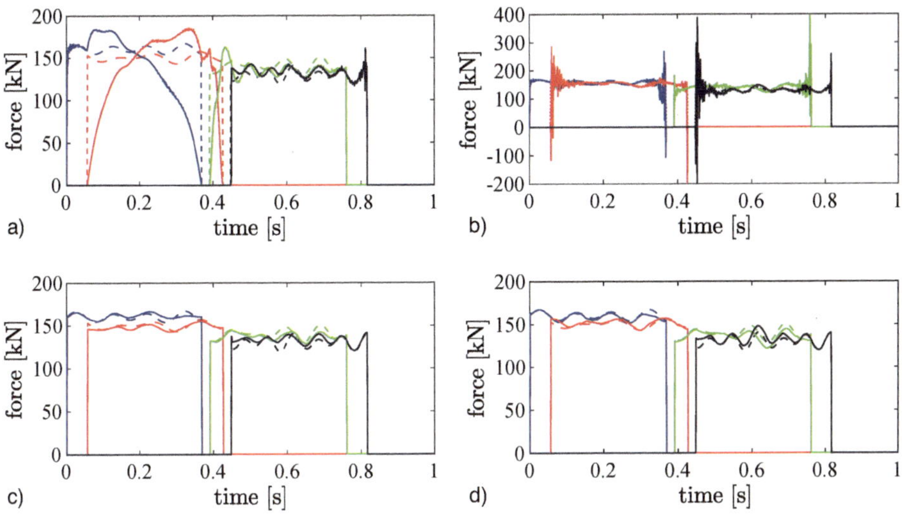

Figure 6.46 Behavior of the solution with violation of the second solvability condition and noisy measurement data (solid lines: identified forces, dashed lines: reference forces): a) results for one acceleration sensor (MP 15) and $w_{p1} = w_{p2} = 0$, $x_{err,b} = 13.74$ %; b) results for two acceleration sensors (MPs 15 and 30) and $w_{p1} = w_{p2} = 0$, $x_{err,b} = 6.33$ %; c) results for one acceleration sensor (MP 15) and $w_{p1} = 10^{-13}$, $w_{p2} = 10^{-8}$, $x_{err,b} = 2.06$ %; d) results for two acceleration sensors (MPs 15 and 30) and $w_{p1} = 10^{-13}$, $w_{p2} = 10^{-8}$, $x_{err,b} = 1.82$ %

confirmed in context of nonzero penalty functions and noisy measurement data. These investigations concerned the load scenario 3 (Section 6.6.1.4 and 6.6.3.3) and implied the identification of up to four forces acting simultaneously on the bridge, while only three vibration modes were considered in the modal superposition (cf. Fig. 6.43).

On the contrary, the violation of the second solvability condition caused a considerable error of the identified forces (cf. Fig. 6.22a). Therefore, the following investigation is concerned with the behavior of the proposed method in case of violated second solvability condition (Section 6.5.2.2) under consideration of noisy data and nonzero penalty functions. In this sense, the example discussed in conjunction with Fig. 6.22 is re-investigated with simulated noisy measurement data ($e_p = 5$ %) and weighting parameters of the penalty functions according to Eq. 6.31. It is recalled that the computations were performed under consideration of two modes and one (Fig. 6.22a) or two (Fig. 6.22b) acceleration sensors. This configuration is also adopted for the present investigation. Fig. 6.46a shows the results obtained under consideration of noisy measurement data from a single acceleration measurement point (MP 15) and $w_{p1} = w_{p2} = 0$. It can be seen that – similar to the noise-free case (Fig. 6.22a) – the identification results are unacceptable, especially for the first two forces. It is obvious that the DFS-SAV property is seriously harmed. By supplementing the measurement data with acceleration signals from MP 30, a significantly better accuracy is obtained (Fig. 6.46b). Nevertheless, the identification results are affected by considerable noise-like components.

Both cases were subsequently re-calculated under consideration of penalty functions weighted according to Eq. 6.31. Fig. 6.46c shows a significant improvement of the identification results with measurement data from a single acceleration sensor. A very satisfactory error of $x_{\mathrm{err,b}} = 2.06$ % was revealed, whereas the DFS-SAV is not violated anymore. Fig. 6.46d illustrates the result obtained using acceleration data from MP 15 and MP 30 and nonzero penalty functions. It is clear that the noise-like components could be suppressed, resulting in a perfectly acceptable error. It can be thus generally stated that the penalty functions are able to successfully compensate the violation of the solvability conditions. Nevertheless, it is generally recommended – if possible – to design the experiments in such a way, that the solvability conditions are fulfilled.

6.7 Influence factors on the identification accuracy

6.7.1 Effect of noise

One of the main issues with respect to the MFI problems in structural dynamics is their sensitivity to the noise components of the measured data. This section investigates the noise effect on the identification accuracy from two points of view: Section 6.7.1.1 focuses on the effects of different noise samples for a given noise level and Section 6.7.1.2 tackles the effect of different noise levels.

6.7.1.1 Different noise samples with the same statistical characteristics

In the following, it will be explored to what extent different noise samples, generated considering the same statistical noise characteristics, influence the accuracy of the MFI. For this, a set of eleven runs with different velocities (80 km/h, 120 km/h, 160 km/h and 200 km/h) is performed for the load scenario 1 and 3, whereas a new set of simulated measurement data according to Eq. 5.24 (for $e_{\mathrm{p}} = 5$ %) is generated for each run[19]. The resulting error values at the end of the backward optimization are shown in Tab. 6.7 for the load scenario 1 and in Tab. 6.8 for the load scenario 3 together with the corresponding average errors and standard deviations. It can be seen in both cases that the proposed method delivers very satisfactory average errors of up to 3 % for the load scenario 1 and of up to 2.7 % for the load scenario 3. The relatively low standard deviations also highlight the robustness of the proposed method. In this context, it is worth mentioning that the standard deviations of the errors obtained for load scenario 3 tend to be lower than for the load case 1. The highest average error could be identified in both cases for the slowest passage (80

[19]The values from Tab. 6.3 were considered to represent the first run, so that only ten additional data sets were generated for each velocity and scenario.

Run no.	Velocity			
	80 km/h	120 km/h	160 km/h	200 km/h
1	3.00 %	2.38 %	0.57 %	1.82 %
2	2.10 %	1.36 %	1.22 %	1.61 %
3	2.23 %	1.84 %	1.10 %	1.19 %
4	2.14 %	1.01 %	0.95 %	1.76 %
5	2.46 %	1.33 %	1.39 %	1.71 %
6	1.82 %	1.11 %	1.35 %	1.71 %
7	0.95 %	0.62 %	1.63 %	1.82 %
8	2.29 %	1.25 %	0.86 %	1.32 %
9	0.89 %	1.65 %	0.95 %	1.38 %
10	2.61 %	1.20 %	1.09 %	1.31 %
11	2.04 %	1.89 %	1.13 %	2.43 %
Average	2.05 %	1.42 %	1.11 %	1.64 %
Standard deviation	0.64 %	0.49 %	0.29 %	0.35 %

Table 6.7 Identification errors $x_{\mathrm{err,b}}$ for different noise samples with the same noise level $e_{\mathrm{p}} = 5\ \%$ under consideration of load scenario 1

km/h). In this respect, it can be observed that the values adopted from Tab. 6.3 (run no. 1 in Tab. 6.7 and 6.8), which were obtained using only a single sample of measurement data, partly represent the worst case scenario of the investigated noise components. This holds, for example, for passages with 80 and 120 km/h of the load scenario 1.

Furthermore, considering that the total computation time was set – independently of the moving force velocity – to 10 seconds for load scenario 3 and 2 seconds for load scenario 1, no clear correlation can be determined between the length of the decay phase included in the signals (e.g. ca. 6.23 s for 200 km/h and only about 2.64 s for 80 km/h) and the identification accuracy. However, the decay phase signals may play an important role in the improvement of the accuracy of the backward optimization in comparison to the forward sliding window approach (cf. Section 6.6.2).

6.7.1.2 Different noise levels

The investigation concerning the influence of different noise levels on the MFI accuracy using the same weighting parameters of the penalty functions is – strictly speaking – not conclusive. The reason for this is that the proposed method implies the determination of the weighting factors w_{p1} and w_{p2} according to Section 6.6.1.3 for a given noise level of the used sensors[20]. As soon as a sensor type is replaced by another with different noise characteristics, a new derivation of w_{p1} and w_{p2} has to be performed. They are then to be kept constant for the identification of all passages over the investigated structure.

[20]The noise level of the a sensor can be determined, for example, by a statistical analysis of a random noise signal recorded during a load free phase.

Run no.	Velocity			
	80 km/h	120 km/h	160 km/h	200 km/h
1	2.42 %	2.46 %	2.13 %	2.63 %
2	2.68 %	2.28 %	2.46 %	2.78 %
3	2.67 %	2.36 %	2.00 %	2.27 %
4	2.51 %	2.48 %	2.37 %	2.48 %
5	2.72 %	2.63 %	2.38 %	2.10 %
6	2.98 %	2.27 %	2.11 %	2.53 %
7	2.87 %	2.22 %	2.07 %	2.33 %
8	2.46 %	2.32 %	2.33 %	2.35 %
9	2.73 %	2.63 %	2.41 %	2.59 %
10	2.65 %	2.23 %	2.11 %	2.31 %
11	2.77 %	2.26 %	2.00 %	2.73 %
Average	2.68 %	2.38 %	2.22 %	2.46 %
Standard deviation	0.17 %	0.15 %	0.17 %	0.21 %

Table 6.8 Identification errors $x_{\mathrm{err,b}}$ for different noise samples with the same noise level $e_\mathrm{p} = 5$ % under consideration of load scenario 3

Despite the aforementioned non-conclusiveness of the results, Tab. 6.9 shows how the proposed MFI method performs, if it is used for analyzing signals with different noise levels. Here, the weighting parameters w_{p1} and w_{p2} are determined according to Section 6.6.1.3 (Eq. 6.31) for $e_\mathrm{p} = 5$ % and kept fixed for all investigated noise levels. It can be clearly seen that the errors are generally very satisfactory for noise levels lower than 5%. This is actually the noise level used for the determination of w_{p1} and w_{p2}. The reason for this is – as already indicated in Section 6.6.1 – the fact that flatter gradients of the error functions are expected for lower noise levels. This means that a wider range of parameters $(w_{\mathrm{p1}}, w_{\mathrm{p2}})$ leads to sufficiently low errors, explaining thus the good behavior of the proposed method

e_p [%]	Load scenario 1		Load scenario 3	
	80 km/h	160 km/h	80 km/h	160 km/h
1	1.65 %	0.68 %	1.78 %	1.73 %
2	2.19 %	0.44 %	2.04 %	1.86 %
3	1.61 %	1.03 %	2.09 %	1.96 %
5	3.00 %	0.57 %	2.42 %	2.13 %
7	1.68 %	1.52 %	3.35 %	2.37 %
10	2.87 %	2.11 %	4.67 %	3.19 %
15	2.94 %	3.14 %	6.51 %	3.61 %
20	4.09 %	2.04 %	8.22 %	5.55 %

Table 6.9 Identification errors $x_{\mathrm{err,b}}$ of load scenario 1 and 3 for different noise levels using fixed weighting parameters of the penalty functions (determined for $e_\mathrm{p} = 5$ %)

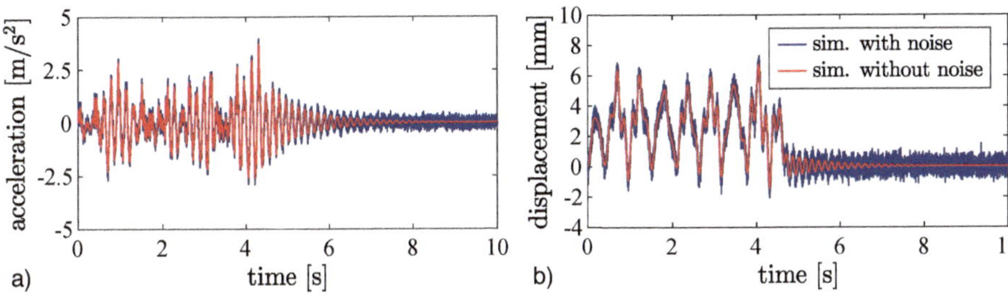

Figure 6.47 Time histories of the simulated measurements at MP 15 resulting from the load scenario 3 under consideration of a noise level of 20 %

for noise levels lower than the reference level used for the determination of w_{p1} and w_{p2}. Values of e_p greater than 5 % obviously lead to higher, partly unacceptable errors. They reach 4.1 % for load scenario 1 and 8.2 % for load scenario 3 for a practically unrealistic level $e_p = 20$ %. In this regard, in order to underline the magnitude of the signal-to-noise ratio for $e_p = 20$ % and thus to support its characterization as "practically unrealistic noise level", Fig. 6.47 illustrates the simulated measurements with and without noise at MP 15. Moreover, Tab. 6.9 indicates that the errors of load scenario 3 are monotonically increasing with the noise level for both investigated velocities. This effect is obviously not present in the results of the load scenario 1. The reason for this is most likely the slightly higher scattering of the results for load scenario 1, which was also observed in Tab. 6.7.

The error values obtained for $e_p = 15$ % and $e_p = 20$ % using the load scenario 3 can be considered as rather unacceptable. However, assuming that the used sensors actually possess such a (practically unrealistic) noise level, the weighting factors of the penalty functions have to be determined according to the corresponding noise level. For example, without going into detail, it is mentioned that a parametric study performed according to Section 6.6.1 for $e_p = 20$ % under consideration of the load scenario 3 delivers an optimal combination of the weighting parameters $(w_{p1}, w_{p2}) = (1 \cdot 10^{-11},\ 1 \cdot 10^{-5})$. The factor w_{p2}, which is responsible for the reduction of the noise component, had to be considerably increased in comparison to the value of 10^{-8} used for $e_p = 5$ %. The corresponding error $x_{err,f}$ obtained at the end of the forward computation is about 3.5 %, confirming that sufficiently accurate results can be obtained even when the measured signals are polluted with very high noise levels.

6.7.2 Effect of computational parameters

This section investigates the influence of certain computational parameters on the accuracy of the identified forces. For this, a set of computations with varying parameters of interest (moving window length, smoothing radius, length of the overlapping interval) was

Window length [samples]	Load scenario 1		Load scenario 3	
	80 km/h	160 km/h	80 km/h	160 km/h
300	4.61 %	1.05 %	3.08 %	2.33 %
350	3.78 %	0.71 %	2.66 %	2.18 %
400	2.69 %	0.63 %	2.71 %	2.00 %
450	3.00 %	0.57 %	2.42 %	2.13 %
500	2.56 %	0.62 %	2.56 %	2.11 %
550	2.20 %	0.53 %	2.31 %	2.09 %
600	2.54 %	0.55 %	2.31 %	2.22 %

Table 6.10 Identification errors $x_{\mathrm{err,b}}$ of load scenario 1 and 3 for different window lengths

performed under consideration of the load scenarios 1 and 3 and two different moving force velocities (80 and 160 km/h). The results are presented in the following.

6.7.2.1 Window length

All computations presented in the previous sections were performed for a window length of 450 samples. In order to analyze how the length of the moving window affects the accuracy of the proposed method, several inverse computations with varying window lengths were performed for the load scenarios 1 and 3. In this sense, a parameter range between 300 and 600 samples was investigated, which is considered as optimal with respect to the computation time. In this respect, it is to be mentioned that window lengths greater than 1000 samples are generally rather less practicable because of the increased computation effort occurring due to the increasing size of the Hessian[21]. The results are shown in Tab. 6.10 and indicate that the window lengths lower than 400 samples may lead to increased errors of the identified forces, whereas the accuracy of the proposed MFI method is rather insensitive to window lengths between 400 and 600 samples.

In general, it is recommended to select the window length so that it contains at least one full cycle of the time varying component of the moving forces with the lowest frequency. Considering that the lowest frequency components of the excitation are 6 Hz for the load scenario 1 (Eq. 6.2) and 3 Hz for the load scenario 3 (Eq. 6.6), the recommended minimal window lengths of about 170 (load scenario 1) and 330 samples (load scenario 3) can be determined for a computation time step of 0.001 s. The default window length of 450 samples used throughout this chapter represents thus a good trade-off, which requires a reasonable computational effort and leads to a satisfactory accuracy.

[21] Assuming that a single moving force is included in a certain window, the number of entries of the Hessian depends quadratically on the window size. For the regular case of a window containing entries assigned to several moving forces, an even steeper increase of the number of the Hessian entries with respect to the window length is achieved (cf. also Fig. 5.1).

r_a	Load scenario 1		Load scenario 3	
[samples]	80 km/h	160 km/h	80 km/h	160 km/h
0	3.38 %	0.92 %	2.96 %	2.72 %
1	2.62 %	0.57 %	2.53 %	2.28 %
2	3.00 %	0.57 %	2.42 %	2.13 %
5	2.74 %	1.80 %	2.75 %	2.74 %
10	6.25 %	4.43 %	4.39 %	4.03 %
20	14.42 %	6.19 %	17.98 %	10.10 %
50	7.34 %	6.80 %	44.25 %	28.68 %
100	7.58 %	5.24 %	52.78 %	65.51 %

Table 6.11 Identification errors $x_{\mathrm{err,b}}$ of load scenario 1 and 3 for different smoothing radii

6.7.2.2 Smoothing radius

A smoothing radius $r_{\mathrm{a}} = 2$ was considered for all computations presented so far. In the following, it is investigated to which extent the smoothing radius affects the accuracy of the force identification. This is done by performing computations under consideration of different smoothing radii in the range between 0 and 100 samples, whereas the same sample of noisy simulated measurement was used for all computations corresponding to a certain scenario. The results are given in Tab. 6.11 and indicate that the proposed method only performs well under consideration of low smoothing radii. The reason for this is that increased smoothing radii tend – due to the more pronounced smoothing effects – to flatten the identified force time histories and are thus not able to correctly retrieve their frequency components. Therefore, it is recommended to use a smoothing radius no greater than 5 samples, whereas even lower values are expected to increase the accuracy.

The selection $r_{\mathrm{a}} = 0$ represents a special case, which cancels the second penalty function. This can be easily proven by evaluating Eq. 5.17 for $r_{\mathrm{a}} = 0$. In this case, the local smoothing matrix is identical to the identity matrix: $\left[\boldsymbol{Q}_{\mathrm{loc},l}\right]_{r_{\mathrm{a}}=0} = \boldsymbol{I}$. This leads to $\boldsymbol{Q}_{\mathrm{p2},l} = \boldsymbol{0}$ and implicitly to $\mathcal{F}_{\mathrm{p2}} = 0$. The effect can be clearly seen in Fig. 6.48, which illustrates the identification results of load scenario 3 for $r_{\mathrm{a}} = 0$ and a velocity of 160 km/h. It is obvious that the solution is seriously affected by noise-like components, especially for the forces situated at the beginning and at the end of the structure. This explains the slightly higher errors given in Tab. 6.11 for $r_{\mathrm{a}} = 0$ than for $r_{\mathrm{a}} = 1$ or $r_{\mathrm{a}} = 2$. Nevertheless, the comparison with the results shown in Fig. 6.18 (similar investigation for $w_{\mathrm{p1}} = w_{\mathrm{p2}} = 0$) confirms that the first penalty function can also suppress – to a certain extent – the effect of noise.

Moreover, considering the findings of Section 6.6.3.2 and Section 6.6.3.3 with respect to the increasing accuracy of the impulse identification for a reduced smoothing effect of the second penalty function (i.e. lower weighting factor w_{p2}), it can be presumed that a lower smoothing radius could also improve the accuracy of the impulse identification. The reason for this is that the smoothness level can be also controlled to a certain extent

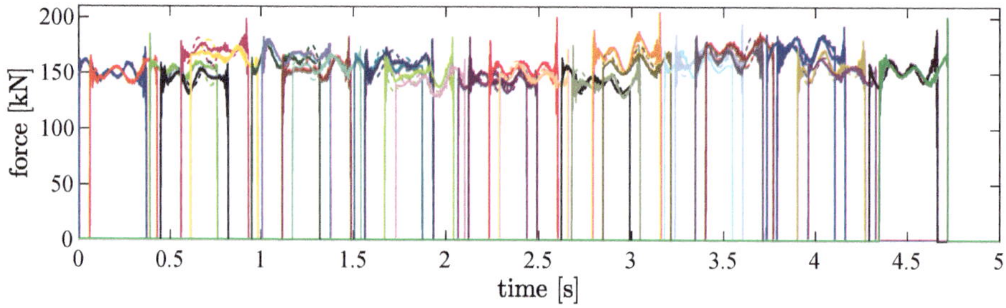

Figure 6.48 Identified and reference force time histories for the load scenario 3 and a smoothing radius $r_a = 0$ samples (dashed lines: reference forces; solid lines: identified forces)

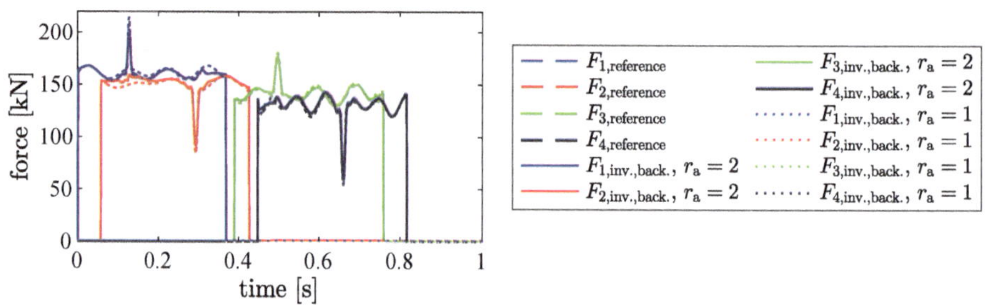

Figure 6.49 Influence of the smoothing radius on the impulse component identification

by adapting r_a. This fact is verified in the following using the load scenario 2 with two smoothing radii: $r_a = 2$ (result already discussed in Section 6.6.3.2) and $r_a = 1$. Fig. 6.49 illustrates the results obtained for $r_a = 1$ ($x_{\text{err,b}} = 1.29$ %) together with the corresponding results obtained for $r_a = 2$ (taken from Fig. 6.41) and the reference force values. Fig. 6.50 shows the same diagrams in detailed views of the intervals around the impulses (grey areas in Fig. 6.49), in order to facilitate the visual comparison of the different results. It is obvious that the consideration of the lower smoothing radius has a positive influence on the identification accuracy of the impulse components. The corresponding errors evaluated according to Eq. 6.32 for the four impulses (with $r_a = 1$) are:

$$\Delta I_1 = -2.86\ \% \quad \Delta I_2 = +1.61\ \% \quad \Delta I_3 = +0.33\ \% \quad \Delta I_4 = +2.74\ \% \tag{6.36}$$

The comparison with the values provided by Tab. 6.4 for the optimal weighting factors $(w_{p1}, w_{p2}) = (10^{-13}, 10^{-8})$ confirms also quantitatively that the reduction of the smoothing radius leads to a better identification of the impulse-like components of the moving forces.

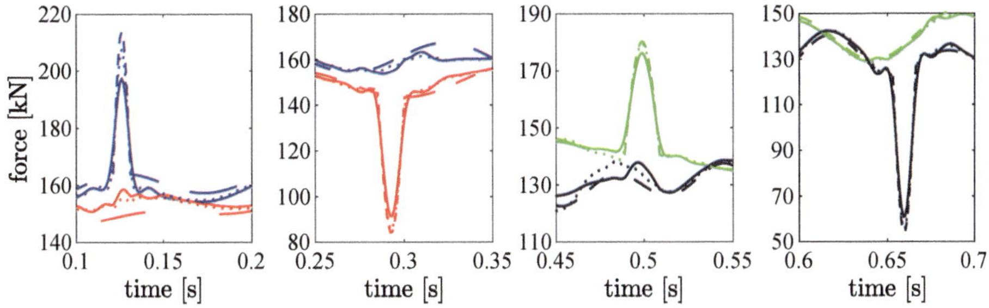

Figure 6.50 Influence of the smoothing radius on the impulse component identification: detailed view (dashed lines: reference forces; solid lines: identified forces with $r_a = 2$; dotted lines: identified forces with $r_a = 1$)

Overlapping interval length [samples]	Load scenario 1		Load scenario 3	
	80 km/h	160 km/h	80 km/h	160 km/h
1	4.43 %	0.54 %	3.05 %	2.63 %
90	2.87 %	0.61 %	2.49 %	2.11 %
113	2.69 %	0.64 %	2.58 %	1.97 %
150	3.00 %	0.57 %	2.42 %	2.13 %
225	3.15 %	0.68 %	2.43 %	2.19 %

Table 6.12 Identification errors $x_{err,b}$ of load scenario 1 and 3 for different lengths of the overlapping intervals

6.7.2.3 Length of the overlapping interval

The length of the overlapping interval is another parameter, which has to be defined by the user prior to the start of the inverse computation. Its effect on the performance of the identification method is briefly discussed in the following by comparing the results obtained for load scenario 1 and 3 under consideration of different overlapping interval lengths. For this, the following number of overlapping samples were investigated: 1, 90 (ca. $1/5$ of the window length), 113 (ca. $1/4$ of the window length), 150 (ca. $1/3$ of the window length) and 225 (ca. $1/2$ of the window length). The results are given in Tab. 6.12 for two different velocities of the moving forces. It can be clearly seen that the accuracy of the method is rather insensitive to the length of the overlapping interval in the investigated range between 90 and 225 samples. On the contrary, the computations with an overlapping interval of only one sample generally deliver considerably higher errors. The reason for this is most likely the fact that the computation with an overlapping interval of a greater length allows to start the iteration process in the subsequent window at a plausible initial point (cf. Section 6.3.2).

Even though it was shown that the length of the overlapping interval has a rather limited influence on the identification accuracy, it represents an important influence factor for the computational time. The reason for this is that it directly affects the number of windows and implicitly the computational effort. The experience of the author indicates that an overlapping interval of $1/3$ of the window length is a good choice in order to obtain a satisfactory performance, both in terms of accuracy and computational effort.

6.7.3 Effect of the amount of modal information

In the following, it will be examined how the number of vibration modes considered in the analysis affects the accuracy of the MFI. In this context, it should be noted that the influence of the number of modes for the noise-free computation without penalty functions has already been studied in Section 6.5.2. Based on the findings, a set of solvability conditions for noise-free problems without regularization were formulated. They give the relations between the number of modes, number of sensors and number of forces acting simultaneously on the structure, which are required to achieve the exact solution.

This section firstly focuses on the case of the noise-free data under consideration of nonzero penalty functions. The purpose of the investigation is to isolate the pure influence of the number of modes from any inaccuracies generated by noise. In this sense, a set of inverse computations with noise-free simulated measurement data containing different numbers of modes was performed for the load scenarios 1 and 3 and two different velocities (80 km/h and 160 km/h). The weighting parameters w_{p1} and w_{p2} of the penalty functions were selected according to Eq. 6.31. It is to be mentioned that this investigation implies the same number of modes used for the generation of the simulated measurement data and for the identification procedure. The number and the locations of the sensors remained unchanged for all trials, as defined in Section 6.3.1. The number of modes was varied between one and five. This means that the first solvability condition is only satisfied for the computations with four and five mode shapes in case of load scenario 3 and for the computations with two to five modes for load scenario 1. On the contrary, the second solvability condition is satisfied for all trials, since a number of five acceleration sensors was used (cf. Fig. 6.9), whereas a maximum number of four forces are acting simultaneously on the bridge (for load scenario 3).

The results of the study with noise-free data and nonzero penalty functions are given in Tab. 6.13. With respect to the fulfilling of the first solvability condition, it would be expected that all results obtained under consideration of 2 to 5 modes for load scenario 1 and of 4 to 5 modes in case of load scenario 3 would reveal the exact solution. However, in comparison to the investigations presented in Section 6.5.2, the case investigated here deals with nonzero penalty functions. These obviously represent a modification of the objective function, which does not allow the reconstruction of the exact solution. Nevertheless, it can be seen that the method delivers stable results with very satisfactory error values. As

Number of modes	Load scenario 1		Load scenario 3	
	80 km/h	160 km/h	80 km/h	160 km/h
1	2.18 %	1.90 %	2.62 %	3.65 %
2	2.29 %	0.81 %	1.90 %	2.36 %
3	2.02 %	0.41 %	1.72 %	1.80 %
4	1.55 %	0.62 %	1.61 %	1.64 %
5	2.55 %	0.58 %	1.67 %	1.58 %

Table 6.13 Identification errors $x_{\mathrm{err,b}}$ of load scenario 1 and 3 for different number of modes considered in the analysis with noise-free data

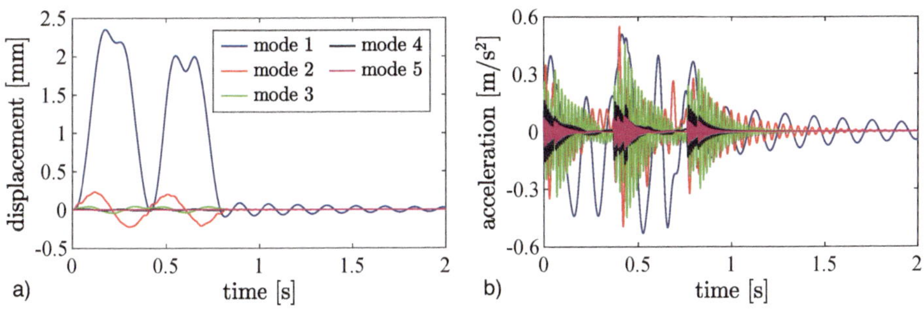

Figure 6.51 First five modal components (without noise) of the simulated displacement (a) and acceleration (b) at MP 10 for load scenario 1

expected, a clear tendency of an increasing accuracy with increasing number of modes can be observed. However, for both load cases, no significant improvement can be observed for computations with more than three modes. The reason for this is most likely the low contribution of the fourth and fifth mode to the total response. This can be explained by the frequencies of the fourth and fifth mode (102.13 Hz and 159.58 Hz), which are much greater that the highest excitation frequency of the moving forces: 15 Hz for load scenario 1 and 8 Hz for load scenario 3 (cf. Eq. 6.2 and 6.6). The clear dominance of the first three modes to the total response is illustrated exemplarily for MP 10 in Fig. 6.51 (load scenario 1) and Fig. 6.52 (load scenario 3).

In the following, it will be investigated, to what extent the number of modes considered in the analysis affects the results of the identification with noisy data and nonzero penalty functions. This represents, in fact, the plausible situation for practical applications. For this, a set of noisy measurement data with $e_{\mathrm{p}} = 5$ % was simulated for each investigated number of modes. The results are shown in Tab. 6.14. It can be seen that the errors are slightly higher than in the noise-free case (Tab. 6.13). The rather small differences between the results from Tab. 6.14 and Tab. 6.13 confirm that the penalty functions fulfill their role of delivering stable and robust solutions, even for data polluted with a rather high noise level. Moreover, when looking at the errors obtained for the load scenario 3, the tendency

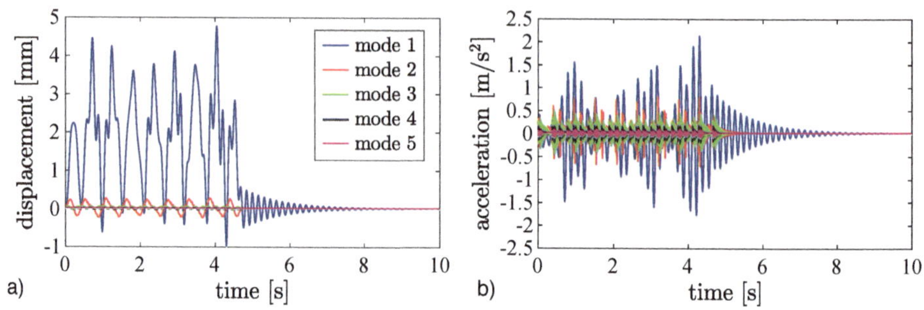

Figure 6.52 First five modal components (without noise) of the simulated displacement (a) and acceleration (b) at MP 10 for load scenario 3

Number of modes	Load scenario 1 80 km/h	Load scenario 1 160 km/h	Load scenario 3 80 km/h	Load scenario 3 160 km/h
1	2.50 %	2.08 %	2.80 %	3.74 %
2	2.76 %	1.45 %	2.58 %	2.39 %
3	3.00 %	0.57 %	2.42 %	2.13 %
4	2.54 %	1.03 %	2.55 %	2.00 %
5	2.26 %	1.21 %	2.43 %	2.16 %

Table 6.14 Identification errors $x_{\mathrm{err,b}}$ of load scenario 1 and 3 for different number of modes considered in the analysis with noisy data ($e_{\mathrm{p}} = 5$ %)

of lower errors for higher number of modes can also be observed in this case. This trend can not be clearly identified in the results of load scenario 1, since in this case a larger noise induced scattering between the results of different runs is expected (cf. Tab. 6.7). In addition, it can be seen also in this case that no significant reduction of the errors is achieved for calculations with more than three modes. Nevertheless, it can be generally stated that a sufficiently accurate force identification can be achieved for both load cases, even for a rather low number of modes, which might possibly violate the first solvability condition.

6.7.4 Effect of the modal truncation

In practice, it is expected that the number of modes contained in the measured responses is larger than the number of vibration modes considered in the inverse analysis. Therefore, the effect of the "modal truncation" on the identification accuracy is studied in the following. For this, a set of simulated measurement data, which includes responses from the first five mode shapes, was generated for load scenario 1 and 3. The identification procedure will be executed under consideration of different number of modes, which ranges from one to five. Similar to the investigations presented in the previous section, computations

Number of modes measurements	Number of modes identification	Load scenario 1		Load scenario 3	
		80 km/h	160 km/h	80 km/h	160 km/h
5	1	5.56 %	4.09 %	4.02 %	8.22 %
5	2	2.09 %	1.90 %	3.09 %	3.34 %
5	3	1.78 %	1.15 %	1.95 %	2.25 %
5	4	1.80 %	0.50 %	1.65 %	1.63 %
5	5	2.25 %	0.58 %	1.67 %	1.58 %

Table 6.15 Identification errors $x_{\mathrm{err,b}}$ of load scenario 1 and 3 for different modal truncation cases with noise-free data

Number of modes measurements	Number of modes identification	Load scenario 1		Load scenario 3	
		80 km/h	160 km/h	80 km/h	160 km/h
5	1	4.40 %	3.62 %	4.04 %	8.86 %
5	2	2.52 %	1.66 %	3.71 %	3.44 %
5	3	1.55 %	1.29 %	2.77 %	2.64 %
5	4	1.51 %	1.24 %	2.43 %	2.08 %
5	5	2.26 %	1.21 %	2.43 %	2.16 %

Table 6.16 Identification errors $x_{\mathrm{err,b}}$ of load scenario 1 and 3 for different modal truncation cases with noisy data ($e_{\mathrm{p}} = 5$ %)

with $e_{\mathrm{p}} = 0$ % and $e_{\mathrm{p}} = 5$ % are performed, in order to allow the isolation of the pure modal truncation effects from the noise influence. The results are given in Tab. 6.15 for the noise-free case and in Tab. 6.16 for computations with noisy data. Firstly, it can be seen that the differences between the results with noise-free and noisy data are rather low, indicating again that the penalty functions can successfully deal with the noise related issues. Moreover, it can be observed in both cases that the error tends to decrease with increasing number of modes considered in the computation. In general, it can be seen that when the number of modal components is no less than three, acceptable results are obtained both for the noisy and noise-free analysis. The reason for this is – as already mentioned above – that the simulated measurement data sets are dominated by the first three modal components. The consideration of additional modes in the inverse computation causes therefore only moderate improvements. However, even though the two investigated load scenarios indicate a similar number of modes that ensures a good accuracy, this can not be generally fixed to a certain value. In this respect, it is recommended that the number of modes considered in the identification procedure is chosen to be at least equal to the number of contributing modes in the measurements, in order to capture all the available information. This has certain implications with regard to the type of measurement sensors. Since the acceleration sensors usually capture higher frequency components than the displacement sensors, it is expected that the force identification from accelerations allows to consider a higher number of modes than the identification from displacements.

6.7.5 Effect of the sensor configuration

The effect of the sensor type and location is studied in the following. For this, a total number of 15 measurement configurations were investigated. They are described in Tab. 6.17 and illustrated in Fig. 6.53. The first four configurations (no. 1 to 4) only include acceleration sensors, while the following four (no. 5 to 8) only consider displacement time histories as simulated measurement data. For the remaining set-ups (no. 9 to 15), a combination of displacement and acceleration sensors was taken into account.

Tab. 6.18 shows the errors $x_{\mathrm{err,b}}$ for all investigated configurations under consideration of load scenario 1 and 3 and two moving force velocities (80 km/h and 160 km/h). It can be observed that the use of only acceleration sensors (no. 1 to 4) does not provide reliable results for the velocity of 80 km/h. This holds especially in case of the load scenario 3, which implies the reconstruction of 32 forces using a total of 32 windows[22]. The reference and identified force time histories for the measurement set-up 2 and a velocity of 80 km/h are shown in Fig. 6.54. It is obvious that the computation seems to fail in reconstructing the quasi-static components of several forces, whereas the error tends to increase towards the

Configuration no.	Measurement points acceleration	Measurement points displacement
1	10, 20, 30, 40, 50	–
2	10, 30, 50	–
3	20, 40	–
4	24	–
5	–	10, 20, 30, 40, 50
6	–	10, 30, 50
7	–	20, 40
8	–	24
9	10, 20, 30, 40, 50	25
10	10, 20, 30, 40, 50	15, 25
11	10, 30, 50	25
12	10, 30, 50	20, 40
13	10, 20, 30, 40, 50	10, 20, 30, 40, 50
14	10, 30, 50	10, 30, 50
15	20, 40	20, 40

Table 6.17 Overview of the investigated sensor configurations within the numerical validation process

[22]The matching numbers of forces and of computational windows is only a coincidence. The formulation of the problem does not impose any correlation between the number of forces and the number of windows.

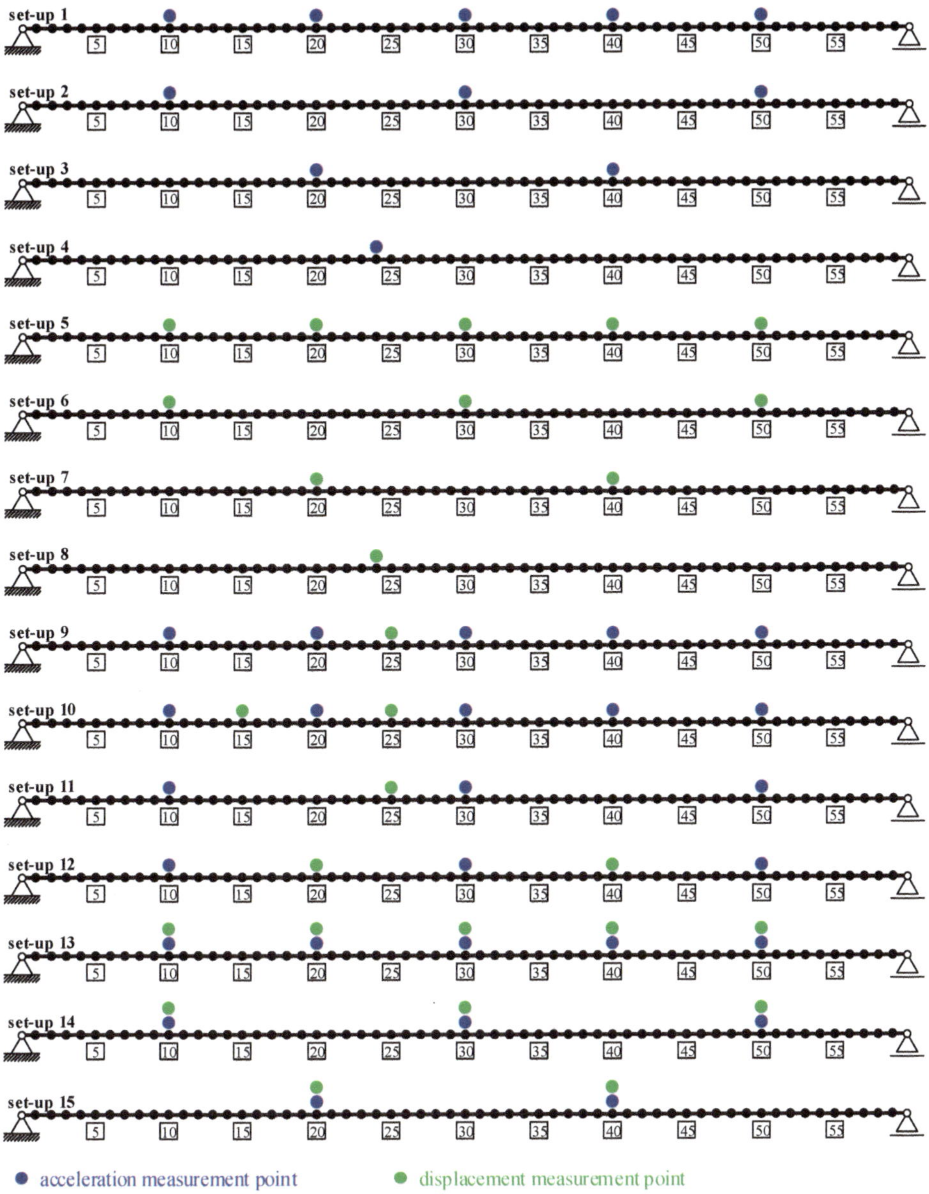

Figure 6.53 Sensor configurations used in the parametric study

end of the time history. Moreover, Fig. 6.54 shows that the identification procedure might even reveal unphysical negative forces, since an unconstrained and unbounded optimization problem is solved (cf. Chap. 5). Nevertheless, despite a completely unsatisfactory force

Set-up no.	Load scenario 1		Load scenario 3	
	80 km/h	160 km/h	80 km/h	160 km/h
1	3.42 %	0.60 %	198.57 %	4.23 %
2	4.25 %	0.70 %	65.80 %	2.62 %
3	3.72 %	1.09 %	45.10 %	6.27 %
4	6.94 %	1.95 %	20.41 %	3.74 %
5	1.91 %	2.29 %	3.16 %	3.24 %
6	1.89 %	2.72 %	3.34 %	3.16 %
7	2.57 %	2.03 %	2.95 %	3.51 %
8	2.40 %	2.42 %	3.04 %	3.68 %
9	3.00 %	0.57 %	2.42 %	2.13 %
10	1.75 %	0.92 %	2.44 %	2.06 %
11	3.18 %	0.86 %	2.55 %	2.28 %
12	1.98 %	0.74 %	2.43 %	2.32 %
13	1.41 %	0.88 %	2.37 %	2.20 %
14	1.34 %	1.26 %	2.54 %	2.22 %
15	1.78 %	0.94 %	2.64 %	3.26 %

Table 6.18 Identification errors $x_{\mathrm{err,b}}$ of load scenario 1 and 3 for different sensor configurations

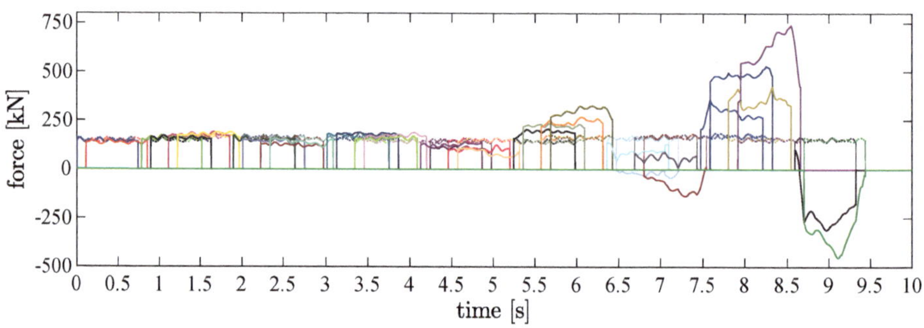

Figure 6.54 Identified and reference force time histories under consideration of load scenario 3, a velocity of 80 km/h and the measurement set-up 2 for a window length of 450 samples (dashed lines: reference forces, solid lines: identified forces)

identification, the re-built accelerations present a very good agreement with respect to the reference values, even at unmeasured locations (cf. Fig. 6.55a). This does not hold for the re-built displacements, which revealed very high deviations with respect to the reference values, especially towards the end of the time history, (Fig. 6.55b).

The results presented above suggest that small identification errors of the first moving forces with respect to their quasi-static component propagate throughout the computation (e.g. by influencing the initial values of the following windows), leading thus to significant

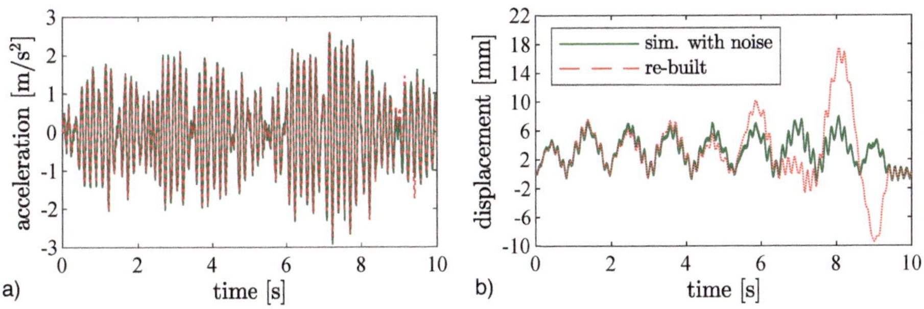

Figure 6.55 Simulated measurements and re-built responses at the unmeasured location MP 22, based on the load scenario 3 and the measurement set-up 2 for a window length of 450 samples

errors and to partly unstable[23] results in the last windows. This possibly also explains the fact that this effect could not be clearly observed for load scenario 1 and a velocity of 80 km/h, which implies only six computational windows. The main cause of the increased errors revealed when using only acceleration data at low speeds is thought to be the fact that the accelerations generally present a low sensitivity with respect to the quasi-static force components. Moreover, it should be noted that the high errors corresponding to set-up no. 1 to 4 in Tab. 6.18 are obtained under consideration of a relatively low amount of regularization, which is directly correlated to the length of the moving window. Slightly erroneous static components determined for the first windows can most likely not be corrected, as the averaging within the penalty functions is performed on a much shorter interval. However, the quasi-static force component only has an insignificant influence on the accelerations, allowing thus to achieve a very low value of the main optimization criterion, which only consists of the minimization of the acceleration differences (cf. Fig. 6.55a). In order to verify this assumption, further computations are performed for the first measurement configuration with window lengths of 700, 1000 and 1500 samples. The error values at the end of the backward computation are $x_{\text{err,b}} = 5.42\ \%$ for a window length of 700 samples, $x_{\text{err,b}} = 2.89\ \%$ for 1000 samples and $x_{\text{err,b}} = 2.62\ \%$ for 1500 samples. In this context, Fig. 6.56 exemplarily illustrates the excellent agreement between the reference and the identified forces for the computation based on a window length of 1500 samples. Hence, it can be clearly confirmed that the proposed moving force reconstruction method is highly sensitive with respect to the amount of regularization and implicitly to the window length, if the measurement set-up only includes acceleration sensors. This holds especially in case of passages with low speeds. However, if such a measurement set-up has to be used for a practical application, it is recommended to tune the optimal window length for identification of low speed passages prior to the start of the experimental investigation. This can be achieved, for example, by means of a numerical study with simulated measurement

[23]By unstable results it is meant that the error would most likely tend to increase beyond all borders, if a hypothetical vehicle with an unlimited number of carriages is assumed.

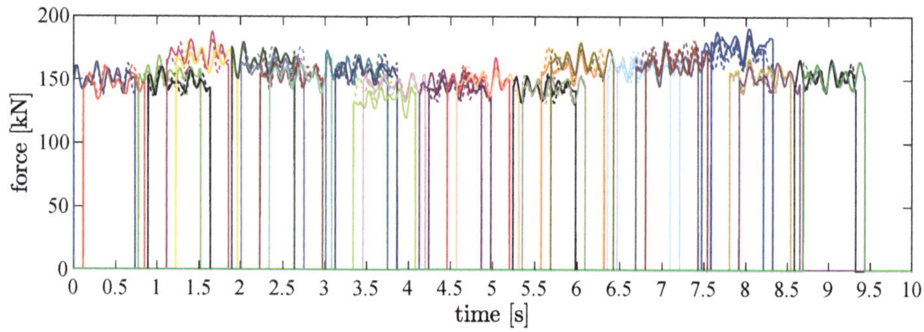

Figure 6.56 Identified and reference force time histories under consideration of load scenario 3, a velocity of 80 km/h and the measurement set-up 1 for a window length of 1500 samples (dashed lines: reference forces, solid lines: identified forces)

data. A rough orientation value with respect to the minimum window length, which is expected to deliver feasible results, is to aim at least the unity ratio between the window length and the passage time of a single force.

On the contrary, consistently stable results were revealed by the computations based on only displacement measurements (no. 5 to 8), even for the basic window length of 450 samples. This holds for both investigated load scenarios and velocities. In a general case, higher errors are expected for poorer content of measurement data. However, this effect can only be clearly identified for load scenario 1 (80 km/h) and load scenario 3 (160 km/h). It is thought that the reason for this is the noise component of the measurement signals, which may slightly affect the computational results. This seems to be a justifiable explanation also for the unexpected behavior of the errors for the measurement set-ups 1 to 4 in case of load scenario 3 with a velocity of 80 km/h. It can be seen that the errors significantly decrease with decreasing amount of measurement information, from about 199 % for five acceleration sensors to only 20 % for a single acceleration sensor. Considering that the problem is in this case – as already mentioned above – prone to become unstable, it is definitely possible that a larger amount of noise (occurring due to additional measurements) combined with the low regularization amount can increase the instability of the results.

The best results are obviously revealed by a combination of displacement and acceleration measurement data (no. 9 to 15). In this respect, one might criticize that direct displacement measurements of bridge structures are quite cumbersome or even impossible due to the lack of a fixed reference point. However, considering the impressing recent advances in the field of non-contact displacement measurements (FIRUS et al., 2016), combining acceleration and displacement measurements within the force reconstruction process on real structures is a justified and practicable approach. The lowest errors are obtained for set-up 13, which involves the use of five acceleration and five displacement sensors. However, all the combined measurement configurations generally reveal quite satisfactory results, which are only slightly different from the minimal values. This indicates that also a more practicable combination of sensors, implying a reduced number of displacement measurements (e.g. set-ups 9-12) can be reliably used with the proposed MFI method.

Related to the previous discussion, it is to be mentioned that the finding of the optimal sensor locations in the force identification is still an open research question. Without going into detail into this topic, it is recommended to seek the acquisition of as many contributing modal components as possible. This means that the sensors are to be placed in such a way, so that they do not match the nodes of the relevant mode shapes contributing to the structural response. In this context, it would be advisable that the accelerometers are located at different positions than the displacement sensors, in order to increase the amount of information. The set-ups 9, 10, and 12 fulfill these conditions. They are able to record components of the first five modes, whereas, in fact, the first three modes already deliver a sufficiently accurate representation of the structural behavior (cf. Fig. 6.52).

6.7.6 Effect of modeling errors

One of the most important aspects of the MFI procedure is the structural model. This is usually formulated based on the finite element method (FEM) and is used for generating the computed motion quantities $u_{h,j}$ and $\ddot{u}_{h,j}$ required by the objective function (Eq. 5.15). Since the proposed method is basically intended to be applied on real existing structures, the corresponding structural models have to be adapted to the real behavior. A widely used approach is to update a set of structural parameters, which minimize the differences between the computed and measured natural frequencies and mode shapes. In this context, it is mentioned that the modal parameters of an existing structure (natural frequencies, damping ratios, mode shapes) are usually obtained by modal testing (PEETERS et al., 1999; REYNDERS et al., 2008). However, even after performing the model updating procedure using the measured data, certain errors are expected to remain in the structural model. In this sense, the present study investigates how different level of modeling uncertainties affect the results of the proposed MFI method. For this, the stiffness of the beam elements and the nodal masses were randomly varied under consideration of a certain disturbance level according to the following rule:

$$
\begin{aligned}
E_{\text{inv},i} &= E_i + E_i \cdot e_{\text{p,E}} \cdot N_{\text{rand},i} && \text{for} \quad i = 1,2,...,60 \\
I_{\text{inv},i} &= I_i + I_i \cdot e_{\text{p,I}} \cdot N_{\text{rand},i} && \text{for} \quad i = 1,2,...,60 \\
m_{\text{inv},i} &= m_i + m_i \cdot e_{\text{p,m}} \cdot N_{\text{rand},i} && \text{for} \quad i = 1,2,...,59
\end{aligned}
\tag{6.37}
$$

where $E_{\text{inv},i}$ describes the Young's modulus and $I_{\text{inv},i}$ the area moment of inertia of the i-th out of 60 beam elements used for the inverse computation, while $m_{\text{inv},i}$ depicts the corresponding nodal mass of the i-th out 59 nodes. E_i, I_i and m_i represent the reference values used for the generation of the simulated measurement data, which are assumed as constant for all elements/nodes. Furthermore, $e_{\text{p,E}}$, $e_{\text{p,I}}$ and $e_{\text{p,m}}$ represent the noise levels defined for disturbing the three parameters. Finally, $\boldsymbol{N}_{\text{rand},i}$ is the i-th entry of the standard distribution vector $\boldsymbol{N}_{\text{rand}}$ with zero mean and unit standard deviation.

A total of seven configurations were investigated, whereas the considered noise levels span from 0 % (reference model) to a worst case scenario of 10 %. It is to be mentioned that the same value e_{model} was assigned for all three noise levels in each configuration:

$$e_{\text{p,E}} = e_{\text{p,I}} = e_{\text{p,m}} = e_{\text{p,model}} \tag{6.38}$$

The resulting stiffness parameters of each element are given in Fig. 6.57 for the Young's modulus and in Fig. 6.58 for the second moment of area. Fig. 6.59 shows the corresponding disturbed nodal masses. The disturbance of the stiffness and mass parameters obviously implies changes of the modal parameters of the system, which are the key quantities in the dynamic analysis. In this sense, the natural frequencies of the disturbed system and the relative errors with respect to the frequencies of the reference model used for the generation of the simulated measurement data (Eq. 6.1) are given in Tab. 6.19, whereas the relative frequency errors are computed as

$$\Delta f_i = \frac{\tilde{f}_i - f_i}{f_i} \cdot 100\ \% \tag{6.39}$$

where $\tilde{f}_i$ represents the i-th natural frequency of the disturbed system and f_i denotes the corresponding natural frequency of the reference system, as given in Eq. 6.1. It can be seen in Tab. 6.19 that the noise levels considered in the disturbance of the structural parameters with values of up to $e_{\text{p,model}} = 3\ \%$ generally only cause slight changes in the natural frequencies. Frequency errors above 1 % are only obtained for noise levels exceeding $e_{\text{p,model}} = 7\ \%$. In this context, it is to be mentioned that Eq. 6.37 is a very simplified assumption for describing the model uncertainty. It is clear that by randomly disturbing both stiffness and mass parameters, a compensating effect will occur with respect to the natural frequencies. Therefore, the interpretation of the model errors with respect to the deviations of the natural frequencies seems to be more appropriate in this case.

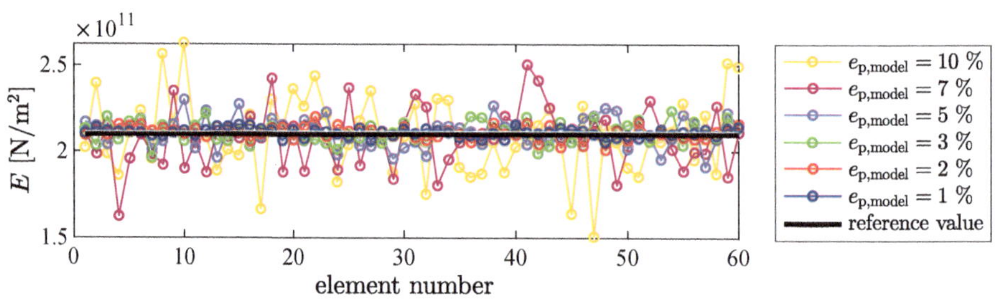

Figure 6.57 Assumed model uncertainties with respect to the Young's modulus of the beam elements

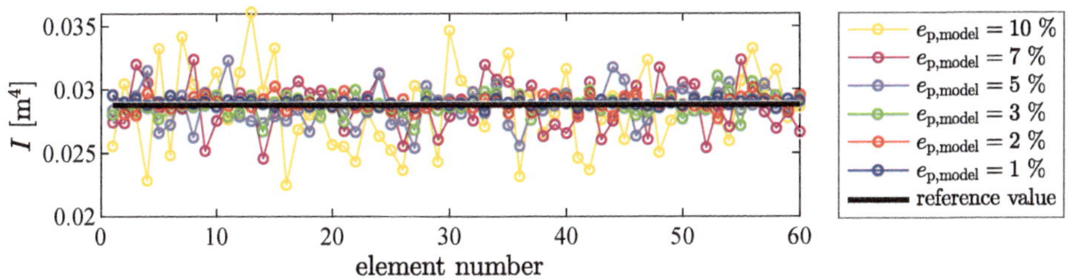

Figure 6.58 Assumed model uncertainties with respect to the second moment of area of the beam elements

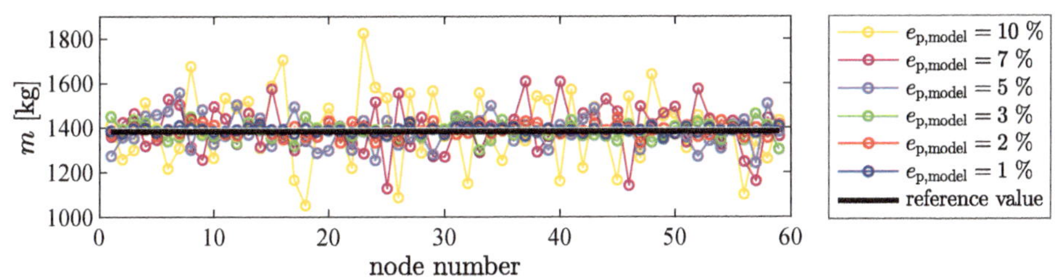

Figure 6.59 Assumed model uncertainties with respect to the nodal masses

The effect of the model errors on the mode shapes can be evaluated by means of the Modal Assurance Criterion (MAC) values between the reference and the perturbed systems. The MAC value represents a measure of the strength of the linear correlation between two vectors and is calculated as follows:

$$MAC_{i,k} = \frac{\left(\phi_i^\top \cdot \tilde{\phi}_k\right)^2}{\left(\phi_i^\top \cdot \phi_i\right) \cdot \left(\tilde{\phi}_k^\top \cdot \tilde{\phi}_k\right)} \tag{6.40}$$

whereas ϕ_i represents the i-th mode shape of the reference system and $\tilde{\phi}_k$ the k-th mode shape of the disturbed system. The MAC values are situated between 0 and 1. They can be arranged in matrix form, obtaining the so called MAC-matrix. In case of an ideal correlation between the mode shapes of the reference and disturbed systems, the MAC-matrix is equivalent to the identity matrix. Tab. 6.19 also includes MAC values corresponding to the diagonal entries of the MAC matrix ($i = k$ in Eq. 6.40) for the first five modes of each disturbed system. They indicate that the mode shapes of the disturbed systems present only negligible differences compared to the unperturbed initial system, even in the case of a rather conservative perturbation level of 10 %. This is in line with the observations of LAW et al. (2011), according to which the eigenvectors should be be rather insensitive to local errors in the model.

Tab. 6.20 gives the force identification errors $x_{err,b}$ determined at the end of the backward computation for all investigated model uncertainty configurations. The results consider the load scenarios 1 and 3 and two moving force velocities (80 km/h and 160 km/h). It can be clearly seen, that the results corresponding to values of $e_{p,model}$ between 1 % to 3 % show rather insignificant changes compared to the reference case $e_{p,model} = 0$ %. This effect can be correlated with the very small changes in the natural frequencies, whereas the fundamental frequency plays a dominant role in dynamic analysis of the present case (cf. Fig. 6.52). As soon as the error of the first natural frequency increases over 0.5 %, corresponding to a perturbation of the structural parameters of $e_{p,model} = 5$ %, a clear drop in the accuracy of the proposed MFI method is revealed. Rather intolerable error values are obtained for noise levels $e_{p,model} = 7$ % and above, corresponding to natural frequency errors greater than 1 %. Therefore, it can be concluded that a very accurate structural model is required, in order to allow a good accuracy of the MFI. In the context of model updating, it should be aimed to achieve an error less than 1 % for all natural frequencies, which are significantly excited by the external forces.

$e_{p,model}$ [%]	f_1 [Hz]	Δf_1 [%]	MAC mode 1
0	6.3831	-	1.000000
1	6.3847	+0.025	0.999999
2	6.3766	-0.102	0.999997
3	6.3751	-0.126	0.999995
5	6.3427	-0.634	0.999989
7	6.3177	-1.025	0.999993
10	6.1470	-3.699	0.999961
$e_{p,model}$ [%]	f_2 [Hz]	Δf_2 [%]	MAC mode 2
0	25.5325	-	1.000000
1	25.5787	+0.180	0.999996
2	25.5393	+0.026	0.999970
3	25.5476	+0.059	0.999990
5	25.5586	+0.102	0.999942
7	25.1638	-1.444	0.999801
10	24.6249	-3.555	0.999537
$e_{p,model}$ [%]	f_3 [Hz]	Δf_3 [%]	MAC mode 3
0	57.4482	-	1.000000
1	57.4789	+0.053	0.999996
2	57.4019	-0.081	0.999974
3	57.4307	-0.031	0.999896
5	57.2307	-0.379	0.999481
7	56.4094	-1.808	0.999746
10	56.3310	-1.945	0.999190

table continued on the next page

$e_{p,model}$ [%]	f_4 [Hz]	Δf_4 [%]	MAC mode 4
0	102.1300	-	1.000000
1	102.1814	+0.050	0.999992
2	102.2334	+0.101	0.999934
3	102.2244	+0.092	0.999879
5	102.0809	-0.048	0.999295
7	100.2465	-1.844	0.999755
10	99.4973	-2.578	0.998560
$e_{p,model}$ [%]	f_5 [Hz]	Δf_5 [%]	MAC mode 5
0	159.5779	-	1.000000
1	159.8810	+0.190	0.999989
2	159.6820	+0.065	0.999881
3	159.4176	-0.100	0.999869
5	158.8181	-0.476	0.998899
7	157.7672	-1.135	0.999374
10	157.9702	-1.007	0.998677

Table 6.19 Natural frequencies of the disturbed systems, the relative errors with respect to the frequencies of the reference system and the corresponding MAC values

$e_{p,model}$ [%]	Load scenario 1		Load scenario 3	
	80 km/h	160 km/h	80 km/h	160 km/h
0	3.00 %	0.57 %	2.42 %	2.13 %
1	2.98 %	0.82 %	2.48 %	2.08 %
2	2.66 %	1.01 %	2.48 %	2.09 %
3	3.26 %	1.00 %	2.56 %	2.33 %
5	3.47 %	2.20 %	3.42 %	3.33 %
7	4.33 %	3.46 %	3.65 %	3.96 %
10	8.22 %	10.14 %	7.93 %	9.13 %

Table 6.20 Identification errors $x_{err,b}$ of load scenario 1 and 3 for different level of modeling errors

It should be mentioned that the previous discussion does not include the effect of an inexact damping ratio. However, this represents in fact a structural parameter that is subjected to high uncertainties in applications on real structures, especially in case of an amplitude dependent damping behavior (REBELO et al., 2008; RAUERT et al., 2010; STOLLWITZER et al., 2020). In addition, the damping has a significant influence on the results of a dynamic analysis of bridge structures, particularly in case of resonance excitation. Therefore, it is expected that the damping assumption also influences to a certain extent the force identification results. In this respect, Section 7.6.3.8 presents a detailed investigation on the effect of an inexact damping ratio on the accuracy of the proposed MFI method both in context of resonant and non-resonant train passages over a real railway bridge.

6.7.7 Effect of an inexact speed

The vast majority of the MFI methods available in the literature consider that the location of the moving forces is known at each time instant. This assumption, which was also postulated for the identification method proposed in this work, seems justified, as an accurate measurement of the vehicle speed is not expected to pose any technical difficulties. There are several state of the art methods that can accomplish this task, e.g. photoelectric sensors (cf. Section 7.3.3), radar devices or classical strain gauges installed at rail level (ŽNIDARIČ et al., 2016; VOSPERNIG et al., 2020). However, in case of identification of pedestrian induced forces, the tracking of the subject over time becomes a little bit more complicated. The reason for this is that both the trajectory and the speed are likely to be subjected to more or less considerable variations during a bridge crossing. This aspect will be discussed in more detail in Section 8.6.4.

The speed of the moving forces determines the coordinates at which each load is situated in each time step. This implicitly means that the time dependent modal loads are directly dependent on the speed (Eq. 3.45). Therefore, it can be stated that the vehicle (or pedestrian) speed is a key aspect in the identification procedure. In order to verify to what extent an uncertain speed assumption affects the identification accuracy, a further set of computations is performed under consideration of different speed perturbations. Two reference speeds were considered (80 km/h and 160 km/h) for the determination of the simulated measurement data, whereas the inverse computation is performed with perturbed speeds, ranging from -3 to $+3$ km/h (with respect to each reference speed). The identification errors $x_{\mathrm{err,b}}$ are shown in Tab. 6.21. It is clear that even the slightest speed perturbation of only 0.1 km/h causes an important loss of accuracy, especially in case of load scenario 3. This effect is even more pronounced for the running speed of 80 km/h than for 160 km/h. The reason for this might be the absolute formulation of the speed perturbation instead of a relative approach. The influence of an absolute perturbation is – from the relative perspective – twice as significant for 80 km/h than for 160 km/h. Speed perturbations above 0.3 km/h consistently delivered unacceptable results for all investigated scenarios. Therefore, it can be generally stated that the proposed MFI procedure requires very accurate speed measurements, in order to achieve a satisfactory accuracy.

6.8 Conclusion

This chapter describes the application of the proposed MFI method on a 1D Bernoulli beam model of a simply supported single-tracked bridge, whereas all computations are based on simulated measurement data. A wide set of investigations and parametric studies were performed, in order to evaluate the influence of various parameters on the solution behavior.

It has been shown that the optimal values of the weighting parameters of the penalty functions w_{p1} and w_{p2} of a given structure are rather insensitive to the magnitude, speed and number of the moving forces. Therefore, they can be determined at the beginning of the investigation by means of a single numerical parametric study and be then kept constant for all trials performed on the respective structure.

Two solvability conditions of the proposed MFI method were formulated for noise-free data and an unregularized objective function. The first one states that the number of modes considered in the inverse analysis has to be at least equal to the maximum number of forces present on the structure at a given time instant, in order to retrieve the exact solution. The second solvability condition requires that the number of sensors included in the data set is at least equal to the number of forces present on the structure in a certain time instant, in order to achieve the exact solution. However, the activation of the penalty functions compensates to a certain extent the violation of the solvability conditions, so that stable and sufficiently accurate solutions can be expected, even when one or both conditions are harmed.

The investigations with simulated measurement data have shown that the method is also able to identify impulse-like force components. However, it has been found that the identification of impulse magnitudes is highly sensitive to the amount of regularization, i.e. high regularization components tend to flatten the impulse-like force components.

The proposed MFI method shows a stable behavior and reveals sufficiently accurate results in relation with different noise samples of simulated measurement data. Regarding

Speed perturbation [km/h]	Load scenario 1		Load scenario 3	
	80 km/h	160 km/h	80 km/h	160 km/h
−3.0	36.87 %	19.93 %	106.63 %	52.33 %
−2.0	28.25 %	14.11 %	76.17 %	42.12 %
−1.0	18.58 %	7.37 %	44.75 %	24.86 %
−0.5	10.59 %	3.70 %	20.43 %	15.89 %
−0.3	7.00 %	2.19 %	13.59 %	11.30 %
−0.1	3.45 %	1.17 %	6.37 %	4.19 %
0.0	3.00 %	0.57 %	2.42 %	2.13 %
+0.1	3.21 %	0.97 %	7.46 %	3.96 %
+0.3	4.78 %	2.99 %	15.60 %	10.90 %
+0.5	7.38 %	5.25 %	19.81 %	15.60 %
+1.0	14.04 %	11.41 %	33.29 %	25.82 %
+2.0	22.43 %	22.97 %	59.95 %	41.40 %
+3.0	31.34 %	33.4 %	89.13 %	51.67 %

Table 6.21 Identification errors $x_{\mathrm{err,b}}$ of load scenario 1 and 3 for different speed perturbations

the noise level, very satisfactory errors were only obtained for noise levels lower than the (very conservative) assumption of 5 %, which was also used for the determination of the weighting parameters.

Furthermore, a good identification accuracy could be obtained for window lengths above 400 samples, smoothing radii below 5 samples and length of the overlapping interval exceeding one fifth of the moving window length. With respect to the amount of modal information, stable and accurate results were obtained for computations under consideration of a number of modes ranging between 1 and 5. A tendency toward higher accuracy for larger number of considered modes could be observed. Nevertheless, no improvements were detected for more than three modes. In addition, investigations on the modal truncation revealed that lower errors could be detected for increasing number of modes considered in the inverse computation.

The accuracy of the force identification is – in case of application with noisy data – very sensitive to the amount of regularization if only acceleration measurements are used in the analysis. If the amount of regularization, which also depends on the window length, is too low, the method fails to reconstruct the quasi-static components of the identified forces, even though the re-built accelerations are in very good agreement with the reference data. The best identification results are obtained for a combination of acceleration and displacement measurements, whereas very satisfactory results can be obtained even for a low number of sensors.

One of the main aspects of the proposed MFI method is that it requires a very accurate model, in order to allow a high identification accuracy. It has been shown that deviations above 1 % with respect to the first natural frequency lead to intolerable errors.

Finally, it has been found that the identification accuracy is highly sensitive to the speed assumed in the inverse computation. It has been shown that even the slightest speed perturbation of 0.1 km/h causes a significant loss of accuracy.

7 Experimental validation: railway bridge in operation

7.1 Framework of the investigation

Infrastructure assets in general, but especially bridge structures, are continuously subjected to new load scenarios during their planned service life. This occurs mainly due to a wide variety of innovations in the vehicle technology. The new generation of high-speed trains of the DB AG, known as the ICE 4 (Intercity Express 4, Fig. 7.1), was commissioned at the end of 2017. The new vehicle impresses with an innovative modular concept, which implies considerably lower maintenance costs. For example, by increasing the length of the car bodies, the number of bogies could be reduced. However, it was found during the planning phase that the modified wheel set arrangement of the ICE 4 (Fig. 7.2) could possibly lead to a different dynamic excitation mechanism on railway bridges. Furthermore, the longer cars of the ICE 4 expectedly imply higher axle loads than in case of previous high-speed train types. The so-called train signature[1] is suitable for a first quick and rough comparison between the dynamic effects of the ICE 4 and the High-Speed-Load-Model A (HSLM-A)

Figure 7.1 New vehicle type ICE 4

[1] The train signature characterizes the dynamic excitation capacity of a particular railway vehicle. It is only dependent on the axle spacing and the axle loads, being thus independent of the structural properties. It is determined by breaking down the load diagram of a vehicle in Fourier series followed by a Taylor series expansion. The equation of the train signature is given in ERRI (1999) together with its detailed derivation.

A. Firus, *A Contribution to Moving Force Identification in Bridge Dynamics*, Mechanik, Werkstoffe und Konstruktion im Bauwesen 65, https://doi.org/10.1007/978-3-658-39838-5_7

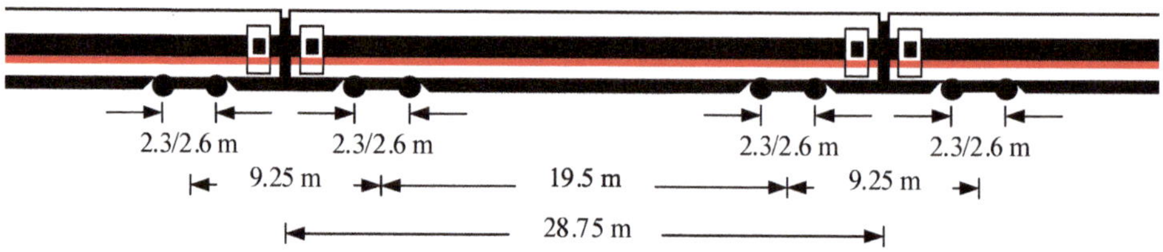

Figure 7.2 Axle arrangement of the ICE 4 (the axle distance within a bogie is 2.3 m for unpowered and 2.6 m for powered cars)

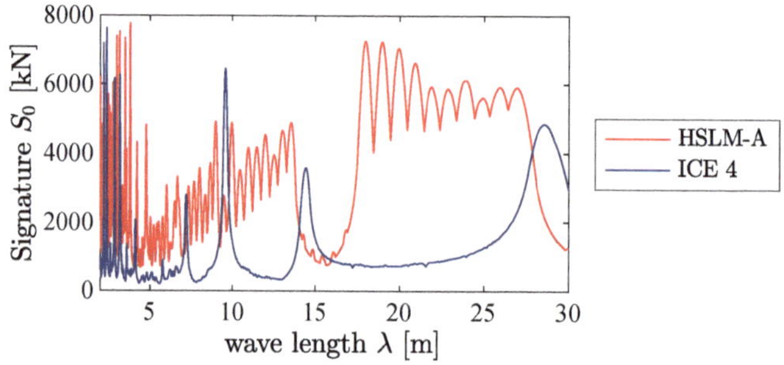

Figure 7.3 Signature of the ICE 4 and envelope signature of the HSLM-A model trains

given in the regulations (RIL 804, DIN EN 1991-2). It represents the excitation capacities of the ICE 4 and the HSLM-A models as functions of the excitation wavelength λ (Fig. 7.3), whereas λ is defined as the ratio between the train velocity v and the first natural frequency of the bridge f_1:

$$\lambda = \frac{v}{f_1} \tag{7.1}$$

Fig. 7.3 clearly indicates that the ICE 4 partly possesses a higher excitation capacity than the HSLM-A, especially at $\lambda \approx 9.6$ m and $\lambda \approx 28.6$ m. This means that for a given natural frequency of the bridge, the train generates resonance at other, even lower velocities than the previous train types, which are generally presumed to be covered by the HSLM-A.

In this context, the DB Netz AG[2] set up the project *Brückenbefahrbarkeit* (english: *bridge trafficability*), which implied the verification of all existing bridges in the railway network of the DB AG with respect to their dynamic responses, in order to ensure a safe operation under excitation of the ICE 4. For this purpose, a multi-stage process was developed and the entire network could be successfully verified in time for the commissioning of the ICE 4. The process is mainly based on an extensive set of database-driven automated calculations, but implies also individual detailed analyses for particular bridge structures

[2] DB Netz AG is a wholly owned subsidiary of the DB AG. It owns and operates – as an infrastructure manager – the vast majoritiy of the German railway network.

with rather complex structural systems (GRUNERT, 2021). Moreover, the reassessment process had to be complemented in several cases by experimental investigations. This step was necessary, as the structural parameters determined through experimental investigations can differ considerably from the parameters resulting from the initial structural analysis based on the design drawings (BAGAYOKO et al., 2008; FIRUS et al., 2018c). Such differences occur mostly due to different assumptions, simplifications and uncertainties within the structural model (REBELO et al., 2008; ZABEL et al., 2008; RAUERT, 2011; REITERER et al., 2019). A total of more than 50 000 structures[3] were analyzed.

Within the framework of the strategic partnership between the Technical University of Darmstadt and the DB AG, the author of this thesis and his supervisor were involved in the experimental investigations of selected bridge structures. One of them was also particularly suitable for validation of the MFI method developed within this work. It will be treated in detail in the following sections. The investigation implied the comprehensive measurement of the dynamic responses of a real structure in operation under excitation of the ICE 4, whereas the vehicle behavior was also recorded at selected points. The main objectives of the measurement were: to verify the excitation mechanism of the ICE 4 (cf. Fig. 7.3), to determine the experimental resonance peak of the structure, to verify the models for dynamic reassessment of bridge structures and, if necessary, to determine safety margins in the evaluation. In addition, the influence of the longitudinal ballast joint and of the moment of resistance generated by the continuity of the ballast bed over the supports on the dynamic behavior of the bridge was to be analyzed. The experimental validation of the MFI method developed within this thesis was integrated as a complementary aim of the investigation. In order to tackle the challenging experiments, an interdisciplinary measurement team was formed from the department for bridge measurement of the DB Netz AG (FRIEBE, 2017) and the Technical University of Darmstadt for measurements of structural responses and DB Systemtechnik[4] for vehicle measurements.

7.2 Investigation object

The investigated bridge (shown in Fig. 7.4) is located in Germany at km 17.316 of the railway route no. 5302 (Augsburg-Ulm), which has a maximum allowable speed of 200 km/h. The bridge has a span of 19.50 m and crosses the river Schmutter[5]. The structure consists of two adjacent, structurally independent identical steel box girders with rectangular

[3] In order to avoid confusion, it is to be mentioned that the mentioned amount of 50 000 structures does not represent the number of investigated bridges. The term *structure* denotes here in fact the substructure of a bridge (from the german *Teilbauwerk*). The railway network of the DB AG includes around 25 200 bridges, consisting of around 50 000 substructures. For example, a simple two tracked single span beam bridge usually consists of two independent substructures, one for each direction.

[4] As a subsidiary of DB AG, DB Systemtechnik is an engineering company, which is mainly responsible for engineering services in the field of rail technology.

[5] Based on the name of the river Schmutter, the investigated structure will be denoted in the following as the *Schmutter bridge*.

Figure 7.4 Investigated bridge over the river Schmutter

section. Each of them consists of two webs (16 mm tick), a lower flange (18 mm thick) and an upper flange (20 mm thick). The width of a box girder is 3.50 m, while the height is 0.973 m. The distance between the inner webs of the two adjacent girders is 0.56 m, leading thus to a width of the main structural system of 7.56 m. Including the width of the additional cantilever systems for the walkways, the total width of the structure becomes 10.0 m (Fig. 7.5).

For the stiffening of the box girders, a set of 31 equally spaced T-beams – consisting of a web 270 x 10 x 3484 mm and a lower flange 130 x 10 x 3484 mm – were disposed inside the box girder, at the upper flange (Fig. 7.5 and 7.6). In addition, two longitudinal 1/2 I-beams (1/2 I400) were installed at the lower flange, each of them at a distance of 0.75 m from the vertical symmetry axis of the cross section. Furthermore, it can be seen in Fig. 7.6 that at six locations along the structure the T-beams at the upper flange and the longitudinal 1/2 I beams at the lower flange are connected by two very stiff trapezoidal beams, as detailed in Fig. 7.5.

The box girders are connected by the ballast bed along a longitudinal ballast joint. This is expected to introduce a coupling effect between the twin decks, even though they are structurally independent. The two substructures are crossed by centrically disposed tracks,

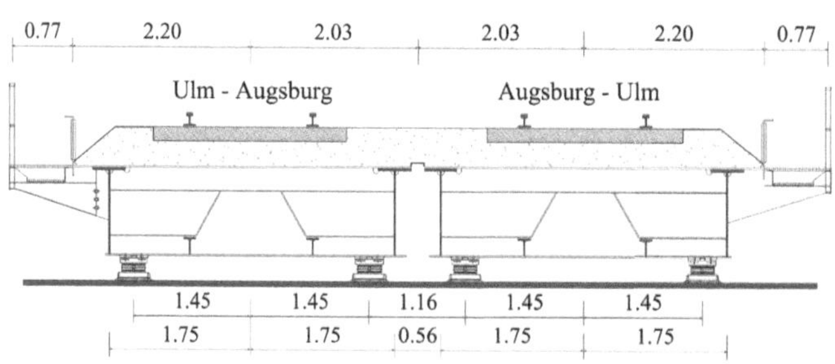

Figure 7.5 Cross-section of the Schmutter Bridge

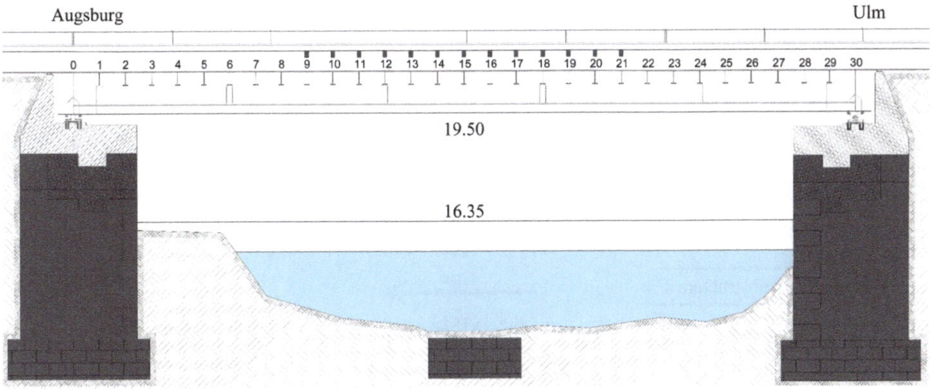

Figure 7.6 Longitudinal section of the Schmutter Bridge

one for each traffic direction. The structural system of the two coupled bridge decks can be described individually and as a whole as a simply supported beam. The corresponding support conditions are achieved using elastomeric bearings.

7.3 Measurement set-up

The various goals of the project required a comprehensive measurement setup consisting of more than 100 accelerometers, displacement sensors and strain gauges. They were arranged at different locations on the superstructure, on the track and at vehicle level. A detailed presentation of the complete measurement set-up for the structure and the vehicle would go far beyond the scope of this thesis. Therefore, only the set-up of the relevant sensors for this work will be discussed in the following. This only concerns sensors installed on the superstructure.

7.3.1 Acceleration measurements

The acceleration is one of the most commonly measured quantities in the railway bridge dynamics. On the one side, the acceleration sensors serve for verifying the absolute maximum deck accelerations, which have to be less than 3.5 m/s^2 for ballast bridges and 5 m/s^2 for bridges with slab tracks (DIN EN 1991-2, RIL 804). On the other side, the accelerations are the main quantity used in the system identification procedures, which represent a necessary step of every model updating procedure. In this sense, the measurement of the Schmutter bridge implied the installation of over 80 acceleration sensors in the box girder and at the track level. However, only the vertical accelerations (in z direction, cf. Fig. 7.8) recorded on the main structural elements (box girders) will be used within the analyses of this thesis. Their positions are shown in Fig. 7.7. The corresponding

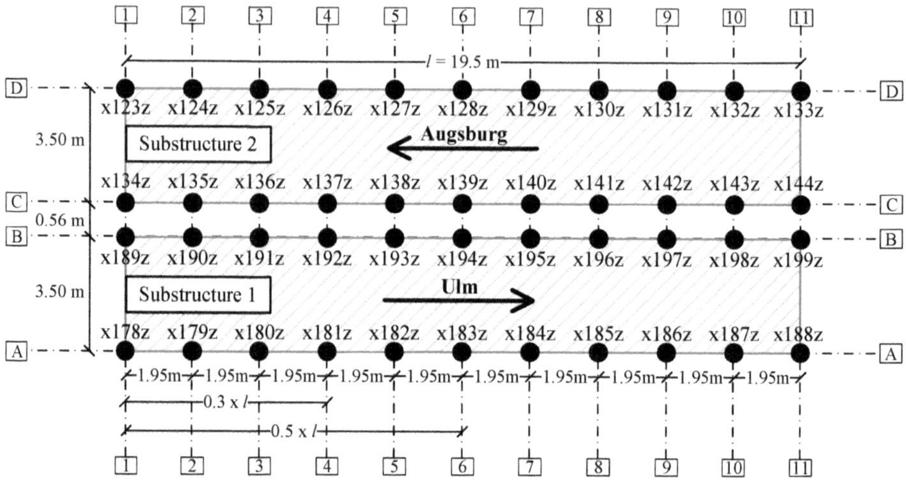

Figure 7.7 Set-up of the acceleration sensors for the Schmutter bridge

sensor types and main technical properties are documented in FIRUS et al. (2018a).

As it can be seen in Fig. 7.7, a very dense array of accelerometers was used. It consists of 44 sensors disposed along four axes of girder webs (A-A to D-D). Each axis contains eleven equally spaced measurement points. On the one side, the dense sensor array is expected to allow a realistic recording of both the global and the local behavior of the coupled structures. On the other side, using the comprehensive array of accelerometers allows an experimental modal analysis in a single set-up, i.e. no rearrangement of the sensors was needed during the measurement campaign. This fact is of great importance, since the measurement was performed under operational conditions and there were only very few and short time slots which allowed undisturbed measurements of ambient vibrations or controlled forced vibrations induced by an impulse hammer. Moreover, a measurement in operational conditions on a route with a very high volume of traffic also considerably limits the possibility of repeated interventions on the measurement set-up during the measurement campaign, which would be required by a multi-setup modal analysis approach.

Fig. 7.8 exemplarily shows the positions of the acceleration sensors inside the box girder for section 6-6. It can be seen that the sensors are mounted on the upper flanges, in the immediate vicinity of the webs. The main idea was to place the sensors at a stiff location of the cross-section, in order to avoid the recording of unwanted local vibrations of individual structural components, e.g. local vibrations of the lower flanges.

7.3.2 Displacement measurements

The novel MFI method developed within this work is based on measured acceleration and displacement data. In this respect, the comprehensive numerical study performed

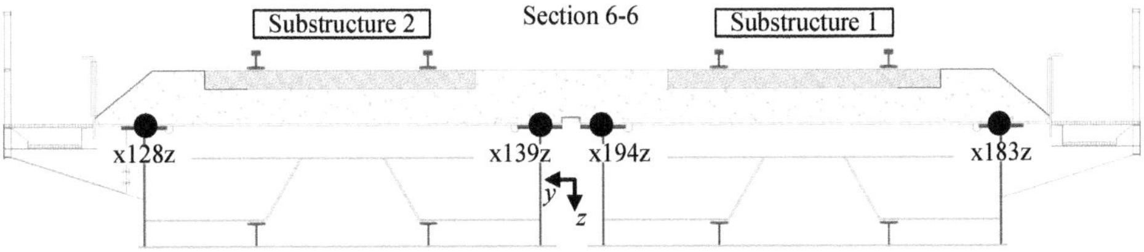

Figure 7.8 Positions of the acceleration sensors inside of the box girder

in Chap. 6 revealed that a combination of acceleration and displacement measurements leads to the best accuracy of the identified forces (cf. Section 6.7.5). It is true that the use of both acceleration and displacement sensors might be criticized at first glance, since it is well known that the displacement measurements employing classical Linear Variable Displacement Transducers (LVDTs) are usually quite cumbersome or even impossible due to the lack of fixed reference points. However, remarkable advances in the field of non-contact displacement measurement methods have been made in the recent decades. Their functioning principles range from rather simple laser vibrometry to more modern technologies like profile scanning or microwave interferometry.

The displacement measurements of the Schmutter bridge had to rely exclusively on non-contact measurement systems. The reason for this that all relevant measurement positions were practically situated above the river, preventing thus the installation of sensors with a fixed reference point. In this sense, all three previously mentioned non-contact measurement techniques were included in the measurement concept. Fig. 7.9a illustrates the set-up of the displacement measurement systems and Fig. 7.9b shows a photo capture of the installed equipment. A total of four laser vibrometers were used, whereas a corner reflector had to be installed for each of them, in order to allow a clear reflection of the laser beams. In addition, a microwave interferometer and a profile scanner were employed. The basics of the profile scanning and some application examples can be found in SCHILL (2018), SCHILL et al. (2019a) and SCHILL et al. (2019b). Insights into the laser vibrometry and some applications examples are are given in NASSIF et al. (2005), MALEKJAFARIAN et al. (2018) and KUTTNER et al. (2019). The microwave interferometry (MI) measurement principle will be discussed in more detail in the following, since it represents a rather new technology with respect to applications in railway bridge dynamics. Moreover, the author of this thesis contributed significantly to a research project dealing with the verification and adaptation of the MI technology for railway-specific applications.

7.3.2.1 Microwave interferometry

Although several researchers have dealt with microwave interferometry measurements of civil engineering structures in the last 15 years and have demonstrated their applicability

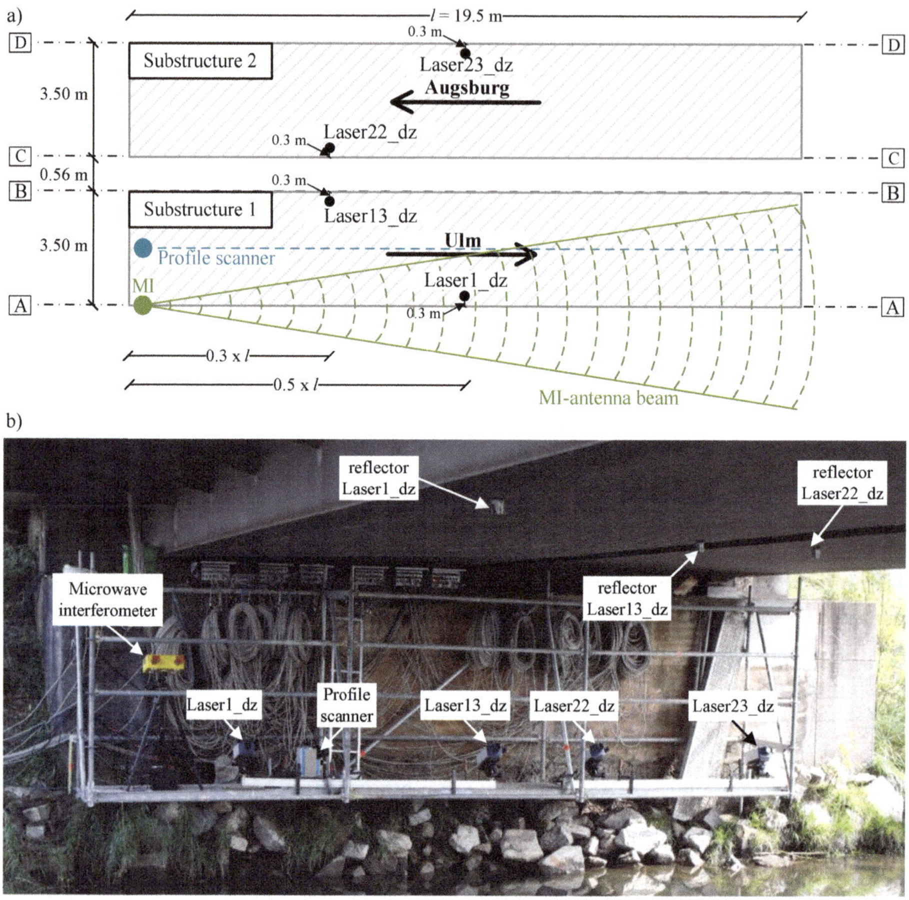

Figure 7.9 Set-up of the displacement sensors for the Schmutter bridge: a) Sketch of the measurement set-up; b) Photo capture of the installed measurement equipment

in individual cases (PIERACCINI et al., 2007; PIERACCINI et al., 2008; GENTILE et al., 2009; PIERACCINI, 2013; NEGULESCU et al., 2013; GENTILE, 2015; ARTESE et al., 2020; NICO et al., 2020), there is a lack of information on the general limitations and boundary conditions for bridge measurements. Therefore, prior to the implementation of the MI as a regular measurement method, it had to be subjected to a comprehensive validation and adaptation process focusing on applications in the railway bridge dynamics. The author of this thesis was significantly involved in this process, which was performed within the cooperation project *STRAMIK - Strukturanalyse mit Mikrowelleninterferometrie*[6] between the Technical University of Darmstadt and the DB Netz AG. It included several MI mea-

[6] English: *Structural analysis using microwave interferometry.*

surements of operational railway bridges of the DB AG and systematic comparisons with results of synchronous conventional measurements. As a result, limitations and boundary conditions for different structural types and materials as well as various recommendations for different measurement situations were formulated. A detailed reproduction of the validation process and its results would go beyond the scope of this thesis. At this point, the author refers to the works of FIRUS et al. (2016), FIRUS et al. (2017) and the research report of BECKER et al. (2017), which summarize the main findings of the validation process. However, considering its novel character for applications in civil engineering and in order to facilitate the understanding of the measurement results presented in the following sections, the basics of the measurement principle will be briefly provided in the following.

The measurement principle of MI allows the non-contact synchronous measurement of structural displacements for a set of distinct points[7] disposed along a structure (GENTILE et al., 2009). The structural behavior can be recorded with a sampling frequency of up to 4 000 Hz, whereas the accuracy specified by the manufacturers lies in the sub-millimeter range. The high sampling frequency also allows the acquisition of dynamic structural responses, which can be used for the straightforward identification of the dynamic structural parameters. As already mentioned in the case of Schmutter bridge, the non-contact character of the MI represents an essential aspect, especially when investigating operational railway bridges with difficult accessibility to the bridge deck. The use of MI does not require any track closure, as it would be the case with conventional sensors. This aspect has a significant implication on the availability of the infrastructure. Moreover, also the measurement costs can be reduced, since a significantly lower expenditure of human labor is required in case of MI measurements.

The MI radar device, denoted also as microwave interferometer, is installed on a tripod and sends electromagnetic waves towards the measurement object (e.g. bridge deck). The waves are reflected as they reach the surface of the target object, whereas any geometrical discontinuities of the structure situated within the antenna beam and exceeding the wave length of the electromagnetic waves ($\lambda = 17.4$ mm) ensure the backscattering of certain signal components. In this respect, the MI functioning principle is based on the acquisition of phase and amplitude of the backscattered signals (BERNARDINI et al., 2007).

The displacement Δr of the target object occurring within a sampling interval is expressed according to Eq. 7.2 as a function of the phase change ϕ and the wavelength λ of the emitted microwaves:

$$\Delta r = -\frac{\lambda}{4 \cdot \pi} \cdot \phi \tag{7.2}$$

The interferometric phase ϕ can be basically determined according to Eq. 7.3:

[7] In fact, as it will be shown later, the MI does not measure the displacements at discrete structural points, but for different resolution cells disposed along the structure. However, in order to facilitate the understanding of the results, the term *measurement points*, which is more common for civil engineers, is used in the following in relation to the integral displacement of a resolution cell.

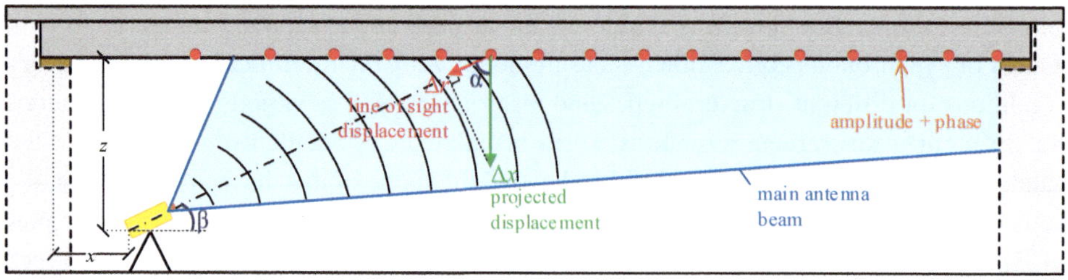

Figure 7.10 Typical MI measurement situation for bridges (the red dots represent the centres of the resolution cells)

$$\phi = \phi_{\text{disp}} + \phi_{\text{atm}} + \phi_{\text{noise}} \pm 2\pi n \tag{7.3}$$

It is obvious that the phase ϕ is influenced by several factors: ϕ_{disp} represents the main phase component and corresponds to the phase difference occurring due to the movement of the reflecting objects along the line of sight of the MI device. Furthermore, ϕ_{atm} is the phase change generated by different atmospheric effects, whereas ϕ_{noise} is used to depict the noise component of the measurement. However, typical MI applications in structural dynamics are characterized by rather short time measurements with a duration of up to few minutes. The atmospheric changes are thus expected to be insignificant over the measurement duration. Therefore, the term ϕ_{atm} in Eq. 7.3 can be usually neglected. On the contrary, the use of MI for long term monitoring requires the consideration of the atmospheric changes (LIU et al., 2015). The term $2\pi n$ describes the wrapping of the interferometric phase between $-\pi$ and $+\pi$, which corresponds to the range $-\lambda/4$ to $\lambda/4$ due to the double travelled distance. Here, n represents the phase ambiguity, since only a certain value of the phase difference can be retrieved and not the integer number of phase cycles. The phase ambiguity can be solved under different assumptions. For example, with respect to the displacement experienced by a bridge deck within a sampling interval, it can be surely assumed that it is lower than a quarter of the wavelength, i.e. $n = 1$ ($\pm\lambda/4 = \pm 4.4$ mm, $\lambda = 17.4$ mm). Assuming a sampling frequency of 200 Hz, the MI would still allow to record a displacement of up to 880 mm within one second. This value obviously lies outside the plausible range for regular bridge structures.

Fig. 7.10 shows a typical measurement set-up for measurements of bridge deck displacements. The MI devices in RAR mode (Real Aperture Radar, suitable for dynamic signal acquisition) can only record one-dimensional displacements Δr, i.e. along the line of sight of the device. These have to be subsequently transformed to the assumed direction of movement Δx, as shown in Fig. 7.10. The corresponding projection relationship is:

$$\Delta x = \frac{\Delta r}{\cos(\alpha)} \tag{7.4}$$

a)  b) 

Figure 7.11 Example of structures with good (a) and poor (b) wave reflectivity characteristics

Therefore, a good understanding of the mechanics of the structure is required, in order to allow realistic assumptions with respect to the real the movement behavior, which has to be taken into account when setting up the MI device.

A further important aspect to be considered when setting up the MI device is that the reflectivity of the targeted objects has a considerable influence on the amplitude of the backscattered signal. In general, any discontinuities of the structure, whose dimensions exceed the wavelength λ, are expected to ensure a good reflectivity. Considering the applications of the microwave interferometry on railway bridges, it could be observed that the steel structures are usually expected to provide a better reflectivity than reinforced or prestressed concrete structures (FIRUS et al., 2016). The reason for this is a typical difference in the constructive appearance of the bottom surface of the bridge deck, which is hit by the antenna beam. While the visible cross-beams and other constructive elements at the bottom side of a steel bridge deck ensure a "discontinous" surface, the concrete bridges usually have a rather plain surface (Fig. 7.11). The latter eventually leads to a dominant forward reflection of the microwaves and thus to a loss of intensity of the backscattered signal (BECKER et al., 2017). Therefore, the quality of the reflected signals has to be verified at the beginning of the measurement. This can be done by analyizing the signal-to-noise ratio (SNR) and the complex representation of signal (FIRUS et al., 2016).

Due to the high spatial resolution and the high accuracy, the microwave interferometers can simultaneously measure several points along the structure under the same time stamp. This represents a decisive advantage over many conventional measuring systems. The stepped frequency continuous wave (SFCW, TAYLOR, 2001) or the frequency modulated continuous wave (FMCW, IGLESIAS et al., 2014) principles are employed, in order to ensure the spatial (range) resolution δ_r along the line of sight. This can be expressed according to Eq. 7.5 as a function of the bandwidth B of the transmitted wave and of the speed of light c:

$$\delta_r = \frac{c}{2 \cdot B} \tag{7.5}$$

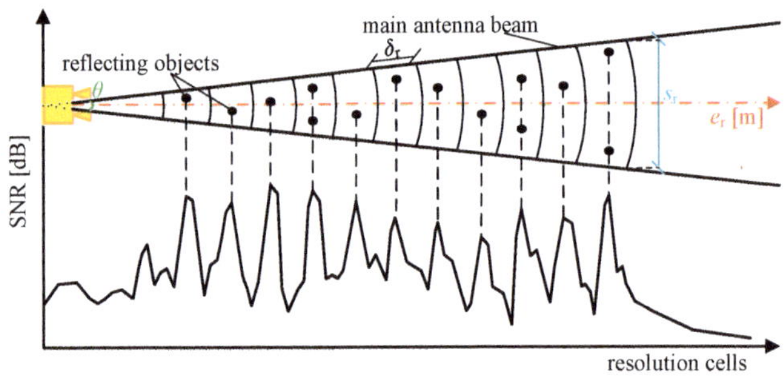

Figure 7.12 Range resolution cells of a microwave interferometer and corresponding SNR values

In Germany, the Federal Network Agency allows a maximum bandwidth of 200 MHz, leading to a minimum range resolution δ_r of 0.75 m (Fig. 7.12).

One of the MI drawbacks with respect to applications in bridge dynamics is the fact that the microwave interferometers deliver a weighted average of the echoes for every resolution cell. In other words, if a resolution cell contains more reflecting targets with independent movements, their corresponding displacements cannot be distinguished (Fig. 7.12). However, this effect can be usually reduced by choosing a proper position for the microwave interferometer, so that each resolution cell contains only one dominant reflector (RÖDELSPERGER et al., 2010). The width s_r of the antenna beam can be used when defining the positioning of the device. It can be roughly estimated according to Eq. 7.6, as a function of the distance e_r and the horizontal opening angle θ of the antenna beam (specified by the manufacturers, see also Fig. 7.12):

$$s_r = 2 \cdot e_r \cdot \tan\left(\frac{\theta}{2}\right) \tag{7.6}$$

However, it is to be mentioned that in reality the antenna beam does not present an exact geometrical delimitation. Besides the main lobe of the radar beam, several side lobes might occur. They can still produce significant echoes in case of strong reflectors (IDS-INGINERIA DEL SISTEMI S.P.A., 2015).

A more detailed overview of the MI is given by BERNARDINI et al. (2007), RÖDELSPERGER et al. (2010) and FIRUS et al. (2016), while several application examples are presented in SCHNEIDER et al. (2015), LIU et al. (2015), NICO et al. (2020), ARTESE et al. (2020) and HUANG et al. (2020).

7.3.3 Axle detection

The MFI method proposed in this work assumes that the location of the forces passing the structure, defined by the travel speed and the axle spacing, is known at each time instant.

Figure 7.13 Photelectric sensors used for the detection of the axle distances and travel speed of the vehicles passing the Schmutter bridge

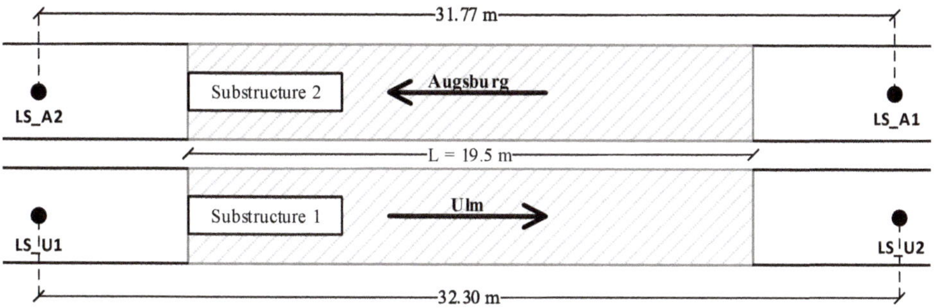

Figure 7.14 Installation positions of the photoelectric sensors

Section 6.7.7 showed that the speed measurement represents crucial aspect, which directly affects the identification accuracy of the moving forces. In this sense, every application of the proposed MFI method with real measurement data has to rely on a sensor based detection of the axle spacing and the vehicle velocities. For the experimental investigation of the Schmutter bridge, a classical solution based on two photoelectric sensors was chosen (Fig. 7.13). Their corresponding set-up is presented in Fig. 7.14. It allows not only the identification of the axle distances, but also of the travel speed and implicitly of the total vehicle length. In order to increase the reliability of the measurements, a complementary measurement of the vehicle velocity was performed using a radar device mounted on the vehicle.

An alternative method for the axle detection of railway vehicles implies the installation of strain gauges on the rail, between two sleepers (ŽNIDARIČ et al., 2016; VOSPERNIG et al., 2016). However, both the solution based on photoelectric sensors and the strain gauge method require a direct access to the track, obviously implying a temporary track closure. This represents an important drawback for both approaches. In this sense, the axle detection based on signals acquired on the structure itself is addressed in some recent research works (LILJENCRANTZ et al., 2007; ŽNIDARIČ et al., 2016). They seek to retrieve the axle information from strains recorded on the main structure. However, only limited success was reported in both cases, since the structural strains at the investigated locations proved to be not sensitive enough to separate the two axles of a single bogie.

For the sake of completeness, it should be mentioned that extensive discussions on axle detection systems for road vehicles, including approaches requiring no sensor installation at pavement level, are given in OBRIEN et al. (2001), OBRIEN et al. (2008), CAPRANI et al. (2013), OJIO et al. (2015), YU et al. (2016) and LYDON et al. (2016).

7.4 Measurement results

This section is firstly concerned with the operational modal analysis performed on the empty structure for identifying the modal parameters of the Schmutter bridge. Subsequently, the measurement results obtained during passages of an ICE 4 vehicle are discussed.

7.4.1 Modal analysis

The Reference-Based Combined Deterministic-Stochastic Subspace Identification (denoted as CSI/ref) procedure – implemented in the software MACEC 3.3 (2014) – was applied, in order to identify the natural frequencies, the mode shapes and the corresponding damping ratios of the empty structure. The description of the mathematical background of the CSI/ref method, even in a brief manner, would go far beyond the scope of this thesis. At this point, the reader is referred to the works of REYNDERS et al. (2008) and PEETERS et al. (1999), which delivered essential contributions towards the application of this method in civil engineering.

In the present case, the system identification by means of CSI/ref implies the use of acceleration responses collected at the 44 measurement points (cf. Fig. 7.7), generated by repeated impulse hammer hits at measurement point x181z over a time interval of about 700 seconds. Fig. 7.15a illustrates the force time histories recorded by the impulse hammer and Fig. 7.15b shows an example of an acceleration time history measured at the point x183z. Both signal types were acquired with a sampling rate of 1200 Hz. The force signals recorded by the impulse hammer represent the deterministic part of the excitation, which is superimposed to a stochastically modeled component, representing the unknown ambient excitation.

Fig. 7.16 shows the first six identified natural frequencies and damping ratios as well as the corresponding mode shapes in vertical direction. The lowest natural frequency of the empty structure of 5.75 Hz occurs – as expected – for the first vertical bending mode of the coupled structure. The second mode arises at 7.59 Hz and describes the first torsional movement of the coupled structure. The third mode (12.06 Hz) represents the first torsional movement (in opposite directions) of the individual structures. Mode 4, occurring at 16.65 Hz, represents in fact a combined bending-torsional mode, which obviously includes the second mode bending component of the coupled structure. The fifth mode shape (23.62

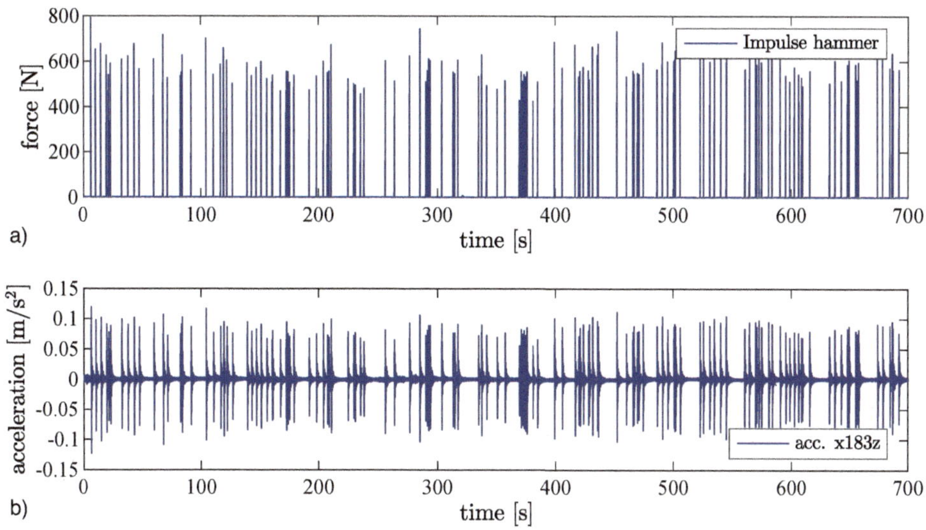

Figure 7.15 Examples of force (a) and acceleration (b) signals used for identifying the modal parameters of the Schmutter bridge

Hz) describes the second torsional movement (in opposite directions) of the individual structures, while mode 6 (26.37 Hz) depicts the individual torsion of the two substructures in the same direction.

The coupling of the mode shapes for the two substructures, illustrated in Fig. 7.16, clearly indicates that the longitudinal ballast joint ensures a strong interaction between the two substructures, even though they are statically uncoupled. However, this statement is for now only valid for the rather low acceleration levels generated by the impulse hammer (cf. Fig. 7.15b). The reason for this is that higher amplitudes in lower modes, as they typically occur during the train passages, are expected to activate additional mechanisms of energy dissipation. These could be, for example, the friction in the supports, the internal friction in the ballast layer (Rebelo et al., 2008; Rauert et al., 2010; Stollwitzer et al., 2020) or damping mechanisms activated by the vertical and horizontal relative displacements between the sleepers and the main bridge structure (Stollwitzer et al., 2020). Hence, an eventual amplitude dependency of the natural frequencies and damping ratios cannot not be accounted for by the modal analysis procedure based on impulse hammer excitation. For example, in the present case, the analysis of the damping ratios using the logarithmic decrement approach (Petersen et al., 2017) from a set of typical acceleration free-decay signals (after train crossing) yields a modal damping of 2.74 %. This value is significantly higher than the results provided in Fig. 7.16 and will be used as a basis for all the calculations performed in the further course of the this work.

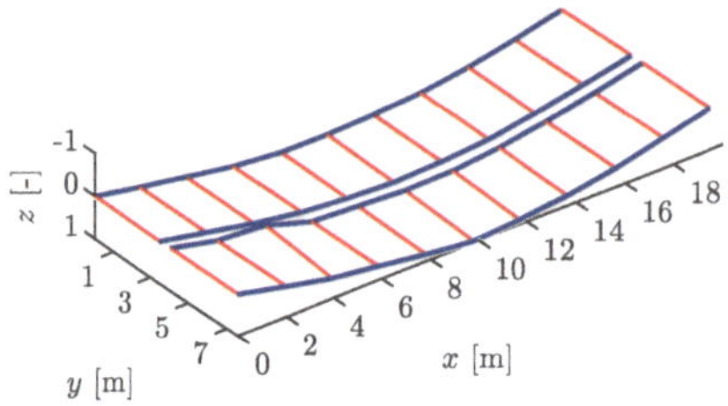

a) $\bar{f}_1 = 5.75$ Hz, $\bar{\xi}_1 = 1.35$ %

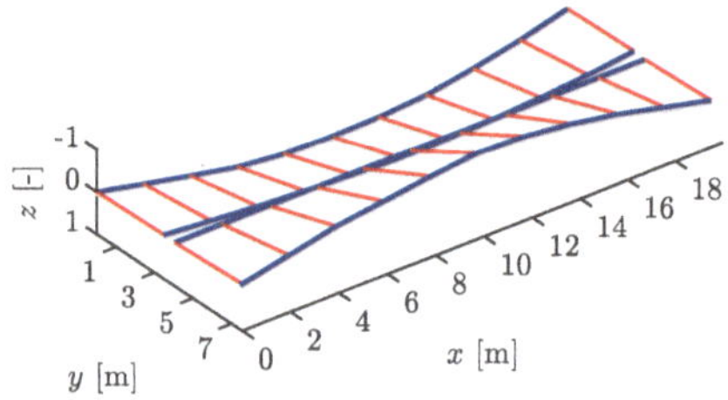

b) $\bar{f}_2 = 7.61$ Hz, $\bar{\xi}_1 = 1.02$ %

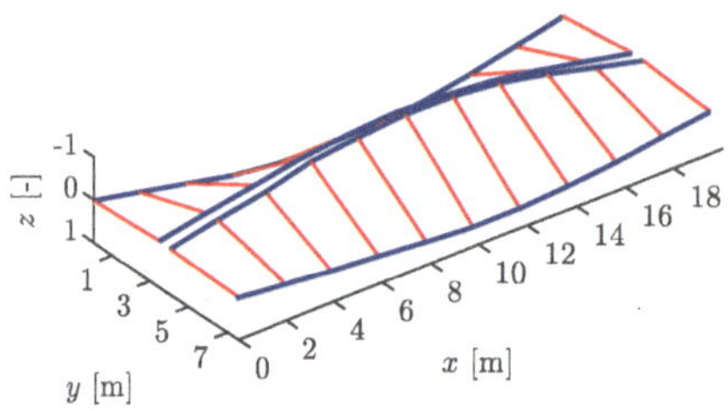

c) $\bar{f}_3 = 12.06$ Hz, $\bar{\xi}_3 = 0.46$ %

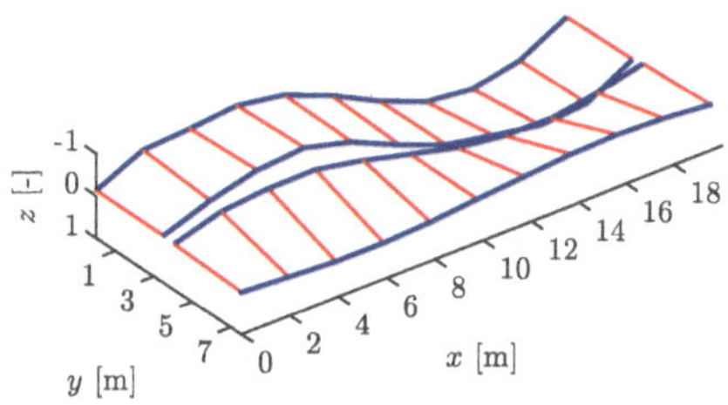

d) $\bar{f}_4 = 16.65$ Hz, $\bar{\xi}_4 = 2.40$ %

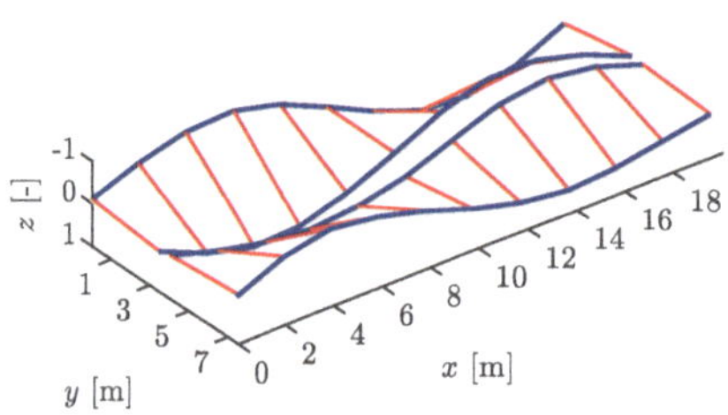

e) $\bar{f}_5 = 23.62$ Hz, $\bar{\xi}_5 = 0.99$ %

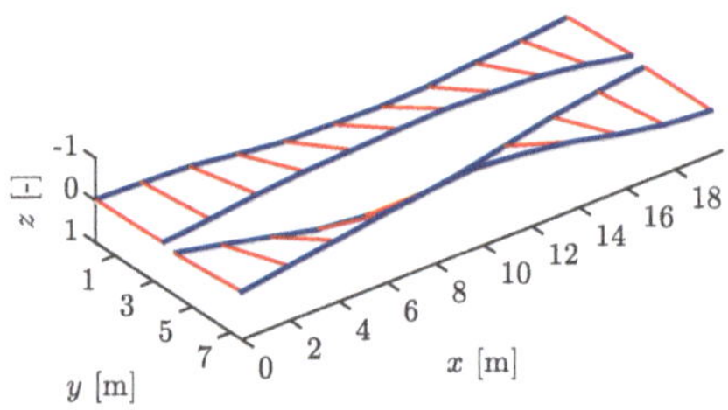

f) $\bar{f}_6 = 26.37$ Hz, $\bar{\xi}_6 = 1.88$ %

Figure 7.16 Experimentally identified mode shapes, natural frequencies and damping ratios of the Schmutter Bridge from tests on the empty structure. The modes are normalized with respect to the maximum absolute value.

7.4.2 Train passages

A total of 52 crossings of the ICE 4 over the Schmutter bridge were recorded, 26 passages in each direction. The passages were performed at predefined train speeds in a range between 10 and 200 km/h. The vehicle was loaded at 80 % of its full capacity using steel weight blocks and sandbags (Fig. 7.17), whereas a weight of approximately 0.8 kN was assigned to each seat. This represents a conservative assumption of the weight of a passenger. After installing the steel weight blocks and the sandbags in all carriages, each axle load of the train

Figure 7.17 Vehicle loading using steel weight blocks

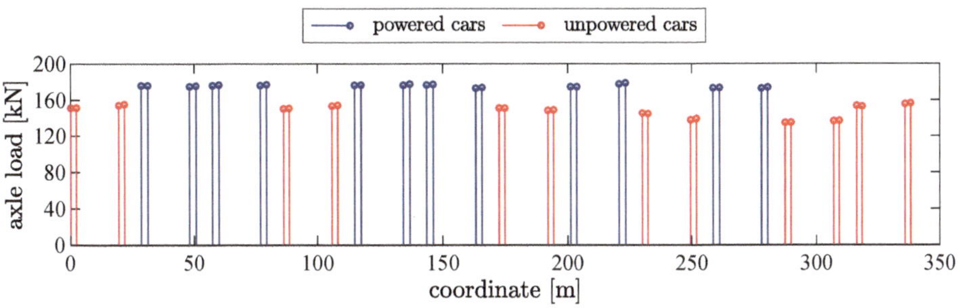

Figure 7.18 Graphical representation of the measured static axle loads for the test configuration of the ICE 4 - train crossings in direction Ulm

was determined by measurements, using a special weighting system for railway vehicles. The measured static loads for all 48 axles are given in Tab. 7.1, whereas the first axle ($i = 1$) is considered as the leading axle for the passages in direction Ulm (over the substructure 1, cf. Fig. 7.7). The resulting total weight of the train is about 7750 kN, corresponding to a total vehicle mass of about 790 t. For the passages in direction Augsburg (over substructure 2), the vehicle drove in reversed configuration, i.e. the leading axle was the 48-th axle in Tab. 7.1. The axle loads given in Tab. 7.1 are plotted against their coordinates with respect to the first axle in Fig. 7.18. In addition, the numbering of the carriages is given in Fig. 7.19 together with the general carriage configuration of the ICE 4. It is obvious that the vehicle consists of six powered cars and six unpowered cars[8]. Carriage 1 was leading for passages in direction Ulm, while carriage 12 lead the passages in direction Augsburg. Fig. 7.18 and Fig. 7.19 also emphasize that – as expected – the axle loads of the powered cars are generally higher than those of the unpowered cars.

[8] The powered and unpowered carriages are also widely denoted in the literature as *traction* and *trailing cars*, respectively.

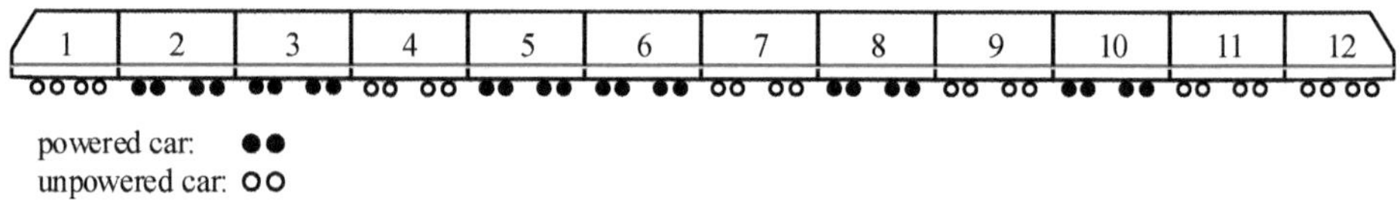

Figure 7.19 Vehicle configuration and carriage numbering of the ICE 4 for the passages in direction Ulm (substructure 1)

As it was already mentioned above, the train crossings over the Schmutter bridge were performed at speeds between 10 km/h and 200 km/h. In fact, only two passages per direction were performed at a low speed of about 10 km/h. This case is considered to generate a rather quasi-static loading and can be used for the static calibration of the numerical model. All other passages were performed with predefined travel speeds in a range between 80 km/h and 200 km/h. They should allow the experimental determination of the resonance curve under action of the ICE 4.

Tab. 7.2 gives the overview of all the recorded passages. It includes the passage number, the corresponding direction, the predefined speed and the measured speeds recorded by the photoelectric sensors (cf. Fig. 7.13) and by a ground radar installed on carriage 8 of the moving vehicle. It can be seen that the measured speeds slightly differ from the predefined values. This effect occurs most likely due to the tolerances of the train control unit with respect to the travel speed. Moreover, it can be seen that even the results of

i	d_i [m]	$F_{\text{st},i}$ [kN]	i	d_i [m]	$F_{\text{st},i}$ [kN]	i	d_i [m]	$F_{\text{st},i}$ [kN]
1	0.00	151.12	17	114.85	176.12	33	230.00	145.19
2	2.30	151.19	18	117.45	176.38	34	232.30	144.52
3	19.50	153.80	19	134.35	176.12	35	249.50	137.55
4	21.80	154.93	20	136.95	177.38	36	251.80	138.63
5	28.60	175.63	21	143.60	176.35	37	258.60	172.92
6	31.20	175.70	22	146.20	176.92	38	261.20	173.34
7	48.10	174.62	23	163.10	172.72	39	278.10	172.71
8	50.70	175.27	24	165.70	173.52	40	280.70	173.85
9	57.35	175.62	25	172.50	150.81	41	287.50	134.35
10	59.95	176.40	26	174.80	150.73	42	289.80	134.50
11	76.85	175.75	27	192.00	148.25	43	307.00	136.08
12	79.45	176.61	28	194.30	148.67	44	309.30	136.46
13	86.25	150.04	29	201.10	174.26	45	316.25	153.44
14	88.55	150.52	30	203.70	174.29	46	318.55	152.63
15	105.75	153.15	31	220.60	177.40	47	335.75	155.41
16	108.05	153.84	32	223.20	178.53	48	338.05	156.22

Table 7.1 Static axle loads measured for the test configuration of the ICE 4

No.	Direction	Speed [km/h] predefined	photoelectric sensor	radar
1	Ulm	140.00	138.71	140.48
2	Augsburg	140.00	140.16	error
3	Ulm	10.00	9.87	9.98
4	Augsburg	10.00	9.75	9.85
5	Ulm	10.00	9.96	10.08
6	Augsburg	10.00	9.75	9.86
7	Ulm	185.00	182.93	185.23
8	Augsburg	185.00	183.64	186.02
9	Ulm	190.00	189.18	191.63
10	Augsburg	190.00	189.80	192.35
11	Ulm	195.00	194.65	197.20
12	Augsburg	195.00	195.00	197.60
13	Ulm	200.00	199.32	201.84
14	Augsburg	200.00	199.81	202.49
15	Ulm	192.00	191.19	193.61
16	Augsburg	192.00	191.35	193.86
17	Ulm	180.00	178.66	180.94
18	Augsburg	180.00	179.63	182.00
19	Ulm	175.00	173.47	175.63
20	Augsburg	175.00	174.11	176.37
21	Ulm	170.00	168.21	170.36
22	Augsburg	170.00	168.92	171.09
23	Ulm	165.00	163.04	165.02
24	Augsburg	165.00	163.58	165.70
25	Ulm	172.00	169.88	172.07
26	Augsburg	172.00	170.95	173.16
27	Ulm	173.00	170.88	173.06
28	Augsburg	173.00	171.37	173.53
29	Ulm	174.00	172.14	174.30
30	Augsburg	174.00	172.44	174.67
31	Ulm	82.00	81.24	82.29
32	Augsburg	173.00	171.97	174.17
33	Ulm	173.00	170.43	172.52
34	Augsburg	173.00	171.83	174.00
35	Ulm	172.00	169.67	171.75
36	Augsburg	172.00	170.90	173.08
37	Ulm	171.00	169.05	171.13
38	Augsburg	171.00	169.70	171.85

table continued on the next page

No.	Direction	Speed [km/h] predefined	photoelectric sensor	radar
39	Ulm	167.00	164.71	166.73
40	Augsburg	167.00	165.59	167.71
41	Ulm	120.00	118.11	119.66
42	Augsburg	120.00	118.99	120.46
43	Ulm	160.00	157.84	159.82
44	Augsburg	160.00	159.02	161.10
45	Ulm	100.00	98.77	100.03
46	Augsburg	100.00	100.39	101.57
47	Ulm	150.00	147.77	149.62
48	Augsburg	150.00	148.97	150.82
49	Ulm	130.00	128.26	129.88
50	Augsburg	130.00	129.37	130.91
51	Ulm	128.00	126.28	127.89
52	Augsburg	128.00	127.38	128.90

Table 7.2 Overview of the ICE 4 passages, listed in chronological order

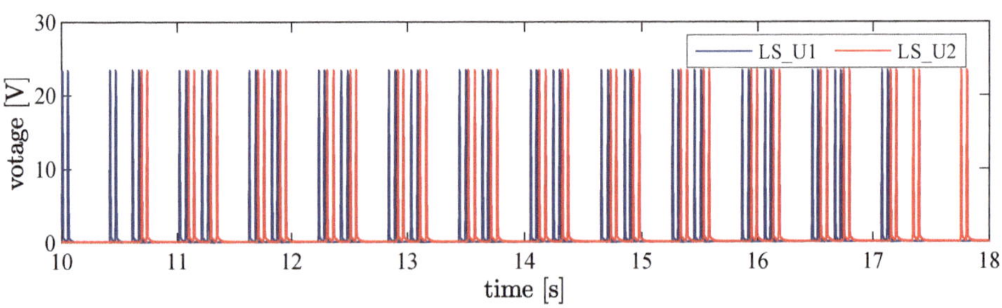

Figure 7.20 Signals of the photoelectric sensors LS_U1 and LS_U2 (cf. Fig. 7.14) for the passage 27 towards Ulm

the two measurement systems (photoelectric sensor and radar) present certain differences. In this respect, the results of the photoelectric sensors will be adopted for all analyses performed in the further course of this thesis. The reason for this is that the photoelectric sensor measurements (exemplarily shown in Fig. 7.20 for passage no. 27) allow – under consideration of known axle distances – a double verification of the speeds determined from the measurement signals. On the one side, the speed can be determined based on the time needed by an axle to cross the known distance between the two photoelectric sensors (cf. Fig. 7.13). On the other side, the time interval between the passing of two subsequent wheels through the same sensor can be used together with the corresponding known axle distance, in order to determine the vehicle velocity. Both approaches based on the photoelectric sensors delivered almost identical results.

7.4.3 Resonance curves of accelerations

As already mentioned in the previous sections, one of the main aims of the investigation performed in cooperation with DB Netz AG was the experimental determination of the resonance curves[9] under action of the ICE 4. Neglecting the slow passages (no. 3 to 6 in Tab. 7.2), a number of 24 passages with speeds between 80 km/h and 200 km/h were performed in each direction.

In order to experimentally determine the resonance curve, it is necessary to have a prior estimation of the resonance speeds. According to RIL 804, the i-th resonance speed $v_{\text{res},i,j,k}$ of the j-th mode can be determined for a given train k as:

$$v_{\text{res},i,j,k} = f_j \cdot \frac{L_{\text{ob},k}}{i} \tag{7.7}$$

where f_j is j-th natural frequency, $L_{\text{ob},k}$ is the length over buffers[10] of the k-th train, while $i = 1,\ 2,\ 3$ or 4. The term:

$$\frac{L_{\text{ob},k}}{i} \tag{7.8}$$

is also denoted as the principal wave length of the excitation frequency. It can be seen that the expression given in Eq. 7.8 is equal to Eq. 7.1, which expresses the excitation wavelength as the ratio between the natural frequency and the vehicle speed.

Substituting the natural frequency $\bar{f}_1 = 5.75$ Hz determined within the operational modal analysis (Section 7.4.1) and the length over buffer of 28.75 m in Eq. 7.7 results – for $i = 3$, corresponding to a wave length of excitation $\lambda \approx 9.6$ m – in an expected resonance speed of about 198 km/h. This value was used to establish the order of the speeds given in Tab. 7.2. After a system calibration run at 140 km/h and two quasi-static passages in each direction, it was aimed to approach the resonance peak at 198 km/h from below, starting with a speed of 185 km/h (passages no. 7 to no. 16). However, as it can be seen in the corresponding resonance curves of the sensors placed at midspan (along the axis 6-6, cf. Fig. 7.7), illustrated in Fig. 7.21, no resonance peak could be identified around the expected speed[11]. Therefore, further passages were performed with speeds below 185 km/h, seeking to approach the – presumably lower – resonance speed from above. This was obviously found at about 171 km/h, delivering a maximal acceleration of about 4.1 m/s^2. In this respect, it is to be mentioned that no influence on the ballast stability could be detected,

[9] The resonance curve consists of the maximal absolute acceleration values at a certain structural point, which are plotted against the vehicle velocity corresponding to each maximum value.

[10] The term *length over buffers* expresses the overall length of a carriage equipped with buffers. It is measured from the outermost surface of the buffer plate to the outermost surface of the buffer plate at the other end of the carriage, whereas the buffers are not under pressure.

[11] In order to facilitate the understanding, the results of the measurement points situated on the substructure crossed by the vehicle are represented in each case using solid lines, while the results of the other superstructure are represented using dashed lines.

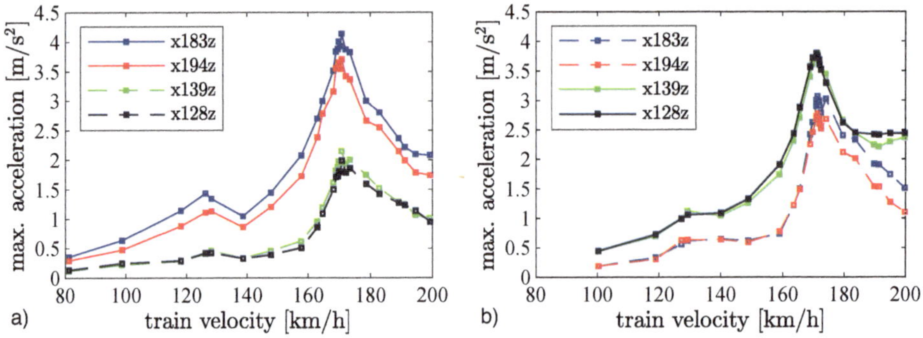

Figure 7.21 Resonance curves of the Schmutter bridge for acceleration channels situated on the axis 6-6 (cf. Fig. 7.7) and low pass filtered signals with a cut-off frequency of 10 Hz: a) train passages in direction Ulm (substructure 1); b) train passages in direction Augsburg (substructure 2)

Figure 7.22 Painted line pattern for the visual inspection of the ballast stability

even though the maximum acceleration level of 3.5 m/s^2 (RIL 804, DIN EN 1991-2) was slightly exceeded. This was checked after each passage by means of visual inspection of a painted line pattern applied on the ballast and sleepers (Fig. 7.22).

Considering the experimentally determined resonance speed of 171 km/h, the corresponding natural frequency for the "loaded" bridge $f_1 = 4.95$ Hz can be determined using Eq. 7.7. This indicates a considerable drop of the first natural frequency during the train passage in comparison to the natural frequency of the empty structure. Hence, the modal parameters identified by the CSI/ref (Section 7.4.1) can not be used for the calibration of the numerical model, which should deliver a realistic representation of the real structural behavior during train passages.

The effect of a lower frequency during the passage was also documented by KIM et al. (2003), LI et al. (2003), REBELO et al. (2008), LIU et al. (2009b), RAUERT et al. (2010), RIGUEIRO et al. (2010), LIU et al. (2014c), VOSPERNIG et al. (2020) and KAROUMI et al. (2021). Basically, there are two main factors which are presumed to contribute to the decrease of the natural frequency. The first is related to the increasing of the total mass of the system. It occurs basically due to the stiff coupling of the unsprung mass of the vehicle (i.e. mass of the wheel sets) to the bridge mass (LI et al., 2003; KIM et al., 2003; LIU et al., 2009b; RIGUEIRO et al., 2010; LIU et al., 2014c). In this regard, it can be stated that the amount of frequency drop is strongly dependent on the ratio between the unsprung mass and the mass of the bridge (LI et al., 2003). On the contrary, the effects of the sprung masses – i.e. masses of the bogies and the car body – on the frequency change of the bridge are usually expected to be rather insignificant. The reason for this is that the dampers and the springs used to connect them with each other and with the wheel-sets shall ensure a dynamic uncoupling, which eventually leads low inertial forces of the car body and the bogies (LI et al., 2003). With respect to the handling of this effect in the engineering practice, it is to be mentioned that the guideline for the dynamic analysis of railway bridges of the Austrian Railways (ÖBB INFRASTRUKTUR, 2011) requires the consideration of an additional mass of 2000 kg/m, disposed along the track axis[12]. This approximately corresponds to the whole mass of a vehicle and contradicts the previous statement, that the inertial masses of the car body are expected to have low values due to the action of the primary and secondary spring and damper levels.

The second factor assumed to be influencing the decrease of the natural frequency during the passage is the nonlinear behavior of the bridge deck. This is characterized by a stiffness decrease when the bridge is subjected to high amplitude levels, i.e. during train passage. The stiffness increases again after the train leaves the bridge and the vibration amplitudes diminish. Since the deformations of the steel structure are expected to be sufficiently low to remain in the elastic range, the only remaining possible source of the nonlinear characteristic is the ballast bed. In this sense, it is presumed that the interlocking forces within the ballast bed cause relative movement between the ballast particles during the train passages, implying thus a drop in the stiffness values (REBELO et al., 2008; RIGUEIRO et al., 2010; RAUERT et al., 2010). This conclusion is also supported by analyzing the evolution of the frequency over time for the acceleration time history recorded at MP x183z during the decay phase of the passage no. 27. The time history of the decay phase is shown in Fig. 7.23 and the corresponding spectrogram[13] for the frequency range between 0 and 20 Hz is given in Fig. 7.24. The increasing of the frequency over time from ca. 5.07 Hz to about 5.7 Hz can be clearly observed.

Furthermore, it is to be mentioned that the signals used for the determination of the resonance curve in Fig. 7.21 were low pass filtered using a Butterworth filter (5th order) with a cut-off frequency of 10 Hz. The reason for the rather low cut-off frequency is that it was aimed – as far as possible – to isolate the resonance phenomena induced solely by the excitation mechanism of the ICE 4. Typical acceleration time histories recorded

---

[12]The requirement is generally valid for bridges with a span length greater than 7 m. For bridges with a span length between 7 and 20 m, an additional computation without additional mass is required. In this way, the two limit situations are investigated, whereas the real behavior is expected to lie in between.

[13]The spectrogram represents the short-time Fourier transform of the original acceleration time history. It implies the division of the acceleration time history into shorter segments (in the present case 500 samples with overlapping interval of 450 samples) and an individual Fourier transformation for each shorter segment.

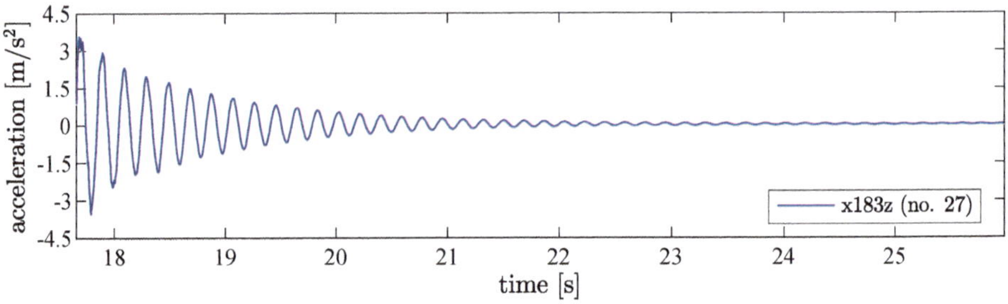

Figure 7.23 Acceleration time history of the decay phase recorded at the measurement point x183z during the passage no. 27 towards Ulm

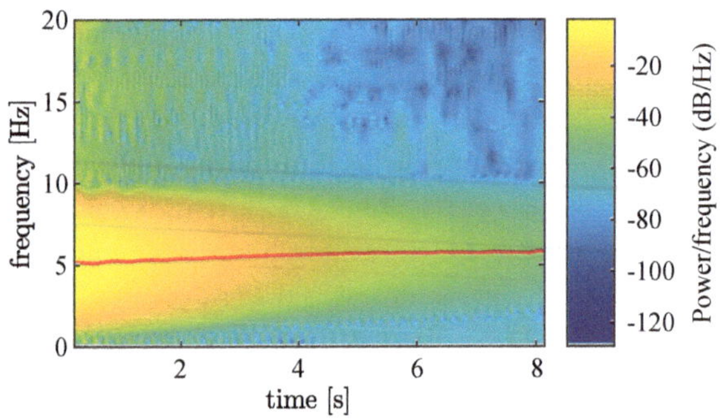

Figure 7.24 Spectrogram of the decay phase for the passage no. 27, sensor x183z

during a passage at resonance speed (passage no. 27 and no. 26 in Tab. 7.2) at MP x183z and x128z are shown in Fig. 7.25a and Fig. 7.26a. The acceleration response is obviously characterized by vibrations at the first natural frequency of 4.95 Hz. Analyzing the amplitudes of the peaks in both diagrams, the effect of the – rather untypical – resonance mechanism of the ICE 4 can be observed. Two subsequent peaks are amplified by the two groups of two axles corresponding to the rear bogie of a carriage and the front bogie of the subsequent carriages. The two amplified peaks are followed by a free vibration. This effect obviously occurs due to the almost integer ratio of the bogie spacing within the car to the bogie spacing of two subsequent carriages (cf. Fig. 7.2): $19.5/9.25 \approx 2.11$.

Fig. 7.27 shows the resonance diagrams based on signals filtered with a cut-off frequency of 20 Hz. No significant changes can be observed in comparison to the curves shown in Fig. 7.21. This confirms that the resonant response of the first mode clearly dominates the total response. This fact is also confirmed by Fig. 7.25b and Fig. 7.26b, which illustrate the acceleration time histories at MP x183z and x128z, which were subjected to a 20 Hz low pass filtering (Butterworth, 5th order). For the sake of completeness, time histories of two non-resonant passages (no. 1 and no. 2 in Tab. 7.2) are illustrated in Fig. 7.28. The signals

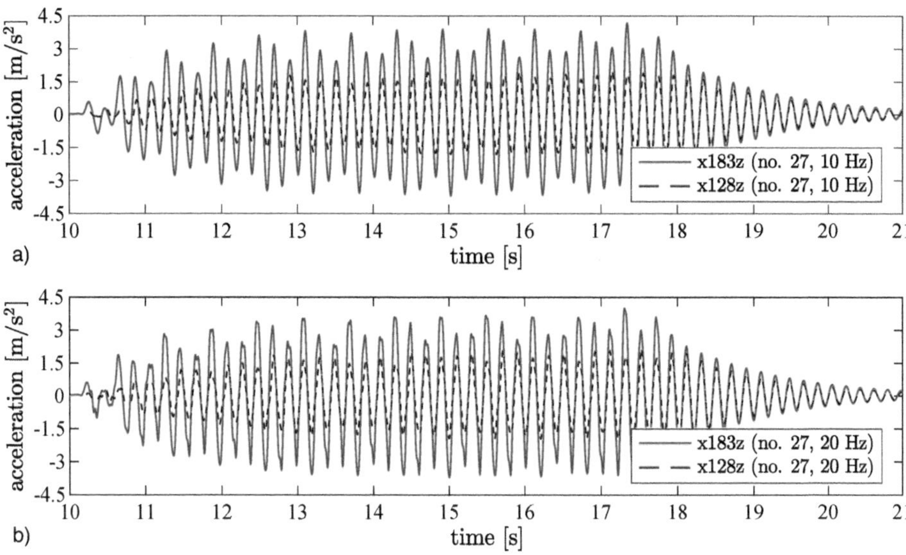

Figure 7.25 Acceleration time histories of of measurement points x183z and x128z recorded during the passage no. 27 towards Ulm: a) signals filtered with a cut-off frequency of 10 Hz; b) signals filtered with a cut-off frequency of 20 Hz;

are clearly dominated by the excitation frequency of the previously discussed excitation mechanism, which results from the groups of two axles corresponding to each bogie. Since the excitation frequency does not match the resonance frequency, the responses decay almost completely before a new group of axles reaches the bridge.

Finally, as it can be seen in Fig. 7.21, the ballast joint ensures that a certain part of the load is transferred from the substructure, which is passed by the vehicle, to the unloaded substructure. This effect was also observed by REBELO et al. (2008), RIGUEIRO et al. (2010), RAUERT et al. (2010) and BIGELOW et al. (2018). It confirms the hypothesis that the longitudinal ballast joint has a significant influence on the dynamic behavior of the structure by increasing its global stiffness. However, the comparison between the resonance curves shown in Fig. 7.21a and Fig. 7.21b reveals that – besides the previously discussed nonlinear behavior of the ballast bed related to the frequency drop – the longitudinal ballast bed causes a further nonlinearity of the investigated system. This refers to the differences between the accelerations of the loaded and the unloaded substructures during train passage. It can be seen that when the train drives towards Ulm (over substructure 1), the substructure 2 experiences accelerations of up about 2 m/s^2 at midspan, reaching thus about 50 % of accelerations of the loaded substructure (ca. 4.1 m/s^2, cf. Fig. 7.21a). On the contrary, the train passages towards Augsburg (Fig. 7.21b) generate accelerations of about 3.8 m/s^2 at midspan of the loaded substructure, while the corresponding measurement points of the unloaded substructure experience maximal accelerations of about 3 m/s^2. This results in

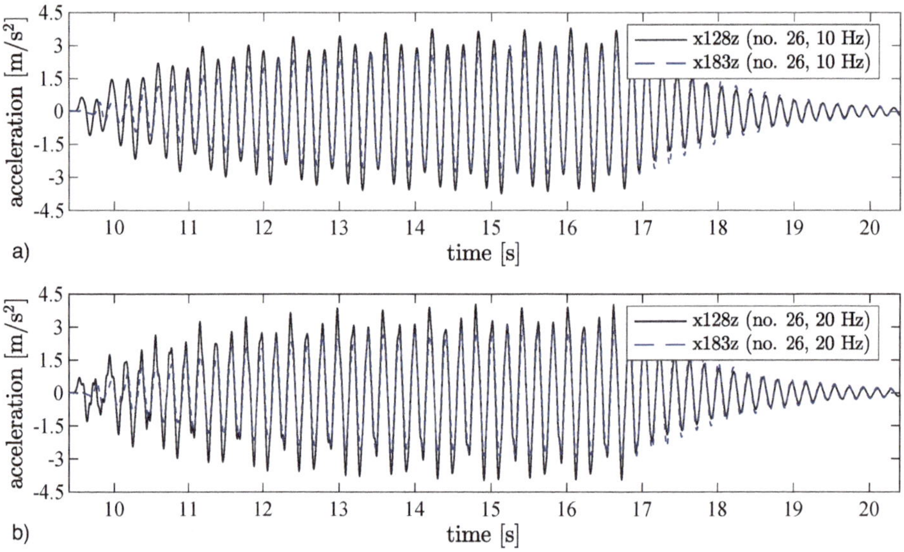

Figure 7.26 Acceleration time histories of of measurement points x183z and x128z recorded during the passage no. 26 towards Augsburg: a) signals filtered with a cut-off frequency of 10 Hz; b) signals filtered with a cut-off frequency of 20 Hz;

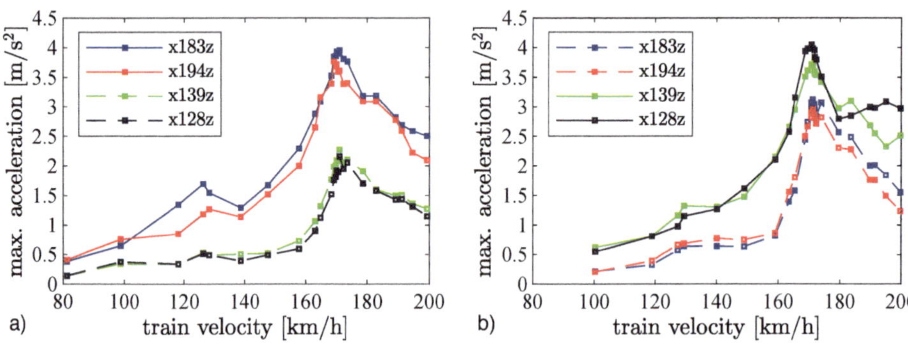

Figure 7.27 Resonance curves of the Schmutter bridge for acceleration channels situated on the axis 6-6 (cf. Fig. 7.7) and low pass filtered signals with a cut-off frequency of 20 Hz: a) train passages in direction Ulm (substructure 1); b) train passages in direction Augsburg (substructure 2)

a ratio of about 80 %. This fact clearly indicates different behaviors of the load carrying mechanism for the two passage directions. Related to this topic, RAUERT et al. (2010) also reported – based on experimental findings – that the shear stiffness of the ballast joint seems to be strongly dependent on the load level and implicitly on the vibration amplitudes. Therefore, it can be stated – also under consideration of the previously discussed effects of the frequency reduction during train passage – that the amplitude dependent stiffness seems

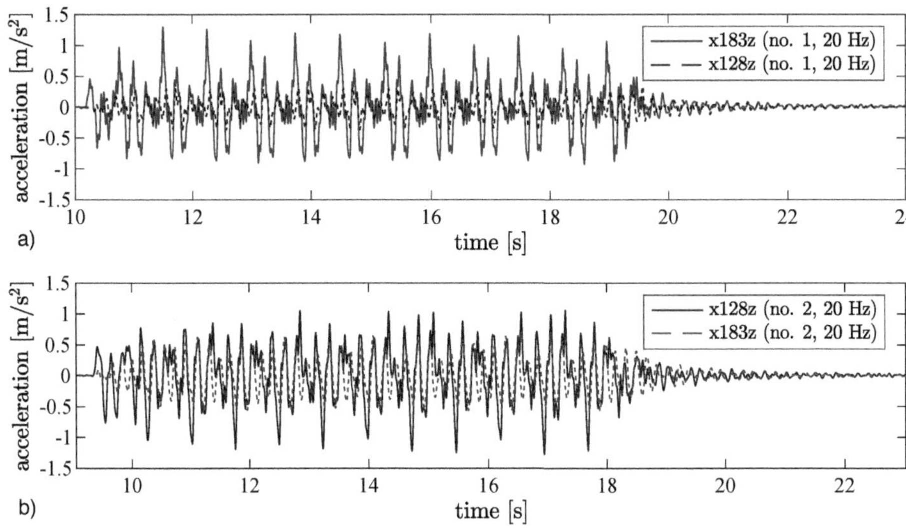

Figure 7.28 Acceleration time histories of measurement points x183z and x128z recorded during the passages no. 1 (a) and 2 (b): signals filtered with a cut-off frequency of 20 Hz;

to be a fundamental property of the ballast bed. The mechanical understanding of this effect and its accurate numerical implementation represents a challenging task, which would go far beyond the scope of this thesis. In this context, it should be mentioned that all numerical investigations carried out in the subsequent parts of this work will refer exclusively to the passages in the direction of Ulm (over substructure no. 1). This means that at least the nonlinearity related to the passage direction can be neglected in the modeling.

7.4.4 Displacement measurements

As mentioned in Section 7.3.2, different non-contact technologies were used for the displacement measurements during the train passages (cf. Fig. 7.9). Their results will be briefly discussed in the following.

Fig. 7.29 shows the time history of the displacements measured using the four laser vibrometers (cf. Fig. 7.9) during the passage no. 27. It can be easily observed that the displacement at the measurement point Laser13_dz (at 0.3 x l) reaches higher values than the displacement Laser1_dz at midspan. This is also valid for the unloaded substructure, i.e. the measurement point Laser22_dz (at 0.3 x l) experiences greater maximal displacements than Laser23_dz (at 0.5 x l). These findings obviously contradict the logical engineering expectation of a maximal displacement at the middle of the span. In this respect, the results of the complementary non-contact measurement methods (MI and profile scanner) can be used, in order to verify the correctness of the laser vibrometer measurements. Fig. 7.30a

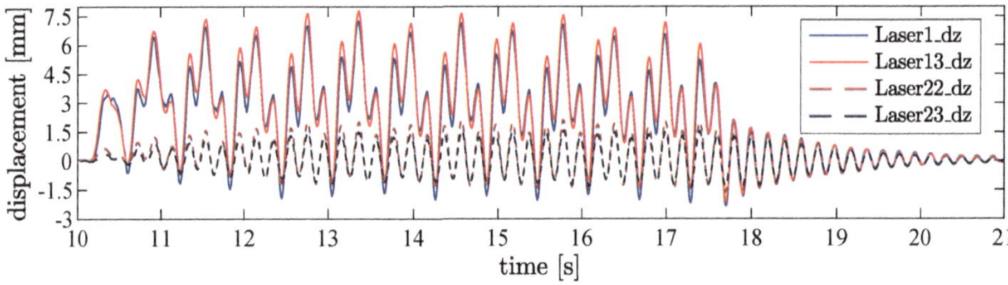

Figure 7.29 Displacement time histories recorded by the four laser vibrometers during the passage no. 27

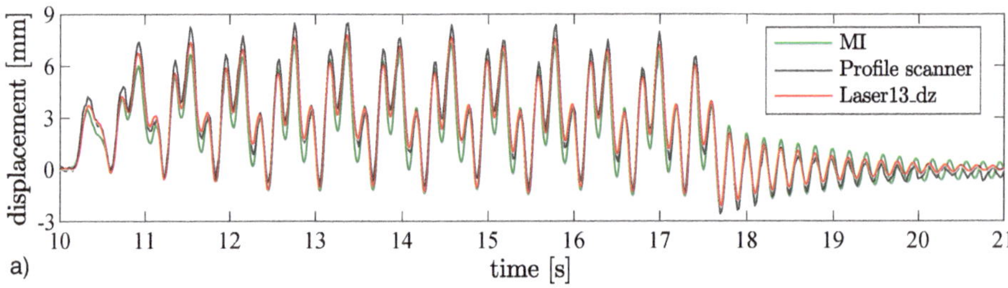

a)

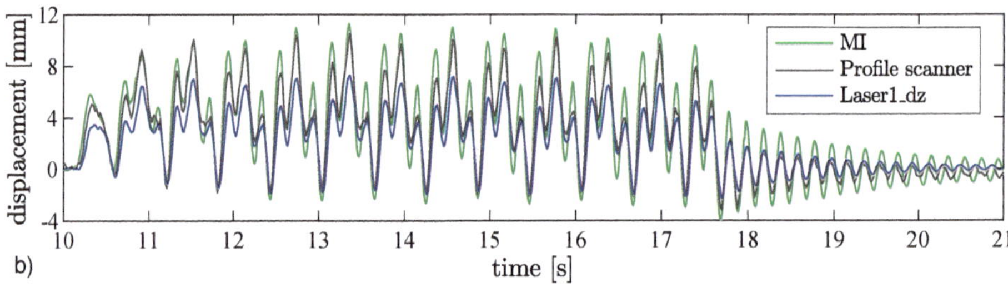

b)

Figure 7.30 Displacement time histories measured with three different non-contact measurement techniques during the passage no. 27: a) results at 0.3 x l (axis 4-4); b) results at 0.5 x l (axis 6-6)

illustrates the comparison of the measured displacements at 0.3 x l, corresponding to the axis 4-4 (cf. Fig. 7.7), which also includes the reflector of MP Laser13_dz. It is obvious, that the three non-contact measurement methods deliver very similar results in this case. In this respect, it is to be noted that the three devices were placed at different y-coordinates (cf. Fig. 7.9). The measurement point Laser13_dz is situated at the inner edge of substructure 1 (axis B-B, cf. Fig. 7.7), the profile scanner acquires the displacements along the longitudinal axis of the same substructure, while the microwave interferometer records displacements of resolution cells, whose centroids are disposed along the outer edge (axis A-A). Therefore, the displacements compared in Fig. 7.30a do not correspond exactly to the same point. However, assuming that no – or very less – profile deformation occurs, the displacements of a substructure at different points with the same x-coordinate are expected to have similar values. This fact is confirmed by the very good agreement of the measured values presented in Fig. 7.30a.

On the contrary, Fig. 7.30b only indicates a good match of the profile scanner and the MI measurements, while the laser vibrometer results are significantly lower. This trend could be confirmed for all recorded passages and indicates a systematic error in the laser vibrometer measurements at midspan. For an additional verification, the time histories of the four laser vibrometers were differentiated two times with respect to the time, in order to obtain the corresponding acceleration time histories. These were subsequently compared with the accelerations recorded by the acceleration sensors installed at the same coordinates (cf. Fig. 7.9 and Fig. 7.7). The results are given in Fig. 7.31. It is obvious that the accelerations determined from laser vibrometer recordings at midspan of both substructures revealed an unsatisfactory match with the corresponding direct acceleration measurements (Fig. 7.31a and d). On the contrary, the comparison of the second derivatives of the Laser13_dz and Laser22_dz signals (both situated at 0.3 x l) with the corresponding accelerations shows an excellent agreement (Fig. 7.31b and c). The systematic error of the measurement points Laser1_dz and Laser23_dz is thus also confirmed by a second, independent approach. The error is thought to occur due to the very low projection angles[14] α_L of the laser vibrometers used for the midspan measurements. The α_L of the four devices are: 14.20° for Laser1_dz, 24.35° for Laser13_dz, 25.50° for Laser22_dz and 14.95° for Laser23_dz. Therefore, both laser vibrometer measurements performed at midspan will not be used in the further analyses of this work.

Having concluded that the signals of the laser vibrometers Laser13_dz and Laser22_dz deliver plausible results, the relative displacement between the two substructures can be analyzed by subtracting the mentioned signals from each other. The results obtained for the passage no. 27 are shown in Fig. 7.32 and reveal a maximum relative displacement of about 5.8 mm. In this respect, it is to be mentioned that the maximum displacement at the point Laser13_dz is about 7.7 mm, while the signal recorded at Laser22_dz reaches values of up to 2.1 mm (Fig. 7.29). These findings emphasize again that the longitudinal ballast joint has a significant contribution to the load transfer between the loaded and the unloaded substructure. In the same context, the relative displacements of the quasi-static passage no. 3 are analyzed, in order to verify if the longitudinal ballast joint also presents the same behavior in the static case. The corresponding results are represented in Fig. 7.33. Maximal displacements of about 5.9 mm at Laser13_dz and 0.6 mm at Laser22_dz were recorded. The corresponding time histories lead to a maximal relative displacement of about 5.5 mm. This confirms that the longitudinal ballast joint also ensures the load transfer between the two substructures in the quasi-static case. The time histories of the quasi-static

[14]The projection angle represents the angle between the line-of-sight (LOS) of the laser vibrometer and the horizontal direction. The projection is needed because – similar to the microwave interferometry (cf. Fig. 7.10 and Eq. 7.4) – the laser vibrometer acquires the displacement in the LOS direction, which has to be projected with respect to the vertical direction by dividing the measured LOS-signal by $\cos(90° - \alpha_L)$.

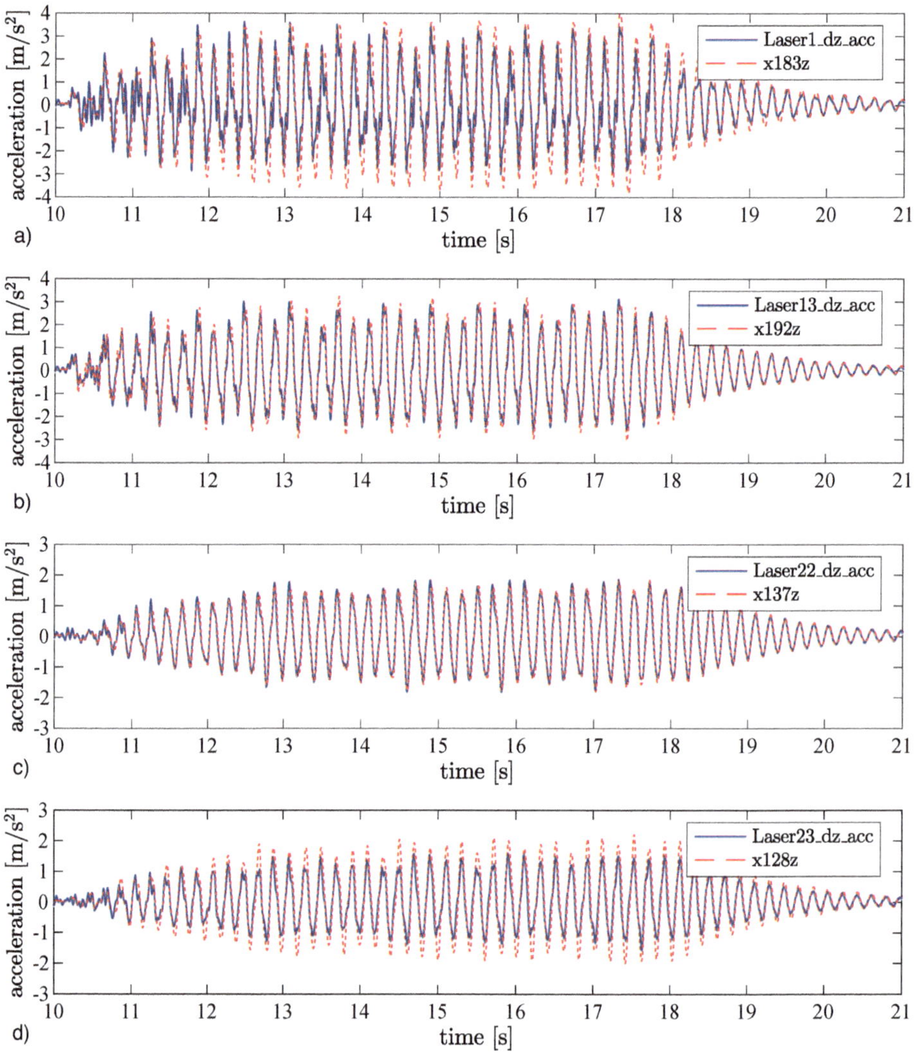

Figure 7.31 Measured acceleration time histories in comparison with the acceleration time histories obtained as second derivatives of the laser vibrometer measurements: a) x183z/Laser1_dz; b) x192z/Laser13_dz; c) x137z/Laser22_dz; d) x128z/Laser23_dz. All signals were filtered using a low pass filter with a cut-off frequency of 20 Hz.

displacements recorded for the passage no. 3 are therefore well suited for the calibration of the numerical model, which has to be able to deliver a realistic prediction for both the static and dynamic behavior of the structure.

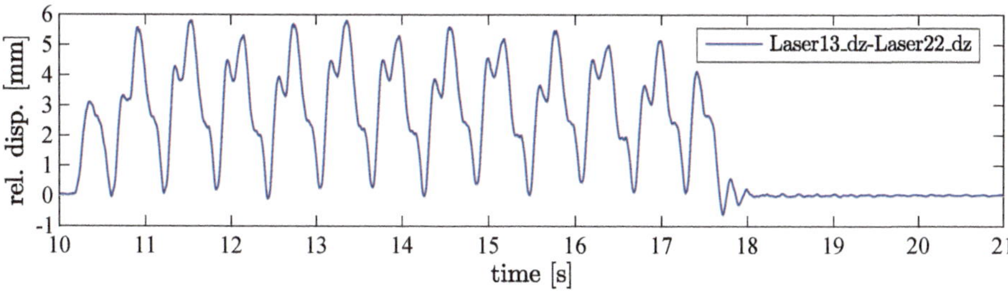

Figure 7.32 Time history of the relative displacement between the two substructures

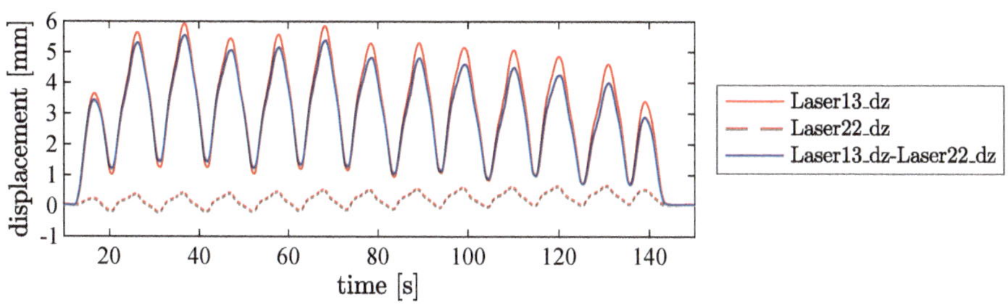

Figure 7.33 Displacement time histories recorded by two laser vibrometers during the passage no. 3 and the resulting relative displacement between the two substructures

7.5 Structural model

The finite element software SOFiSTiK 2020 was used to create a detailed numerical model for the Schmutter Bridge according to the design plans. The steel structure was idealized as a three-dimensional folded-plate model using shell elements, as shown in Fig. 7.34. The material parameters of the steel components were assumed to be deterministic and were assigned typical values from the engineering practice: $\rho = 7850\ \text{kg/m}^3$, $E = 2.1 \cdot 10^8\ \text{kN/m}^2$ and $\nu = 0.3$. Four bearing supports were defined at each side of the bridge according to the design plans (Fig. 7.35), whereas all of them were considered free to rotate. The average depth of the ballast bed over the bridge deck is about 0.44 meters. This was only considered in the model as an added mass, defined under consideration of a density of $1.9\ \text{t/m}^3$. It is to be noted that this density was assigned to the whole ballast body, which also includes the volume of the concrete sleepers (cf. Fig. 7.5). In order to account for the density difference of the volume occupied by the slippers and to also consider the mass of the rails, an additional mass of about 230 kg/m was defined in the finite element model for each substructure. Moreover, further additional masses corresponding to the two walkway constructions disposed along the outer edges of the two substructures were included in the model. They were estimated according to the design plans to about $900\ \text{kg/m}$ for each substructure.

As it can be seen in Fig. 7.34, the model consists in fact of the two substructures, which are connected by vertical springs and present a partial moment restriction at the supports. The latter was defined at each end of the two substructures using a set of horizontal springs disposed along the upper flanges of the girder boxes. In this context, the spring $k_{\mathrm{h},1,x}$ was assigned to substructure 1 and $k_{\mathrm{h},2,x}$ to substructure 2. They were included in the modeling approach, since the ballast track and the rails present a continuity over the supports, which is expected to generate a certain flexural stiffness. For simplicity, the same values of $k_{\mathrm{h},1,x}$ and $k_{\mathrm{h},2,x}$ were assumed for both ends of the substructure 1 and substructure 2, respectively. Furthermore, it is to be mentioned that the vertical springs $k_{\mathrm{v},12,z}$ connecting the two girder boxes are included in the model to account for the shear stiffness of the ballast, which – as already shown in the previous sections – obviously ensures the coupled behavior of the two substructures.

In the previous sections it was shown that the ballast bed presents a strong nonlinear behavior, which presumably leads to a reduced stiffness during the train passage and thus to a drop of the natural frequencies. However, since the proposed MFI method is only valid for linear calculations, a linearized model of the Schmutter bridge must be used in this work. This implies that a linear behavior has to be assumed for all spring elements. In this sense, despite a set of orientation values for $k_{\mathrm{h},1,x}$, $k_{\mathrm{h},2,x}$ and $k_{\mathrm{v},12,z}$ provided in the literature (REBELO et al., 2008; RIGUEIRO et al., 2010; RAUERT et al., 2010; BIGELOW et al., 2018), the spring parameters used in the present investigation were determined by performing a model calibration. For this, an optimization problem is formulated. Its main idea is to find the optimal stiffness parameters $k_{\mathrm{h},1,x}$, $k_{\mathrm{h},2,x}$ and $k_{\mathrm{v},12,z}$, which minimize the differences between the computed and measured first natural frequency during train passage ($\bar{f}_1 = 4.95$ Hz, cf. Section 7.4.3) and between the measured and computed maximal quasi-

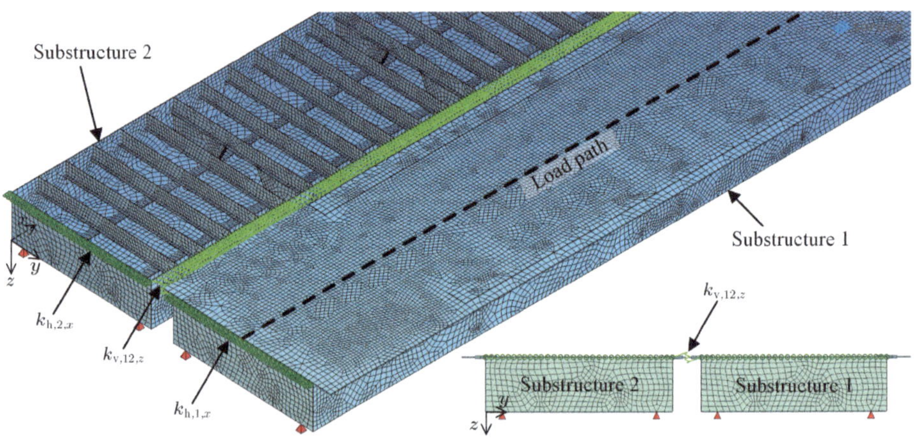

Figure 7.34 Finite element model of the Schmutter Bridge (the deck plate of the substructure 2 was hidden, in order to allow the representation of the structural elements located inside the girder box)

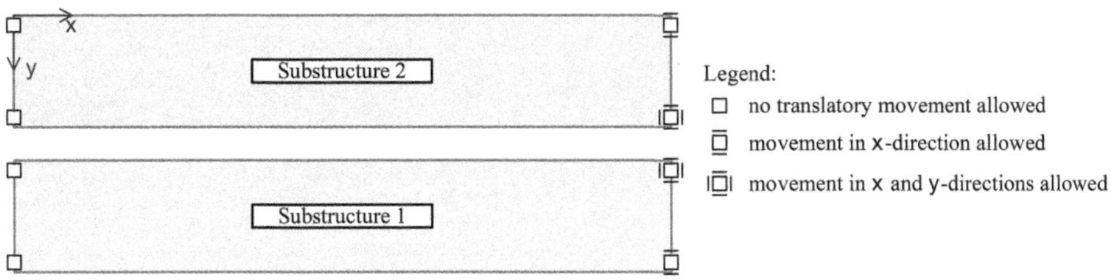

Figure 7.35 Sketch of the bearing support configuration of the Schmutter bridge

static displacements at the measurement points Laser13_dz and Laser22_dz, respectively (cf. Fig. 7.33). The corresponding objective function is expressed as:

$$\min_{\boldsymbol{k}} \mathcal{F}(\boldsymbol{k}) = w_{\text{f}} \cdot (f_1 - \bar{f}_1)^2 + w_{\text{u13}} \cdot (u_{\text{max,L13}} - \bar{u}_{\text{max,L13}})^2 + w_{\text{u22}} \cdot (u_{\text{max,L22}} - \bar{u}_{\text{max,L22}})^2 \tag{7.9}$$

where

$$\boldsymbol{k} = \begin{pmatrix} k_{\text{h},1,x} \\ k_{\text{h},2,x} \\ k_{\text{v},12,z} \end{pmatrix} \tag{7.10}$$

is the vector of the design variables. Furthermore, f_1 represents the computed and $\bar{f}_1$ the experimentally identified first natural frequency, while $u_{\text{max,L13}}$ and $\bar{u}_{\text{max,L13}}$ denote the computed and the measured maximal absolute displacements at the location of Laser13_dz for the quasi-static passage no. 3. A similar notation is also used for the absolute maximal displacements at Laser22_dz ($u_{\text{max,L22}}$ and $\bar{u}_{\text{max,L22}}$ in Eq. 7.9). At this point, it is to be mentioned that no normalization or weighting of the three components of the objective function was used, i.e. $w_{\text{f}} = w_{\text{u13}} = w_{\text{u22}} = 1$. The reason for this is that similar orders of magnitude can be reached for all terms, if the units [Hz] and [mm] are used for the natural frequencies and displacements, respectively.

The optimization problem expressed in Eq. 7.9 was solved using the sequential quadratic programming (SQP) strategy (NOCEDAL et al., 2006; HARZHEIM, 2019). A single constraint imposing the positivity of the stiffness parameters was considered in the present case. The corresponding results are:

$$\begin{aligned} k_{\text{h},1,x} &= 250.8 \cdot 10^3 \quad \text{kN/m/m} \\ k_{\text{h},2,x} &= 247.6 \cdot 10^3 \quad \text{kN/m/m} \\ k_{\text{v},12,z} &= 0.46 \cdot 10^3 \quad \text{kN/m/m} \end{aligned} \tag{7.11}$$

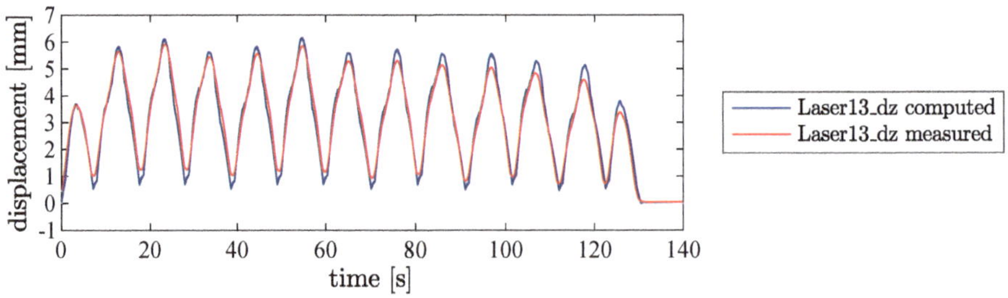

Figure 7.36 Computed and measured displacement time histories of the measurement point Laser13_dz for the quasi-static passage no. 3

It can be clearly observed that the optimization procedure delivered quite similar values for the horizontal springs $k_{\mathrm{h},1,x}$ and $k_{\mathrm{h},2,x}$. The vertical spring $k_{\mathrm{v},12,z}$, ensuring the coupling of the two substructures, obviously has a significantly lower value. This can be explained by the rather high relative displacement between the nodes corresponding to the measurement points Laser13_dz and Laser22_dz (cf. Fig. 7.33). Here, it is worth mentioning that – partly due to the linearization assumption of the system – the values obtained by solving the optimization problem do not necessarily have to be physically justifiable. They simply represent a set of equivalent parameters, which are expected to ensure that the finite element model approximates the real structural behavior as closely as possible. In this respect, it has to be mentioned that the solution of Eq. 7.10 was not quite straightforward. Due to the presumably non-convex behavior of the objective function, the optimization firstly landed in a local minimum with unsatisfactory values of the components of the objective function. Therefore, the last steps of the model updating had to be manually tuned, in order to obtain the solution given in Eq. 7.11. With regard to the sensitivity of the model to the three components of the objective function, it should be mentioned that all three criteria must be considered, in order to achieve a reasonable solution. For example, an optimization that only takes into account the minimization of the frequency term is not sensitive at all with respect to the vertical spring. The reason for this is that the first vertical spring is not active in the first mode of vibration (cf. Fig. 7.16). This means that more or less arbitrary values of the displacements at Laser13_dz and Laser22_dz can be retrieved, which are dependent on the initial value of $k_{\mathrm{v},12,z}$.

Fig. 7.36 shows the displacement time history of the structural node corresponding to Laser13_dz, computed for the quasi-static passage no. 3 ($v = 9.87$ km/h) under consideration of the optimal stiffness values provided by Eq. 7.11. In addition, Fig. 7.36 also includes the measured time history of the displacement at the same location, in order to allow a direct visual comparison between the model results and the real measurements. It can be clearly seen, that the numerical model achieves an excellent agreement with respect to the reference displacement generated by quasi-static train passage no. 3. The maximal computed displacement value is 6.1 mm, which only lies about 3 % above the measured

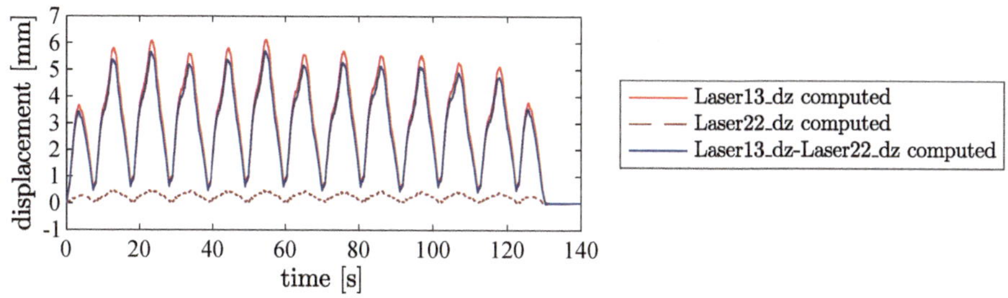

Figure 7.37 Displacement time histories computed for the passage no. 3 at MP Laser22_dz and Laser13_dz and the resulting relative displacement

value of 5.9 mm. For the sake of completeness, Fig. 7.37 shows the computed time histories at the measurement points Laser13_dz and Laser22_dz as well as the resulting relative displacement between the two substructures. The maximal displacement computed at the Laser22_dz is about 0.5 mm and lies thus only a tenth of millimeter below the measured value of about 0.6 mm (cf. Fig. 7.33). The resulting value of the maximal relative displacement between the two substructures is 5.7 mm, which is in very good agreement with the experimentally determined value of 5.5 mm. It can be therefore clearly stated that the model calibration with respect to the static behavior is successful.

As a following step, the optimization criterion regarding the first natural frequency is verified. In this sense, Fig. 7.38 shows the computed mode shapes[15] with significant vertical components[16] and the corresponding natural frequencies below 30 Hz. The first computed natural frequency is 4.95 Hz, indicating that an exact agreement with the target value has been achieved. In this context, it is to be mentioned that neither the values of the higher natural frequencies nor the qualitative figures of the corresponding mode shapes can be directly compared with experimental results. The reason for this is that the resonance curves only provide clear information regarding the first natural frequency during the train passages (cf. Fig. 7.27). Hence, the structural model should be also able to realistically represent the dynamic structural behavior, at least for the first mode. This delivers, in fact, the main contribution to the total response (cf. Section 7.4).

A closer look at the mode shapes illustrated in Fig. 7.38 clearly reveals that the computed mode shapes always occur in pairs of symmetric and asymmetric modes with respect to the vertical contact plan between the two substructures[17]. The mode no. 1 (Fig. 7.38a) obviously represents the first symmetric bending mode. Since both substructures experience similar movements in the same direction, the vertical springs disposed along the longitudinal

[15]In order to facilitate the visual representation, each mode shape shown in Fig. 7.38 was normalized with respect to its maximal absolute value, i.e. scaling to a maximum value of 1.

[16]The computed mode shapes are only illustrated based on the values of the vertical components at the 44 structural nodes corresponding to the acceleration measurement points. This facilitates a rough visual comparison with the experimentally identified mode shapes of the empty structure, shown in Fig. 7.16.

[17]A similar effect was also noticed by RIGUEIRO et al. (2010).

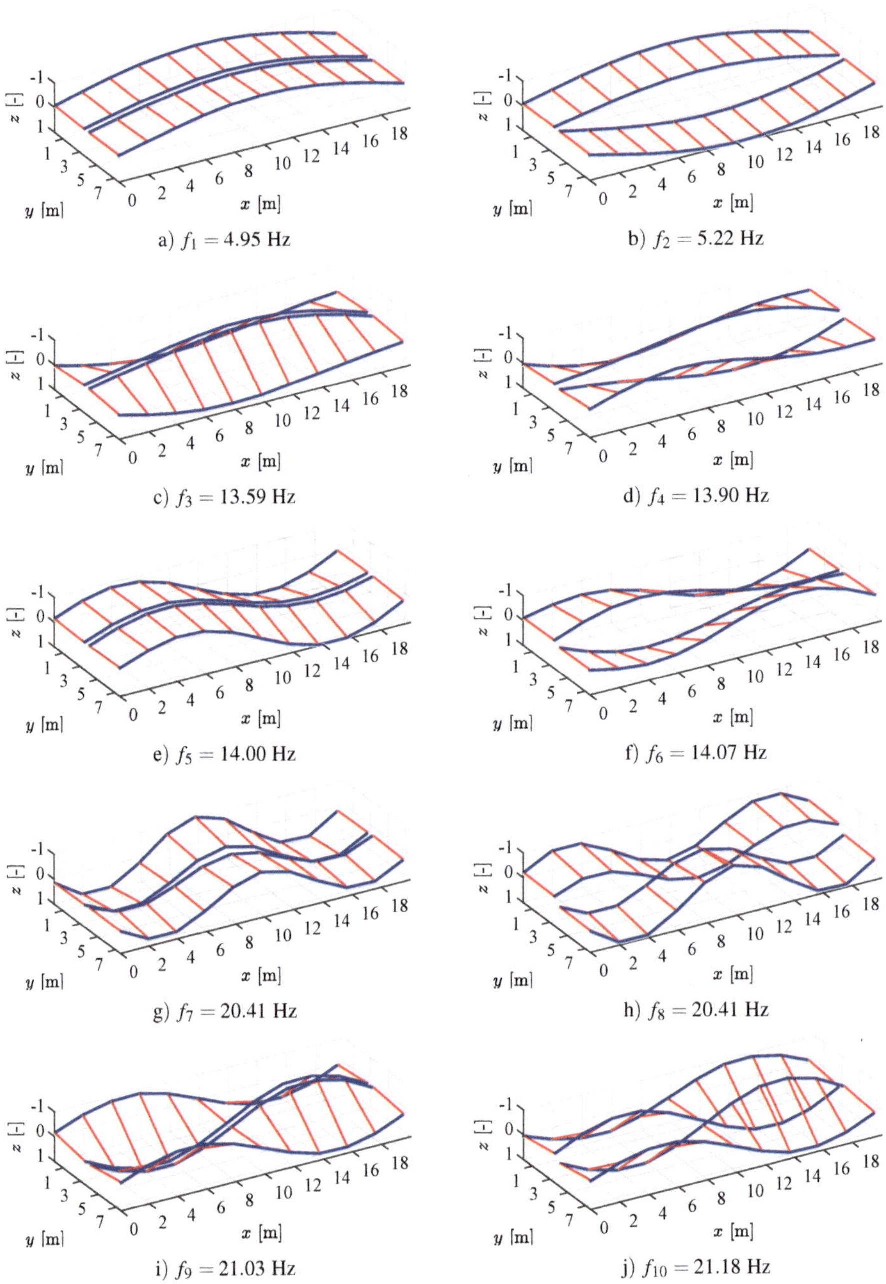

Figure 7.38 Computed mode shapes and natural frequencies of the Schmutter bridge

ballast joint are not active and have thus a neglectable influence on the natural frequency and on the qualitative figure of the mode shape. For this reason, the mode shape seems to be qualitatively similar to the first bending mode identified by the experiments on the empty structure (Fig. 7.16a). The second mode (Fig. 7.38b) represents the corresponding asymmetric bending mode. Since it obviously presents a significant relative displacement between the substructures, the vertical springs are activated. The corresponding natural frequency of 5.22 Hz is about 5.5 % higher than the value for the symmetric mode. In fact, the separation between the natural frequencies of the symmetric and asymmetric modes is determined by the value of the vertical springs. Considering the lower limit case $k_{\mathrm{v},12,z} = 0\,\mathrm{kN/m/m}$, the two substructures are expected to vibrate independently from each other, leading to a duplication of the first natural frequency[18] of 4.95 Hz. On the contrary, if $k_{\mathrm{v},12,z}$ tends to a very high value, the corresponding mode shape converts to a torsional mode of the (stiffly) coupled structure, similar to the one shown in Fig. 7.16b. The corresponding natural frequency obtained for a presumably stiff vertical spring ($k_{\mathrm{v},12,z} = 10^{11}\,\mathrm{kN/m/m}$) is 7.37 Hz.

The previous discussion with respect to the symmetric and asymmetric modes and their frequency separation is clearly valid for all further mode shapes: the first symmetric and asymmetric torsional modes (Fig. 7.38c and d), the second symmetric and asymmetric bending modes (Fig. 7.38e and f), the third symmetric and asymmetric bending modes (Fig. 7.38g and h) as well as the second symmetric and asymmetric torsional modes (Fig. 7.38i and h). In this context, it is to be mentioned that the separation of the natural frequencies between the symmetric and asymmetric modes seems to decrease with increasing order of the mode shapes.

These findings represent in fact an additional confirmation of the hypothesis discussed in Section 7.4, which states that the real structure has a much higher stiffness at low vibration amplitudes (e.g. generated by the impulse hammer excitation) than during the train passage. This remark is also valid for the coupling springs, which obviously tend to a very stiff value at low amplitudes.

7.6 Force identification with real measurement data

7.6.1 Computational parameters

In the following, the application of the developed MFI method with real measurement data is discussed in detail for the ICE 4 passage no. 27 (cf. Tab. 7.2). For this, different combinations of measured acceleration and displacement time histories will be considered in the analysis (Tab. 7.3). Typical acceleration time histories at the points x183z and x128z

[18] It is to be noted that in the present case the duplication of the first natural frequency is not expected to be mathematically exact, since the horizontal springs disposed at the supports of the two substructures ($k_{\mathrm{h},1,x}$ and $k_{\mathrm{h},2,x}$) have slightly different values (cf. Eq. 7.11).

were already shown in Fig. 7.25. The force identification will be firstly tackled by only considering the first two modes in the inverse computation (cf. Fig. 7.38). In this respect, the measured data was lowpass filtered with a cut-off frequency of 10 Hz, in order to adapt the frequency content of the measured time histories to the frequency content of the computations. A time step of the computation $\Delta t = 0.003333$ s was used, which allows a sufficiently fine time discretization of the considered frequency content. This also required a resampling of the measured signals, which were originally sampled at a frequency of 1200 Hz. The window length of the computation was selected to 1000 samples, whereas a smoothing radius $r_\mathrm{a} = 2$ samples was adopted (cf. Section 6.7.2.2). With respect to the window overlapping interval, a value corresponding to one third of the window length was used (cf. Section 6.7.2.3). In addition, both the computed and measured vectors of the motion quantities at each location are normalized with respect to the absolute maximum values of the measured vectors, according to Eq. 5.13. This was achieved by defining the weighting parameters $w_{\mathrm{d},h}$ and $w_{\mathrm{a},h}$ according to Eq. 5.14.

Furthermore, it is to be noted that a validation based on comparisons with experimentally determined dynamic forces according to Eq. 5.23 is not possible, since no system for direct acquisition of reference dynamic axle loads was available. Therefore, the comparisons have to be carried out on the level of the re-built signals[19] or of the static components. The latter are defined as:

$$x_{\mathrm{err,st,tot}} = \frac{\left\| \boldsymbol{x}_{\mathrm{identified,st}} - \boldsymbol{x}_{\mathrm{reference,st}} \right\|_1}{\left\| \boldsymbol{x}_{\mathrm{reference,st}} \right\|_1} \cdot 100\ \% \tag{7.12}$$

The reference vector $\boldsymbol{x}_{\mathrm{reference,st}}$ in Eq. 7.12[20] contains n_F values $x_{\mathrm{reference,st},l}$ (one for each of the n_F axles with $l = 1, 2, ..., n_\mathrm{F}$). They were determined by weighting the vehicle prior to the investigation (cf. Tab. 7.1). The identified static component vector $\boldsymbol{x}_{\mathrm{identified,st}}$ also consists of n_F entries $x_{\mathrm{identified,st},l}$. These were approximated for each force l – under consideration of the DFS-SAV property of traffic loads on bridges – as an average of all identified force values assigned to each of the n_F axles:

$$x_{\mathrm{identified,st},l} = \frac{1}{n_{\mathrm{tp},l}} \cdot \sum_{i=1}^{n_{\mathrm{tp},l}} x_{l,i} \tag{7.13}$$

[19] It is to be mentioned that an error evaluation based on the re-built signals according to Eq. 6.11 has to be interpreted with caution. By substracting each entry of the reference vectors from the corresponding entries of the re-built vectors, it is expected that even slight phase shifts between the time histories of the real measurement data and of the re-built vectors can erroneously indicate poor results, even in case of a good match of amplitudes and frequency content. Therefore, the comparison performed on the level of re-built results implying real measurement data will be basically only discussed from qualitative point of view.

[20] In the following, the term $x_{\mathrm{err,st,tot}}$ in Eq. 7.12 will be referred to as $x_{\mathrm{err,f,st,tot}}$ for the error obtained at the end of the forward optimization and as $x_{\mathrm{err,b,st,tot}}$ for the final error revealed by the backward optimization. Similarly, also the notations $x_{\mathrm{err,st},l}$ in Eq. 7.14 and $x_{\mathrm{err,st,avg}}$ in Eq. 7.15 can be separately defined for the forward and backwards computation, i.e. $x_{\mathrm{err,f,st},l}/x_{\mathrm{err,b,st},l}$ and $x_{\mathrm{err,f,st,avg}}/x_{\mathrm{err,b,st,avg}}$, respectively.

No.	MP accelerations	MP displacement
1	x192z, x194z, x196z, x181z, x183z, x185z	Laser13_dz
2	x194z, x183z	Laser13_dz
3	x192z, x194z, x196z, x181z, x183z, x185z	–
4	x194z, x183z	–
5	x192z, x194z, x196z, x181z, x183z, x185z	Laser1_dz, Laser13_dz
6	x194z, x183z	Laser1_dz, Laser13_dz
7	–	Laser1_dz, Laser13_dz
8	–	Laser13_dz
9	x191z to x197z and x180z to x186z	Laser13_dz
10	x192z, x194z, x196z, x181z, x183z, x185z, x126z, x128z, x130z, x137z, x139z, x141z	Laser13_dz
11	x128z, x139z, x194z, x183z	Laser13_dz, Laser22_dz
12	x126z, x128z, x130z, x137z, x139z, x141z	Laser22_dz

Table 7.3 Measurement configurations used for the MFI of the Schmutter bridge

The isolated error $x_{\mathrm{err,st},l}$ for a certain axle l can be determined as:

$$x_{\mathrm{err,st},l} = \frac{\left|x_{\mathrm{identified,st},l} - x_{\mathrm{reference,st},l}\right|}{\left|x_{\mathrm{reference,st},l}\right|} \cdot 100\ \% \tag{7.14}$$

Based on Eq. 7.14, the average static axle load error is defined as:

$$x_{\mathrm{err,st,avg}} = \frac{1}{n_{\mathrm{F}}} \cdot \sum_{l=1}^{n_{\mathrm{F}}} x_{\mathrm{err,st},l} \tag{7.15}$$

Finally, it is to be mentioned that the total computation time was firstly set to 7.5 s, which approximately corresponds to the total passage time of the complete train:

$$t_{\mathrm{out}} = \frac{l + d_{48}}{v} = \frac{19.5\ [\mathrm{m}] + 338.05\ [\mathrm{m}]}{170.88\ [\mathrm{km/h}]} = 7.53\ [\mathrm{s}] \tag{7.16}$$

This means that the decay phase is firstly not considered at all in the inverse computation. The reason for this is the previously discussed nonlinear character of the structure (cf. Section 7.4). As it was shown in Fig. 7.24, the natural frequency of the bridge changes during the decay phase. However, the numerical model used within this work was defined as linear and is therefore not able to represent the amplitude dependent frequency variation. This means that a large and variable phase shift between the measured and the re-built results is expected for the decay phase. This has a direct influence on the value of the

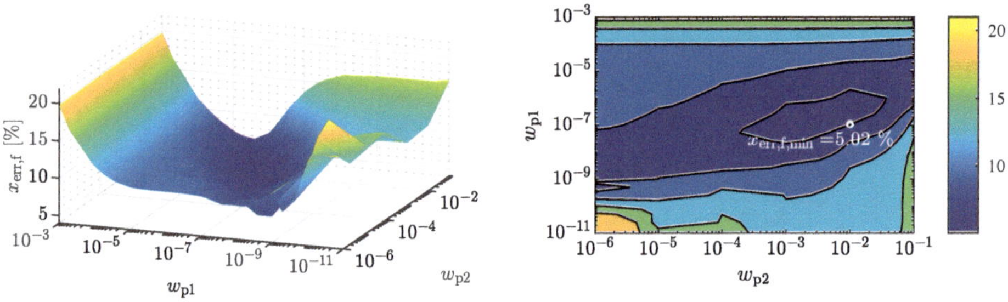

a) b)

Figure 7.39 Parametric study for the determination of the weighting factors of the penalty functions w_{p1} and w_{p2} for the Schmutter bridge: a) 3D plot; b) contour plot

objective function (Eq. 5.15). In order to avoid any difficulties related to this aspect, the decay phase is firstly neglected within the computation. Nevertheless, the effect of the nonlinear frequency change in the decay phase will be briefly addressed in Section 7.6.3.

7.6.2 Weighting factors of the penalty functions

Before proceeding to the solution of the optimization problem, the weighting factors w_{p1} and w_{p2} still have to be determined. For this, a parametric study according to Section 5.1.4, i.e. solely based on numerical analyses, was performed. The simulated measurement data for the parametric study was generated using the reference load scenario 3 (cf. Section 6.2.3 and Eq. 6.6) under consideration of a rather conservative noise level $e_p = 5$ %, whereas the sensor configuration no. 1 (cf. Tab. 7.3) was adopted. The result, obtained by only employing the forward optimization, is shown in Fig. 7.39 and reveals a minimum error $x_{err,f} = 5.02$ % at $w_{p1} = 5 \cdot 10^{-7}$ and $w_{p2} = 1 \cdot 10^{-2}$. However, it can be noticed – as in all cases discussed in Section 6.6.1 – that the error curve presents a rather flat region (in logarithmic representation) around the absolute minimum point, indicating that a wider range of weighting parameters is expected to ensure a good accuracy of the force identification.

7.6.3 Results and discussion of the experimental validation

The identification results obtained under consideration of the first measurement configuration (Tab. 7.3) are illustrated in Fig. 7.40 for the forward and in Fig. 7.41 for the backward computation approach. In order to allow a rough visual comparison, the reference static axle loads are also included in the figures. The convergence criteria of both computation approaches were selected according to Section 6.3.3. At the first sight, a rather poor match between the identified and the reference results can be observed both for the forward and the backward computation. The corresponding overall errors of the identified static axle loads with respect to the measured axle weights, computed according to Eq. 7.12, are

$x_{\text{err,f,st,tot}} = 5.69\ \%$ and $x_{\text{err,b,st,tot}} = 5.77\ \%$. This means – without a detailed discussion of the axle specific results at this stage (performed later) – that the proposed method retrieved in the investigated case the total weight of the vehicle with an accuracy of around 6 %. This value lies in the acceptable range.

With respect to the corresponding re-built acceleration and displacement time histories, a very good agreement with the measurements was clearly revealed. This can be exemplarily seen in Fig. 7.42, which shows the re-built and reference accelerations and displacements for MP x183z and Laser13_dz, respectively. These two sensors were also provided as measured data for the inverse problem (cf. Tab. 7.3). Moreover, even the re-built responses at "unmeasured" locations[21] show a very good match with respect to the reference (measured) values (Fig. 7.43). This is valid for "unmeasured" locations of both the loaded substructure (Fig. 7.43a and b) and the unloaded substructure[22] (Fig. 7.43c and d), even though only measured time series of the loaded substructure were considered in the inverse computation (cf. Tab. 7.3). It is thus again indicated that the modeling approach of the longitudinal ballast joint seems to ensure a sufficiently realistic representation of the real structural behavior for the investigated resonant passage. Moreover, the results shown in Fig. 7.41 and 7.43 indicate that an accurate prediction of the responses at unmeasured locations seems to be easier to achieve than an accurate force identification.

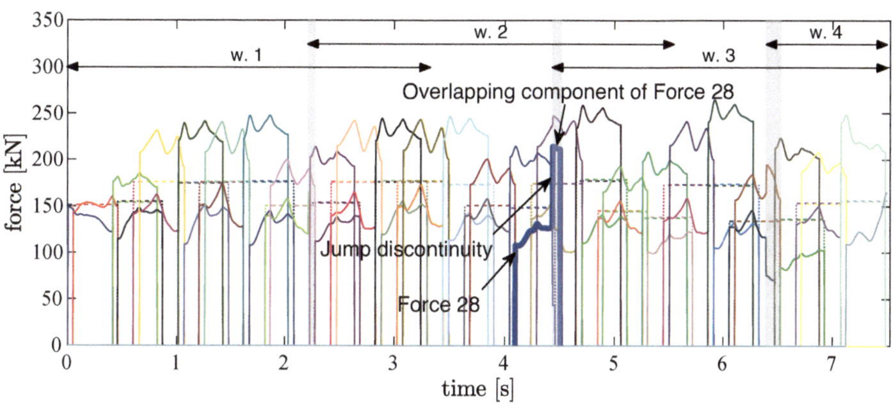

Figure 7.40 Identified dynamic force time histories (solid lines) and reference static axle loads (dashed lines) for the passage no. 27 over the Schmutter bridge – results obtained for the forward solution approach under consideration of the weighting factors $w_{\text{p1}} = 5 \cdot 10^{-7}$ and $w_{\text{p2}} = 1 \cdot 10^{-2}$

[21]The points of the Schmutter bridge, which were instrumented with accelerometers or displacement sensors but were not provided as measurement data for the inverse problem are denoted as "unmeasured" locations. They allow the verification of the proposed method by means of independent data sets, which were not included in the definition of the objective function.

[22]It is to be mentioned that the reference displacement time history used in Fig. 7.43b (measurement position Laser1_dz) was recorded using the microwave interferometer. The laser vibromenter results at midspan could not be used due to the inaccuracies of the measurements discussed in Section 7.4.4.

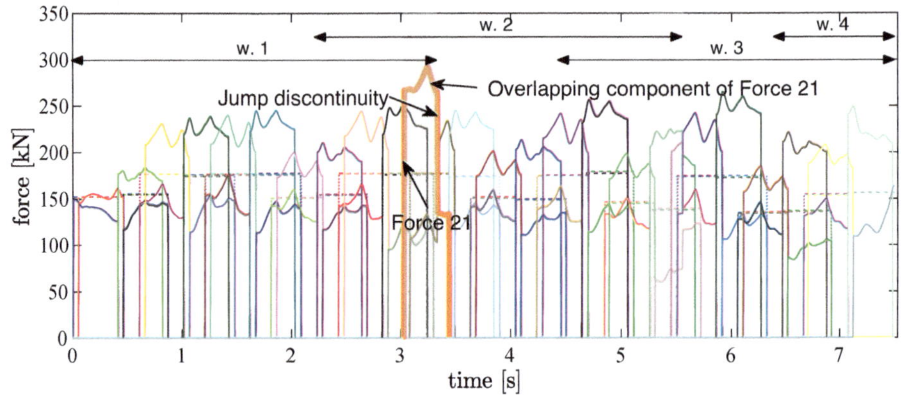

Figure 7.41 Identified dynamic force time histories (solid lines) and reference static axle loads (dashed lines) for the passage no. 27 over the Schmutter bridge – results obtained for the backward solution approach under consideration of the weighting factors $w_{p1} = 5 \cdot 10^{-7}$ and $w_{p2} = 1 \cdot 10^{-2}$

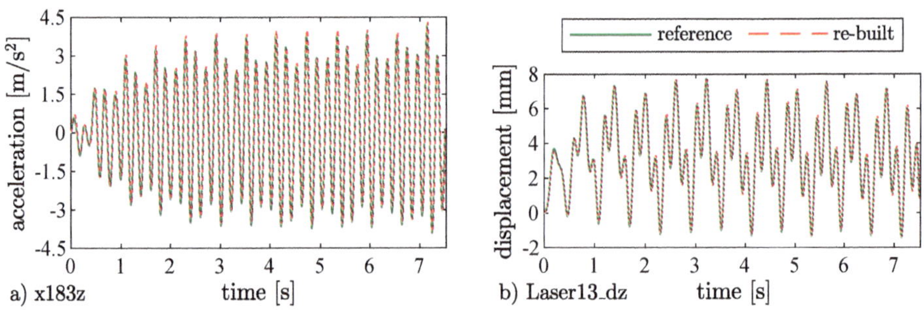

Figure 7.42 Reference and re-built time histories at measured locations for the passage no. 27 and measurement configuration 1 based on the identified forces from Fig. 7.41: a) acceleration at MP x183z; b) displacement at MP Laser13_dz

However, even though the re-built responses are in very good agreement with the measured results, Fig. 7.40 and Fig. 7.41 clearly show, that the identified forces are basically affected by two effects, which lead to high inaccuracies with respect to the static components of the individual axles:

- A set of jump discontinuities in the time histories of some identified axle loads can be seen in the results of both the forward and the backward computation (see grey areas in Fig. 7.40 and Fig. 7.41).

- Significant differences between the two static components associated with the two axles of a bogie can be observed for the majority of the bogies.

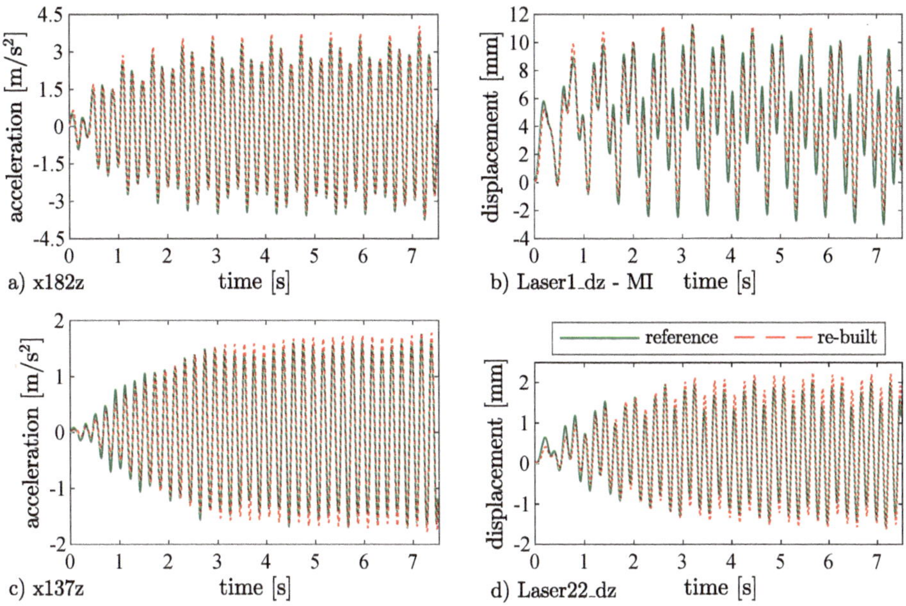

Figure 7.43 Reference and re-built time histories at "unmeasured" locations for the passage no. 27 and measurement configuration 1 based on the identified forces from Fig. 7.41: a) acceleration at MP x182z; b) displacement at MP Laser1_dz (measured using the microwave interferometer); c) acceleration at MP x137z; d) displacement at Laser22_dz

There is no doubt that both previously mentioned effects are unrealistic. On the one hand, a jump discontinuity in the force diagram would mean that the static component of the axle load would suddenly change during the train passage. This is obviously not possible. On the other hand, the bogies are constructed in reality in such a way, that the static load is nearly evenly distributed on both axles. This can be also confirmed by the results of the weighting protocols given in Tab. 7.1. In the following, the two effects will be discussed in more detail together with corresponding strategies to suppress them.

7.6.3.1 Jump discontinuities of the identified forces

As it can be seen in Fig. 7.40 and Fig. 7.41, the jump discontinuities of the identified forces always occur either at the end or at the beginning of a computational window. Actually, it can be observed that in the forward optimization, the jump discontinuities only appear at the beginning of the windows, i.e. at the beginning of the overlapping intervals. On the contrary, the effect is only observed at the end of the time windows for the backward optimization. In fact, only the force time histories, which are split by the initial or final point of a window, can be affected by this type of discrepancy. It can be therefore stated that this effect is of pure numerical nature and only appears due to the solution approach based on moving windows. This implies that the force time histories divided by the lower or upper limit of a window contain a certain number of force values that are optimized twice. For example, the force 28 (highlighted in Fig. 7.40) is firstly optimized over its whole length within the second window. However, due to the overlapping interval, the time history of force 28 is split by the initial point of the third window, so that a rather short component of force 28 (highlighted grey area) will be subjected to a new optimization in window 3. The same aspects are valid also for the backward sliding direction, as exemplified in Fig. 7.41 for the force 21. Indeed the first penalty function from Eq. 5.15 makes sure that the identified dynamic forces of an axle vary around a constant average value. However, this condition is valid within a single window. This means that if any force is split by the window boundaries, the algorithm – in its present form – makes sure that the values of the respective force time history vary in each window around a constant average value. The average values of different windows are uncorrelated, so that different static components can be determined for the same force.

Interestingly, this behavior was not explicitly observed within the numerical validation process based on simulated measurement data (Chap. 6). The reason for this is thought to be the fact that the simulated measurements include – due to the perfect mathematical excitation of the considered modes – a larger amount of information, which is perfectly compatible with the direct analysis results (computed motion quantities) used within the optimization process. This seems to largely suppress the discrepancies that appear in the form of jump discontinuities in the identified forces.

Nevertheless, a rather simple strategy for suppressing this effect can be defined. It can be seen that the results of the forward and backward computations from Fig. 7.40 and Fig. 7.41 are generally very similar. Exceptions are only the forces divided by any window boundaries (see grey areas). Assuming that the backward computation usually only generates a slight improvement of the results (with respect to the forward computation), it seems justifiable to adopt the solution of the backward computation as a basis solution and to modify it by replacing the identified force time histories, which are affected by the effect of jump discontinuities, with the corresponding solutions from the forward computation. The result obtained for the case discussed above is shown in Fig. 7.44.

As already mentioned, the effect of the jump discontinuities occurs due to the sliding window formulation. It is therefore expected, that it does not appear, if a timely expensive computation with only one window is performed. In this sense, Fig. 7.44 also includes the results obtained at the end of the backward computation (without any correction) for a window length of 2252 samples, i.e. all time steps are comprised within a single window. On the one side, it reveals – as expected – that the jump discontinuity effects are not present in the single window computation. On the other side, it can be seen that the approach proposed for suppressing the jump discontinuity effects achieves in the investigated case – in terms of identified forces – an almost exact match with respect to the results of the single window computation. The corresponding errors are $x_{\mathrm{err,b,st,tot}} = 5.47\ \%$ for a moving

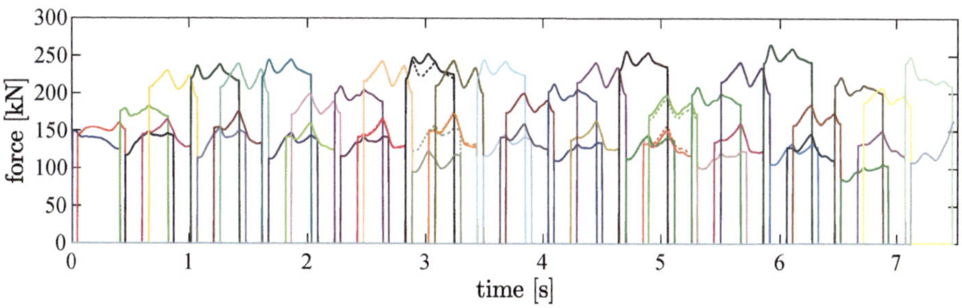

Figure 7.44 Identification results for the passage no. 27 over the Schmutter bridge with suppressed effects of the jump discontinuities in the identified force time histories – results obtained for window lengths of 2252 (dashed lines) and 1000 (solid lines) samples and the weighting factors $w_{p1} = 5 \cdot 10^{-7}$ and $w_{p2} = 1 \cdot 10^{-2}$

window of 1000 samples and $x_{\mathrm{err,b,st,tot}} = 5.80\ \%$ for the single window solution. Moreover, the re-built responses at the same locations used in Fig. 7.43 are illustrated in Fig. 7.45. They indicate that a very good agreement with respect to the reference data can be also achieved using of the corrected identified force time histories. Almost no difference between the results of Fig. 7.45 and Fig. 7.43 can be visually observed. Therefore, the correction method will be used as default in the further analyses.

7.6.3.2 Significant differences of the static axle loads within a bogie

After solving the issue of the jump discontinuities, the considerable differences between the identified static components of the two axles of a single bogie are addressed. Fig. 7.46 obviously shows that this effect occurs for the majority of the bogies, whereas one axle load generally tends to be considerably higher and the other one considerably lower than the reference values. It seems that the mathematical algorithm of the proposed MFI method in its original form is – in case of applications with real measurement data – not able to properly deal with the closely spaced axles of a bogie (cf. Fig. 7.2). In this respect, it is interesting to notice that this effect was not explicitly observed during the investigations based on simulated measurements. This behavior seems to be an identifiability problem, which might be partly explained by the rather low curvatures of the two mode shapes considered in the computations. It can be seen in Fig. 7.47 that the load paths of the first two mode shapes can be approximately described by a half sine wave over the bridge length. This leads to more or less comparable mode shapes values for the two closely spaced axles of a bogie in each time step. Considering the expression of the modal forces given in Eq. 3.45, it is obvious that in case of similar values of the mode shapes assigned to the two axles of the i-th bogie[23], i.e.:

[23] Generally, the two axle of the i-th bogie have the indexes $2i-1$ and $2i$.

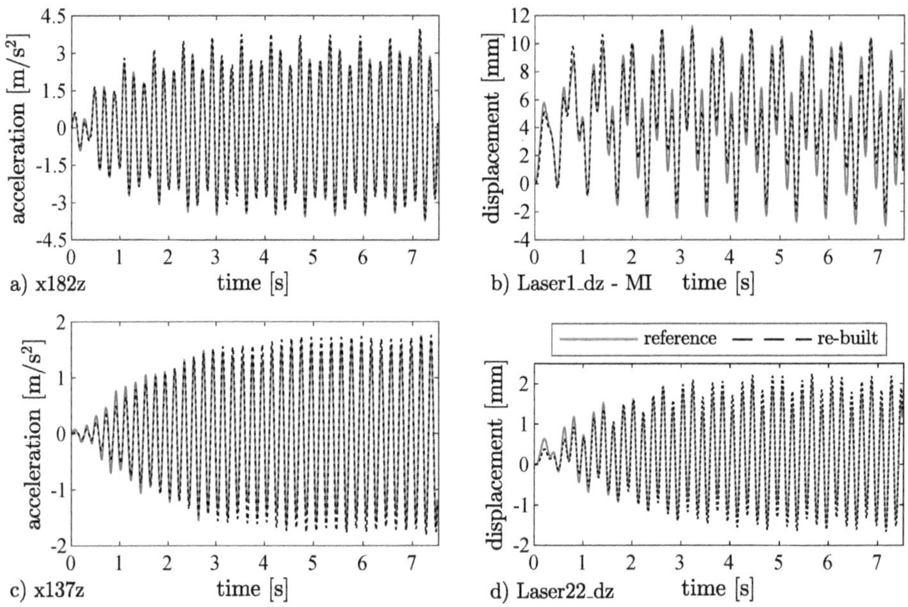

Figure 7.45 Reference and re-built time histories at "unmeasured" locations for the passage no. 27 and measurement configuration 1 based on the identified forces from Fig. 7.44 (window length 1000 samples): a) acceleration at MP x182z; b) displacement at MP Laser1_dz (measured using the microwave interferometer); c) acceleration at MP x137z; d) displacement at Laser22_dz

$$\phi_{1,2i-1} \approx \phi_{1,2i} \text{ and } \phi_{2,2i-1} \approx \phi_{2,2i} \tag{7.17}$$

the contribution of the i-th bogie to each modal load, expressed as:

$$\begin{aligned} &\phi_{1,2i-1} \cdot F_{2i-1} + \phi_{1,2i} \cdot F_{2i} \text{ for mode 1 and} \\ &\phi_{2,2i-1} \cdot F_{2i-1} + \phi_{2,2i} \cdot F_{2i} \text{ for mode 2} \end{aligned} \tag{7.18}$$

can be achieved by different axle force combinations (F_{2i-1}, F_{2i}). Without going into detail, it is mentioned that this effect does not occur in case of an inverse analysis of the Schmutter bridge model using noisy simulated measurement data and the same computation parameters. The reason for this is most likely the fact that the simulated responses are – due to the clear mathematical excitation – more sensitive to the closely spaced axle distances than the measured responses. This hypothesis is fully justified considering that this work is based in fact on a simplified structural model, which does not include the track characteristics, e.g. the ballast load distribution. These are very likely to affect the responses measured in the box girder (LIU et al., 2009b; RIGUEIRO et al., 2010).

In order to deal with the issue of unevenly distributed identified static axle load components within a bogie, the realistic condition of equal static axle forces of the bogie is

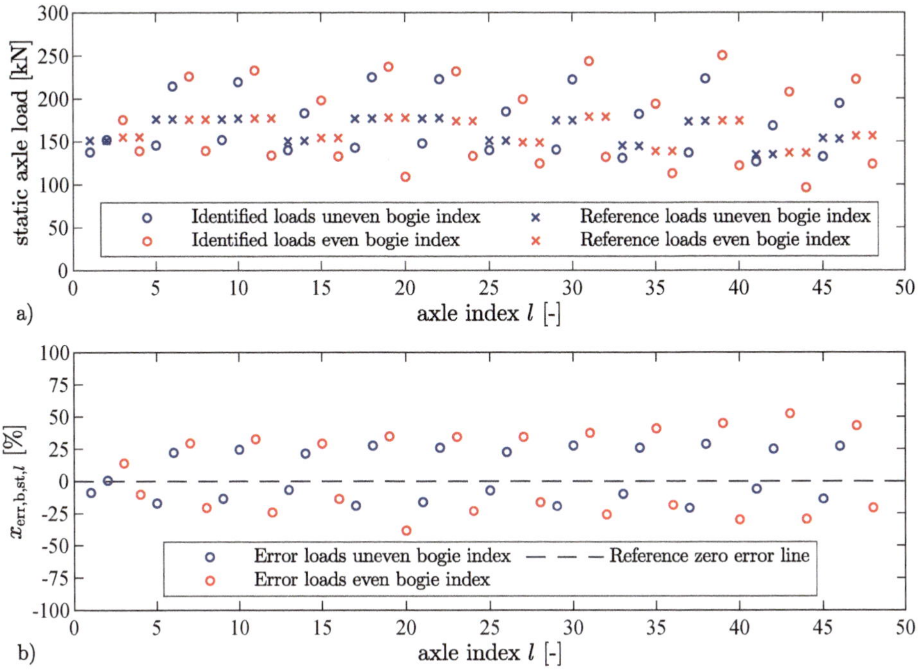

Figure 7.46 Identified and reference static axle loads for the passage no. 27 under consideration of suppressed jump discontinuity effects: a) comparison of the force values; b) relative errors of the identified static forces with respect to the corresponding measured results

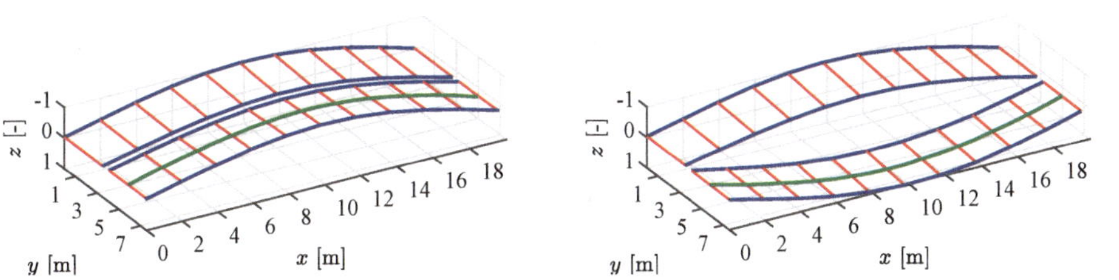

a) $f_1 = 4.95$ Hz b) $f_2 = 5.22$ Hz

Figure 7.47 Computed mode shapes and natural frequencies of the Schmutter bridge considered within the inverse analysis – results including the loading path (green line). The modes are normalized with respect to the maximum absolute value.

enforced to the force identification results. This means that the identification results of the 48 forces illustrated in Fig. 7.44 are shifted along the vertical axis in such a way, that the static components of every two axles corresponding to the same bogie have the same value. This is achieved by determining the average identified static force value for each bogie and by subsequently shifting the time histories of the identified forces corresponding to the respective bogie along the vertical axis, so that the static component of each axle load (computed according to Eq. 7.13) matches the computed average force of the bogie axles. This means that both dynamic force time histories of the identified forces are shifted by the same absolute value, one in positive and the other one in negative direction.

The result for the previously discussed case is represented in Fig. 7.48. The error determined according to Eq. 7.12 is $x_{\text{err,b,st,tot}} = 5.42\ \%$. The corresponding re-built time histories of displacements and accelerations at unmeasured locations are shown in Fig. 7.49. It can be clearly observed that also in this case an excellent match between the re-built and the reference responses was achieved. This confirms that the shift of the dynamic forces along the vertical axis causes rather neglectable modifications in the re-built responses. The reason for this is that the modal forces are expected – in case of low differences between the mode shape values at the positions of the closely spaced axles – to experience almost no change due to the shifting of the identified axle loads. Therefore, this correction will be used for all analyses performed in the further course of this work.

Furthermore, it is to be noted that the overall error of the static forces (i.e. error of the total vehicle mass) did not experience a significant improvement after applying this adaptation – $x_{\text{err,b,st,tot}} = 5.42\ \%$ for the corrected case (Fig. 7.48) and $x_{\text{err,b,st,tot}} = 5.47\ \%$ for the corresponding result from Fig. 7.44. The reason for this is – as already mentioned above – that the two axles of a bogie are shifted along the vertical axis by the same amount. This step is expected to generate nearly no change in the total average of the force time history. On the contrary, a significant improvement was achieved on the level of the individual static forces. Fig. 7.50 shows that the forces of each bogie lie now much closer to their measured reference values, whereas the identified values tend to be generally higher than the reference values. The absolute maximum error value $x_{\text{err,b,st},l} \approx 11\ \%$ is achieved for the static force corresponding to the axle no. 43. The average of the absolute error values

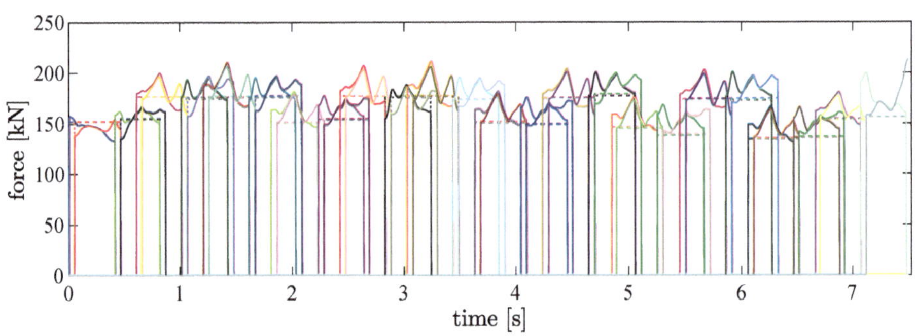

Figure 7.48 Identified dynamic force time histories (solid lines) and reference static axle loads (dashed lines) for the passage no. 27 over the Schmutter bridge under consideration of suppressed jump discontinuity effects and an even distribution of the static loads within a bogie – results obtained for the weighting factors $w_{\text{p1}} = 5 \cdot 10^{-7}$ and $w_{\text{p2}} = 1 \cdot 10^{-2}$

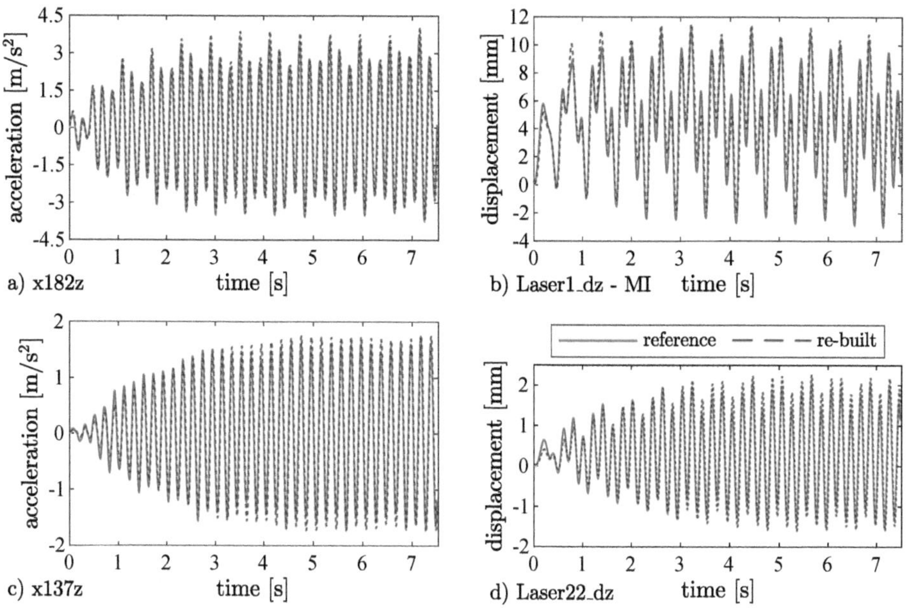

Figure 7.49 Reference and re-built time histories at "unmeasured" locations for the passage no. 27 and measurement configuration 1 based on the identified forces from Fig. 7.48: a) acceleration at MP x182z; b) displacement at MP Laser1_dz (measured using the microwave interferometer); c) acceleration at MP x137z; d) displacement at Laser22_dz

for the 48 static forces is $x_{\text{err,b,st,avg}} \approx 6.1\ \%$ with a standard deviation $\sigma_{\text{err,b,st,avg}} \approx 2.8\ \%$. This result is considered as satisfactory, especially under consideration of the important uncertainties of the numerical model.

7.6.3.3 Dynamic force components

With respect to the dynamic components of the identified forces, no reference data is available for their validation. Nevertheless, a statistical description of the dynamic force for each axle has to be performed in order to gain at least a rough insight into the extent of the dynamic force variation during the train passage. In this sense, the root-mean-square error (RMSE) of the identified dynamic forces for each axle l is evaluated with respect to the identified static axle load $x_{\text{identified,st},l}$:

$$x_{\text{rmse},l} = \sqrt{\frac{1}{n_{\text{tp},l}} \cdot \sum_{i=1}^{n_{\text{tp},l}} \left(x_{l,i} - x_{\text{identified,st},l}\right)^2} \tag{7.19}$$

whereas $x_{\text{identified,st},l}$ is determined according to Eq. 7.13 as average of the identified dynamic force values. Furthermore, in order to allow a more general comparison between different

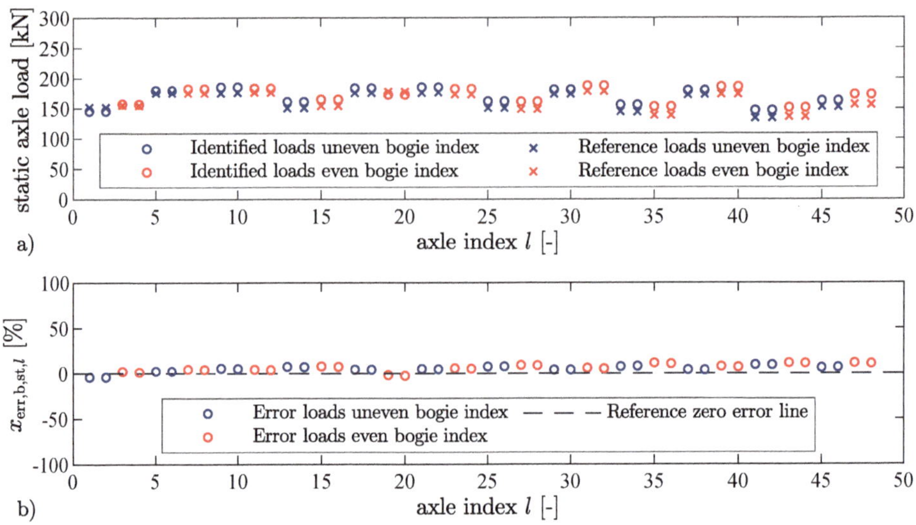

Figure 7.50 Identified and reference static axle loads for the passage no. 27 under consideration of suppressed jump discontinuity effects and an even distribution of the static loads within a bogie: a) comparison of the force values; b) relative errors of the identified static forces with respect to the corresponding measured results

passages, the average RMSE error of a whole passage is defined as mean value of all axle-specific RMSE values $x_{\mathrm{rmse},l}$:

$$x_{\mathrm{rmse,avg}} = \frac{1}{n_{\mathrm{F}}} \cdot \sum_{l=1}^{n_{\mathrm{F}}} x_{\mathrm{rmse},l} \tag{7.20}$$

Fig. 7.51 shows the RMSE values for the $n_{\mathrm{F}} = 48$ identified dynamic axle loads. It can be seen that they vary between 3 and 15 kN with an average $x_{\mathrm{rmse,avg}} = 8.9$ kN and a standard deviation of about $\sigma_{\mathrm{rmse,avg}} = 2.4$ kN. However, it is to be mentioned that the RMSE value computed according to Eq. 7.19 is not to be used for the quantitative performance evaluation of the proposed MFI method, since the direct comparison of the dynamic forces with the corresponding static axle loads is not justified from physical point of view. It delivers, however, a good insight into the variation of the dynamic force components around their average value. This characteristic can be used for a rough comparison of dynamic components of different computations with each other.

The previously discussed results obviously indicate that the identified forces include significant dynamic components (cf. Fig. 7.48 and 7.51). However, since no reference dynamic forces could be measured, the occurrence source of the dynamic force components can not be clearly isolated. This means that it can not be determined, to which extent the identified dynamic force components correspond to the real dynamic forces generated by

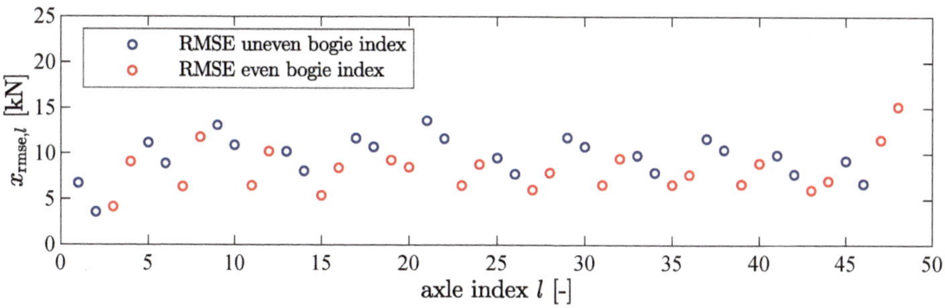

Figure 7.51 Root-mean-square error of the identified forces for the passage no. 27 under consideration of suppressed jump discontinuity effects and an even distribution of the static loads within a bogie

the train-structure interaction or by possible track irregularities and out-of-round wheels[24] and to which extent they are numerically generated by the optimization algorithm, without a direct correlation to the real interaction forces. An attempt to better understand this effect is made by closely analyzing a detailed view of the identified forces in the range $0-3$ s (see Fig. 7.52). It is immediately noticeable that the maximum points of the dynamic forces usually occur around the times when another axle arrives on or leaves the bridge (see vertical dashed black lines in Fig. 7.52). This behavior might suggest that the main source of dynamic components lies most likely in the mathematical algorithm. A possible reason for this effect could be possible inaccuracies regarding the speed of the vehicle. On the one hand, a certain speed measurement error could occur and, on the other hand, the real speed could deviate slightly from the perfectly constant speed assumed in the computation. If a train axle arrives on the structure slightly later in the numerical model than in the real measurements, its contribution to the structural response in the missing time steps has to be compensated by the axles, which are already on the bridge at that time point. The same is valid for an axle leaving the bridge earlier in the numerical model than in reality. However, this statement can not be clearly confirmed without reference dynamic forces and is therefore to be rather interpreted as a plausible hypothesis. This effect could also not be detected in the investigations based on simulated measurement data and a beam model under consideration of perturbed speeds in the inverse analysis (Section 6.7.7). Nevertheless, it should be noted that the identified dynamic forces seem to accurately represent the effect of the (unknown) real dynamic forces, even though their occurrence source is not clearly determinable. This fact has already been proven by means of the re-built responses (Fig. 7.49), which showed a very good agreement with the measurement data.

[24]Considering that a more or less brand new vehicle was used within the test campaign, the presence of out-of-round wheels is rather unlikely.

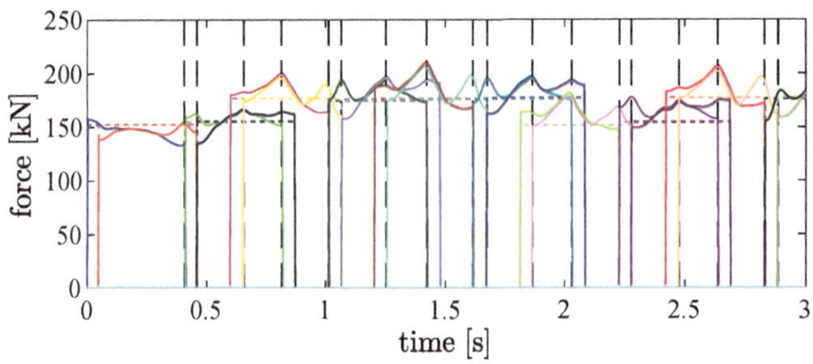

Figure 7.52 Detailed view in the range $0-3$ s of the identified dynamic force time histories (solid colored lines) and reference static axle loads (dashed colored lines) for the passage no. 27 over the Schmutter bridge under consideration of suppressed jump discontinuity effects and an even distribution of the static loads within a bogie – results obtained for the weighting factors $w_{p1} = 5 \cdot 10^{-7}$ and $w_{p2} = 1 \cdot 10^{-2}$. The vertical dashed black lines highlight time points, at which certain axles arrive on or leave the bridge

7.6.3.4 Improved weighting factors of the penalty functions by means of calibration passages

It is to be mentioned that most direct methods for static axle force identification (e.g. WIM or B-WIM, cf. Chap. 2) usually imply a calibration run employing a test vehicle with known static axle loads. The MFI method proposed in this work is expected to deliver – under consideration of a successful adaptation of the numerical model to the real structural behavior – a sufficient accuracy of the identified forces, even without a calibration run. However, the quality of the results is expected to increase if a calibration run is included as additional information in the determination of the optimal weighting parameters of the penalty functions. As the Schmutter bridge investigation implied several passages of the ICE 4 vehicle with known static axle loads, it is possible to refine the search of the optimal parameters w_{p1} and w_{p2} based on the relative error of static components (cf. Eq. 7.12) obtained using real measurement data.

In this sense, further parameter sets lying in the rather flat region (around the absolute minimum point) of the initial error curve (Fig. 7.39) are used in the analysis with real measurement data (see red markers in Fig. 7.53). The purpose is to check whether some other parameter set is able to deliver a better identification of the static components of the axle loads. The investigation implying a total of 12 additional parameter combinations revealed a considerable increase of the accuracy for $w_{p1} = 5 \cdot 10^{-5}$ and $w_{p2} = 1 \cdot 10^{-1}$. The identification result is shown in Fig. 7.54. Its corresponding relative error is $x_{\text{err,b,st,tot}} = 1.78$ %, clearly indicating an excellent agreement between the identified and measured static axle loads. A very good match can also be observed in Fig. 7.55, which depicts the comparison between the re-built and reference accelerations and displacements at some selected unmeasured locations. The identified and reference static forces and the corresponding relative errors are shown in Fig. 7.56 for each axle in particular. It can be

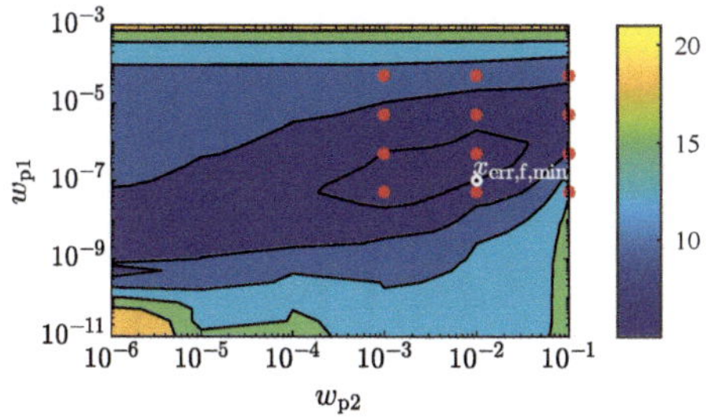

Figure 7.53 Contour plot of the error function used for the determination of the weighting factors w_{p1} and w_{p2}. The red markers highlight the parameter array defined for the refining of w_{p1} and w_{p2}

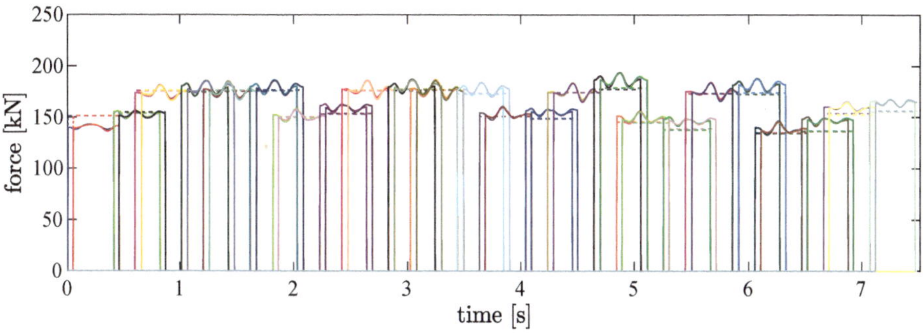

Figure 7.54 Identified dynamic force time histories (solid lines) and reference static axle loads (dashed lines) for the passage no. 27 over the Schmutter bridge under consideration of suppressed jump discontinuity effects, an even distribution of the static loads within a bogie and refined weighting parameters of the penalty functions ($w_{p1} = 5 \cdot 10^{-5}$ and $w_{p2} = 1 \cdot 10^{-1}$)

clearly seen that the identified static axle loads lie in this case very close to the reference values. The absolute maximum value of the error is $x_{err,b,st,l} = 7.7$ %, corresponding – also in this case – to the axle no. 43. The average of the absolute error values is $x_{err,b,st,avg} = 2.8$ %, confirming thus that excellent results are achieved also for the individual axles. Furthermore, Fig. 7.57 shows the RMSE results for all 48 axles. It can be observed that the RMSE values lie between 1.8 kN and 4.9 kN with an average of $x_{rmse,avg} = 3.4$ kN and a standard deviation of $\sigma_{rmse,avg} = 0.94$ kN. The variation of the dynamic forces around the average value is obviously lower for the refined weighting parameters ($w_{p1} = 5 \cdot 10^{-5}$ and $w_{p2} = 1 \cdot 10^{-1}$) than in the case of the originally determined values ($w_{p1} = 5 \cdot 10^{-7}$ and $w_{p2} = 1 \cdot 10^{-2}$). The reason for this is the increased value of the weighting factor w_{p1} of the first penalty function. Its role is to ensure that the dynamic components of each force do not lie too far from their average value. It makes therefore sense that the increasing weight of this criterion imposes a lower variability of the dynamic components. The increased weighting of the second penalty function (from $1 \cdot 10^{-2}$ to $1 \cdot 10^{-1}$), which should suppress the noise

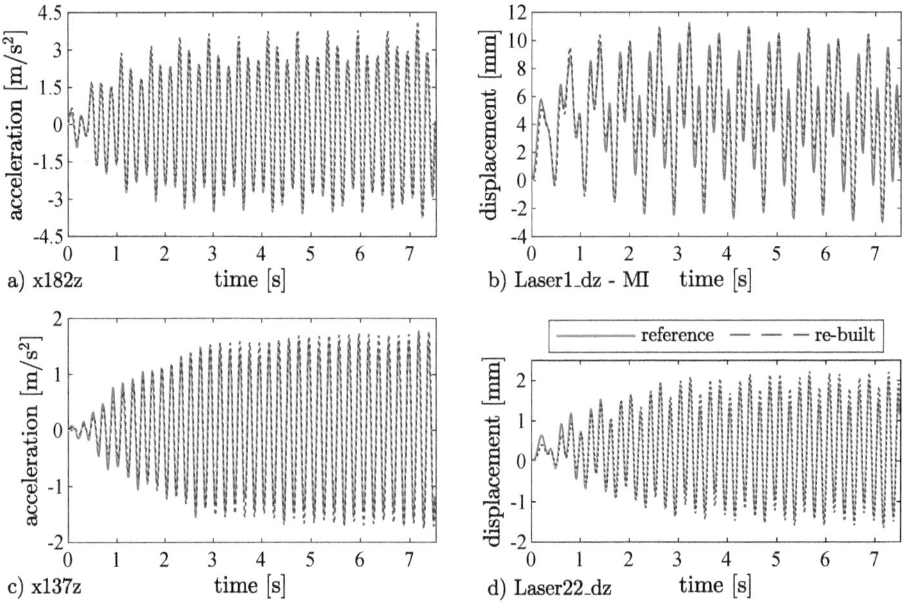

Figure 7.55 Reference and re-built time histories at "unmeasured" locations for the passage no. 27 and measurement configuration 1 based on the identified forces from Fig. 7.54: a) acceleration at MP x182z; b) displacement at MP Laser1_dz (measured using the microwave interferometer); c) acceleration at MP x137z; d) displacement at Laser22_dz

effects, can be explained by the fact that the sensors used in the real measurement have a much lower noise level than the value $e_{\mathrm{p}} = 5\ \%$ assumed for the simulated measurements in the original determination of the weighting factors (cf. Section 7.6.2). Therefore, the refined parameter set $w_{\mathrm{p1}} = 5 \cdot 10^{-5}$ and $w_{\mathrm{p2}} = 1 \cdot 10^{-1}$ will be used – unless otherwise specified – as default in the following investigations.

7.6.3.5 Effect of the decay phase on the identification accuracy

It has already been mentioned in Section 7.6.1 that the decay phase is not considered by default in the inverse computation. The reason for this is the hypothesis of large phase shifts between the computed and the re-built responses in the decay phase, occurring due to the (amplitude dependent) nonlinearity of the system (cf. Section 7.4). In order to verify whether this assumption holds or not, a computation implying the passage no. 27 was performed for a total computation time of 10 s instead of 7.5 s (as in Fig. 7.54).

The result of the force identification is shown in Fig. 7.58, whereas selected re-built responses are illustrated in Fig. 7.59. The re-built responses obviously confirm the expectation of a significant phase shift in the decay phase. This can be seen by analyzing the interval between 7.5 and 10 s, highlighted by the grey areas in Fig. 7.59. Considering

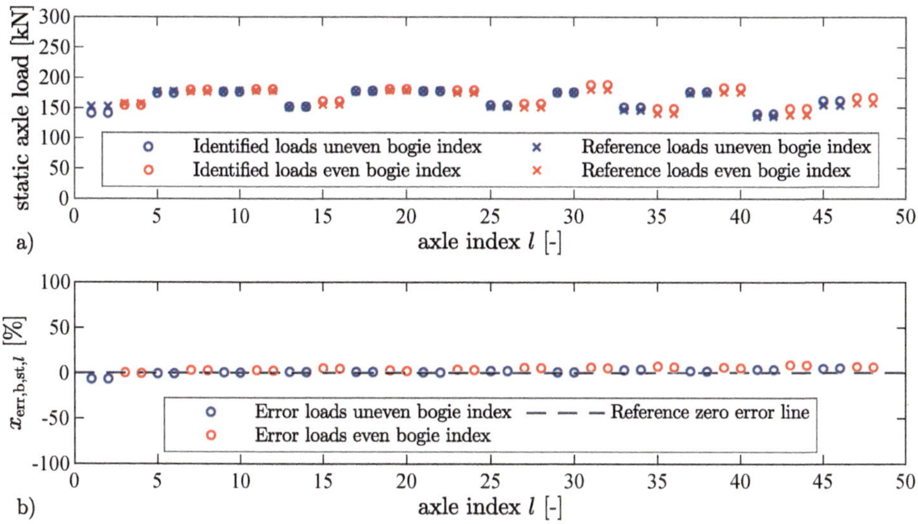

Figure 7.56 Comparison of the identified static axle loads with the reference measured values for the passage no. 27 under consideration of suppressed jump discontinuity effects, an even distribution of the static loads within a bogie and refined weighting factors of the penalty functions: a) force values; b) relative errors of the identified static forces with respect to the corresponding measured results

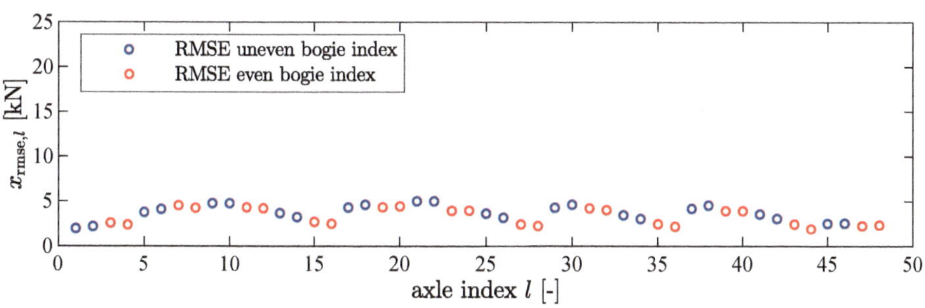

Figure 7.57 RMSE of the identified forces for the passage no. 27 under consideration of suppressed jump discontinuity effects, an even distribution of the static loads within a bogie and refined weighting parameters of the penalty functions

that the main optimization criterion (Eq. 5.15) implies the minimization of the distance between the measured and computed time histories, it is obvious that there is no feasible solution that fulfills it in the decay phase. This means that by considering the decay phase in the computation, an additional error source is deliberately introduced into the solution process. Nevertheless, it has to be noted that the error values caused by the decay phase shift for the investigated case (Fig. 7.58) are still acceptable. The total relative error of the static fore components is $x_{\mathrm{err,b,st,tot}} = 2.16$ %. This value is slightly higher than the result

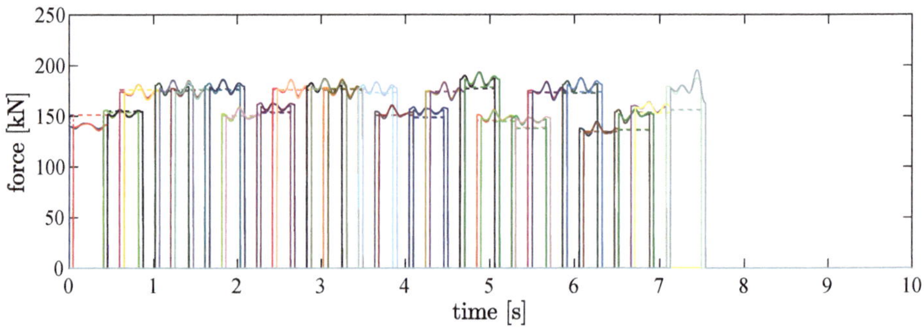

Figure 7.58 Identified dynamic force time histories (solid lines) and reference static axle loads (dashed lines) for the passage no. 27 and measurement configuration 1 under consideration of the decay phase in the measurement data set

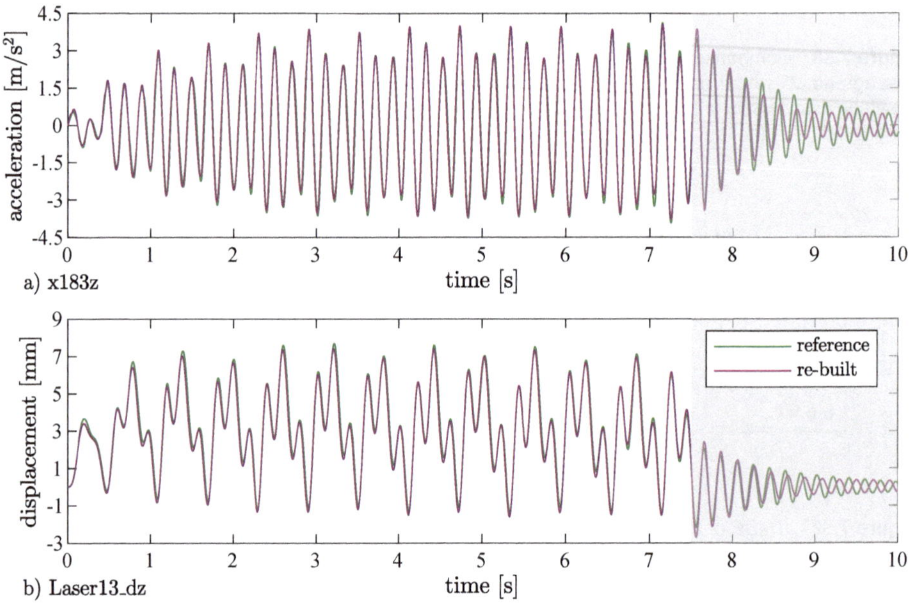

Figure 7.59 Reference and re-built time histories at measured locations for the passage no. 27 and measurement configuration 1 under consideration of the decay phase in the measurement data set: a) acceleration at MP x183z; b) displacement at MP Laser13_dz

$x_{err,b,st,tot} = 1.78$ % corresponding to Fig. 7.54. The average error value for the individual forces is $x_{err,b,st,avg} = 3.16\%$, lying as well above the value of $x_{err,b,st,avg} = 2.8\%$ obtained for the identifications without decay phase (Fig. 7.54). In this respect, it is recommended to perform the MFI without taking the decay phase into account if the investigated structure has amplitude-dependent stiffness and implicitly variable natural frequencies.

7.6.3.6 Application of the method on further train passages

The previously presented results were all obtained based on the ICE 4 passage no. 27 (cf. Tab. 7.2), performed at resonance speed (171 km/h, cf. Eq. 7.7). This is characterized by rather smooth time histories of structural responses, which are clearly dominated by the first modal component (cf. Fig. 7.25). In this case, it was shown that the proposed MFI method revealed excellent results. However, in real practical applications, the pure resonance excitation of a railway bridge is rather an exception. Moreover, the refining of the weighting parameters of the penalty functions by means of a "calibration" run was also performed solely based on the passage no. 27. Therefore, in order to avoid the bias effects related to the calibration of the weighting parameters on the same measurement used in the identification and to also verify the performance of the proposed method with different regular passages, three further out-of-resonance passages (no. 13, 45 and 49 in Tab. 7.2) and one additional near-resonant trial (no. 29 in Tab. 7.2) were subjected to the inverse computation procedure. The additional validation passages were chosen in such a way, so that the whole investigated speed range between 100 and 200 km/h is covered[25]. The corresponding results will be presented in the following, whereas the quantitative comparison will be carried out using the following parameters:

- the total relative error of the static axle forces $x_{\mathrm{err,b,st,tot}}$, determined according to Eq. 7.12. This value represents in fact the relative error of the total vehicle weight.
- the mean relative error $x_{\mathrm{err,st,avg}}$ determined according to Eq. 7.15 as the average of all individual static component errors (i.e. the axle specific errors $x_{\mathrm{err,st},l}$ according to Eq. 7.14). In addition, the corresponding standard deviation $\sigma_{\mathrm{err,st,avg}}$ of all n_{F} errors $x_{\mathrm{err,st},l}$ is evaluated. These parameters are considered to be the most meaningful within the results analysis, since they provide a quick insight into the axle specific errors, which might behave different than the total weight error. The latter can reach reasonable values, even if the individual errors are not acceptable (cf. Fig. 7.41).
- the average root-mean-square error $x_{\mathrm{rmse,avg}}$ determined according to Eq. 7.20 and the corresponding standard deviation $\sigma_{\mathrm{rmse,avg}}$ of all $x_{\mathrm{rmse},l}$ values. These parameters are

[25] It is to be mentioned that the passage no. 31 with a travel speed of about 81 km/h could not be included in the set of inverse computations. The reason for this is, on the one hand, that the ICE 4 encountered another vehicle coming from opposite direction, making it difficult to identify exactly which components of the motion quantities correspond to the actual ICE 4 excitation. The acceleration values considered for this passage in Fig. 7.21 and Fig. 7.27 were assumed as the maximum values occurring before the second vehicle reached the structure. On the other hand, the initially predefined speed for the passage no. 31 was 174 km/h. This could not be achieved due to a malfunctioning of the manual control system, resulting in a much slower passage with slightly non-constant speed.

Weighting parameter set	Error type	Passage no. 13	27	29	45	49
1	$x_{err,b,st,tot}$ [%]	1.79	1.78	3.81	1.45	3.46
	$x_{err,b,st,avg}$ [%]	2.79	2.77	4.19	4.01	4.24
	$\sigma_{err,b,st,avg}$ [%]	2.41	2.27	3.29	2.56	2.02
	$x_{rmse,avg}$ [kN]	4.44	3.39	3.77	6.91	3.93
	$\sigma_{rmse,avg}$ [kN]	1.36	0.94	1.00	2.91	0.91
2	$x_{err,b,st,tot}$ [%]	2.61	5.42	10.04	5.39	6.15
	$x_{err,b,st,avg}$ [%]	3.86	6.13	10.12	6.63	6.56
	$\sigma_{err,b,st,avg}$ [%]	2.99	2.75	5.99	4.31	3.64
	$x_{rmse,avg}$ [kN]	16.79	8.88	11.79	18.47	14.95
	$\sigma_{rmse,avg}$ [kN]	6.55	2.45	2.81	3.66	4.27

Table 7.4 Identification errors for different passages of the ICE 4 over the Schmutter bridge

only used for a rough insight into the amount of dynamic components included in the identified forces.

Fig. 7.60 illustrates the identified dynamic forces for the four additional passages in comparison to the reference values of the static axle loads. In addition, Fig. 7.61 graphically shows the achieved static axle loads and the corresponding axle specific identification errors $x_{err,b,st,l}$. At first sight of Fig. 7.60 and 7.61a, it can be observed, that the proposed MFI method delivers – from the qualitative point of view – very good results, independent on the vehicle velocity. In general, the identified static axle loads tend to be slightly higher than the reference values (Fig. 7.61a).

The quantitative analysis of the identified forces can be performed using Tab. 7.4, which includes the results determined for both previously discussed combinations of w_{p1} and w_{p2}. Parameter set 1 represents the default parameter set defined using the calibration passage ($w_{p1} = 5 \cdot 10^{-5}$ and $w_{p2} = 1 \cdot 10^{-1}$, cf. Section 7.6.3.4) and the parameter set 2 denotes the initial values determined solely based on the numerical investigation ($w_{p1} = 5 \cdot 10^{-7}$ and $w_{p2} = 1 \cdot 10^{-2}$, cf. Section 7.6.2). With respect to the parameter set 1, Tab. 7.4 shows that all relative errors $x_{err,b,st,tot}$ lie below 4 % and are thus considered as very satisfactory. Moreover, the axle specific errors present as well excellent average values $x_{err,st,avg} \leq 4.2$ % and standard deviations $\sigma_{err,st,avg} \leq 3.3\%$. The latter indicates that the error values of the individual axles generally vary within a rather narrow interval (see also 7.61b).

The good agreement between the identified and the reference forces is generally also supported by the comparisons between the reference and the re-built motion quantities. These can be seen in Fig. 7.62 for a set of measured locations and in Fig. 7.63 for selected "unmeasured" locations[26], whereas the results correspond to the slowest passage (no. 45). In general, a good accuracy was only revealed for the measured locations and for the

[26] It is to be mentioned that the reference displacement time history used in Fig. 7.63b (measurement position Laser1_dz) was recorded using the profile scanner as the laser vibromenter results at midspan could not be

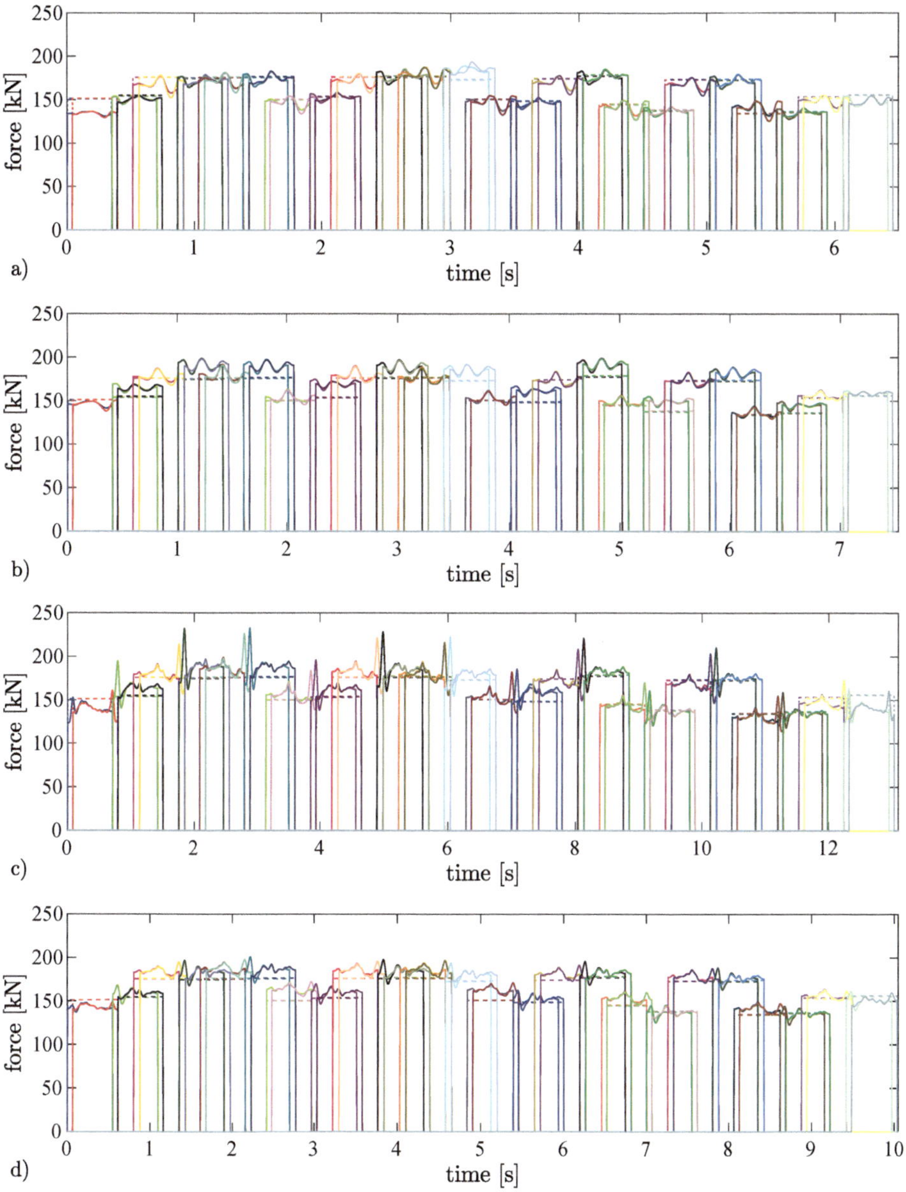

Figure 7.60 Identified dynamic force time histories (solid lines) and reference static axle loads (dashed lines) for different regular passages obtained under consideration of the weighting parameters $w_{p1} = 5 \cdot 10^{-5}$ and $w_{p2} = 1 \cdot 10^{-1}$: a) passage no. 13 (199 km/h); b) passage no. 29 (172 km/h); c) passage no. 45 (99 km/h); d) passage no. 49 (128 km/h)

---

used due to the inaccuracies of the measurements discussed in Section 7.4.4 and this specific passage was not recorded by the microwave interferometer.

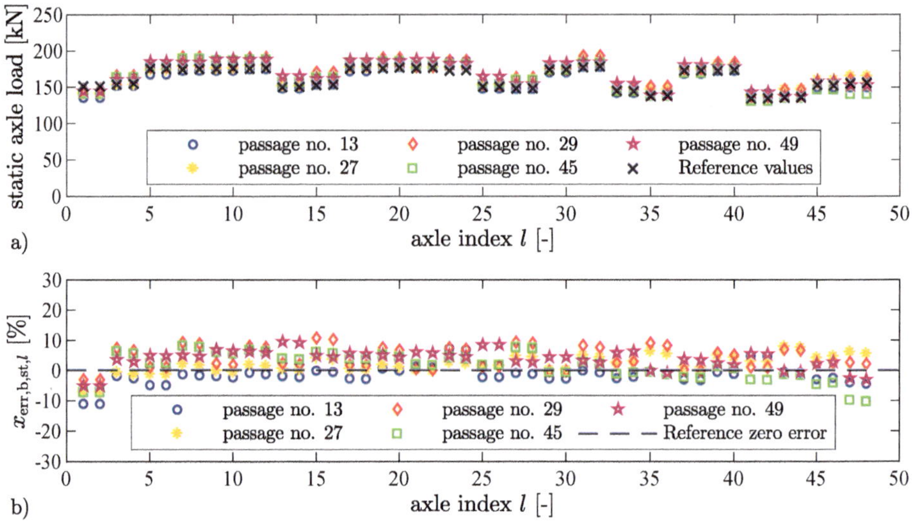

Figure 7.61 Comparison of the identified static axle loads with the reference measured values for different regular passages: a) force values; b) relative errors of the identified static forces with respect to the corresponding measured results

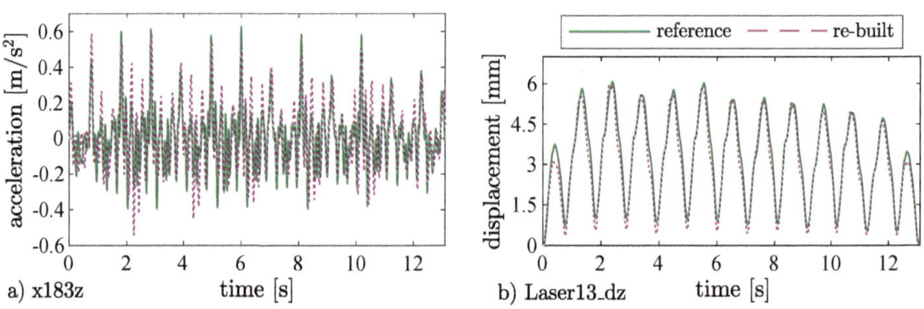

Figure 7.62 Reference and re-built time histories at measured locations for the passage no. 45 and measurement configuration 1 under consideration of refined weighting parameters of the penalty functions: a) acceleration at MP x183z; b) displacement at MP Laser13_dz

"unmeasured" points situated on the loaded substructure (Fig. 7.63a and b). On the contrary, the re-built responses at "unmeasured" locations on the unloaded substructure (Fig. 7.63c and d) present rather moderate accuracies[27]. These results contradict to some extent the findings obtained for the passage no. 27, which indicated a very good accuracy of the

[27] In this respect it is also to be noted that the measured displacement time history of the "unmeasured" location shown in Fig. 7.63d presents a linear drift. Considering that the displacement is lower than 1 mm, the drifting effect might indicate that the structural displacement of the point Laser22_dz lies close to the limits of the measurement device in terms of accuracy.

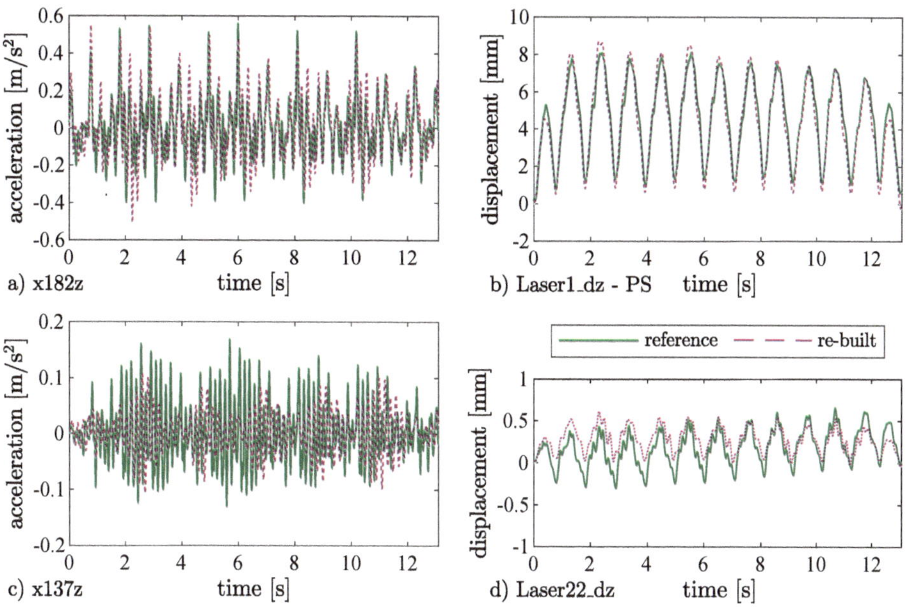

Figure 7.63 Reference and re-built time histories at "unmeasured" locations for the passage no. 45 and measurement configuration 1 under consideration of the refined weighting parameters of the penalty functions: a) acceleration at MP x182z; b) displacement at MP Laser1_dz (measured using the profile scanner); c) acceleration at MP x137z; d) displacement at Laser22_dz

re-built responses for "unmeasured" locations situated on both the loaded and the unloaded structure (Fig. 7.55). In order to clarify this effect, the re-built responses of the remaining passages (no. 13, 29 and 49) are generated and plotted in Appendix B.1. Their analysis leads to the conclusion that the motion quantities of the points assigned to the loaded structure can always be re-built with a high accuracy. On the contrary, the results corresponding to the unloaded structure only present a satisfactory accuracy in case of resonant passages (no. 27 and no. 29). The main reason for this is thought to be the approach used for the model calibration, which implied the fitting of the natural frequency corresponding to a resonance excitation. The responses of the resonant passages are obviously dominated by the first mode, decreasing thus the influence of model errors related to the effect of the longitudinal ballast joint in the higher modes. In addition, the amplitude of the responses recorded both on the loaded and on the unloaded structures are much lower in case of out-of-resonance passages than in case of resonance excitation, as already seen in Fig. 7.63. Considering the amplitude dependent natural frequency of the structure, it is expected that the model calibrated at the resonant frequency presents increased errors for computations of non-resonant load cases. Moreover, eventual amplitude-independent measurement errors of the laser vibrometers have a higher influence for low signal amplitudes. They could also contribute to the higher inaccuracies of the re-built responses for non-resonant passages observed in the investigated cases.

Regarding the dynamic components of the identified forces, it can be stated that the investigated train passages generally present a similar behavior with RMSE values of around 4 kN, excepting the passage no. 45 with a travel speed of 99 km/h. In this case, the mean RMSE almost reaches 7 kN. The reason for this can be seen in the corresponding plot (Fig. 7.60c), which shows a set of identified high-frequency components, affecting almost all axles. There is no doubt that this effect can not be related to any physical occurrence. A closer look at Fig. 7.60c indicates that the high peaks always occur around the times when another axle either arrives or leaves the structure, similar to the discussion on Fig. 7.52. However, the maximum force values in this case are more pronounced than in Fig. 7.52. The reason for this might be the lower travel speed, which obviously implies a longer passage time, being thus more susceptible to slight variations in the vehicle speed.

Without going into the same of level of detail, the results obtained for the second set of weighting parameters, also provided in Tab. 7.4 and illustrated in Appendix B.2, will be briefly discussed. It can be seen that the results concerning the static components of the axle loads achieve higher errors than in case of parameter set 1. Both the error of the total vehicle weight and the mean error of the individual static axle loads reach values of up to 10 %. Under consideration of the rather high standard deviations of the axle specific error values, it can be stated that the results partly lie outside the acceptable range (e.g. passage 29). The tendency of higher values of the identified static force components is generally observed also in this case (cf. Fig. B.8). Regarding the dynamic components of the identified forces, it is obvious that the lower weighting parameter of the first penalty function (w_{p1}) generally allows a higher variation of the identified results around their mean value. The average RMSE values reach up to 19 kN for the passage no. 45. Even though the general appearance of the results might indicate a plausible solution, the high dynamic variations of the axle specific results are not justifiable from the physical point of view, especially for the out-of-resonance passages (no. 13, 45 and 49). These seem to retrieve significantly higher RMSE values than the resonant passages (no. 27 and 29). All these findings clearly emphasize the importance of a calibration run to determine an appropriate set of weighting parameters for the two penalty functions.

7.6.3.7 Effect of the measurement configuration

All previously discussed results were generated using the measurement configuration 1, defined in Tab. 7.3. It includes a set of six acceleration sensors and one displacement measurement device, representing thus a practicable measurement set-up with respect to the real applications. Nevertheless, a set of further measurement configurations defined in Tab. 7.3 are investigated in order to find out if a sufficient accuracy of the force identification can be achieved with even fewer sensors or if the consideration of additional sensor recordings leads to a significant increase of the result quality. In addition to the description provided in Tab. 7.3, the 12 measurement configurations are graphically illustrated in Fig. 7.64, which should facilitate a better understanding and a quicker visual interpretation of the results presented in Tab. 7.5[28] and illustrated in Appendix B.3.

Measurement configuration	Error type	Passage no. 13	27	29	45	49
1	$x_{err,b,st,tot}$ [%]	1.79	1.78	3.81	1.45	3.46
	$x_{err,b,st,avg}$ [%]	2.79	2.77	4.19	4.01	4.24
	$\sigma_{err,b,st,avg}$ [%]	2.41	2.27	3.29	2.56	2.02
	$x_{rmse,avg}$ [kN]	4.44	3.39	3.77	6.91	3.93
	$\sigma_{rmse,avg}$ [kN]	1.36	0.94	1.00	2.91	0.91
2	$x_{err,b,st,tot}$ [%]	1.69	3.20	6.19	3.18	5.28
	$x_{err,b,st,avg}$ [%]	2.72	3.85	6.20	5.21	5.87
	$\sigma_{err,b,st,avg}$ [%]	2.98	1.88	2.21	3.65	4.51
	$x_{rmse,avg}$ [kN]	3.65	3.40	3.72	6.46	4.18
	$\sigma_{rmse,avg}$ [kN]	0.97	0.87	0.95	2.33	0.89
3	$x_{err,b,st,tot}$ [%]	17.08	14.56	17.92	23.00	16.54
	$x_{err,b,st,avg}$ [%]	17.10	15.19	17.89	22.84	16.58
	$\sigma_{err,b,st,avg}$ [%]	3.97	8.72	10.32	11.31	7.09
	$x_{rmse,avg}$ [kN]	3.02	2.43	2.25	6.13	2.98
	$\sigma_{rmse,avg}$ [kN]	0.99	0.60	0.53	2.51	1.11
4	$x_{err,b,st,tot}$ [%]	13.87	17.15	19.20	28.93	19.30
	$x_{err,b,st,avg}$ [%]	13.91	17.86	19.25	28.72	19.38
	$\sigma_{err,b,st,avg}$ [%]	3.58	9.99	10.46	11.79	7.60
	$x_{rmse,avg}$ [kN]	2.38	1.93	1.82	4.85	2.63
	$\sigma_{rmse,avg}$ [kN]	0.65	0.48	0.48	1.79	0.84
5	$x_{err,b,st,tot}$ [%]	–	1.81	3.80	3.09	6.10
	$x_{err,b,st,avg}$ [%]	–	2.76	3.78	4.07	6.12
	$\sigma_{err,b,st,avg}$ [%]	–	2.40	2.01	2.36	3.79
	$x_{rmse,avg}$ [kN]	–	3.57	3.38	7.07	4.36
	$\sigma_{rmse,avg}$ [kN]	–	0.98	0.90	2.96	0.97
6	$x_{err,b,st,tot}$ [%]	–	2.74	2.90	4.12	7.30
	$x_{err,b,st,avg}$ [%]	–	3.03	2.90	6.45	7.35
	$\sigma_{err,b,st,avg}$ [%]	–	2.80	1.41	5.14	5.90
	$x_{rmse,avg}$ [kN]	–	3.48	3.17	6.57	4.39
	$\sigma_{rmse,avg}$ [kN]	–	0.85	0.82	2.37	0.83

table continued on the next page

[28] It is to be mentioned that the configurations 5, 6 and 7 could not be investigated for the passage no. 13, since they imply the consideration of the measurement point Laser1_dz in the analysis. As shown in Section 7.4.4, the laser vibrometer signals at this point are erroneous and could not be used in the further analyses. Moreover, no profile scanner or microwave interferometer measurements were performed for the passage no. 13, so that no usable signals were available at this point.

Measurement configuration	Error type	Passage no. 13	27	29	45	49
7	$x_{\text{err,b,st,tot}}$ [%]	–	3.90	2.00	7.72	9.72
	$x_{\text{err,b,st,avg}}$ [%]	–	4.61	2.53	9.03	9.80
	$\sigma_{\text{err,b,st,avg}}$ [%]	–	3.86	1.57	7.71	4.63
	$x_{\text{rmse,avg}}$ [kN]	–	2.56	2.35	1.76	1.63
	$\sigma_{\text{rmse,avg}}$ [kN]	–	0.76	0.65	0.32	0.18
8	$x_{\text{err,b,st,tot}}$ [%]	7.19	4.61	8.58	9.94	9.23
	$x_{\text{err,b,st,avg}}$ [%]	9.33	4.88	8.60	10.83	9.15
	$\sigma_{\text{err,b,st,avg}}$ [%]	10.17	2.24	2.16	8.55	5.65
	$x_{\text{rmse,avg}}$ [kN]	1.39	2.69	2.92	1.14	1.28
	$\sigma_{\text{rmse,avg}}$ [kN]	0.31	0.77	0.81	0.15	0.20
9	$x_{\text{err,b,st,tot}}$ [%]	4.28	0.06	1.16	0.26	1.38
	$x_{\text{err,b,st,avg}}$ [%]	4.63	3.58	5.63	5.51	2.80
	$\sigma_{\text{err,b,st,avg}}$ [%]	2.93	2.45	2.41	3.31	2.33
	$x_{\text{rmse,avg}}$ [kN]	4.70	3.59	4.07	7.46	4.44
	$\sigma_{\text{rmse,avg}}$ [kN]	1.31	1.04	1.10	3.07	1.00
10	$x_{\text{err,b,st,tot}}$ [%]	0.72	1.82	0.84	11.40	1.17
	$x_{\text{err,b,st,avg}}$ [%]	1.77	2.23	4.09	14.12	4.76
	$\sigma_{\text{err,b,st,avg}}$ [%]	1.50	1.94	1.80	10.17	3.42
	$x_{\text{rmse,avg}}$ [kN]	3.30	3.95	4.13	10.72	4.22
	$\sigma_{\text{rmse,avg}}$ [kN]	1.07	1.17	1.11	4.61	1.32
11	$x_{\text{err,b,st,tot}}$ [%]	2.24	3.15	0.83	10.70	0.15
	$x_{\text{err,b,st,avg}}$ [%]	3.60	3.24	2.81	13.35	4.65
	$\sigma_{\text{err,b,st,avg}}$ [%]	2.56	1.60	1.52	8.42	3.03
	$x_{\text{rmse,avg}}$ [kN]	2.52	3.98	4.18	7.04	4.32
	$\sigma_{\text{rmse,avg}}$ [kN]	0.88	1.13	1.06	2.50	1.34
12	$x_{\text{err,b,st,tot}}$ [%]	20.72	6.46	30.37	25.74	6.07
	$x_{\text{err,b,st,avg}}$ [%]	25.58	10.70	30.08	38.62	22.96
	$\sigma_{\text{err,b,st,avg}}$ [%]	19.87	7.47	14.75	22.63	16.35
	$x_{\text{rmse,avg}}$ [kN]	9.07	2.80	1.93	9.72	6.32
	$\sigma_{\text{rmse,avg}}$ [kN]	2.61	1.31	0.77	2.87	1.85

Table 7.5 Identification errors for different measurement configurations of the Schmutter bridge

Analyzing the results of the configuration 2 in Tab. 7.5, it can be seen that the reduction of the measurement set-up to only two acceleration sensors and a single displacement sensor generally causes an increase of the identification errors. However, the total error of the vehicle weight and the mean error of the individual static axle loads are still in an acceptable range. With respect to the dynamic components of the results, no significant change can be observed with respect to the configuration 1. At first sight, these findings might indicate that increasing the number of sensors is associated with a decrease of the identification errors.

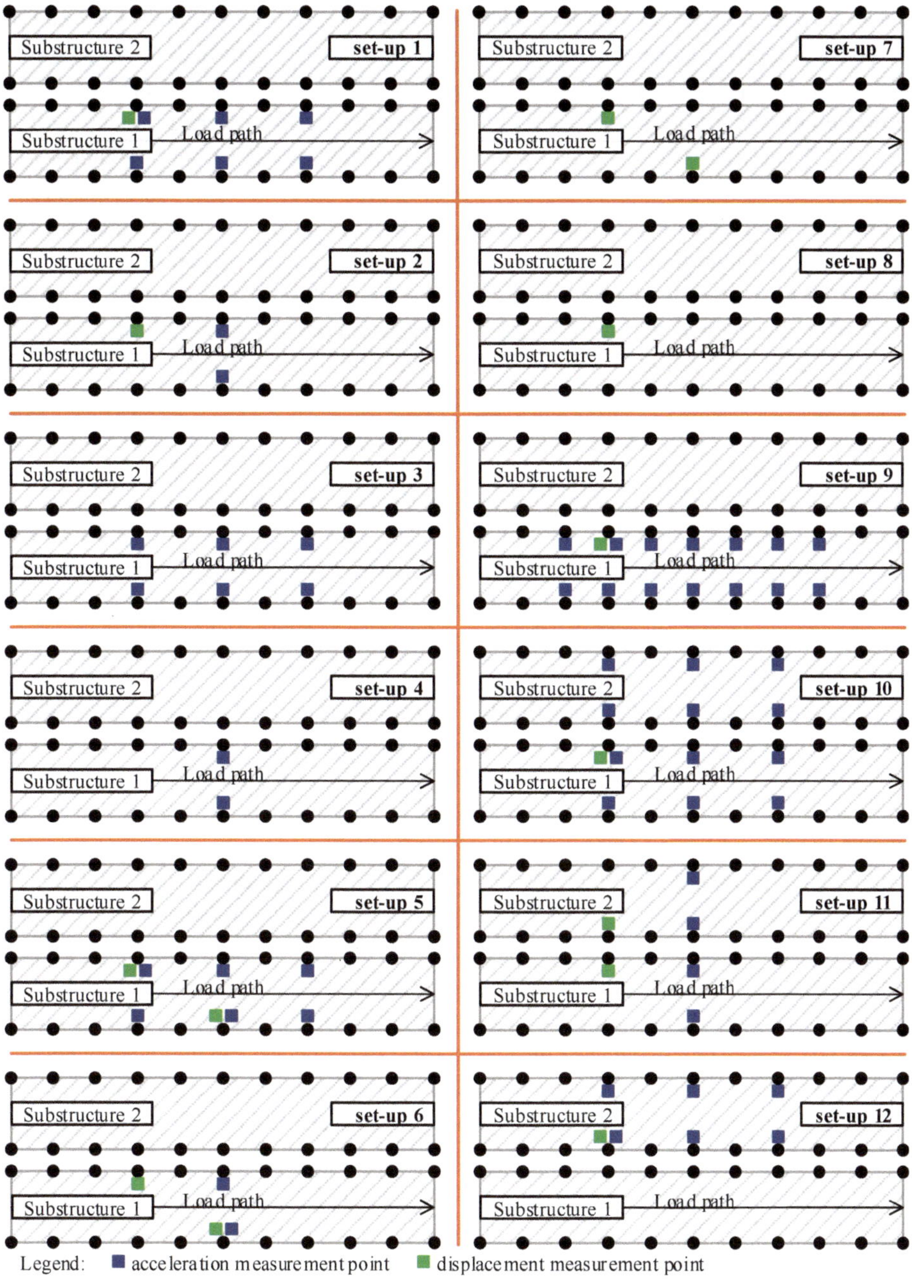

Figure 7.64 Investigated measurement configurations of the Schmutter bridge

However, it is to be mentioned that this can not be generally postulated. For example, the configurations 5 and 6 are similar to the set-ups 1 and 2, but include an additional displacement sensor (Laser1_dz). The consideration of the additional displacement measurement point obviously leads to an improvement of the results for the resonant passages. However, higher errors are revealed for the non-resonant passages. A similar tendency can also be observed for the configuration 9, which implies a very comprehensive and rather impracticable measurement set-up consisting of 14 accelerometers and one displacement sensor. It was found that the higher number of accelerometers generally improves the error of the total vehicle weight (exception: passage no. 13) but leads to increased average errors of the individual static axle forces for almost all passages (except for passage no. 49). The effect of lower identification accuracy for additional sensors considered in the analysis clearly indicates the presence of modeling errors. A similar behavior has been also observed for other related applications, e.g. the strain estimation of an offshore monopile wind turbine (MAES et al., 2016a).

The set-ups 10 and 11 consider measurement data from both the loaded and the unloaded structure. Comparing the results with those of configurations 1 and 2, it is clear that the additional sensors partly improve and partly reduce the accuracy of the results. The largest error increase is observed for the slow passage no. 45, thus reconfirming the findings from Section 7.6.3.6, which revealed a very poor accuracy of the re-built responses of the unloaded structure for the out-of-resonance passages. According to these results, it is expected that considering measurement points located only on the unloaded structure will lead to even lower identification accuracy. This situation was investigated with the set-up 12, which indeed yielded unacceptable errors. This emphasizes again the need for a very accurate model for achieving an accurate MFI, especially in case of twin bridges sharing a common ballast bed. If the moving forces of both tracks of a twin bridge are to be identified, it is recommended to install sensors on both substructures, in order to minimize the influence of the expectedly higher modeling errors of the longitudinal ballast joint.

The configurations 3 and 4 only consider acceleration sensors in the analysis. The corresponding errors are significantly higher, lying by far outside the acceptable range. The reason for this is that – in case of only acceleration signals – the method fails to reconstruct the static component of the force. This effect matches the findings obtained for the investigation with simulated measurement data and a low regularization amount[29] (cf. Section 6.7.5). It can can be explained by the fact that the accelerations are less sensitive to the quasi-static axle forces than the displacements. The analog cases, implying only displacement sensors (set-up 7 and 8) revealed only a moderate accuracy with a slight tendency to better results for the resonant passages (no. 27 and 29), confirming thus that the model seems to be more accurate in the near-resonant band. In any case, the static load components revealed a better match than in the case of a MFI based only on acceleration

[29] The analysis with simulated measurements and an increased regularization amount, achieved by using a higher window length, lead to a very good accuracy, even when only acceleration sensors are used.

sensors (configurations 3 and 4). This confirms the hypothesis of higher sensitivity of the displacement sensors with respect to static force components compared to the sensitivity of the accelerations. In addition, it is to be mentioned that the identified forces based on only displacement sensors present lower average values of the dynamic components, highlighting that the acceleration measurements are more sensitive with respect to dynamic forces than the displacement measurements.

In general, the highest accuracy could be found for a combination of acceleration and displacement sensors, whereas sufficiently accurate results could be obtained even for a low number of sensors, which mainly have to be placed on the loaded structure.

7.6.3.8 Effect of the damping ratio

The previously presented results were obtained with a damping ratio $\xi = 2.74$ % obtained by applying the logarithmic decrement approach on several free-decay signals. However, it is to be mentioned that this value was in fact obtained for the first mode and assumed in the computation also for the higher modes. The reason for this is that the investigated set of measurements did not contain clear decay phases for the higher modes, which are required for a reliable estimation of the corresponding damping ratios. In this respect, it is recalled that a set of further attempts to determine the modal damping ratios by means of the CSI/ref method were not conclusive because of the amplitude dependency of both the damping and the natural frequency (cf. Section 7.4.1 and 7.4.3).

Even though the previously shown identification results revealed that an excellent accuracy can be achieved with the uniform damping ratio $\xi = 2.74$ %, the simplifying assumption of the same damping ratio for all modes is still associated with high uncertainties. It is well known that the modal damping ratio of bridge structures is not expected to have a constant value over all modes (LIU et al., 2009b; MAGALHÃES et al., 2010; LIU et al., 2014c; CANTIENI et al., 2016; PETERSEN et al., 2017). In this sense, a short study on the effect of the damping ratio on the MFI accuracy is performed in the following. For this, the five investigated passages are subjected to further computations under consideration of different damping ratios, which were assigned to both modes considered in the analysis. The damping ratio was varied by ± 0.5 % and ± 1.0 %, resulting in the following new samples: 1.74 %, 2.24 %, 3.24 % and 3.74 %.

The results are given in Tab. 7.6 and illustrated in Appendix B.4. They show that the damping ratio only has an influence on the results of the resonant passages, which is justifiable from the mechanical point of view. Analyzing the resonant passages (no. 27 and 29), it can be seen that lower damping ratios are associated with an increased identification accuracy of the static axle loads. This effect occurs due to the fact that the identification procedure revealed so far – under consideration of the default value $\xi = 2.74$ % – a tendency towards slightly higher identified static force components than the corresponding reference values (cf. Fig. 7.61). In this situation, it is obvious that the decrease of the damping ratio requires lower forces for the fulfilling the main optimization criterion.

Damping ratio ξ	Error type	Passage no. 13	27	29	45	49
1.74 %	$x_{\text{err,b,st,tot}}$ [%]	2.02	0.06	2.35	1.40	3.28
	$x_{\text{err,b,st,avg}}$ [%]	3.08	1.65	2.95	3.99	4.18
	$\sigma_{\text{err,b,st,avg}}$ [%]	2.23	1.70	2.57	2.56	2.00
	$x_{\text{rmse,avg}}$ [kN]	4.64	4.22	4.54	6.93	4.15
	$\sigma_{\text{rmse,avg}}$ [kN]	1.28	0.94	0.98	2.91	1.04
2.24 %	$x_{\text{err,b,st,tot}}$ [%]	1.91	0.95	3.10	1.43	3.37
	$x_{\text{err,b,st,avg}}$ [%]	2.93	2.16	3.56	4.00	4.21
	$\sigma_{\text{err,b,st,avg}}$ [%]	2.29	1.92	2.96	2.56	2.00
	$x_{\text{rmse,avg}}$ [kN]	4.54	3.76	4.13	6.92	4.03
	$\sigma_{\text{rmse,avg}}$ [kN]	1.32	0.95	1.01	2.91	0.97
2.74 %	$x_{\text{err,b,st,tot}}$ [%]	1.79	1.78	3.81	1.45	3.46
	$x_{\text{err,b,st,avg}}$ [%]	2.79	2.77	4.19	4.01	4.24
	$\sigma_{\text{err,b,st,avg}}$ [%]	2.41	2.27	3.29	2.56	2.02
	$x_{\text{rmse,avg}}$ [kN]	4.44	3.39	3.77	6.91	3.93
	$\sigma_{\text{rmse,avg}}$ [kN]	1.36	0.94	1.00	2.91	0.91
3.24 %	$x_{\text{err,b,st,tot}}$ [%]	1.67	2.59	4.51	1.48	3.55
	$x_{\text{err,b,st,avg}}$ [%]	2.69	3.43	4.80	4.03	4.27
	$\sigma_{\text{err,b,st,avg}}$ [%]	2.55	2.66	3.62	2.56	2.04
	$x_{\text{rmse,avg}}$ [kN]	4.35	3.12	3.47	6.89	3.83
	$\sigma_{\text{rmse,avg}}$ [kN]	1.41	0.93	0.95	2.91	0.87
3.74 %	$x_{\text{err,b,st,tot}}$ [%]	1.55	3.38	5.21	1.50	3.64
	$x_{\text{err,b,st,avg}}$ [%]	2.68	4.11	5.45	4.04	4.30
	$\sigma_{\text{err,b,st,avg}}$ [%]	2.64	3.05	3.89	2.57	2.06
	$x_{\text{rmse,avg}}$ [kN]	4.27	2.93	3.21	6.88	3.74
	$\sigma_{\text{rmse,avg}}$ [kN]	1.45	0.92	0.89	2.91	0.83

Table 7.6 Effect of the damping ratio on the accuracy of the force identification

Generally, it can be stated that – in case of regular train passages with non-resonant speed – the accuracy of the force identification is rather insensitive to estimation errors of the damping ratios. Nevertheless, the damping plays an important role when dealing with resonant or near-resonant excitation. Even though the resonant excitation of railway bridges occurs in rather isolated cases, a realistic estimation of the damping should be aimed at, in order to eliminate an additional possible source of error.

7.6.3.9 Effect of the number of modes considered in the analysis

In the following, the effect of the number of modes considered in the analysis will be investigated. The previously presented results were generated under consideration of the first two modes. For this, all measurement data sets were low pass filtered using a

Number of modes	Error type	Passage no. 13	27	29	45	49
1	$x_{err,b,st,tot}$ [%]	80.65	74.62	73.51	79.49	86.64
	$x_{err,b,st,avg}$ [%]	80.62	74.61	73.41	79.34	86.48
	$\sigma_{err,b,st,avg}$ [%]	4.25	3.97	4.43	7.49	5.18
	$x_{rmse,avg}$ [kN]	4.30	6.38	7.92	9.39	5.67
	$\sigma_{rmse,avg}$ [kN]	1.69	1.61	1.63	3.88	1.79
2	$x_{err,b,st,tot}$ [%]	1.79	1.78	3.81	1.45	3.46
	$x_{err,b,st,avg}$ [%]	2.79	2.77	4.19	4.01	4.24
	$\sigma_{err,b,st,avg}$ [%]	2.41	2.27	3.29	2.56	2.02
	$x_{rmse,avg}$ [kN]	4.44	3.39	3.77	6.91	3.93
	$\sigma_{rmse,avg}$ [kN]	1.36	0.94	1.00	2.91	0.91
3	$x_{err,b,st,tot}$ [%]	6.56	0.58	1.23	6.64	2.57
	$x_{err,b,st,avg}$ [%]	6.47	2.81	3.62	6.89	3.55
	$\sigma_{err,b,st,avg}$ [%]	2.47	2.64	2.71	4.50	2.66
	$x_{rmse,avg}$ [kN]	4.45	3.77	4.14	9.38	9.35
	$\sigma_{rmse,avg}$ [kN]	1.38	1.08	0.97	2.69	1.86
4	$x_{err,b,st,tot}$ [%]	9.59	5.48	3.99	12.67	8.30
	$x_{err,b,st,avg}$ [%]	9.53	5.90	7.00	12.72	8.33
	$\sigma_{err,b,st,avg}$ [%]	2.20	4.49	4.08	3.83	3.67
	$x_{rmse,avg}$ [kN]	5.78	4.12	4.77	9.25	11.27
	$\sigma_{rmse,avg}$ [kN]	1.61	1.27	1.09	2.37	2.89
5	$x_{err,b,st,tot}$ [%]	13.04	5.92	4.43	17.58	11.12
	$x_{err,b,st,avg}$ [%]	12.88	5.70	5.11	17.77	11.16
	$\sigma_{err,b,st,avg}$ [%]	3.84	3.69	3.60	5.31	3.72
	$x_{rmse,avg}$ [kN]	5.77	4.03	4.91	9.73	9.69
	$\sigma_{rmse,avg}$ [kN]	1.38	1.33	1.16	1.80	2.97

Table 7.7 Effect of the number of modes considered in the computation on the accuracy of the force identification

Butterworth filter (5th order) with a cut-off frequency of 10 Hz. In this section, four additional computation sets will performed, which imply the consideration of 1, 3, 4 or 5 modes. In this respect, it is to be mentioned that the computational modes 3 to 5 posses natural frequencies between 10 and 20 Hz (Fig. 7.38). Therefore, the measurement data used for these cases was filtered under consideration of a cut-off frequency of 20 Hz.

The results are summarized in Tab. 7.7, whereas their graphical representation is provided in Appendix B.5. It can be seen that the use of only one mode leads to a significant overestimation of the static components of the identified forces (cf. Fig. B.26). The reason for this is that the first two modes are almost identical along the loading path. This can be seen in Fig. 7.65, which represents the two modes normalized with respect to a modal mass of 1 t. Two nearly identical mode shapes obviously imply nearly equal

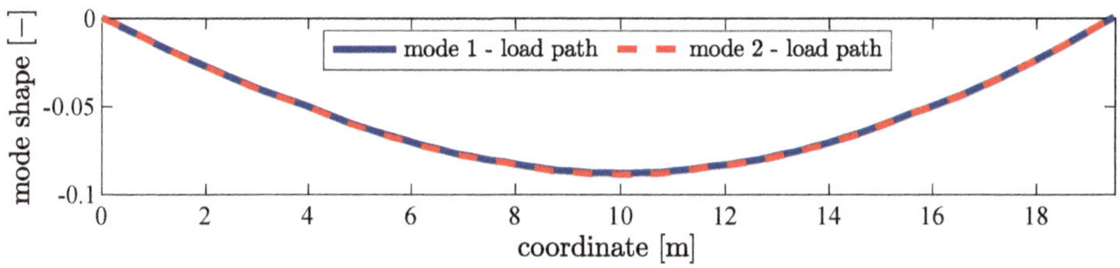

Figure 7.65 Computed mode shapes of the Schmutter bridge along the load path – modes normalized with respect to the unit modal mass (1 t)

modal loads (cf. Eq. 3.45) and therefore a similar contribution of the two modes to the total structural response. When considering only one mode within the identification procedure, the effect of the second mode has to be compensated by the first one. This can only be achieved by a considerable increase in the identified loads.

The computations with a higher number of modes revealed that increasing the number of modes does not necessarily imply a better identification accuracy. This contradicts both the engineering expectations and the findings from the investigations with simulated measurement data (Section 6.7.3). In fact, a clear trend of increasing errors for higher number of modes can be observed, whereas greater discrepancies are revealed for the out-of-resonance passages. These findings can be mainly explained by the modeling errors. It was presented in Section 7.5, that the structural model was calibrated with respect to the resonant frequency occurring during the resonant passages and to the displacements acquired during a quasi-static passage. For this reason, on the one hand, the model is expected to present certain inaccuracies for passage speeds outside the resonant range, which are associated to lower structural responses. This is thought to be the cause for the higher deviations of the identified forces in case of passage 13, 45 and 49. On the other hand, by updating the dynamic behavior of the numerical model with respect to only the first natural frequency determined using the resonance curve (Fig. 7.21), it is highly uncertain to which extent the higher computational modes match the real situation. In this regard, the results given in Tab. 7.7 for 3 to 5 modes indicate that the structural model of the Schmutter bridge presents considerable errors for the higher modes, leading to a substantial decreasing of the identification accuracy. In order to reduce the influence of the modeling errors related to the higher modes as far as possible, it is recommended to perform the force identification only under consideration of the certainly identifiable modes. In the same context, the frequency content of the measurement data has to be limited according to the highest natural frequency to be considered. However, this approach completely neglects the eventual influence of out-of-band modes. It has been indeed reported, that the disregard of the quasi-static contribution of the out-of-band modes might lead in certain situations to increased force identification errors (Maes et al., 2019). Nevertheless, the investigations performed in this work revealed a very robust behavior and sufficiently accurate results with

Speed assumption acc. to Tab. 7.2	Error type	Passage no. 13	27	29	45	49
Photoelectric sensors	$x_{\mathrm{err,b,st,tot}}$ [%]	1.79	1.78	3.81	1.45	3.46
	$x_{\mathrm{err,b,st,avg}}$ [%]	2.79	2.77	4.19	4.01	4.24
	$\sigma_{\mathrm{err,b,st,avg}}$ [%]	2.41	2.27	3.29	2.56	2.02
	$x_{\mathrm{rmse,avg}}$ [kN]	4.44	3.39	3.77	6.91	3.93
	$\sigma_{\mathrm{rmse,avg}}$ [kN]	1.36	0.94	1.00	2.91	0.91
Radar device	$x_{\mathrm{err,b,st,tot}}$ [%]	17.42	14.13	7.41	10.61	8.52
	$x_{\mathrm{err,b,st,avg}}$ [%]	23.55	17.95	13.75	23.32	21.70
	$\sigma_{\mathrm{err,b,st,avg}}$ [%]	25.92	20.53	16.26	23.20	19.48
	$x_{\mathrm{rmse,avg}}$ [kN]	12.46	11.38	9.70	14.86	13.17
	$\sigma_{\mathrm{rmse,avg}}$ [kN]	7.51	5.71	4.92	7.52	7.41

Table 7.8 Effect of an inexact train speed on the accuracy of the force identification

respect to the modal truncation. Therefore, it is not expected that neglecting the influence of out-of-band modes will significantly affect the accuracy of the proposed MFI method for typical structures such as those studied in the present work.

7.6.3.10 Effect of an inexact train speed

The investigation presented in Section 6.7.7 showed that the vehicle speed assumed in the computation is one of most sensitive parameter for the accuracy of the proposed MFI method. As a consequence, a strong requirement for an accurate speed measurement was formulated. In order to emphasize this aspect in the context of applications with real measured data, a further set of computations is performed for the five investigated passages under consideration of perturbed speeds. In this respect, it is recalled that the speed measurements for the ICE 4 passages were realized using photoelectric sensors and a radar device installed at the level of carriage 8 of the vehicle (cf. Section 7.3.3 and 7.4.2). The speed results obtained from both systems are given in Tab. 7.2. It can be seen that the values of the photoelectric sensors and the radar device mainly differ by about 1 to 2 km/h. However, the values of the photoelectric sensors were adopted for all computations performed in this chapter. The reason for this is that the photoelectric measurements allowed – due to the known axle distances of the ICE 4 – a double verification of the speed results (cf. Section 7.4.2). In this way, it could be proven that these are the correct values.

The additional computations performed in this section take into account the speed values retrieved by the radar device (cf. Tab. 7.2). The results are given in Tab. 7.8 in comparison with the values obtained under consideration of the photoelectric sensor speeds. It can be clearly seen that the consideration of an inexact speed in the inverse computation leads to unacceptable identification errors. This is valid for all passages. The corresponding graphical illustration of the results, which is provided in Fig. B.31, indicates a similar trend for all investigated passages. Rather low errors can be observed at the beginning of the time histories, where the phase shift between the computed and the measured signals is not yet significant. As the phase shift increases, steadily higher error values can be observed. An intolerable identification accuracy is revealed towards the end of the time history. Given the findings from Section 6.7.7, the results presented in this section are not surprising. Nevertheless, they have been included in the discussion to raise awareness with respect to the importance of accurate speed measurements. It has been shown, that even widely technically accepted speed measurement solutions can lead to highly erroneous identification results.

7.7 Conclusion

This chapter describes a first comprehensive experimental validation process of the proposed MFI method. It is based on investigations on a real railway bridge and implies acceleration and displacement measurements during passages of an ICE 4 vehicle. For this, a detailed folded plate FE model was defined according to the drawing plans. It was calibrated with respect to first resonant frequency and to the quasi-static displacements determined within the experimental investigations.

The first identification results have shown that the proposed MFI method fails to properly identify the static components of the individual forces, even though the total vehicle weight and the re-built responses are in very good agreement with the measured values. This leads to the conclusion that it is easier to achieve an accurate prediction of the responses at unmeasured locations than to correctly identify the real forces. In this context, two main effects were identified that lead to the high axle-specific errors of the quasi-static force components. On the one side, a set of jump discontinuities appear in the time histories of some forces due to the moving window formulation. On the other side, significant differences between the two static components of the two closely spaced axles of a bogie can be observed. This issue is thought to occur mainly due to modeling errors, e.g. by neglecting the load-distributing characteristics of the ballasted track, in combination with the low differences between the values of mode shapes corresponding to the positions of the two axles. Nevertheless, strategies for the correction of both previously mentioned issues in the post-processing have been proposed. They lead to a very satisfactory accuracy with regard to both the total weight of the vehicle and the individual static axle loads.

Furthermore, it has been shown that in case of applications with real measurement data, the weighting parameters of the penalty functions can be improved by including a calibration run in their determination. A total of five passages with speeds between 100 km/h and 200 km/h were investigated under consideration of refined weighting parameters. Excellent results with respect to relative error of the total weight were achieved. All values lie below 4 %. The individual static axle loads could be identified with average accuracy values below 4.2 %. This confirms that the proposed method can identify the static axle loads with very satisfactory accuracy. In general, a very good accuracy of the re-built responses at "unmeasured" locations can only be observed for the points situated on the loaded substructure. On the contrary, a rather poor match of the re-built responses of the unloaded substructure could be detected. However, this does not entirely hold for the resonant passages. The reason for this is that the model was calibrated with respect to the resonant frequency. This implicates that the model errors are most likely higher for a non-resonant excitation. With respect to the accuracy of the identified dynamic force components, no clear statement can be formulated, since no reference dynamic force measurements were performed. However, it has been observed that the maxima of the dynamic force components usually occur around the time steps where another axle arrives on or leaves the bridge. This effect might indicate that the dynamic components are strongly influenced by numerical artifacts.

Similar to the investigations with simulated measurement data (Chap. 6), the highest identification accuracy could be achieved for a combination of acceleration and displacement time histories. However, the sensors should be always installed on the loaded substructure. In this context, it is highlighted that dynamic displacement time histories recorded by a microwave interferometer were used in the identification process. Contrary to the results of the simulated measurement data, an accurate force identification based on only acceleration sensors is not possible. The reason for this is most likely that the real measured accelerations are rather insensitive to the static force components. It has also been shown that increasing the number of sensors does not necessarily increase the accuracy of the identified static axle loads.

Finally, a parametric study with varying damping ratios revealed that the damping only plays a determinant role in case of resonant passages. Therefore, the requirement of a highly accurate model should also refer to the damping ratio. Moreover, contrary to the findings obtained for simulated measurement data, a higher number of modes considered in the computation does not necessarily imply an increased identification accuracy. The reason for this are the modeling errors. By only updating the model with respect to its first natural frequency, it is highly uncertain, to which extent the higher frequencies and mode shape values match the reality.

8 Experimental validation: pedestrian bridge

8.1 Framework of the investigation

Chap. 7 was concerned with the validation and application of the proposed MFI method in the field of railway bridges. The present chapter deals with an additional experimental validation of the force identification procedure. It implies the reconstruction of the pedestrian induced loads on a full-scale experimental footbridge. It should be mentioned that this experiment was conducted as part of another research project on human-structure interaction in which the author of this paper was significantly involved. Therefore, the experiment was not ideally designed and conducted for the validation of the developed MFI technique. Nevertheless, an available set of appropriate measurement data could be identified and used within the scope of this work. The results should underline – based on validations with real measurement data – the wide range of possible applications for the proposed MFI method: from small structures excited by rather low pedestrian forces up to the "heavy category" of a complete train passing a railway bridge in operation. Before proceeding to the presentation of the actual experimental validation, a brief introduction into the topic of the direct and indirect measurement of the pedestrian induced load, also denoted in the following as ground reaction force (GRF), is provided in the following.

8.2 Ground reaction forces: problem statement

In the current context of increasingly stringent architectural demands, particularly in terms of span, slenderness and use of innovative lightweight material combinations, many new footbridges are susceptible to a high level of vibrations. This eventually leads to a discomfort of the pedestrians that has to be avoided to the greatest extent possible. Current methods for the dynamic analysis of footfall induced vibrations represent the pedestrian action as a force moving over the structure at a constant speed (RACIC et al., 2009; ŽIVANOVIĆ, 2016; CAPRANI et al., 2016; PETERSEN et al., 2017). The parameters describing the time dependent force induced by the pedestrian to the structure – denoted as ground reaction force (GRF) – are usually adopted from literature. The most common approach is to describe the GRF time history as a Fourier series with up to four harmonic

A. Firus, *A Contribution to Moving Force Identification in Bridge Dynamics*, Mechanik, Werkstoffe und Konstruktion im Bauwesen 65, https://doi.org/10.1007/978-3-658-39838-5_8

components (BACHMANN, 1988; KERR et al., 2001; SEILER et al., 2004; ŽIVANOVIĆ et al., 2005), whereas these are usually determined based on experimental investigations on rigid (non-vibrating) floors. However, it has been reported in several cases that this approach often leads to an overestimation of the real structural responses (ŽIVANOVIĆ, 2016; SHAHABPOOR et al., 2016; AHMADI et al., 2018a; FIRUS et al., 2018b). There is increasing evidence that this might be partly a result of the mechanical interaction between the structure and the pedestrian, widely known as the human-structure-interaction (HSI) effect. This is usually expected to occur in case of structures experiencing a high level of vibrations (CAPRANI et al., 2016; ŽIVANOVIĆ, 2016; SHAHABPOOR et al., 2016; AHMADI et al., 2017; VAN NIMMEN et al., 2017; AHMADI et al., 2021). The HSI effect can be also analyzed in terms of ground reaction forces. In this context, it can be noted that the HSI affects the contact forces arising at the interface between the pedestrian foot and the surface of underlying structure, i.e. the GRF induced by a pedestrian walking on a vibrating structure is expectedly lower than the GRF exerted during walking on a rigid surface (BAUMANN et al., 1988; DANG et al., 2016; FIRUS et al., 2018b; AHMADI et al., 2018b).

The classical approach for a direct measurement of GRFs on rigid (non-vibrating) structures implies the use of either biomechanical force plates (KISTLER, 2017) or treadmills instrumented with force transducers (BELLI et al., 2001). These methods are commonly incorporated in biomechanical studies of the human gait. The foot pressure insoles, i.e. instrumented shoes (FORNER CORDERO et al., 2004; FONG et al., 2008), represent a rather new alternative for the direct GRF measurement. First applications in civil engineering have already been reported by SEILER et al. (2004), AHMADI et al. (2018a) and AHMADI et al. (2018b). The main advantage of the pressure insoles measurement is that the subjects are allowed to perform their natural gait, without having any locomotion constraints, such as targeting the force plates (RACIC et al., 2013). A further main category of GRF measurements on rigid ground available in the literature are the so called indirect approaches. These can also be divided – depending on the respective underlying measurement principle – in the following subcategories: methods based on 3D recordings of the human body kinematics (RACIC et al., 2010; VAN NIMMEN et al., 2014; VAN NIMMEN et al., 2018), approaches based on inertial measurement units (BOCIAN et al., 2016; SHAHABPOOR et al., 2018; BROWNJOHN et al., 2018; ANCILLAO et al., 2018; REVI et al., 2020) or even recently published methods implying smartphone recordings (CHEN et al., 2016; MARTINELLI et al., 2020). It is to be noted that these indirect measurement methods of GRFs based on the acquisition of the human body motion will be denoted in the following as "human-centred". They are usually characterized by a rather limited accuracy, especially with respect to the absolute peak-GRFs (ANCILLAO et al., 2018). In this context, the required assumption regarding the inertial properties of each body segment represents one of the main uncertainty sources, since these are expected to present a large scattering between different subjects. Moreover, a justifiable tendency towards using a low, practicable number of sensors for the GRF reconstruction based on human body recordings can be observed in most of the published works. However, the low number of sensors is expectedly associated with higher identification errors (ANCILLAO et al., 2018). Further error sources are the drift between the sensors and the body segment they are attached to and the inaccurate separation of the accelerometer data to the spatial directions (BROWNJOHN et al., 2018). Extensive literature overviews on the measurement methods of GRFs on rigid ground are given in RACIC et al. (2009), RACIC et al. (2013) and ANCILLAO et al. (2018).

Regarding the measurement of GRFs induced by pedestrians on vibrating structures, only few direct approaches are feasible in terms of costs and effort required for installation. The direct installation of biomechanical force plates on a vibrating structure is indeed possible. First applications in the field of pedestrian induced structural vibrations have already been reported (COMER et al., 2013; FIRUS et al., 2018b; AHMADI et al., 2018b). However, such experimental investigations imply high costs and can only be realized in laboratory conditions. Furthermore, in case of experimental laboratory structures, the GRFs can also be measured by load cells installed at the supports of the test structures (FIRUS et al., 2018b; AHMADI et al., 2018b). In this respect, it is to be mentioned that both the use of biomechanical force plates and the measurements using load cells at the supports might be subjected to high measurement errors because of the inertial forces arising due to the accelerations of the forces plates or of the bridge mass (WERKLE et al., 2016; AHMADI et al., 2018b). In order to suppress these errors to the greatest extent possible, some additional post-processing steps are required. They essentially involve determining the inertial force components and eliminating them from the original signals (AHMADI et al., 2018b). Another viable option for the GRF measurement on vibrating structures are the aforementioned in-shoe pressure sensors. However, they show limited accuracy of the force amplitudes obtained by integrating the distributed pressure over the foot insole area (RACIC et al., 2013). VAN HAWERMEIREN et al. (2020) presents a human-centered indirect approach for identification of the contact forces on a vibrating footbridge. The method uses laboratory measurements of GRF and accelerations of the human body at lower back level performed during walking on a rigid surface. These are subsequently incorporated into an updated structural model of a pedestrian bridge (VAN DEN BROECK et al., 2020) with HSI, in order to determine the crowd contact forces on the vibrating structure. The performance of the method is analyzed on the level of reconstructed structural responses. They show that only a limited accuracy can be achieved, partly due to the high signal-to-noise ratio of the interaction human body accelerations with respect to the technical properties of the used acceleration sensors.

In contrast to the drawbacks of the previously introduced methods for GRF measurement on flexible (vibrating) structures, the structural responses can be usually acquired with high accuracy and reasonable effort. In this sense, the main idea of this investigation is to verify if the MFI method proposed in this work is suitable for investigations of GRFs on flexible structures, especially pedestrian bridges. In case of a successful validation, the indirect GRF measurement on flexible structures could be realized without use of costly and sensitive biomechanical equipment. In the following, the indirect measurement of GRFs based on structural responses will be referred to as "structure centred". To the best knowledge of the author, only the study of OBATA et al. (2006) and the very recent work of CHEN et al. (2021) address the structure-centered MFI in the context of pedestrian bridge applications. Both works use genetic algorithms in order to identify the optimal parameter describing the GRF. The measured data consists either of only accelerations (OBATA et al., 2006) or a combination of acceleration and displacement (CHEN et al., 2021) time histories. On the one hand, OBATA et al. (2006) aims to identify the parameters describing a force model (contact time, heel period, tip of toe period, heel impact, tip of toes impact and heel-tip of toe impact), which is characterized by successive identical single steps. On the other hand, the approach of CHEN et al. (2021) seeks to retrieve the parameters describing the equivalent Fourier representation of the total GRF, which results from the linear addition of the forces induced by the left and the right foot. Moreover, CHEN et al. (2021) also deals with the problem of pedestrian groups. In this sense, an additional optimization variable is defined as the equivalent number of pedestrians. However, since the GRF identification is only tackled by means of a parameter set describing a generalized single step or a Fourier force model, the methods of OBATA et al. (2006) and CHEN et al. (2021) seem to be unable to identify eventual differences between different steps. These are usually observed in GRF measurements (VAN NIMMEN et al., 2014) and might provide important insights towards a better understanding of the HSI effects. On the contrary, the formulation of the proposed MFI problem (cf. Chap. 5) considers the GRF values in each time step as optimization variables. This is expected to allow the identification of eventual differences between the GRFs of the individual step sequences.

Before proceeding to the presentation of the experimental campaign and to the discussion of the results, it is to be mentioned that the procedures used in the experimental work were approved by the ethical committee of of the Technical University of Darmstadt and the participants gave written informed consent prior to participation.

8.3 Investigation object

In the following, the MFI method developed in this work will be subjected to an additional experimental validation procedure based on real measurement data from an experimental pedestrian bridge. The structure is shown in Fig. 8.1 and will be denoted in the following as the *HUMVIB-Bridge*. It was built on the campus of the Technical University of Darmstadt (Germany), mainly for research purposes in the field of HSI.

The static system of the HUMVIB-Bridge is a simply supported beam with a span of $l = 13.24$ m. The bearing structure mainly consists of two longitudinal steel beams (steel type S235JR) with HEB240 cross section and a reinforced concrete deck. The latter is composed of 13 separate precast elements with dimensions $250 \times 100 \times 12$ cm (Fig. 8.2).

Figure 8.1 HUMVIB experimental pedestrian bridge (the middle support was removed during the experimental investigations)

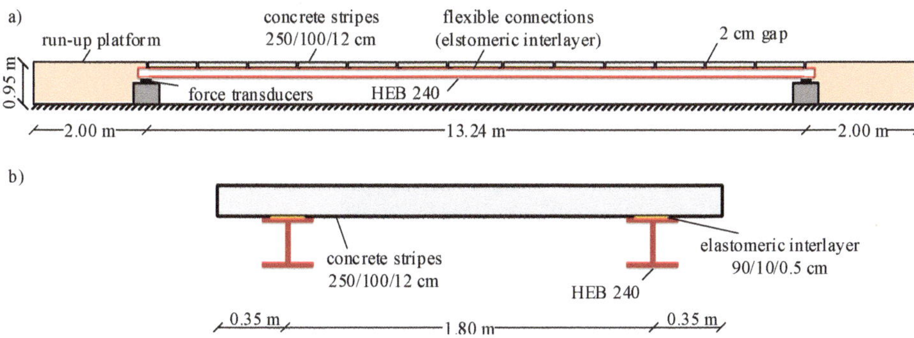

Figure 8.2 Sketch of the experimental pedestrian bridge HUMVIB: a) side view; b) cross section

The approach of using separate concrete elements was preferred instead of an uninterrupted deck slab, in order to avoid – to the greatest extent possible – a composite action of the concrete deck. This means that the concrete elements are supposed to contribute to the global structural behavior only as an additional mass. The reason for this is that it was aimed to achieve a structure having a first natural frequency in a range, which is excitable by the human walking, i.e. between 1.5 and 2.5 Hz according to BACHMANN (1988) or between 1.6 and 2.2 Hz according to HAWRYSZKÓW (2011). As the concrete elements do not influence the global stiffness of the structure, their mass could be used for a fine tuning of the natural frequency. Moreover, the use of the concrete stripes also reduces the modeling uncertainties related to the scattering of material properties of concrete. The gap between the concrete elements is 2 cm wide, small enough to ensure a safe locomotion of the subjects but large enough to avoid the composite action of the concrete to the global behavior. Finally, it is to be noted that the precast concrete elements and the steel beams are separated by an elastomeric interlayer with a thickness of 5 mm (cf. Fig. 8.2b). The

HUMVIB-Bridge has a total mass of about 12 tons. This leads to a first natural frequency of 2.04 Hz, which is a good match of the mean values for the typical walking frequency distributions provided in BACHMANN (1988) and HAWRYSZKÓW (2011).

8.4 Measurement set-up

For the experimental investigations performed on the HUMVIB-Bridge, an extensive measurement set-up was implemented. It was initially designed for the investigation of the HSI effect, both from biomechanical and structural point of view. The set-up is pictured in Fig. 8.3 and schematically illustrated in Fig. 8.4. On the one hand, the measurement set-up implied the acquisition of structural responses in vertical direction at eight measurement points (MP) disposed along the steel beams. Both accelerometers and displacement sensors

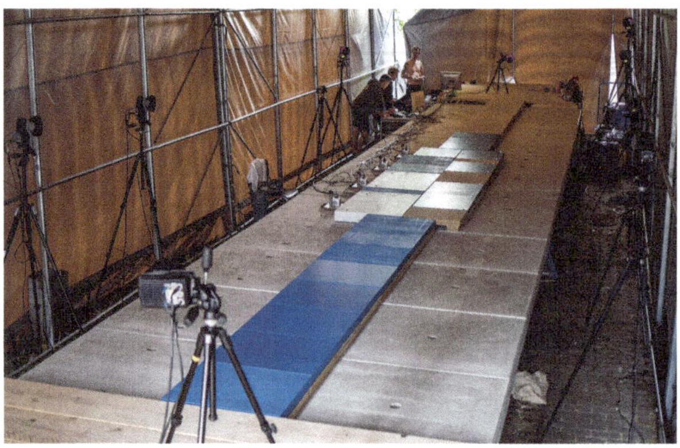

Figure 8.3 Photo caption of the measurement set-up used in the experimental investigation of the HUMVIB-Bridge

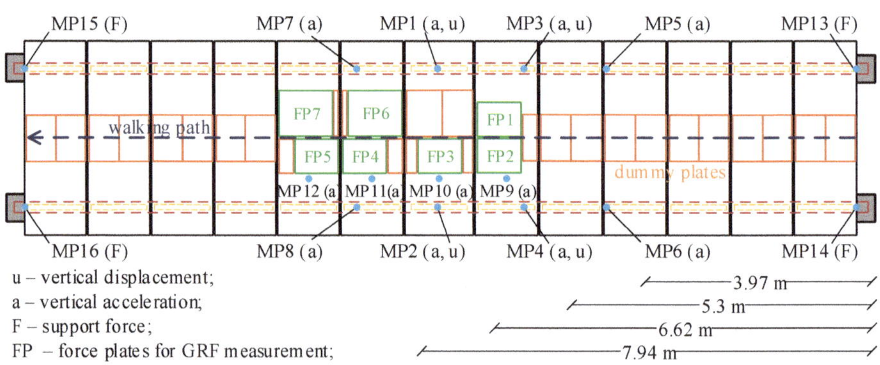

Figure 8.4 Measurement set-up of the HUMVIB-Bridge

were installed at the locations MP 1 to MP 4, while only accelerations were recorded at MP 5 to MP 8. The accelerations were recorded using 1D seismic accelerometers of type PCB-TLD393B04 for MP 1 to MP 4 and 3D acceleration sensors PCB-TLD356A17 for MP 5 to MP 8. All displacement time histories were measured using LVDTs of type HBM 1-WA/50MM-L. Moreover, the dynamic reaction forces at the four supports (MP 13 to MP 16) were recorded using a set of load cells of type Induk UC350.

On the other hand, a set of biomechanical measurement devices were incorporated in the investigation. The recording of the GRFs of the walking pedestrians was realized using five force plates (FP 1 to FP 5) of type KISTLER 9260AA6 (dimensions: 600 × 500 × 50 mm) and two force plates (FP 6 and FP 7) of type KISTLER 9287CA (dimensions: 900 × 600 × 100 mm). The corresponding force plate masses specified by the manufacturer are 8.6 kg and 25 kg, respectively (KISTLER, 2017). The arrangement of the force plate array was designed in such a way, so that as many steps as possible can be recorded. The configuration illustrated in Fig. 8.4 allowed the acquisition of up to four individual steps per trial. The forces recorded by the force plates will serve as reference values within the validation process of the proposed MFI method. In addition, it is to be mentioned that the walking path had to be leveled using dummy plates with a thickness of 100 mm, in order to compensate the different heights of the forces plates. In this way, proper conditions for a natural gait are ensured and the tripping hazard of the subjects is reduced.

As already mentioned in Section 8.2, the accuracy of the force plates might be affected by vibrations of the structure on which they are installed (WERKLE et al., 2016). Therefore, the original force plate signals acquired during the walking trials have to be corrected in the post-processing. This is achieved by substracting the inertial forces generated by the acceleration of the force plate mass from the total force plate signals. However, the inertial forces of the force plates can only be determined if the acceleration time histories at the location of each force plate are recorded. In this sense, a set of additional 1D accelerometers was installed in the middle of each concrete plate, which contained at least a force plate (MP 9 to MP 12). Finally, it is to be noted that the biomechanical force plates present – in case of proper installation – natural frequencies of about 200 Hz (9260AA6) to 500 Hz (9287CA) (KISTLER, 2017). They lie thus far above the relevant vibration frequencies of the HUMVIB-Bridge. Therefore, it can be presumed that both internal vibrations of the plates and relative vibrations of the force plates with respect to the concrete elements can be neglected.

8.5 Structural model

A detailed FE model of the HUMVIB-Bridge was created according to the as-built plans using the finite element software ANSYS 2020 R1 (Fig. 8.5). The two longitudinal steel beams were defined using beam elements (BEAM188), whereas the precast concrete elements were modeled using shell elements (SHELL181). With respect to the material parameters of the steel and concrete elements, it is to be mentioned that they were assumed to be deterministic. In this respect, typical values from the engineering practice were adopted (Tab. 8.1). Furthermore, the elastomeric interlayer elements situated between the steel beams and the concrete deck components were idealized using a group of three translational springs ($k_{t,2,x}$, $k_{t,2,y}$ and $k_{t,2,z}$ along the x, y and z axis, respectively) and one rotational spring ($k_{r,2,x}$ around the longitudinal x-axis). Their definition is illustrated in Fig. 8.6. It can be seen that a set of four spring elements was defined at four points of each plate (two on each side), whereas the connections were assumed to lie 20 cm from the ends of the plate. Hence, a number of 16 spring elements (COMBIN14) for each plate was achieved. The supports of the main longitudinal beams were assumed to be rigid in the vertical z direction, whereas their horizontal movement (in x and y directions) is permitted. They were defined – according to the as-built situation – at the lower flanges of the HEB 240 profiles. The horizontal movement capability at the supports was ensured by defining a set of two translational springs, $k_{t,1,x}$ and $k_{t,1,y}$ at each support (Fig. 8.5). The finite element model presents a total number of 15 550 nodes, 13 760 elements and 89 974 DOFs.

Figure 8.5 Finite element model of the HUMVIB-Bridge

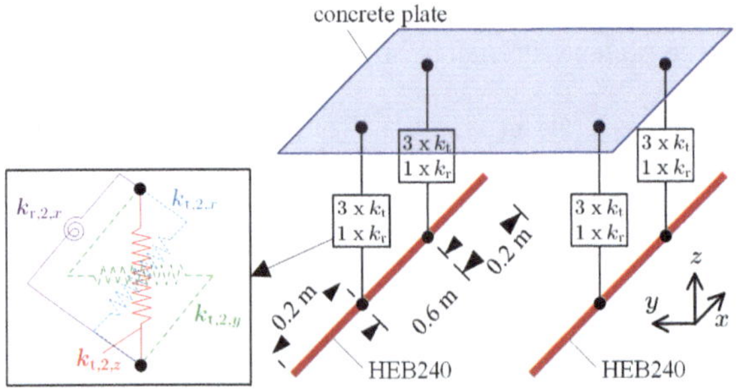

Figure 8.6 Idealized connection between the steel beams and concrete plates of the HUMVIB-Bridge

Structural element	Finite Element type	density ρ [kg/m^3]	Young's modulus E [kN/m^2]	Poisson's ratio ν [−]
Steel beams	BEAM188	7850	$2.1 \cdot 10^8$	0.3
Concrete elements	SHELL 181	2500	$3 \cdot 10^7$	0.2

Table 8.1 Material parameters used for the definition of the FE model of the HUMVIB-Bridge

Since the rotational and translational springs used for the idealization of the support conditions and of the connection between concrete and steel components represent only some equivalent elements, which seek to approximate the real – rather more sophisticated – coupling and support behavior, no typical values of the corresponding stiffness coefficients could be directly defined. Their values had to be determined by solving an optimization problem, which aims to minimize the sum of squared differences between the measured and computed natural frequencies for seven modes[1] (first three vertical bending, first two torsional and first two lateral bending modes):

$$\min_{\boldsymbol{k}} \mathcal{F}(\boldsymbol{k}) = \sum_{i=1}^{7} w_i \cdot \left(f_i - \bar{f}_i\right)^2 \tag{8.1}$$

Before the optimization procedure was carried out, an experimental system identification was performed using – as in case of the Schmutter bridge – the reference based combined deterministic-stochastic subspace identification algorithm (REYNDERS et al., 2008). Fig. 8.7 shows the first five experimentally identified vertical mode shapes together with their corresponding natural frequencies. The optimization problem was solved using the SQP strategy (HARZHEIM, 2019) and delivered the following set of stiffness coefficients:

$$\begin{aligned}
k_{t,1,x} &= 12.64 \cdot 10^3 \quad \text{kN/m}\\
k_{t,1,y} &= 10.72 \cdot 10^3 \quad \text{kN/m}\\
k_{t,2,x} &= 11.14 \cdot 10^3 \quad \text{kN/m}\\
k_{t,2,y} &= 2.52 \cdot 10^3 \quad \text{kN/m}\\
k_{t,2,z} &= 18.22 \cdot 10^3 \quad \text{kN/m}\\
k_{r,2,x} &= 11.95 \cdot 10^3 \quad \text{kNm}
\end{aligned} \tag{8.2}$$

whereas the following weighting parameters w_i were used:

$$w_1 = w_3 = w_5 = 10; \quad w_2 = 1000; \quad w_4 = w_6 = 1; \quad w_7 = 0.1 \tag{8.3}$$

[1] It can be seen that the objective function defined in Eq. 8.1 does not include any terms corresponding to the mode shapes values. The reason for this is that an optimization based on only natural frequencies was firstly performed. Since this already revealed an excellent agreement between the computed and measured data for both natural frequencies and mode shapes (cf. Tab. 8.2), there was no need to extend the objective function to include mode shape terms. In this way, a cross validation of the identified parameters was also achieved.

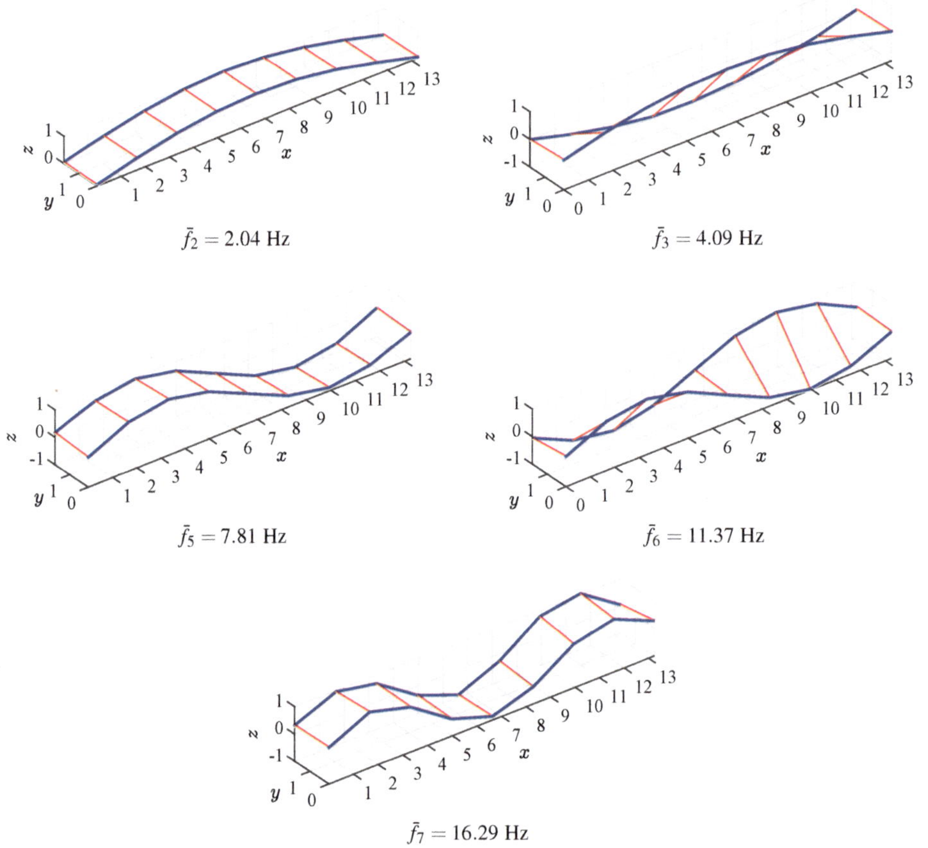

Figure 8.7 Experimentally identified vertical mode shapes of the HUMVIB-Bridge and the corresponding natural frequencies

It is to be mentioned that the first and the fourth summation terms of Eq. 8.2 are associated with the first and the second horizontal bending modes, whereas the rest represent the vertical mode shapes shown in Fig. 8.7. Following the findings from Section 6.7.6, the optimization procedure aimed to achieve – to the greatest extent possible – an exact match of the frequency $\bar{f}_2$ associated with the first vertical bending mode. For this, a very high weighting parameter w_2 was selected (cf. Eq. 8.3), which ensured an accuracy of the frequency f_2 to three decimal places ($f_2 = 2.040$ Hz). However, such an accuracy might indeed be interpreted as slightly exaggerated with respect to practical applications. In this respect, it is to be mentioned that a significant reduction of $w_2 = 1000$ to $w_2 = 100$ also delivered perfectly acceptable results with a computed natural frequency $f_2 = 2.046$ Hz. Nevertheless, according to the high ratio between w_2 and all other weighting parameters

Nr.	Mode type	$\bar{f}_i$ [Hz]	$\bar{\xi}_i$ [%]	f_i [Hz]	Δf_i [%]	MAC mode i [-]
1	1st lateral bending	1.69	2.83	1.63	-3.55	0.9906
2	1st vertical bending	2.04	0.23	2.04	≈ 0	0.9998
3	1st torsional	4.09	2.50	4.09	≈ 0	0.9981
4	2nd lateral bending	4.38	1.77	4.96	+13.24	0.9115
5	2nd vertical bending	7.81	0.23	7.80	-0.12	0.9933
6	2nd torsional	11.37	1.25	11.34	-0.26	0.9959
7	3rd vertical bending	16.29	0.50	16.70	+2.52	0.9618

Table 8.2 Experimentally identified modal characteristics of the HUMVIB-Bridge (natural frequencies $\bar{f}_i$ and damping ratios $\bar{\xi}_i$) of the modes considered in the FE updating process, the corresponding results predicted by the calibrated structural model, the calculated MAC values and the relative frequency deviation Δf_i

(Eq. 8.3), it can be noted that the natural frequency of the first natural bending mode seems to be the determinant quantity for the model updating. Furthermore, it is worth mentioning that a mode tracking had to be implemented, in order to account for eventual mode swaps during the iteration process (HARZHEIM, 2019). This was achieved by checking the maximum MAC values (cf. Eq. 6.40) between the measured and the calculated mode shapes in each iteration step and by repositioning the summation terms in Eq. 8.1 accordingly.

Tab. 8.2 shows the results of the model updating in comparison to the results of the experimental identification. It includes the values of the identified natural frequencies $\bar{f}_i$ and damping ratios $\bar{\xi}_i$ (adopted also for the numerical investigation), the calculated natural frequencies f_i, the relative frequency error Δf and the corresponding MAC values determined according to Eq. 6.40. In addition, Fig. 8.8 illustrates the first five computed mode shapes with relevant vertical components and the corresponding natural frequencies. In general, an excellent agreement between the experimentally identified and the computed frequencies can be observed. The highest absolute errors are 3.55 % and 13.23 % for the first and second horizontal bending mode, respectively. However, only the modes with dominant vertical components are relevant for the following numerical investigations, i.e. the vertical bending and the torsional modes shown in Fig. 8.8. The corresponding maximal relative error is about 2.5 %, whereas all MAC values are close to 1. Hence, it can be stated that the model is sufficiently accurate for analyses under vertical excitation and can be reliably used in the following for the GRF identification.

However, since the extensive biomechanical equipment and the dummy plates installed on the HUMVIB-Bridge (see Fig. 8.3) have a non-neglectable contribution to the total system mass, a further (slight) model adaptation had to be performed for the validation process with real measurement data. In this respect, it is to be noted that unfortunately no experimental system identification was carried out on the bridge instrumented with biomechanical equipment. The reason for this is – as already mentioned before – that the experiments illustrated in Fig. 8.3 and 8.4 were initially designed and executed with

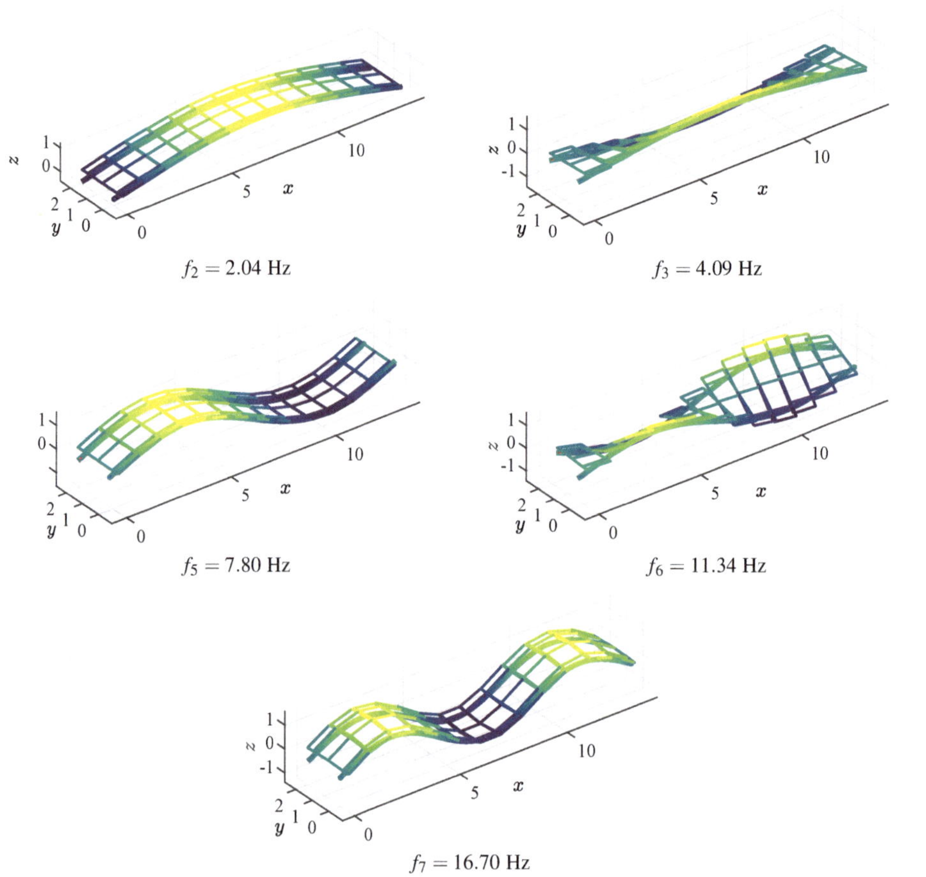

Figure 8.8 Computed mode shapes and natural frequencies of the HUMVIB-Bridge

other scientific aims, which did not require a proper experimental system identification. Nevertheless, the force and the dummy plates were installed in such a way, so that they only act as additional mass, without contributing to the global stiffness. Therefore, an additional mass of about 60 kg/m was included in the previously described structural model. This resulted in an updated computed natural frequency of the first bending mode $f_2 = 1.98$ Hz, confirmed also by FFT-analyses of the free decay phases recorded during different trials. The computed natural frequencies of the higher vertical modes are: $f_3 = 3.97$ Hz (1st torsional mode), $f_5 = 7.56$ Hz (2nd bending mode), $f_6 = 11.02$ Hz (2nd torsional mode) and $f_7 = 16.02$ Hz (3rd bending mode)[2].

[2] An additional representation of the mode shapes for the refined model under consideration of the mass of the biomechanical equipment is not provided, since no difference can be visually observed with respect to the modes illustrated in Fig. 8.8.

All the computations for MFI with real measurement data will rely on the model updated under consideration of the mass of the biomechanical equipment. The original model calibrated with respect to the empty structure will only be used in the parametric study needed for the determination of the weighting parameters w_{p1} and w_{p2} (Section 8.6.2). In this sense, the vertical components of the mode shapes 2, 3, 5, 6 and 7 along the walking path and at the coordinates of the measurement points were exported from the FE software together with the corresponding modal masses and natural frequencies. They were subsequently integrated within the MFI identification procedure implemented in the computer algebra system Matlab (THE MATHWORKS INC., 2019).

8.6 Force identification with real measurement data

8.6.1 Computational parameters

In the following investigations, a moving window size of 400 samples, an overlapping interval of $1/3$ of the window length and a smoothing radius $r_a = 2$ were used. The default measurement configuration considered in the analysis consists of eight accelerometers (MP1 to MP8) and four displacement sensors (MP1 to MP4). Unless otherwise specified, this set-up will be used both for the applications with simulated measurement data (i.e. for the determination of the weighting parameters w_{p1} and w_{p2} in Section 8.6.2) and for the validation process using real measurement data (Section 8.6.4). The corresponding weighing factors of the main optimization criterion are $w_{a,h} = 1$ and $w_{d,h} = 5 \cdot 10^5$ (for all measurement points h). In this respect, it is to be mentioned that the identification procedure was carried out in this case under consideration of the original measured and computed displacement and acceleration vectors. This means that no normalization (e.g. according to Eq. 5.11 or 5.13) was used. By using the original vectors, significant differences between the orders of magnitude of the two terms of the main optimization criterion will occur[3]. This explains the very high value of $w_{d,h}$, which seeks in fact to balance the order of magnitude of the displacement and acceleration terms. The initial points and the convergence criteria of the optimization were formulated according to Section 6.3.2 and 6.3.3, respectively.

Furthermore, it is to be mentioned that – unless otherwise specified – the first five vertical modes (as represented in Fig. 8.8) will be considered in the modal superposition. However, the assumption of the force moving along the shear axis actually prevents the contribution of the torsional modes to the total response. Besides that, a total computation time of 10 s and a computational time step of $\Delta t = 0.003333$ s were used. The latter allows a sufficiently accurate representation of the vibrations in the fifth mode with a computational frequency of 16.70 Hz.

[3] It is to be mentioned that the displacement and acceleration vectors were introduced in the objective function in their SI units, namely $[\mathrm{m}]$ and $[\mathrm{m/s^2}]$, respectively.

8.6.2 Weighting factors of the penalty functions

The weighting parameters w_{p1} and w_{p2} of the penalty functions were determined according to the procedure described in Section 5.1.4. It relies on a numerical parametric study, which is based on a reference time history of a moving force and a set of simulated measurement data. The latter is determined by applying the assumed reference force to the FE model of the empty structure and evaluating the response by a direct analysis, performed according to Sections 3.3.3 and 5.1.2. Subsequently, the results of the direct computation are polluted with artificial white noise, defined according to Eq. 5.24 for a – rather unrealistic – noise level $e_p = 5\ \%$.

As it is recommended to define the amplitude of the reference force in a plausible range of the expected forces, a typical approach for representation of pedestrian loads is assumed. It considers a single force time history, which consists of the superposition of the forces associated to the left and the right foot. This is usually formulated as a Fourier series with limited number of harmonic components (BACHMANN, 1988; KERR et al., 2001; SEILER et al., 2004; ŽIVANOVIĆ et al., 2005), implying thus a perfectly periodic force function. The force time history used in this work is defined according to BACHMANN (1988) as a summation of three Fourier terms, which are considered to be sufficient for dynamic analyses of low frequency structures, as it is the case for the HUMVIB-Bridge:

$$F(t) = G + \sum_{i=1}^{3} G \cdot \alpha_i \cdot \sin(2 \cdot \pi \cdot i \cdot f_s \cdot t - \varphi_i) \tag{8.4}$$

Within the direct analysis, the force $F(t)$ is assumed to move along the walking path shown in Fig. 8.4 with a predefined subject velocity $v_s = 1.5$ m/s. Fig. 8.9 illustrates the reference force time history used within the present investigation. It was obtained under consideration of a subject weight $G = 747$ N and a step frequency of $f_s = 2$ Hz, whereas the Fourier coefficients and the corresponding phase angles were adopted from BACHMANN (1988), as follows: $\alpha_1 = 0.4$, $\alpha_2 = \alpha_3 = 0.1$, $\varphi_1 = 0$, $\varphi_2 = \varphi_3 = \pi/2$. Fig. 8.9 emphasizes that the idealization of the pedestrian load as a Fourier series implies only identical strides. This is a justified assumption for numerical analyses. However, it usually does not entirely hold in real cases, where a set of successive steps are expected to present certain differences (VAN NIMMEN et al., 2014).

The error curve resulting from the parametric study of the HUMVIB-Bridge is shown in Fig. 8.10. The minimum error value is found for $w_{p1} = 5.5 \cdot 10^{-8}$ and $w_{p2} = 1.0 \cdot 10^{-4}$. However, also in this case a nearly flat region of the curve (in logarithmic representation) can be observed around the minimum point. This indicates that a good accuracy of the identified forces is expected for a wider parameter range. In order to allow a quick (visual) plausibility check, the identification results obtained for the aforementioned optimal weighting parameter set is shown in Fig. 8.11. The relative error evaluated according to Eq. 5.23 at the end of the backward optimization is $x_{err,b} = 4.72\ \%$. It can be stated that a very

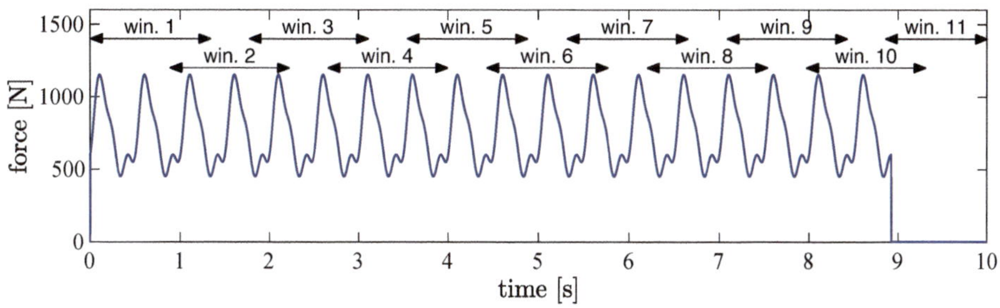

Figure 8.9 Reference force time history used for simulating the measurement data on the HUMVIB-Bridge and window division for a window length of 400 samples

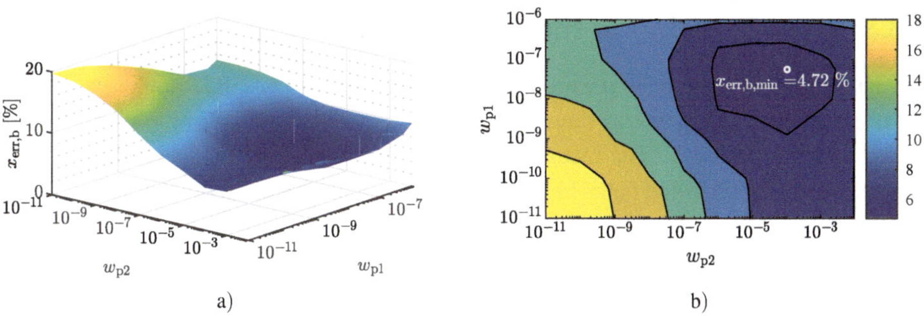

Figure 8.10 Parametric study for determination of the weighing factors w_{p1} and w_{p2} for the HUMVIB-Bridge: a) 3D plot; b) contour plot

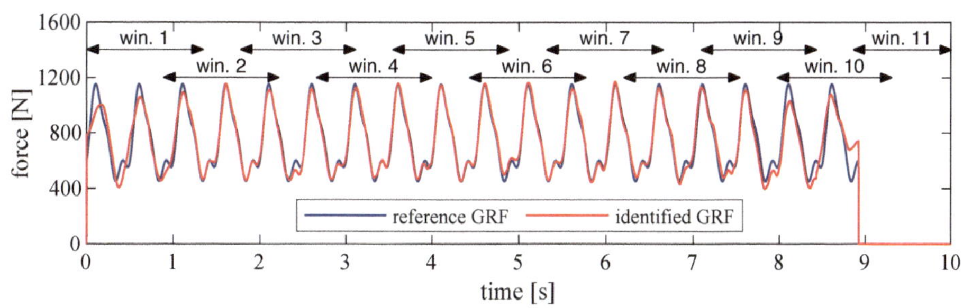

Figure 8.11 Reference and identified force time histories obtained for the HUMVIB-Bridge using simulated measurement data and the optimal set of weighting factors w_{p1} and w_{p2}

good global agreement between the reference and the identified GRF time histories was achieved. A slightly higher deviation of the identified forces with respect to the reference values can be only observed for a limited number of time steps, in which the force is located in the proximity of the supports. The main reason for this effect is thought to be the low sensitivities of acceleration and displacement at the measurement points with respect to forces acting at the beginning and the end of the bridge. However, the penalty functions ensure stable and sufficiently accurate results even at these locations.

Finally, it is to be mentioned that the parametric study presented in this section was based on a structural model, which was calibrated with respect to the empty structure, i.e. without the biomechanical equipment illustrated in Fig. 8.3 and 8.4. It has already been mentioned in Section 8.6.1 that the investigation based on real measurement data will rely on an updated structural model, which includes the effect of the additional equipment mass. However, since the two models (with and without additional mass of the biomechanical equipment) only present very slight differences, it is justifiable to use the same weighting factors in both cases.

8.6.3 Preprocessing of the measurement data

The experimental validation of the proposed MFI method based on GRFs implies the recording of structural responses and biomechanical quantities according to the measurement set-up shown in Fig. 8.4. The measurement scenario was a single subject ($G = 747$ N) walking along the specified path (Fig. 8.4) with a predefined step frequency $f_s = 2$ Hz. A first preprocessing step was concerned with the low-pass filtering of the measurement data under consideration of a cut-off frequency of 18 Hz (Butterworth filter, 5th order). This should ensure the compatibility between the frequency content of the computation and the measured data. In addition, the selected computational time step of $\Delta t = 0.003333$ s also required a resampling of the measured signals, which were initially sampled at 1200 Hz.

Furthermore, the measurement data of the load cells (MP 13 to MP 16) and biomechanical force plates (FP 1 to FP 7) were used as reference values within the experimental validation procedure. However, as already mentioned, the load cells and the force plate signals include both the pedestrian induced loads and the inertial forces occurring due to the accelerations of the bridge and the force plate masses, respectively (WERKLE et al., 2016; AHMADI et al., 2018b). Therefore, in order to obtain reliable reference force data for the validation procedure, the inertial effects had to be removed from the original measurement data, using the following relations, provided by AHMADI et al. (2018b):

$$
\begin{aligned}
F_{\mathrm{GRF,LC}}(t) &= F_{\mathrm{m,LC}}(t) - a_{\mathrm{m,LC}}(t) \cdot M_{\mathrm{I,LC}} \\
F_{\mathrm{GRF,FP}}(t) &= F_{\mathrm{m,FP}}(t) - a_{\mathrm{m,FP}}(t) \cdot M_{\mathrm{I,FP}}
\end{aligned}
\tag{8.5}
$$

where $F_{\mathrm{GRF,LC}}$ denotes the GRF component of the load cell signals, $F_{\mathrm{GRF,FP}}$ represents the pure GRF component of a single force plate, while $F_{\mathrm{m,LC}}$ is the sum of the four load cell signals and $F_{\mathrm{m,FP}}$ is an original force plate reading. Furthermore, $a_{\mathrm{m,LC}}$ is used for determining the inertial components of the bridge and denotes the acceleration time history at the midspan, evaluated as an average of the accelerations recorded at MP 1 and MP 2. $a_{\mathrm{m,FP}}$ describes an acceleration time history of MP 9 to MP 12 (used for the inertia of the

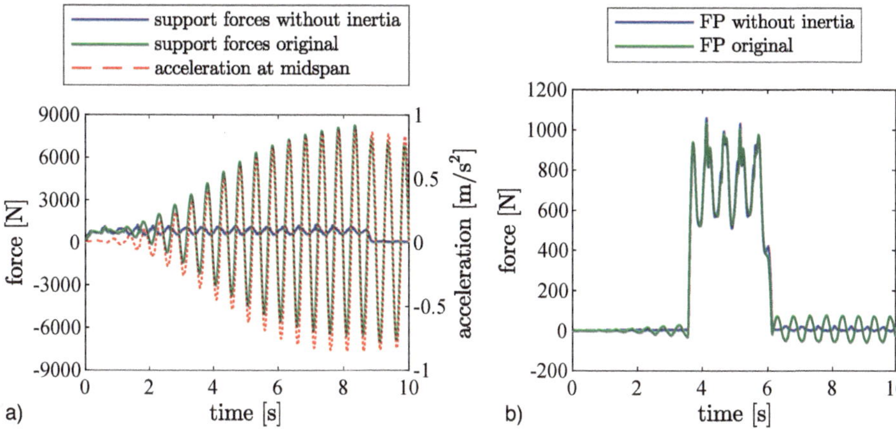

Figure 8.12 Removing of the inertia from the force measurements on the HUMVIB-Bridge: a) signals of the load cells and b) force plates recordings

force plates). $M_{\mathrm{I,LC}}$ and $M_{\mathrm{I,FP}}$ represent the "inertial mass" of the bridge or of a single force plate, respectively. $M_{\mathrm{I,LC}}$ and $M_{\mathrm{I,FP}}$ can be determined using acceleration and force signals recorded in a time window without external loading. The decay phase recorded after the subject walked off the bridge is very well suited for this purpose. In this free-vibration phase, the original sensor readings ($F_{\mathrm{m,LC}}$ or $F_{\mathrm{m,FP}}$) coincide with the respective inertial forces, since the externally induced forces $F_{\mathrm{GRF,LC}}$ and $F_{\mathrm{GRF,FP}}$ are zero, leading to:

$$\begin{aligned} M_{\mathrm{I,LC}} &= \frac{F_{\mathrm{m,LC}}(t)}{a_{\mathrm{m,LC}}(t)} \\ M_{\mathrm{I,FP}} &= \frac{F_{\mathrm{m,FP}}(t)}{a_{\mathrm{m,FP}}(t)} \end{aligned} \tag{8.6}$$

$M_{\mathrm{I,LC}}$ and $M_{\mathrm{I,FP}}$ from Eq. 8.6 are expected to result in rather constant values, which can be then inserted in Eq. 8.5, eventually leading to the isolated GRF components. Fig. 8.12 shows the original measurement signals and the results obtained after suppressing the inertial effects for both the load cells and the force plate readings[4]. It is to be mentioned that only the first bending mode was considered in the determination of the inertial forces, whereas the computed inertial masses are: $M_{\mathrm{I,LC}} \approx 8370$ kg, $M_{\mathrm{I,FP}} \approx 25$ kg for FP 6 and FP 7 and $M_{\mathrm{I,FP}} \approx 8.2$ kg for FP 1 to FP 5. They are in a very good agreement with the values specified by the manufacturer, see Section 8.4 and KISTLER (2017).

In addition, it is to be mentioned that the support forces without inertial effects also served for the identification of the time points for entering and leaving the bridge. This is a

[4] In order to obtain a single force time history containing the contributions of both the left and the right foot, the analysis was performed here using the sum of the four load cells or of the seven force plates, respectively.

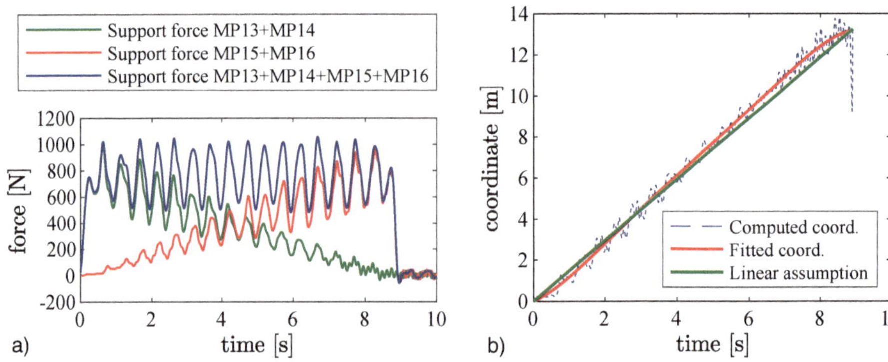

Figure 8.13 Determination of the time dependent location of the subject walking over the HUMVIB-Bridge: a) support forces; b) subject location resulting from the support forces

very important aspect when dealing with the identification of GRFs, since a pedestrian is more likely to present variations of the passage speed than a vehicle driving over a bridge. For example, it is possible that the subject maintains the prescribed step frequency[5], but unconsciously varies the step length. This has a direct influence on the subject speed, which can be determined as:

$$v_{\mathrm{s}} = f_{\mathrm{s}} \cdot l_{\mathrm{s}} \tag{8.7}$$

Any error related to the pedestrian speed directly influences the calculation of the modal loads and influences therefore the identification results. For this reason, the issue of the pedestrian location during the passage has to be addressed before proceeding to the identification process.

In order to reduce the uncertainties related to the effect of a variable pedestrian speed, the time-dependent location of the subject can be approximated using the support forces measured at the two ends of the bridge. In this regard, Fig. 8.13a illustrates the support forces at each end of the bridge and the total support force as a sum of the four load cells. Here, it is to be noted that the force time histories shown in Fig. 8.13a were low-pass filtered, so that they only include frequencies of up to 5 Hz. The reason for this is that it was aimed to to analyze signals – as far as possible – free of dynamic components. Under consideration of a structural behavior similar to a simply supported beam, the location can be approximately determined by means of static equilibrium:

$$x(t) = \frac{F_{\mathrm{sup},2}(t) \cdot l}{F_{\mathrm{sup,tot}}(t)} \tag{8.8}$$

[5] In the present investigation, a metronome was used to allow the subject to tune his pacing rate.

where $F_{\mathrm{sup},2}(t)$ denotes the force resulting from the summation of the load cells signals acquired at MP 15 and MP 16. $F_{\mathrm{sup,tot}}(t)$ represents the total GRF, evaluated as a sum of the measured signals from all load cells. The pedestrian location over time, determined according to Eq. 8.8, is shown in Fig. 8.13b. It can be observed that the computed time dependent coordinates are slightly polluted by some high-frequency components. These are most likely numerical artifacts generated by some residual inertial effects of the higher modes, which are still present in the force time histories showed in Fig. 8.13a. Therefore, a curve fitting was performed, in order to achieve a smooth monotonically increasing location curve, which is also shown in Fig. 8.13b. Its slight nonlinearity confirms the hypothesis of an inconstant walking speed. In order to allow a quick qualitative analysis of extent of the nonlinearity, Fig. 8.13b also includes the coordinates determined under assumption of a constant speed. Alternatively, the location of the subject can be tracked using, for example, video recordings or motion tracking systems from the biomechanics, which are also applicable on real structures in operation (MEJIACRUZ et al., 2021).

8.6.4 Results and discussion of the experimental validation

In the following, the results of the GRF identification based on real measurement data of the HUMVIB-Bridge will be presented. It is to be mentioned on beforehand that this investigation is only intended to provide a quick additional experimental validation of the proposed MFI method and to identify its potential for the applications in the research field of HSI. Therefore, the investigations do not reach the same depth and level of detail as for the Schmutter bridge (Section 7.6.3).

Besides the default measurement configuration mentioned in Section 8.6.1 (denoted in the following as set-up no. 1), three further sensor combinations will be investigated (Tab. 8.3). The second configuration consists as well of a combination of acceleration and displacement sensors. However, it includes a more practicable number of sensors. Set-ups no. 3 and no. 4 consider only accelerations or only displacements as measurement data.

Fig. 8.14 illustrates the identification results for the configuration no. 1 together with the reference GRFs. The latter were determined according to Eq. 8.5 and already shown in Fig. 8.12. A first visual analysis reveals that the identified GRF results seem to be in

Configuration no.	Measurement points acceleration	Measurement points displacement
1	MP1 to MP8	MP1 to MP4
2	MP3, MP4	MP3, MP4
3	MP3, MP4	–
4	–	MP3, MP4

Table 8.3 Overview of the investigated sensor configurations for the HUMVIB-Bridge

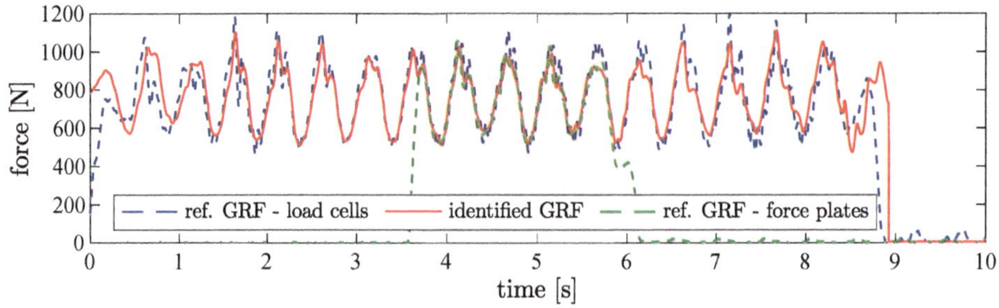

Figure 8.14 Identified and reference force time histories of a subject walking over the HUMVIB-Bridge for the measurement configuration 1

very good agreement with reference GRFs ($F_{\mathrm{GRF,LC}}$) corresponding to the load cells. The effect of a lower accuracy for the force values located in the proximity of the supports can also be clearly seen in this case. This was initially observed in the results obtained with simulated measurement data (Fig. 8.11) and occurs most likely due to the low sensitivities of the acceleration and displacements to forces acting towards the beginning or the end of the bridge. The quantitative analysis of the results is performed based on the relative error $x_{\mathrm{err,LC}}$, which is obtained according to Eq. 5.23, whereas the measured GRF components of the load cells $F_{\mathrm{GRF,LC}}$ are considered as reference term $\boldsymbol{x}_{\mathrm{reference}}$. The corresponding result determined for the configuration 1 is $x_{\mathrm{err,LC}} = 10.18\ \%$. Considering that there are still various uncertainties in the problem definition (e.g. with respect to the modeling, subject velocity, measurement inaccuracies, etc.), the result can be interpreted – from quantitative point of view – as very satisfactory. However, the quantitative evaluation of the MFI accuracy by means of the relative error $x_{\mathrm{err,LC}}$, which is determined using load cell signals, has to be interpreted with high caution. The reason for this is that the error $x_{\mathrm{err,LC}}$ is directly affected by any errors occurring in the process of eliminating the inertial forces from the measurement load cell signals (cf. Section 8.6.3). A close look at the force time histories associated with the decay phase, i.e. after the subject left the structure, reveals the inertial forces could not be completely eliminated (Fig. 8.14, blue dashed curve in the interval between 9 and 10 s). The force signals are expected to be exactly zero, if no external load is present on the bridge. This effect can indeed be only isolated during the decay phase but it actually affects the whole time history and might thus lead to erroneous relative errors $x_{\mathrm{err,LC}}$. Nevertheless, the load cell signals provide a valuable tool for the general comparison of the GRFs along the bridge.

In contrast to the reference GRFs determined from the load cell recordings, the results obtained from the force plate signals ($F_{\mathrm{GRF,FP}}$) are expected to be more accurate due to their more suitable measurement range and to the low "inertial masses" of the force plates. These eventually lead to low inertial force components and thus to lower errors related to their elimination from the original signals. In order to achieve a more reliable validation of the identified GRF, the relative error $x_{err,FP}$ of the identified force with respect to the reference GRFs determined from the force plate signals is introduced. Obviously, it can only be evaluated for the interval, in which the subject passes the force plate array (grey area in Fig. 8.14). $x_{err,FP}$ is determined using Eq. 5.23, whereas the measured GRF component of the force plate signals $F_{GRF,FP}$ (green dashed curve within the grey area of Fig. 8.14) is considered as the reference force vector $\boldsymbol{x}_{reference}$. The result obtained for the first sensor combination is $x_{err,FP} = 3.20$ %, indicating an excellent agreement between the identified and the reference forces. Furthermore, the subject weight can be approximated according to Eq. 7.13 as average of all force values of the identified force time history. The result obtained for the identified force from Fig. 8.14 is 764 N. This lies only 2 % appart from the reference value of 747 N, obtained by static weighting of the subject and confirmed by averaging the force plate signals.

A further analysis can be made based on the general appearance of the identified and the measured GRF time histories from Fig. 8.14. It can be seen that the GRFs obtained from the load cells signals consist in fact of a set of imperfect (unequal) step sequences. In the middle time interval – i.e. between 3 and 6.5 seconds, associated with a pedestrian location around the middle of the bridge (cf. Fig. 8.13b) – a tendency towards lower amplitudes and narrower frequency content of the identfied forces is observed compared to the signals at the beginning and the end of the passage time. Fig. 8.14 clearly shows that this trend could be successfully reproduced by the proposed MFI procedure. The ability to identify the difference between successive step sequences represents a great achievement of the proposed method in regard to the arising possibility of an enhanced understanding of the HSI effects. Considering that the presented MFI method is structure-centered, i.e. it is based only on measured structural responses, it directly reveals the contact forces at the interface bridge-pedestrian, including thus all passive and active HSI effects. Hence, it can be stated that a clear potential towards future applications for a detailed analysis of the HSI effects was identified.

The results of the measurement configuration 1 showed a high accuracy and stability of the GRF identification. However, the number of sensors (eight accelerations and four displacements) might seem inadequate with respect to real applications. In this sense, the results of the configuration 2, implying a more practicable sensor set-up (two accelerometers and two displacements sensors), are discussed in the following. Fig. 8.15 illustrates the results in comparison to the reference force time histories. The corresponding errors are $x_{err,LC} = 10.13$ % and $x_{err,FP} = 3.48$ %. Hence, only very slight differences could be observed in comparison to the results of the measurement configuration 1. This confirms that the method is also able to deliver sufficiently accurate results for a reduced number of sensors. In the same context, a forward analysis was performed using the identified force time history showed in Fig. 8.15. The re-built displacement and accelerations at MP 2, which were not considered in the analysis implying the configuration no. 2 (i.e. it is assumed as an "unmeasured" location), are illustrated in Fig. 8.16 together with the

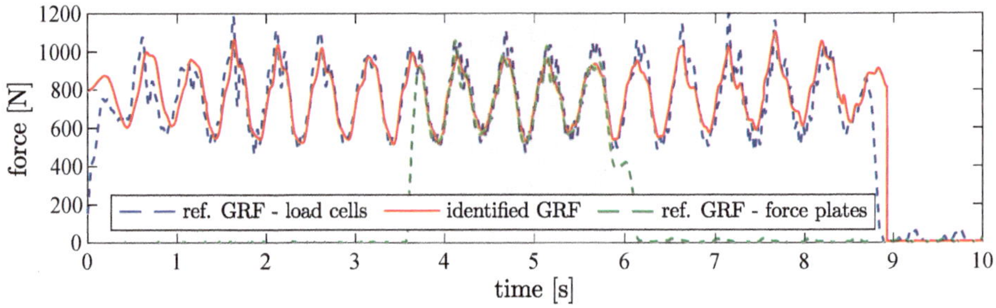

Figure 8.15 Identified and reference force time histories of a subject walking over the HUMVIB-Bridge for the measurement configuration 2

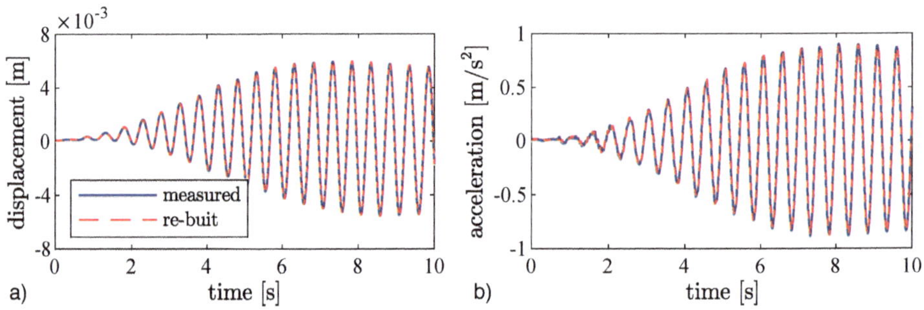

Figure 8.16 Measured and re-built responses at the "unmeasured" location MP 2 of the HUMVIB-Bridge (measurement configuration 2): a) displacement time history; b) acceleration time history.

corresponding measurement time histories. The measured and re-built data are obviously in a nearly perfect agreement. Since there is almost no phase shift between the two time histories, the relative errors of the re-built responses can be evaluated according to Eq. 6.11. The results are $u_{\text{err}} = 1.20$ % and $\ddot{u}_{\text{err}} = 2.41$ %. They indicate that the MFI method can be also used to predict structural responses at unmeasured locations generated by the pedestrian locomotion on a vibrating structure[6].

The identification results for the configurations 3 and 4 are shown in Fig. 8.17 and 8.18, respectively. The corresponding errors are given in Tab. 8.4, which provides an overview over the results for the four investigated sensor combinations. The worst case scenario is obviously the measurement configuration 3, which only implies measurement data from acceleration sensors. The errors lie by far outside the acceptable range. Analyzing the identified force time history from Fig. 8.17, it becomes clear that the proposed MFI method fails to reconstruct the static component of the GRF, even though more or less plausible

6 As already shown within the numerical validation (Chap. 6) and based on the applications on the Schmutter bridge (Chap. 7), the accurate prediction of structural responses at unmeasured locations was found to be even easier to achieve than the correct reconstruction of the external forces.

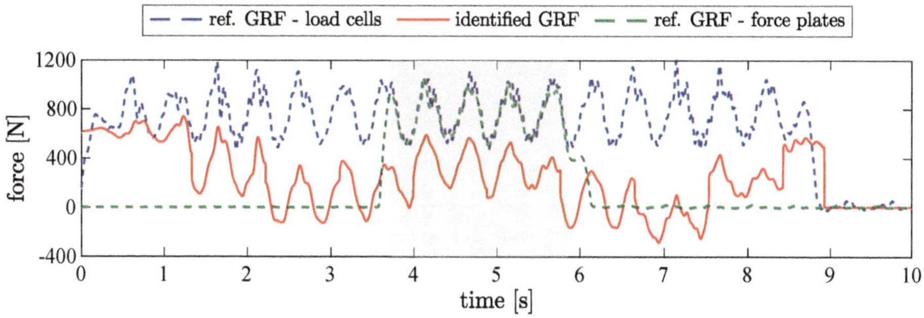

Figure 8.17 Identified and reference force time histories of a subject walking over the HUMVIB-Bridge for the measurement configuration 3

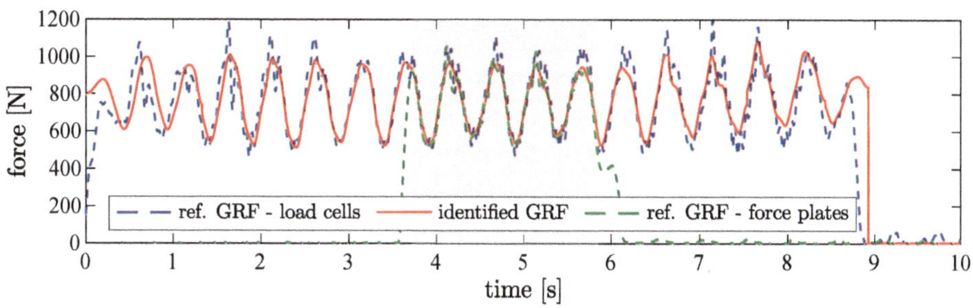

Figure 8.18 Identified and reference force time histories of a subject walking over the HUMVIB-Bridge for the measurement configuration 4

transient components seem to be retrieved. This effect is consistent with the findings obtained in the numerical validation (Section 6.7.5) and within the experimental validation of the Schmutter bridge (Section 7.6.3). It can be explained by the fact that acceleration measurements are less sensitive to the static force component than the displacements signals. However, eventual applications of the GRF reconstruction for investigating HSI effects are generally intended to be carried out in laboratory conditions. This implies that the mass of the test subjects and thus the global average of the unknown force time history is known. In this respect, it is expected that introducing this additional a priori information into the problem formulation, i.e. replacing the global average in the first penalty function by the subject weight, would also allow the accurate GRF reconstruction based on only acceleration sensors. However, a detailed discussion on this topic would go beyond the scope of this work.

Contrary to the results obtained for the identification based on acceleration measurements, the errors obtained for measurement data consisting of only displacement signals present similar values as the results of the combined measurement configurations. Never-

Measurement configuration	Error type	Error value
1	$x_{\mathrm{err,LC}}$ [%]	10.18
	$x_{\mathrm{err,FP}}$ [%]	3.20
2	$x_{\mathrm{err,LC}}$ [%]	10.13
	$x_{\mathrm{err,FP}}$ [%]	3.48
3	$x_{\mathrm{err,LC}}$ [%]	68.98
	$x_{\mathrm{err,FP}}$ [%]	61.01
4	$x_{\mathrm{err,LC}}$ [%]	10.21
	$x_{\mathrm{err,FP}}$ [%]	3.45

Table 8.4 Identification errors for different measurement configurations of the HUMVIB-Bridge

theless, the visual analysis of the reconstructed force time history (Fig. 8.18) reveals that some higher frequency force components could not be successfully retrieved, indicating – as expected – that the displacements are less sensitive to the dynamic components of the external forces. Considering that almost equal errors were retrieved for the configurations 2 and 4, it seems that the error approach based on the l_1 norm (Eq. 5.23) is not sensitive enough to the very small changes of the higher frequency components.

The previous discussion emphasizes that the proposed MFI method delivers accurate and robust results with respect to the identification of GRFs induced by a pedestrian during walking over a footbridge. This is even valid for as few as four sensors, whereas the best results are achieved for a combination of acceleration and displacement sensors. The qualitative comparison between the performances of the proposed structure-centered approach and of the recently published alternative structure-centered method for GRF identification of CHEN et al. (2021) suggests that the method proposed in the present thesis retrieves more accurate results in case of identifying GRF from real measurement data. The reason for this is thought to be that the GRFs for successive strides of a pedestrian are not perfectly identical. It was shown that the proposed method can successfully handle the reconstruction of "imperfect" steps, while CHEN et al., 2021 seeks to reconstruct only the Fourier parameters of the load time history, implying thus that the GRF consists of a succession of identical, perfectly periodic steps sequences. Nevertheless, the work CHEN et al. (2021) presents an important additional approach for dealing with pedestrian groups, which opens the way to further applications.

In the following, three additional computations are used to highlight whether the accuracy of the GRF identification is affected by further parameters: inconstant speed of the subject over time, modeling errors and an uncertain trajectory of the subject.

Effect of an inexact subject speed All previous results for GRF reconstruction were generated under consideration of the realistic speed variation over time, which was determined according to the load cell signals (cf. Fig. 8.13). In order to verify the influence

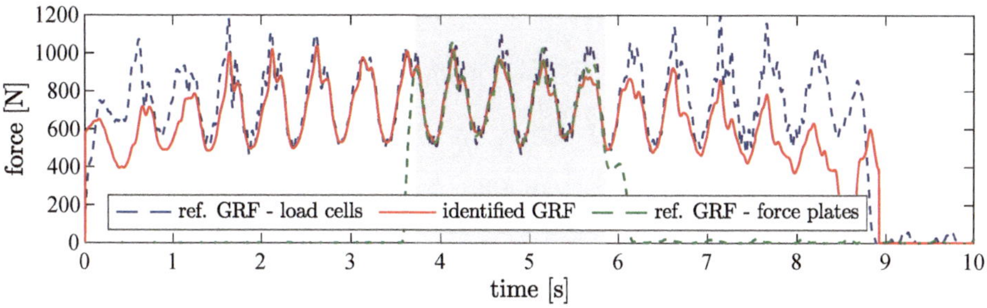

Figure 8.19 Identified and reference force time histories of a subject walking over the HUMVIB-Bridge for the measurement configuration 1 and a perfectly constant speed

of an inexact subject speed, a computation is made under the assumption that the subject perfectly keeps the prescribed frequency at a perfectly constant step length. This means that the location curve of the subject is assumed to vary linearly in time (see green curve in Fig. 8.13). The result is shown in Fig. 8.19. It can be clearly observed that the identified force time history presents considerable deviations at the beginning and at the end of the passage time window. The corresponding relative error $x_{\text{err,LC}}$ increases in this case to 16.61 %. In this respect, it can be observed that the length of the inaccurate time intervals correspond well to the intervals of the highest nonlinearity of the location curve shown in Fig. 8.13b, i.e. approximately between 0 and 2 s and between 5.5 and 9 s. The consideration of an erroneous location leads to erroneous modal forces, which directly propagate to the identification accuracy. In contrast to the intervals at the beginning and the end of the time history, the interval in the middle part of the time history (approximately between 2 and 5.5 s) seems to be characterized by more accurate identification results. A quite satisfactory value of 3.26 % is achieved for the error $x_{\text{err,FP}}$[7]. The effect of a better reconstruction in the middle of time history can be possibly explained by the almost linear variation of the pedestrian speed in the considered interval, approximately between 2 and 5.5 s. This can be observed by the nearly overlapping green and red curves in Fig. 8.13b in the specified time frame. It can be thus stated that the identification accuracy is very sensitive to the subject speed considered in the computation.

Effect of modeling errors The modeling errors were identified as a further sensitive parameter affecting the GRF identification accuracy. This can be easily demonstrated by performing an additional calculation with real measured data, but considering the original structural model calibrated with respect to the empty structure instead of the model containing the mass contribution of the biomechanical equipment. The natural frequency

[7] It is to be noted that the sole analysis of $x_{\text{err,FP}}$ might lead in this case to erroneous interpretations with respect to the identification accuracy. The reason for this is that $x_{\text{err,FP}}$ is only evaluated for the middle part of the force time history, which includes signals of the biomechanical force plates.

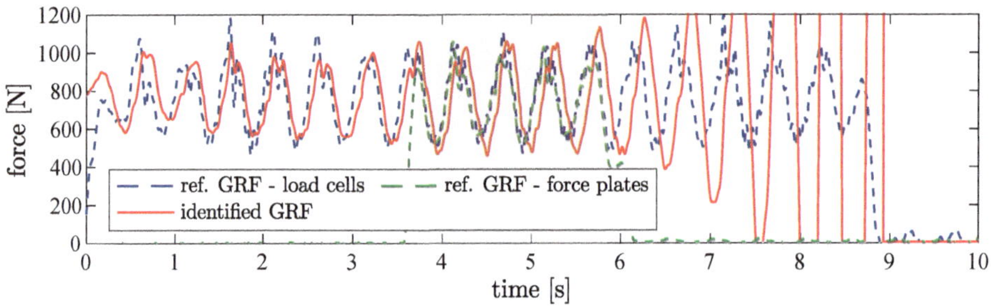

Figure 8.20 Identified and reference force time histories of a subject walking over the HUMVIB-Bridge for the measurement configuration 1 and an insufficiently calibrated structural model

for the first vertical bending mode of the model of the empty structure is $f_2 = 2.04$ Hz (cf. Fig. 8.8). The consideration of an additional mass of 60 kg/m, corresponding to the biomechanical equipment, causes a drop of the frequency of about 3 %, resulting in an adapted model with $f_2 = 1.98$ Hz. The results computed using real measurement data and the original model (without additional mass, $f_2 = 2.04$) are shown in Fig. 8.20. It can be seen that the deviations of the identified forces with respect to the reference values are rather low at the beginning of the time history. However, they increase over time and the result tends to a divergent behavior towards the end of the considered time window. The corresponding errors are $x_{\text{err,LC}} = 70$ % and $x_{\text{err,FP}} = 20.8$ % and lie thus by far in an unacceptable range. This result is consistent with the findings from Section 6.7.6 and justify once again the requirement of a very accurate model for achieving satisfactory GRF identification results.

Effect of an erroneous pedestrian trajectory The trajectory of the moving force is a parameter without any relevance in case of applications on railway bridges (Chap. 7). The reason for this is that the train strictly follows the rail, allowing no lateral variation of the loading path. However, the locomotion of a pedestrian over a footbridge is prone to deviations from the perfectly linear trajectory or to slight lateral variations with respect to a prescribed load path. Moreover, a certain amount of error is in any case deliberately included in the solution process by assuming that the left and right foot are both perfectly situated on the same path, i.e. the step width is considered to be zero. Within the present experiment, the walking surface defined by force plates array and the dummy plates (Fig. 8.3) was at least 60 cm wide. The subjects were asked to follow the middle axis (cf. Fig. 8.4), which was, however, not marked. Therefore, any trajectory deviations can not be excluded, both with respect to the lateral position of the loading path and to the linearity of the locomotion trajectory.

A rough insight into the sensitivity of the GRF identification to the assumed trajectory can be gained by the following computation, which considers a lateral deviation of the

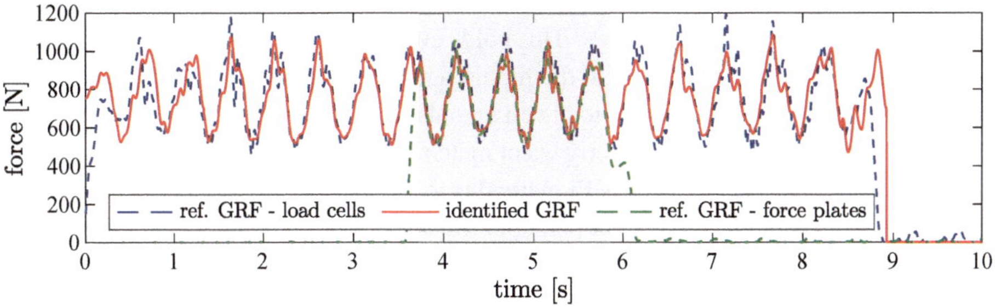

Figure 8.21 Identified and reference force time histories of a subject walking over the HUMVIB-Bridge for the measurement configuration 1 and a lateral deviation of the pedestrian trajectory

loading path of 20 cm. The results are shown in Fig. 8.21. At first sight, a very good match between the identified and the reference force time histories can be generally observed. However, the activation of the torsional modes partly causes slightly more pronounced transient components. The corresponding errors are $x_{\mathrm{err,LC}} = 10.41\ \%$ and $x_{\mathrm{err,FP}} = 4.53\ \%$. These indicate that an erroneous consideration of the subject trajectory might lead – in the present case – to a rather moderate increase of identification errors. Nevertheless, the 3D recording of the pedestrian trajectory is nowadays technically achievable and recommended in case of future research applications on the HSI effect, in order to eliminate an additional error source.

8.7 Conclusion

This chapter describes a second experimental validation process of the proposed MFI method. It is concerned with the identification of the ground reaction forces induced by a pedestrian on an experimental pedestrian bridge. For this, a FE model was firstly defined according to the as-built situation. It was calibrated with respect to the first seven experimentally identified natural frequencies. The load scenario is a subject walking over the bridge with a step frequency of 2 Hz. In order to allow the verification of the identified forces, an array of seven biomechanical force plates was used to record up to four steps in the middle area of the bridge. In addition, the support forces were acquired using four load cells installed at the bridge supports. However, in order to obtain the reference GRF components, the inertial forces of the bridge and of the force plates need to be subtracted from the original sensor readings of the load cells and of the biomechanical force plates, respectively. Moreover, the pedestrians are likely to fluctuate their speed during the passage over a vibrating structure. In this sense, a method for determination of the time dependent subject location was proposed. As expected, the studied case showed a slight non-linear variation of the location in time.

In general, very accurate GRF identification results were obtained for a combination of acceleration and displacement sensors. This holds even for as low as four sensors. The error values lie around 3 % with respect to the highly accurate force plate readings and around 10 % related to the more uncertain load cell signals. In addition, the re-built responses at unmeasured locations revealed a nearly exact match with respect to the reference values. A great achievement of the proposed MFI method is that it enables the evaluation of imperfect step sequences. This indicates a great potential for applications dealing with the HSI effects.

Similar to the investigations on the railway bridge (Chap. 7), it is not possible to retrieve accurate forces if only acceleration sensors are used. The reason for this is the rather low sensitivity of the measured accelerations with respect to quasi-static force components. The identification based on only displacement sensors generally revealed a good accuracy. Nevertheless, some higher frequency force components could not be correctly retrieved because the displacement sensors are less sensitive to high frequency dynamic load components.

It has been also shown that an inexact subject speed has a significant influence on the identification accuracy. In addition, the effect of modeling errors was investigated under consideration of a perturbed structural model, whose first natural frequency differs by 3 % with respect to the calibrated model. This led to a partially divergent behavior of the identified forces and unacceptable identification errors. Finally, the sensitivity of the identification accuracy with respect to an inexact pedestrian trajectory was also studied. It has been shown that perturbing the lateral position of the walking path by 20 cm causes only a moderate increase in the identification errors. This can be possibly explained by the activation of the torsional modes. They are responsible for some high frequency components of the identified forces, which are not present in the reference time histories.

9 Conclusions and outlook

9.1 Summary and conclusions

The present work discusses the development and experimental validation of a method for moving force identification (MFI) on bridge structures. Its basic idea is to find the optimal forces that minimize the difference between the computed and the measured time histories of structural responses for a limited number of measurement points. For this, an optimization problem was formulated in time domain, which considers the magnitude of each moving force in each time instant as an unknown optimization variable. The problem formulation takes into account acceleration and displacement measurement data. The direct computation of the structural responses is based on the modal superposition approach, whereas the modal equations of motion are solved using the time integration method of Newmark. This formulation allows the application of the proposed method with any finite element (FE) model with arbitrary level of detail.

The regularization of the inverse problem is addressed by introducing two weighted penalty terms, which define physical a priori information on the solution. The first one ensures the discrete finite signal with stable average value (DFS-SAV) property of traffic forces on bridges, while the second one seeks to suppress the noise influence on the identified force time histories. The solution of the optimization problem is addressed by means of the trust region optimization strategy. This presents an optimal convergence behavior if it is supplied with an exact gradient and Hessian of the objective function. These were determined in the present work by analytical differentiation, whereas the time stepping method of Newmark was used to solve a set of differential equations occurring in the implicit differentiation process.

In addition, the solution method is formulated using a forward-backward sliding window approach, in order to allow the application of the method for arbitrarily large problems. These are typical, for example, when dealing with the axle load identification of a complete railway vehicle, which usually presents a large number of axles and needs several thousands of time steps to cross the structure. The sliding window method implies the division of the total computational time into a set of separate time windows of given length with a certain overlap. Firstly, the optimization problem is formulated and solved separately in each window, whereas the progress from one window to the next is based on the forward sliding direction. Subsequently, a second optimization for a backward sliding window is performed.

A. Firus, *A Contribution to Moving Force Identification in Bridge Dynamics*, Mechanik, Werkstoffe und Konstruktion im Bauwesen 65, https://doi.org/10.1007/978-3-658-39838-5_9

This allows the consideration of a more extensive set of measurement information, which also includes the decay phase of the signals, eventually leading to an improved accuracy.

The proposed method was firstly tested by means of numerical investigations with simulated measurement data and a simplified 1D beam model. These were complemented by two experimental validation processes, which used detailed 3D FE models. The first one implies an ICE 4 vehicle traveling at predefined speeds over a railway bridge in regular operation. In this case, the validation of the identification could only be performed based on the static components of the forces and on the re-built responses, since no dynamic axle load measurements was possible. The second experimental validation campaign was concerned with the identification of ground reaction forces (GRF) induced by a subject during walking on a pedestrian bridge with a predefined step frequency. The validation of the identified forces could be performed using reference force measurements from a set of biomechanical force plates and four load cells installed at the bridge supports. The pedestrian bridge presents a clear linear behavior and its structural model could be straightforwardly calibrated with respect to the measured natural frequencies. On the contrary, the railway bridge is characterized by an amplitude dependent stiffness, which leads to a frequency drop during the train passage of about 14 % compared to the result of the empty structure. However, a linearized structural model had to be used, since the proposed MFI method is based on the modal superposition approach and can thus be only used in conjunction with linear models.

One of the key aspects of the proposed MFI method is the determination of appropriate weighting parameters for the penalty functions. For this, a robust strategy was proposed. It implies a pure numerical parametric study, which is based on noisy simulated measurement data. It has been shown that the optimal weighting parameters for a certain structure are rather insensitive to the magnitude, speed and number of the moving forces. They can therefore be determined at the beginning of the investigation using a single parameter study and then kept constant for all tests on the given structure. This fact represents a great achievement with respect to real applications. Nevertheless, the experimental validation performed on the railway bridge revealed that even more reliable regularization parameters can be obtained if a calibration run is included in their determination.

Two solvability conditions of the proposed MFI method were defined for the case of noise-free data and an unregularized objective function. The first one states that the number of modes considered in the inverse analysis has to be at least equal to the maximum number of forces present on the structure at a given time instant, in order to retrieve the exact solution. The second one requires that the number of sensors included in the data set is at least equal to the number of forces present on the structure in a certain time instant, in order to achieve the exact solution. Nevertheless, it has been shown that the activation of the penalty functions compensates to a certain extent the violation of the solvability conditions, so that stable and sufficiently accurate solutions can be expected, even if one or both solvability conditions are violated. This is equally valid for the application with noise-free or noisy simulated measurement data and for the investigations with real measurement data.

The first identification results obtained using simulated acceleration and displacement measurement data were very satisfactory with respect to both the static and the dynamic components of the identified forces. Error values below 3 % were obtained for different load scenarios and travel speeds. On the contrary, the first results of the experimental validation on the railway bridge revealed that the proposed method failed – under consideration of both acceleration and displacement measurements – to properly identify the static components of the individual axle forces, even though the total vehicle weight and the re-built responses at unmeasured locations were in a very good agreement with the reference values. This clearly indicates that an accurate prediction of the structural responses at unmeasured locations is generally easier to achieve than the identification of the correct forces.

The identified forces of the railway bridge are basically affected by two effects. On the one side, a set of jump discontinuities appear in the time histories of some moving forces, whereas they always correspond to a moving window boundary. On the other side, significant differences between the two static axle forces of a bogie could be detected, whereas one identified value was always considerably higher and the other one considerably lower than the reference value. The first issue is of pure numerical nature and occurs due to the formulation based on the moving window approach. The second one arises because the proposed method seems not to be able to deal with closely spaced axle loads in case of applications with real measurement data. The reason for this are the modeling errors in combination with low differences between the values of the mode shapes corresponding to the positions of the two axles of a bogie. It is very likely that the measurement data is rather insensitive to the close spacing between the wheel-sets of a bogie, e.g. due to the load-distributing characteristics of the ballast bed, which was neglected in the structural model. Nevertheless, strategies for suppressing both effects in the post-processing were proposed. Considerably lower errors were revealed after implementing the corrections. A total of five passages with speeds between 100 and 200 km/h were investigated. The identified total ICE 4 weight consistently delivered errors below 4 %, while the static components of the individual loads were identified with an excellent average accuracy below 4.2 %.

The previously mentioned effects for the railway bridge did not appear in the investigations of pedestrian bridge. The GRF of the pedestrian could be retrieved very accurately, even for as low as two displacement and two acceleration sensors. Errors of approx. 3 % and 10 % were obtained with respect to the accurate biomechanical force plate recordings and to the highly uncertain load cell readings, respectively. In this respect, a great achievement of the proposed method is that it is able to reconstruct imperfect step sequences successfully. This indicates a clear potential of the MFI method towards applications in the field of human-structure-interaction (HSI). In the same context, it has been shown that the speed of the pedestrian is one of the most sensitive parameters with respect to the accuracy of the force identification. Therefore, the GRF reconstruction had to consider the slight variations of the pedestrian speed over time, which might possibly occur due to slight variations of the step length at a constant step frequency. The importance of the speed measurement was also emphasized in the context of the numerical investigations and within the experimental validation on the Schmutter bridge. The latter showed that even the speed values retrieved by a classical radar device can lead to unacceptable force identification errors. Therefore, the speed measurement has to be performed with great care, in order to allow stable and accurate identification results.

The re-built responses at unmeasured locations were generally retrieved with a very high accuracy by all investigations. However, an exception was detected in case of the Schmutter bridge. It was revealed, that only the re-built responses at points situated on the loaded substructure could be consistently accurately predicted. On the contrary, the re-built responses of the unloaded structure only delivered sufficiently accurate results for the resonant passages. This is a consequence of the modeling errors. By updating the structural model with respect to a single natural frequency obtained from the experimental resonance curve, it is very likely that the model errors are higher for a non-resonant excitation.

In all three investigations, the best identification results were obtained for a combination of acceleration and displacement sensors. In this context, it was shown that even a very low and thus practicably feasible number of displacement and acceleration sensors leads to sufficiently accurate results. Furthermore, both experimental investigations indicated that the use of only acceleration sensors leads to highly erroneous static components of the identified forces. The reason for this is the low sensitivity of the accelerations with respect to the static components of the moving forces. On the contrary, the numerical investigation showed that an accurate force identification is possible even if only acceleration measurements are considered. Nevertheless, the identification based on only acceleration data is in this case very sensitive to the amount of regularization.

In general, the numerical investigations indicate a tendency towards a better identification accuracy for an increased number of sensors. However, the experimental validations have shown that increasing the number of sensors does not necessarily lead to lower identification errors. This effect was detected for the investigation on the Schmutter bridge, where the consideration of additional displacement sensor only lead to improved accuracy for the resonant passages, while higher errors were retrieved for all non-resonant passages. This confirms again the presence of the modeling errors, especially for the non-resonant excitation. In the same context, an increased number of modes considered in the computation generally lead in case of analyses with simulated measurement data – as expected – to lower identification errors. However, the investigation of the Schmutter bridge revealed exactly the opposite behavior. In this case, a clear trend of higher errors for higher number of modes considered in the computation could be observed. This is also a clear consequence of the modeling errors. By only updating the model with respect to the first natural frequency, it is highly uncertain, to what extent the higher frequencies and mode shape values match the real structural behavior.

Moreover, the investigations based on simulated measurements have shown that the proposed method is rather insensitive to the noise components of the measurement data, revealing a sufficient accuracy, even for very conservative noise levels. With respect to the computation parameters related to the inverse problem formulation, it was revealed that a good accuracy is expected for moving window lengths over 400 samples, smoothing radii below 5 samples and lengths of the overlapping interval greater than one fifth of the moving window length. These conditions were respected for both the numerical and the experimental investigations performed in this work.

Regarding the validation of the identified dynamic force components, no clear statement can be formulated for the investigation on the Schmutter bridge, since no reference values were available. However, an interesting effect was observed. The maximum points of the dynamic components usually occur around the time steps where another axle reaches or leaves the bridge. This might indicate that the dynamic force components obtained for the Schmutter bridge are influenced by numerical effects. On the contrary, the dynamic components of the forces could be very accurately identified for the experimental investigations on the HUMVIB-Bridge and for the numerical study of the 1D beam model with simulated measurement data. In the latter case, it has been also shown that the proposed method is even able to retrieve impulse-like components of the unknown forces. This is possible due to the formulation of the optimization problem, which considers the value of each force at each time instant as an unknown optimization variable. Nevertheless, the identification of the impulse magnitudes is highly dependent on the regularization amount. High regularization components tend to flatten the impulse-like components. In this respect, a strategy for a more accurate determination of the impulse magnitudes was proposed. However, the impulse identification could not be validated by experiments. In this context, it should be mentioned that – considering the previously mentioned modeling errors – there is considerable doubt that high frequency impulse components can be retrieved for a real ballasted railway bridge in operation.

As it has already been suggested before, the modeling errors are one of the most relevant parameters affecting the accuracy of the proposed MFI method. This could be confirmed both for the numerical investigations using simulated measurement data and for the two experimental validations. Therefore, the most important requirement of the proposed MFI method is an accurate structural model. The required model accuracy refers, above all, to the global structural behavior. In this respect, it has been shown that even a slight model deviation of 1 % with respect to the first natural frequency can lead to a divergent behavior of the identified forces and thus to a completely intolerable accuracy. Therefore, the match of the computed and measured first natural frequencies of the models used in the experimental validations was considered as a primary calibration criterion. Moreover, the damping ratios play an important role in the MFI from resonant passages. This is a relevant scenario for railway bridge applications. Furthermore, it is presumed that using a more detailed structural model and calibrating it with respect to further aspects of the local structural behavior, e.g. the load-distributing effect of the ballast bed, are expected to have a positive influence on the results.

Finally, it is concluded that the proposed MFI method could be successfully experimentally validated. Despite a set of certain inevitable modeling errors included in the investigated structural models, the method consistently revealed a robust and stable behavior and delivered very accurate identification results, even under consideration of a low number of sensors. In this context, different issues, particularities and boundary conditions associated with applications on real structures were discussed and corresponding solutions were proposed. The successful experimental validation based on two different structures underlines the wide range of possible applications of the proposed MFI approach: from small-scale structures excited by low pedestrian forces to the "heavy category" of a complete train passing a railway bridge in regular operation. The reconstruction of all axles of an ICE 4 vehicle is the ultimate proof that the method can handle arbitrarily large problems. Therefore, an important gap with respect to the applications of the MFI concept for real structures in operation has been closed. On the one side, this opens the way towards a continuous load monitoring on railway bridges based on structural responses. On the other side, the small-scale applications on pedestrian bridge could be potentially used for a better understanding of the HSI effects and a corresponding improvement of the design codes for dynamic analysis of pedestrian induced vibrations.

9.2 Outlook

The foregoing discussions on the proposed MFI method allow to indicate a few points, where the approach can be refined:

- The effect of unevenly distributed static axle load components within a bogie – detected in the experimental validation of the Schmutter bridge – was previously corrected in the post-processing. However, a mathematically more consistent solution of this issue would imply the consideration of the realistic condition of evenly distributed static axle loads within a single bogie as a constraint for the total objective function. In this way, the optimal result is expected to be directly retrieved by the optimization algorithm, without any need for post-processing.
- Even though the current formulation of the problem allows the solution of large problems with reasonable computational effort, setting-up of the Hessian is the most computationally expensive step of the identification process. In this respect, it would be interesting to explore, whether low-rank approximations or the consideration of eventual sparsity characteristics of the Hessian could lead to an improved computational time without loss of accuracy.
- The current formulation of the MFI method considers accelerations and displacements as measurement data. However, despite the recent advances in the field of non-contact

displacement measurement technologies, they are not yet widely applied for bridge measurements. As the method can not retrieve acceptable identification results from only acceleration data, the measurement of a further quantity that reacts sensitively to the static force components would be helpful. In this respect, the extension of the problem formulation for consideration of strain measurements seems to be a justifiable choice.

- The determination of the weighting parameters for the two penalty functions is a key aspect of the identification procedure. Even though it has been shown that they present a robust behavior and can be kept constant for a certain structure, their determination is quite expensive from computational point of view. This does not represent a major problem considering that the corresponding parametric study only has to be performed once, at the beginning of the investigation. Nevertheless, it is believed that further studies, e.g. based in artificial intelligence algorithms, could lead to alternative methods for a faster determination of appropriate weighting factors.
- Even though the method has been experimentally validated on two structures, the sensors used in the measurements do not cover all possible noise behaviors of typical sensors employed in civil engineering. For the sake of completeness, it would make sense to perform further numerical analyses under consideration of worst-case noise characteristics (e.g. amplitude dependent and colored noise), in order to make sure that the proposed MFI is robust and accurate over the whole range of possible noise properties. However, it is to be mentioned that this aspect is not expected to reveal any significant reduction of the performance.
- The previous investigations on real railway bridges were only concerned with the identification of the ICE 4 loads and a single railway bridge. Prior to a reliable long-term practical application, further investigations should be made on other structures under consideration of arbitrary vehicles.

Bibliography

Literature

AHMADI, E., CAPRANI, C. C., ŽIVANOVIĆ, S. and EVANS, N. (2018a): A framework for quantification of human-structure interaction in vertical direction, in: *Journal of Sound and Vibration* vol. 432, pp. 351–372.

AHMADI, E., CAPRANI, C. and HEIDARPOUR, A. (2017): An equivalent moving force model for consideration of human-structure interaction, in: *Applied Mathematical Modelling* vol. 51, pp. 526–545.

AHMADI, E., CAPRANI, C., ŽIVANOVIĆ, S. and HEIDARPOUR, A. (2018b): Vertical ground reaction forces on rigid and vibrating surfaces for vibration serviceability assessment of structures, in: *Engineering Structures* vol. 172, pp. 723–738.

AHMADI, E., CAPRANI, C., ZIVANOVIC, S. and HEIDARPOUR, A. (2021): Experimental validation of moving spring-mass-damper model for human-structure interaction in the presence of vertical vibration, in: *Structures* vol. 29, pp. 1274–1285.

AHMARI, S. and YANG, M. (2013): Impact location and load identification through inverse analysis with bounded uncertain measurements, in: *Smart Materials and Structures* vol. 22, pp. 275–284.

AL-QADI, I., WANG, H., OUYANG, Y., GRIMMELSMAN, K. and PURDY, J. (2016): *LTBP Program's Literature Review on Weigh-in-Motion Systems*, tech. rep., McLean, Virginia, United States of America: U.S. Department of Transportation - Federal Highway Administration.

ANCILLAO, A., TEDESCO, S., BARTON, J. and O'FLYNN, B. (2018): Indirect Measurement of Ground Reaction Forces and Moments by Means of Wearable Inertial Sensors: A Systematic Review, in: *Sensors* vol. 18.2564.

ARTESE, S. and NICO, G. (2020): TLS and GB-RAR Measurements of Vibration Frequencies and Oscillation Amplitudes of Tall Structures: An Application to Wind Towers, in: *Applied Sciences* vol. 10.

ASNACHINDA, P., PINKAEW, T. and LAMAN, J. A. (2008): Multiple vehicle axle load identification from continuous bridge bending moment response, in: *Engineering Structures* vol. 30, pp. 2800–2817.

AUCEJO, M. and DE SMET, O. (2017): A multiplicative regularization for force reconstruction, in: *Mechanical Systems and Signal Processing* vol. 85, pp. 730–745.

A. Firus, *A Contribution to Moving Force Identification in Bridge Dynamics*, Mechanik, Werkstoffe und Konstruktion im Bauwesen 65, https://doi.org/10.1007/978-3-658-39838-5

AUERSCH, L. (2015): Realistic axle-load spectra from ground vibrations measured near railway lines, in: *International Journal of Rail Transportation* vol. 3.4, pp. 180–200.

AZAM, S. E., CHATZI, E. and PAPADIMITRIOU, C. (2015): A dual Kalman filter approach for state estimation via output-only acceleration measurements, in: *Mechanical Systems and Signal Processing* vol. 60-61, pp. 866–886.

BACHMANN, H. (1988): Schwingungsprobleme bei Fußgängerbauwerken, in: *Bauingenieur* vol. 63, pp. 67–75.

BAGAYOKO, L., KOCH, E. and PATZ, R. (2008): Dynamik von Eisenbahnbrücken, in: *Stahlbau-Kalender*, pp. 647–693.

BATHE, K.-J. (2014): *Finite Element Procedures*, 2nd ed., Upper Saddle River, USA: Prentice Hall.

BAUMANN, K. and BACHMANN, H. (1988): *Durch Menschen verursachte dynamische Lasten und deren Auswirkungen auf Balkentragwerke*, tech. rep., Zürich, Switzerland: ETH Zürich.

BECKER, M., SCHNEIDER, J., FIRUS, A., LEINEN, S. and PULLAMTHARA, J. J. (2017): *Strukturanalyse mit Mikrowelleninterferometrie. Teil A: Forschungsanteil*, tech. rep., Darmstadt: Technical University of Darmstadt, unpublished.

BELLI, A., BUI, P., BERGER, A. and GEYSSANT, H. (2001): A treadmill ergometer for three-dimensional ground reaction forces measurement during walking, in: *Journal of Biomechanics* vol. 34, pp. 105–112.

BERNAL, D. and USSIA, A. (2015): Sequential deconvolution input reconstruction, in: *Mechanical Systems and Signal Processing* vol. 50-51, pp. 41–55.

BERNARDINI, G., DE-PASQUALE, G., BICCI, A., MARA, A., COPPI, F., RICCI, P. and PIERACCINI, M. (2007): "Microwave interferometer for ambient vibration measurement on civil engineering structures: 1. Principles of the radar technique and laboratory tests", in: *EVACES '07 - Experimental Vibration Analysis for Civil Engineering Structures*, Porto, Portugal.

BIGELOW, H., RAUERT, T., HOFFMEISTER, B. and FELDMANN, M. (2018): Steifigkeitsbeiträge durchgängiger Schotterschichten auf Eisenbahnbrücken mit Längsfugen - Interaktion zwischen baulich getrennten Überbauten durch ein gemeinsames Schotterbett, in: *Stahlbau* vol. 87.3, pp. 215–224.

BJÖRCK, Å. (2015): Linear Least Squares Problems, in: *Numerical Methods in Matrix Computations*, Springer International Publishing Switzerland, pp. 211–430.

BOCIAN, M., BROWNJOHN, J., RACIC, V., HESTER, D., QUATTRONE, A. and MONNICKENDAM, R. (2016): A framework for experimental determination of localised vertical pedestrian forces on full-scale structures using wireless attitude and heading reference systems, in: *Journal of Sound and Vibration* vol. 376, pp. 217–243.

BRANCH, M. A., COLEMAN, T. F. and LI, Y. (1999): A subspace, interior, and conjugate gradient method for large-scale bound-constrained minimization problems, in: *SIAM Journal on Scientific Computing* vol. 21.1, pp. 1–23.

BRAUNE, W. (1977): Beitrag zur Feststellung der Belastbarkeit von Ingenieurbauten durch die Bestimmung von Einflußlinien aus Dehnungsmessungen unter beliebigen wandernden Lasten, in: *Bautechnik* vol. 54.5, pp. 145–149.

BRONSTEIN, I. N., SEMENDJAJEV, K. A., MUSIOL, G. and MÜHLIG, H. (2012): *Taschenbuch der Mathematik*, 8th ed., Frankfurt am Main: Harri Deutsch, p. 1235.

BROWNJOHN, M. J. W., CHEN, J., BOCIAN, M., RACIC, V. and SHAHABPOOR, E. (2018): Using inertial measurement units to identify medio-lateral ground reaction forces due to walking and swaying, in: *Journal of Sound and Vibration* vol. 426.May, pp. 90–110.

BUSBY, H. R. and TRUJILLO, D. M. (1997): Optimal regularization of an inverse dynamics problem, in: *Computers and Structures* vol. 63.2, pp. 243–248.

BYRD, R. H., SCHNABEL, R. B. and SCHULTZ, G. A. (1988): Approximate solution of the trust region problem by minimization over two-dimensional subspaces, in: *Mathematical Programming* vol. 40, pp. 247–263.

CANTERO, D., ARVIDSSON, T., OBRIEN, E. and KAROUMI, R. (2015): Train-track-bridge modelling and review of parameters, in: *Structure and Infrastructure Engineering* vol. 2015.

CANTIENI, R. (1992): *Dynamic Behavior of Highway Bridges Under the Passage of Heavy Vehicles*, tech. rep., Swiss Federal Laboratories of Materials Testing and Research.

CANTIENI, R., BAJRI, A. and BRINCKER, R (2016): Experimentelle Bestimmung der Dämpfung eines Bauwerks am Beispiel einer Fußgängerbrücke - Worauf man achten sollte, in: *Bauingenieur* vol. 91, pp. 2–9.

CANTIENI, R. and KREBS, W. (2000): *EMPA Test Report No. 153'031. OECD IR 6 DIVINE Project - Dynamic Interaction Between Vehicle and Infrastructure Experiment. Element 6, Bridge Research*, tech. rep., Duebendorf, Switzerland: Swiss Federal Laboratories for Materials Testing and Research.

CAPRANI, C. and AHMADI, E. (2016): Formulation of human-structure interaction system models for vertical vibration, in: *Journal of Sound and Vibration* vol. 377, pp. 346–367.

CAPRANI, C. and OBRIEN, E. (2013): "Vision Systems for Analysis of Congested Traffic", in: *Proceedings of the IABSE conference*, Rotterdam.

CEBON, D. (1987): "Vehicles, Assessment of the dynamic wheel forces generated by heavy road", in: *Symposium on Heavy Vehicle Suspension and Characteristics*, Canberra, Australia: Australian Road Research Board.

CHAN, T. H. T., LAW, S. S. and YUNG, T. H. (2000a): Moving force identification using an existing prestressed concrete bridge, in: *Engineering Structures* vol. 22, pp. 1261–1270.

CHAN, T. H. T., LAW, S. S., YUNG, T. H. and YUAN, X. R. (1999): An interpretive method for moving force identification, in: *Journal of Sound and Vibration* vol. 219.3, pp. 503–524.

CHAN, T. H. T., YU, L., LAW, S. S. and YUNG, T. H. (2001a): Moving Force Identification Studies, I: Theory, in: *Journal of Sound and Vibration* vol. 247.1, pp. 59–76.

CHAN, T. H. T., YU, L., LAW, S. S. and YUNG, T. H. (2001b): Moving force identification studies, II: comparative studies, in: *Journal of Sound and Vibration* vol. 247.1, pp. 77–95.

CHAN, T. H. T. and YUNG, T. H. (2000b): A theoretical study of force identification using prestressed concrete bridges, in: *Engineering Structures* vol. 23, pp. 1529–1537.

CHEN, J., TAN, H. and PAN, Z. (2016): Experimental validation of smartphones for measuring human-induced loads, in: *Smart Structures and Systems* vol. 18.3, pp. 625–642.

CHEN, J., WANG, P. and WANG, H. (2021): Pedestrian-Induced Load Identification from Structural Responses Using Genetic Algorithm with Numerical and Experimental Validation, in: *Journal of Bridge Engineering* vol. 26.3.

CHEN, T.-C. and LEE, M.-H. (2008): Research on Moving Force Estimation of the Bridge Structure using the Adaptive Input Estimation Method, in: *Electronic Journal of Structural Engineering* vol. 8, pp. 20–28.

CHOI, H. G., THITE, A. N. and THOMPSON, D. J. (2007): Comparison of methods for parameter selection in Tikhonov regularization with application to inverse force determination, in: *Journal of Sound and Vibration* vol. 304, pp. 894–917.

CHOPRA, A. K. (2014): *Dynamics of Structures*, 4th ed., Upper Saddle River, New Jersey: Pearson/Prentice Hall, p. 944.

CLOUGH, R. W. and PENZIEN, J. (1995): *Dynamics of structures*, 3rd ed., Berkley, USA: Computers & Structures, Inc.

COLEMAN, T. F. and YUYING, L. (1994): On the convergence of interior-reflective Newton methods for nonlinear minimization subject to bounds, in: *Mathematical Programming* vol. 67, pp. 189–224.

COLEMAN, T. F. and YUYING, L. (1996): An Interior Trust Region Approach for Nonlinear Minimization Subject to Bounds, in: *SIAM Journal on Optimization* vol. 6.2, pp. 418–445.

COMER, A. J., BLAKEBOROUGH, A and WILLIAMS, M. S. (2013): Rhythmic crowd bobbing on a grandstand simulator, in: *Journal of Sound and Vibration* vol. 332.2, pp. 442–454.

DANG, H. and ŽIVANOVIĆ, S. (2016): Influence of Low-Frequency Vertical Vibration on Walking Locomotion, in: *Journal of Structural Engineering* vol. 142, pp. 1–12.

DENG, L. U. and CAI, C. S. (2010): Identification of Dynamic Vehicular Axle Loads: Theory and Simulations, in: *Journal of Vibration and Control* vol. 16.14, pp. 2167–2194.

DENG, L. U. and CAI, C. S. (2011): Identification of Dynamic Vehicular Axle Loads: Demonstration by a Field Study, in: *Journal of Vibration and Control* vol. 17.2, pp. 183–195.

DENNIS JR., J. E. and SCHABEL, R. B. (1996): *Numerical Methods for Unconstrained Optimization and Nonlinear Equations*, Englewood Cliffs, USA: Prentice Hall, p. 378.

DEUTSCHE BAHN AG (2019): *Deutsche Bahn - Daten & Fakten 2018*, tech. rep., Deutsche Bahn AG.

DING, Y., LAW, S. S., WU, B., XU, G. S., LIN, Q., JIANG, H. B. and MIAO, Q. S. (2013): Average acceleration discrete algorithm for force identification in state space, in: *Engineering Structures* vol. 56, pp. 1880–1892.

DJAMAA, M. C., OUELAA, N., PEZERAT, C. and GUYADER, J. L. (2007): Reconstruction of a distributed force applied on a thin cylindrical shell by an inverse method and spatial filtering, in: *Journal of Sound and Vibration* vol. 301, pp. 560–575.

DOMÉNECH, A., MUSEROS, P. and MARTÍNEZ-RODRIGO, M. (2014): Influence of the vehicle model on the prediction of the maximum bending response of simply-supported bridges under high-speed railway traffic, in: *Engineering Structures* vol. 72, pp. 123–139.

DOYLE, J. F. (2004): *Modern Experimental Stress Analysis*, Chichester, UK: John Wiley & Sons, Ltd.

ERRI (1999): *ERRI D214/RP6: Eisenbahnbrücken für Geschwindingkeiten > 200 km/h. Berechnung einfach gelagerter Brücken bei der Durchfahrt eines Zugverbandes*, tech. rep., Utrecht, Netherlands: European Railway Research Institute (ERRI).

ENGL, H. W., HANKE, M. and NEUBAUER, A. (2000): *Regularization of Inverse Problems*, ed. by K. A. PRESS, Dordrecht, The Netherlands, p. 321.

FAURE, C., ABLITZER, F., ANTONI, J. and PÉZERAT, C. (2017): Empirical and fully Bayesian approaches for the identification of vibration sources from transverse displacement measurements, in: *Mechanical Systems and Signal Processing* vol. 94, pp. 180–201.

FENG, D. and FENG, M. Q. (2017): Identification of structural stiffness and excitation forces in time domain using noncontact vision-based displacement measurement, in: *Journal of Sound and Vibration* vol. 406, pp. 15–28.

FENG, D., SUN, H. and FENG, M. Q. (2015): Simultaneous identification of bridge structural parameters and vehicle loads, in: *Computers and Structures* vol. 157, pp. 76–88.

FIRUS, A., BERTHOLD, H. and SCHNEIDER, J. (2018a): *Experimentelle und numerische Untersuchungen an der Eisenbahnüberführung über die Schmutter*, tech. rep., Technical University of Darmstadt, unpublished.

FIRUS, A., SCHNEIDER, J., BERTHOLD, H., SEYFARTH, A. and ALBINGER, M. (2018b): "Parameter identification of a biodynamic walking model for human-structure interaction", in: *Proceedings of the 9th International Conference on Bridge Maintenance, Safety and Management (IABMAS 2018)*, Melbourne, Australia.

FIRUS, A., BERTHOLD, H., SCHNEIDER, J. and GRUNERT, G. (2018c): "Untersuchungen zum dynamischen Verhalten einer Eisenbahnbrücke bei Anregung durch den neuen ICE 4", in: *VDI-Berichte 2321, 6. VDI-Conference on Structural Dynamics*, Verein Deutscher Ingenieure, Würzburg, pp. 233–248.

FIRUS, A., SCHNEIDER, J., BECKER, M., GRUNERT, G. and PULLAMTHARA, J. J. (2016): Dynamische Verformungsmessungen an Eisenbahnbrücken mittels Mikrowelleninterferometrie, in: *Bautechnik* vol. 93.10, pp. 701–710.

FIRUS, A., SCHNEIDER, J., BECKER, M. and PULLAMTHARA, J. J. (2017): "Microwave Interferometry Measurements for Railway-Specific Applications", in: *COMPDYN 2017, 6th ECCOMAS Thematic Conference on Computational Methods in Structural Dynamics and Earthquake Engineering*, ed. by M. PAPADRAKAKIS and M. FRAGIADAKIS, Rhodes Island, Greece.

FLETCHER, R. (1987): *Practical Methods of Optimization*, 2nd ed., West Sussex, England: John Wiley & Sons, Ltd, p. 436.

FONG, D. T.-P., CHAN, Y.-Y., HONG, Y., YUNG, P. S.-H., FUNG, K.-Y. and CHAN, K.-M. (2008): Estimating the complete ground reaction forces with pressure insoles in walking, in: *Journal of Biomechanics* vol. 41, pp. 2597–2601.

FORD, W. (2015): *Numerical Linear Algebra with Applications*, 1st ed., San Diego, USA: Elsevier Science Publishing Co Inc, p. 628.

FORNER CORDERO, A., KOOPMAN, H. J. F. M. and VAN DER HELM, F. C. T. (2004): Use of pressure insoles to calculate the complete ground reaction forces, in: *Journal of Biomechanics* vol. 37, pp. 1427–1432.

FRIEBE, R. (2017): Die Fachstelle Brückenmessung der DB Netz AG, in: *Stahlbau* vol. 86, pp. 1089–1097.

FRÝBA, L. (2001): A rough assessment of railway bridges for high speed trains, in: *Engineering Structures* vol. 23, pp. 548–556.

FRÝBA, L. (1972): *Vibration of solids and structures under moving loads*, 1st ed., Groningen, The Netherlands: Noordhoff International Publishing.

GEISSLER, K. and QUOOS, V. (1998): Vorschlag für die Bewertung der Restnutzungsdauer stählerner Eisenbahnbrücken, in: *Eisenbahn-Ingenieur-Kalender*.

GENTILE, C. (2015): Dynamic investigation of a suspension footbridge using accelerometers and microwave interferometer, in: *MATEC Web of Conferences* vol. 24.

GENTILE, C. and BERNARDINI, G. (2009): An interferometric radar for non-contact measurement of deflections on civil engineering structures: laboratory and full-scale tests, in: *Structure and Infrastructure Engineering* vol. 6.5, pp. 521–534.

GFRERER, H. (1987): An A Posteriori Parameter Choice for Ordinary and Iterated Tikhonov Regularization of Ill-Posed Problems Leading to Optimal Convergence Rates, in: *Mathematics of Computation* vol. 49.180, pp. 523–542.

GHAJARI, M., ALIABADI, M. H. and APICELLA, A. (2013): Identification of impact force for smart composite stiffened panels, in: *Smart Materials and Structures* vol. 22, pp. 1–20.

GILLIJNS, S. and DE MOOR, B. (2007): Unbiased minimum-variance input and state estimation for linear discrete-time systems with direct feedthrough, in: *Automatica* vol. 43, pp. 934–937.

GLATZ, B. and FINK, J. (2021): A redesigned approach to the additional damping method in the dynamic analysis of simply supported railway bridges, in: *Engineering Structures* vol. 241.

GOLUB, G. H. and VON MATT, U. (1997): Tikhonov Regularization for Large Scale Problems, in: *Scientific Computing*, ed. by G. H. GOLUB, S. H. LUI, F. T. LUK and R. J. PLEMMONS, Berlin: Springer, pp. 3–26.

GOLUB, G. H., HEATH, M. and WAHBA, G. (1979): Generalized Cross-Validation as a Method for Choosing a Good Ridge Parameter, in: *Technometrics* vol. 21.2, pp. 215–223.

GONZÁLES, A., ROWLEY, C. and OBRIEN, E. J. (2008): A general solution to the identification of moving vehicle forces on a bridge, in: *International Journal for Numerical Methods in Engineering* vol. 75, pp. 335–354.

GÓRAL, G., BYDOŃ, S. and UHL, T. (2002): Intelligent transducers of in-operational loads in construction fatigue monitoring, in: *Machine Dynamics Problems* vol. 2-3.73-88.

GRUNERT, G. (2021): Data and evaluation model for the description of the static-dynamic interface between trains and railway bridges, to be published.

GUNAWAN, F. E. (2012): Levenberg-Marquardt iterative regularization for the pulse-type impact-force reconstruction, in: *Journal of Sound and Vibration* vol. 331.25, pp. 5424–5434.

GUNAWAN, F. E., HOMMA, H. and MORISAWA, Y (2008a): Impact-Force Estimation by Quadratic Spline Approximation, in: *Journal of Solid Mechanics and Materials Engineering* vol. 2.8, pp. 1092–1103.

GUNAWAN, F. E. and HOMMA, H. (2008b): A Solution of the Ill-Posed Impact-Force Inverse Problems by the Weighted Least Squares Method, in: *Journal of Solid Mechanics and Materials Engineering* vol. 2.2, pp. 188–198.

GUNAWAN, F. E., HOMMA, H. and KANTO, Y. (2006): Two-step B-splines regularization method for solving an ill-posed problem of impact-force reconstruction, in: *Journal of Sound and Vibration* vol. 297, pp. 200–214.

GUNAWAN, F. E. and HOMMA, H. (2004): Efficient Iterative Solution for Large Elasto-Dynamic Inverse Problems, in: *JSME International Journal Series A: Solid Mechanics and Material Engineering* vol. 47.2, pp. 130–137.

GUPTA, S, DEGRANDE, G and LOMBAERT, G (2009): Experimental validation of a numerical model for subway induced vibrations, in: *Journal of Sound and Vibration* vol. 321, pp. 786–812.

HADAMARD, J. (1923): *Lectures on Cauchy's Problem in Linear Partial Differential Equations.* New Haven: Yalw University Press.

HAN, S. L. and KINOSHITA, T. (2012): Investigation of a Stochastic Inverse Method to Estimate an External Force: Applications to a Wave-Structure Interaction, in: *Mathematical Problems in Engineering* vol. 2012.February, pp. 1–25.

HANSEN, P. C. (1987): The truncated SVD as a method for regularization, in: *BIT* vol. 27, pp. 534–553.

HANSEN, P. C. (1992): Analysis of discrete ill-posed problems, in: *SIAM Review* vol. 34.4, pp. 561–580.

HANSEN, P. C. (1998): *Rank-Deficient and Discrete Ill-Posed Problems Mathematical Modeling and Computation*, 1st ed., Philadelphia: SIAM.

HANSEN, P. C. (2001): The L-curve and its use in the numerical treatment of inverse problems, in: *Computational inverse problems inelectrocardiology*, ed. by P. JOHNSTON, Southampton: WIT Press, pp. 119–142.

HARZHEIM, L. (2019): *Strukturoptimierung. Grundlagen und Anwendungen*, 2nd ed., Haan-Gruiten: Europa-Lehrmittel.

HAWRYSZKÓW, P. (2011): "Analysis of dynamical sensitivity and comfort of footbridges", in: *4th International Conference Footbridge 2011*, Wroclaw, Poland, pp. 1037–1046.

HEBDEN, M. D. (1973): *An algorithm for minimization using exact second derivatives*, tech. rep., Harwell, England: Atomic Energy Research Establishment (A.E.R.E), pp. 1–21.

HUANG, C.-H. (2001a): A non-linear inverse vibration problem of estimating the external forces for a system with displacement-dependent parameters, in: *Journal of Sound and Vibration* vol. 248.5, pp. 789–807.

HUANG, C.-H. (2001b): An inverse non-linear force vibration problem of estimating the external forces in a damped system with time-dependent system parameters, in: *Journal of Sound and Vibration* vol. 242.5, pp. 749–765.

HUANG, C.-H. (2005): A nonlinear inverse problem in estimating simultaneously the external forces for a vibration system with displacement-dependent parameters, in: *Journal of the Franklin Institute* vol. 342, pp. 793–813.

HUANG, C.-H., SHIH, C.-C and KIM, S. (2009): An inverse vibration problem in estimating the spatial and temporal-dependent external forces for cutting tools, in: *Applied Mathematical Modelling* vol. 33.6, pp. 2683–2698.

HUANG, Q., WANG, Y., LUZI, G., CROSETTO, M., MONSERRAT, O., JIANG, J., ZHAO, H. and DING, Y. (2020): Ground-Based Radar Interferometry for Monitoring the Dynamic Performance of a Multitrack Steel Truss, in: *Remote Sensing* vol. 12, pp. 1–19.

IDS-INGINERIA DEL SISTEMI S.P.A. (2015): IBIS Surveyor v 01.01 - User Manual, Pisa.

IGLESIAS, R., AGUASCA, A., FABREGAS, X., MALLORQUI, J., MODELLS, D., LÓPEZ-MARTÍNEZ, C. and PIPIA, L. (2014): Ground-Based Polarimetric SAR Interferometry for the Monitoring of Terrain Displacement Phenomena–Part I: Theoretical Description, in: *IEEE Journal of Selected Topics in Apllied Earth Observations and Remote Sensing* vol. 8, pp. 980–993.

JACOB, B. and DOLCEMASCOLO, V. (1998): "Dynamic Interaction Between Instrumented Vehicles and Pavements", in: *International Symposium on Heavy Vehicle Weights and Dimensions*, Queensland, Australia, pp. 142–160.

JACQUELIN, E., BENNANI, A. and HAMELIN, P. (2003): Force reconstruction: analysis and regularization of a deconvolution problem, in: *Journal of Sound and Vibration* vol. 265, pp. 81–107.

JIA, Y., YANG, Z. and SONG, Q. (2015): Experimental study of random dynamic loads identification based on weighted regularization method, in: *Journal of Sound and Vibration* vol. 342, pp. 113–123.

JIANG, C., HAN, X., GUAN, F. J. and LI, Y. H. (2007): An uncertain structural optimization method based on nonlinear interval number programming and interval analysis method, in: *Engineering Structures* vol. 29, pp. 3168–3177.

JIANG, R. J., AU, F. T. K. and CHEUNG, Y. K. (2003): Identification of masses moving on multi-span beams based on a genetic algorithm, in: *Computers and Structures* vol. 81, pp. 2137–2148.

KABANIKHIN, S. I. (2011): *Inverse and Ill-posed Problems. Theory and Applications*, Berlin: Walter de Gruyter, p. 459.

KAROUMI, R., CALÇADA, R., ANDERSSON, A., ALLAHVIRDIZADEH, R., KAMALI, A. Z., SILVA, A., RIBEIRO, D., FERREIRA, G., MONTENEGRO, P. and MUSEROS, P. (2021): *In2Track2: Proactive Bridge and Tunnel Assessment, Repair and Upgrade*, tech. rep., Stockholm/Porto/Valencia: KTH/FEUP/UPV.

KEMMLER, R. (2014): Strukturoptimierung – Theorie und Anwendung. Lecture Notes, University of Applied Sciences HTWG Konstanz, Konstanz, Germany.

KERN, M. (2016): *Numerical Methods for Inverse Problems*, London, UK and Hoboken, USA: ISTE Ltd and John Wiley & Sons, Inc., p. 215.

KERR, S. and BISHOP, N. (2001): Human induced loading on flexible staircases, in: *Engineering Structures* vol. 23, pp. 37–45.

KIM, C.-Y., JUNG, D.-S., KIM, N.-S., KWON, S.-D. and FENG, Q. (2003): Effect of vehicle weight on natural frequencies of bridges measured from traffic-induced vibration, in: *Earthquake Engineering and Engineering Vibration* vol. 2.1, pp. 109–115.

KISTLER (2017): Product catalogue: Force plates and accessories. Biomechanics measuring systems for performance diagnostics, gait analysis, rehabilitation, and ergonomics, Winterthur, Switzerland.

KOULOCHERIS, D. (2014): Evaluation of the Vertical Wheel Loads of a Heavy Vehicle with Two Different Experimental Methods, in: *Austin Journal of Robotics & Automation* vol. 1.1, pp. 1–5.

KUTTNER, T. and ROHNEN, A. (2019): *Praxis der Schwingungsmessung*, 2nd ed., Wiesbaden, Germany: Springer Vieweg.

LAI, T., YI, T.-H. and LI, H.-N. (2016): Parametric study on sequential deconvolution for force identification, in: *Journal of Sound and Vibration* vol. 377, pp. 76–89.

LAW, S. S., BU, J. Q., ZHU, X. Q. and CHAN, S. L. (2004): Vehicle axle loads identification using finite element method, in: *Engineering Structures* vol. 26, pp. 1143–1153.

LAW, S. S. and CHAN, T. H. T. (1999): Moving Force Identification-A Frequency and Time Domains Analysis, in: *Journal of Dynamic System, Measurement and Control, ASME* vol. 12.3, pp. 394–401.

LAW, S. S., CHAN, T. H. T. and ZENG, Q. H. (1997): Moving force identification: a time domain method, in: *Journal of Sound and Vibration* vol. 201.1, pp. 1–22.

LAW, S. S., CHAN, T. H. T., ZHU, X. Q. and ZENG, Q. H. (2001a): Regularization in moving force identification, in: *Journal of Engineering Mechanics* vol. 128, pp. 989–1000.

LAW, S. S. and FANG, Y. L. (2001b): Moving force identification: optimal state estimation approach, in: *Journal of Sound and Vibration* vol. 239.2, pp. 233–254.

LAW, S. S. and LI, J. (2010): Updating the reliability of a concrete bridge structure based on condition assessment with uncertainties, in: *Engineering Structures* vol. 32, pp. 286–296.

LAW, S. S., WU, S. Q. and SHI, Z. Y. (2008): Moving Load and Prestress Identification Using Wavelet-Based Method, in: *Journal of Applied Mechanics* vol. 75.2, pp. 141–147.

LAW, S. S. and ZHU, X. Q. (2000): Study on different beam models in moving force identification, in: *Journal of Sound and Vibration* vol. 234.4, pp. 661–679.

LAW, S. S. and ZHU, X. Q. (2011): *Moving Loads - Dynamic Analysis and Identification Techniques*, ed. by D. M. FRANGOPOL, 1st ed., London: CRC Press/Balkema.

LEBLANC, P. A., WOODROOFFE, J. H. and PAPAGIANNAKIS, A. T. (1992): "Comparison of the Accuracy of Two Types of Instrumentation for Measuring Vertical Wheel Load", in: *3rd Int. Symposium on Heavy Vehicles Weigths and Dimensions*, Cambridge: Thomas Telford, London, pp. 86–94.

LEVENBERG, K. (1944): A method for the solution of certain non-linear problems in least squares, in: *Quarterly of Applied Mathematics* vol. 2, pp. 164–168.

LI, J., SU, M. and FAN, L. (2003): Natural Frequency of Railway Girder Bridges under Vehicle Loads, in: *Journal of Bridge Engineering* vol. 8.4, pp. 199–203.

LI, Q. and LU, Q. (2016): Impact localization and identification under a constrained optimization scheme, in: *Journal of Sound and Vibration* vol. 366, pp. 133–148.

LI, Q. and LU, Q. (2017): Force localization and reconstruction using a two-step iterative approach, in: *Journal of Vibration and Control* vol. 24.17, pp. 3830–3841.

LI, Q. and LU, Q. (2019): A revised time domain force identification method based on Bayesian formulation, in: *International Journal for Numerical Methods in Engineering* vol. 118, pp. 411–431.

LI, X. Y. and LAW, S. S. (2009): Identification of structural damping in time domain, in: *Journal of Sound and Vibration* vol. 328, pp. 71–84.

LI, X. Y. and LAW, S. S. (2010): Adaptive Tikhonov regularization for damage detection based on nonlinear model updating, in: *Mechanical Systems & Signal Processing* vol. 24.6, pp. 1646–1664.

LILJENCRANTZ, A., KAROUMI, R. and OLOFSSON, P. (2007): Implementing bridge weigh-in-motion for railway traffic, in: *Computers and Structures* vol. 85, pp. 80–88.

LIU, J., HAN, X., JIANG, C., NING, H. M. and BAI, Y. C. (2011): Dynamic Load Identification for Uncertain Structures Based on Interval Analysis and Regularization Method, in: *International Journal of Computational Methods* vol. 8.4, pp. 667–683.

LIU, J., MEG, X., JIANG, C., HAN, X. and ZHANG, D. (2016): Time-domain Galerkin method for dynamic load identification, in: *International Journal for Numerical Methods in Engineering* vol. 105, pp. 620–640.

LIU, J., SUN, X., HAN, X., JIANG, C. and YU, D. (2014a): A novel computational inverse technique for load identification using the shape function method of moving least square fitting, in: *Computers and Structures* vol. 144, pp. 127–137.

LIU, K., DE ROECK, G. and LOMBAERT, G. (2009a): The effect of dynamic train-bridge interaction on the bridge response during a train passage, in: *Journal of Sound and Vibration* vol. 325.1-2, pp. 240–251.

LIU, K., LAW, S. S., ZHU, X. Q. and XIA, Y. (2014b): Explicit form of an implicit method for inverse force identification, in: *Journal of Sound and Vibration* vol. 333.3, pp. 730–744.

LIU, K., REYNDERS, E., DE ROECK, G. and LOMBAERT, G. (2009b): Experimental and numerical analysis of a composite bridge for high-speed trains, in: *Journal of Sound and Vibration* vol. 320.1-2, pp. 201–220.

LIU, K., LOMBAERT, G. and DE ROECK, G. (2014c): Dynamic Analysis of Multispan Viaducts with Weak Coupling between Adjacent Spans, in: *Journal of Bridge Engineering* vol. 19.1, pp. 83–90.

LIU, X., TONG, X., DING, K., ZHAO, X., ZHU, L. and ZHANG, X. (2015): Measurement of Long-Term Periodic and Dynamic Deflection of the Long-Span Railway Bridge Using Microwave Interferometry, in: *IEEE Journal of Selected Topics in Apllied Earth Observations and Remote Sensing* vol. 8.9, pp. 4531–4538.

LOURENS, E., PAPADIMITRIOU, C., GILLIJNS, S., REYNDERS, E., DE ROECK, G. and LOMBAERT, G. (2012a): Joint input-response estimation for structural systems based on reduced-order models and vibration data from a limited number of sensors, in: *Mechanical Systems and Signal Processing* vol. 29, pp. 310–327.

LOURENS, E., REYNDERS, E., DE ROECK, G., DEGRANDE, G. and LOMBAERT, G. (2012b): An augmented Kalman filter for force identification in structural dynamics, in: *Mechanical Systems and Signal Processing* vol. 27, pp. 446–460.

LU, Z. R. and LAW, S. S. (2006): Force Identification Based on Sensitivity in Time Domain, in: *Journal of Engineering Mechanics* vol. 132.10, pp. 1050–1056.

LU, Z. R. and LAW, S. S. (2007): Features of dynamic response sensitivity and its application in damage detection, in: *Journal of Sound and Vibration* vol. 303, pp. 305–329.

LU, Z. R. and LIU, J. K. (2011): Identification of both structural damages in bridge deck and vehicular parameters using measured dynamic responses, in: *Computers and Structures* vol. 89, pp. 1397–1405.

LUDESCHER, H. and BRÜHWILER, E. (2009): Dynamic Amplification of Traffic Loads on Road Bridges, in: *Structural Engineering International* vol. 2.

LYDON, M., TAYLOR, S. E., ROBINSON, D. and OBRIEN, E. J. (2016): Recent developments in bridge weigh-in-motion (B-WIM), in: *Journal of Civil Structural Health Monitoring* vol. 6.1, pp. 69–81.

MACEC 3.3 (2014): A Matlab Toolbox for experimental and operational modal analysis – User Manual. KU Leuven – Structural Mechanics Section.

MAES, K, ILIOPOULOS, A, WEIJTJENS, W, DEVRIENDT, C and LOMBAERT, G (2016a): Dynamic strain estimation for fatigue assessment of an offshore monopile wind turbine using filtering and modal expansion algorithms, in: *Mechanical Systems and Signal Processing* vol. 76-77, pp. 592–611.

MAES, K and LOMBAERT, G (2019): The influence of out-of-band modes in system inversion, in: *Mechanical Systems and Signal Processing* vol. 115, pp. 173–187.

MAES, K., LOURENS, E., VAN NIMMEN, K. and REYNDERS, E. (2015): Design of sensor networks for instantaneous inversion of modally reduced order models in structural dynamics, in: *Mechanical Systems and Signal Processing*, pp. 628–644.

MAES, K., SMYTH, A. W., DE ROECK, G. and LOMBAERT, G. (2016b): Joint input-state estimation in structural dynamics, in: *Mechanical Systems and Signal Processing* vol. 70-71, pp. 445–466.

MAES, K., VAN NIMMEN, K., LOURENS, E., REZAYAT, A. and GUILLAUME, P. (2016c): Verification of joint input-state estimation for force identification by means of in situ measurements on a footbridge, in: *Mechanical Systems and Signal Processing* vol. 75, pp. 245–260.

MAGALHÃES, F., CUNHA, Á., CAETANO, E. and BRINCKER, R. (2010): Damping estimation using free decays and ambient vibration tests, in: *Mechanical Systems & Signal Processing* vol. 24, pp. 1274–1290.

MALEKJAFARIAN, A., MARTINEZ, D. and OBRIEN, E. J. (2018): The Feasibility of Using Laser Doppler Vibrometer Measurements from a Passing Vehicle for Bridge Damage Detection, in: *Shock and Vibration* vol. 2018, pp. 1–10.

MARQUARDT, D. W. (1963): An algorithm for least-squares estimation of nonlinear parameters, in: *SIAM Journal of the Society for Industrial and Applied Mathematics* vol. 11.2, pp. 431–441.

MARTINELLI, L., RACIC, V., ALBERTO, B., LAGO, D. and FOTI, F. (2020): Testing Walking-Induced Vibration of Floors Using Smartphones Recordings, in: *Robotics* vol. 9.37, pp. 1–22.

MEJIACRUZ, Y., FRANCO, J., HAINLINE, G., FRITZ, S., JIANG, Z., CAICEDO, J. M., DAVIS, B. and HIRTH, V. (2021): Walking Speed Measurement Technology: a Review, in: *Current Geriatrics Reports* vol. 10, pp. 32–41.

MENSINGER, M., HACKER, A., RADLBECK, C. and LANGEN, T. (2014): "Bestimmung der Restnutzungsdauer historischer Stahlbrücken", in: *12. Fachtagung Baustatik-Baupraxis*, Munich, pp. 141–150.

MEYER, L. (1998): Formoptimierung in der Strukturdynamik, PhD Thesis, Universität Hannover.

MIDDLETON, J. and RHODES, A. H. (1994): The Effect of Dynamic Loading on Road Pavement Wear: A Study of the Relationship Between Road Profiles and Pavement Wear on an Instrumented Test Road, in: *Proceedings of the Institution of Civil Engineers* vol. 105.

MINER, M. A. (1945): Cumulative damage in fatigue, in: *Journal of Applied Mechanics* vol. 12, pp. 149–164.

MORE, J. J. (1977): "The Levenberg-Marquardt Algorithm: Implementation and Theory", in: *Conference on Numerical Analysis*, Dundee, Scotland.

MORÉ, J. J. and SORENSEN, D. (1983): Computing a Trust Region Step, in: *SIAM Journal of Scientific and Statistical Computing* vol. 4.3, pp. 553–572.

MOROZOV, V. A. (1984): *Methods for Solving Incorrectly Posed Problems*, New York, USA: Springer.

MOSES, F. (1979): Weigh-In-Motion System Using Instrumented Bridges, in: *Journal of Transportation Engineering* vol. 105, pp. 233–249.

MOTTERSHEAD, J. E. and FOSTER, C. D. (1991): On the treatment of ill-conditioning in spatial parameter estimation from measured vibration data, in: *Mechanical Systems & Signal Processing* vol. 5.2, pp. 139–154.

NASSIF, H. H., GINDY, M. and DAVIS, J. (2005): Comparison of laser Doppler vibrometer with contact sensors for monitoring bridge deflection and vibration, in: *NDT&E International* vol. 38.April, pp. 213–218.

NEGULESCU, C., LUZI, G., CROSETTO, M., RAUCOULES, D., ROULLÉ, A., MONFORT, D., PUJADES, L., COLAS, B. and DEWEZ, T. (2013): Comparison of seismometer and radar measurements for the modal identification of civil engineering structures, in: *Engineering Structures* vol. 51, pp. 10–22.

NEWMARK, N. M. (1959): A Method of Computation for Structural Dynamics, in: *A.S.C.E., Journal of Engingeering Mechanics Division* vol. 85, pp. 67–94.

NICO, G., PREZIOSO, G., MASCI, O. and ARTESE, S. (2020): Dynamic Modal Identification of Telecommunication Towers Using Ground Based Radar Interferometry, in: *Remote Sensing* vol. 12.

NOCEDAL, J. and WRIGHT, S. J. (2006): *Numerical Optimization*, 2nd ed., New York, USA: Springer Science+Business Media, LLC.

NORDSTRÖM, L. J. (2006): A dynamic programming algorithm for input estimation on linear time-variant systems, in: *Computer Methods in Applied Mechanics and Engineering* vol. 195.44-47, pp. 6407–6427.

OBRIEN, E. J., KAROUMI, R. and QUILLIGAN, M. (2006): Calculating an Influence Line from Direct Measurements, in: *Proceedings of the Institution of Civil Engineers: Bridge Engineering* vol. 159.1, pp. 31–34.

OBRIEN, E. J., ROWLEY, C., GONZÁLEZ, A. and GREEN, M. F. (2009): A regularised solution to the bridge weigh-in-motion equations, in: *International Journal of Heavy Vehicle Systems* vol. 16.3, pp. 310–327.

OBRIEN, E. J., ŽNIDARIČ, A., BAUMGÄRTNER, W., GONZÁLEZ, A. and MCNULTY, P. (2001): *Weighing-In-Motion of Axles and Vehicles for Europe (WAVE) WP1.2: Bridge WIM Systems*, tech. rep., Dublin: University College Dublin.

OBRIEN, E. J., ŽNIDARIČ, A. and OJIO, T. (2008): "Bridge weigh-in-motion – latest developments and applications world wide", in: *Proceedings of the 5th International Conference on WI; (HVParis'2008)*, Paris, pp. 39–56.

O'CONNOR, C. and CHAN, T. H. T. (1987): Dynamic Wheel Loads From Bridge Strains, in: *Journal of Structural Engineering* vol. 114, pp. 1703–1723.

OBATA, T. and MIYAMORI, Y. (2006): Identification of a human walking force model based on dynamic monitoring data from pedestrian bridges, in: *Computers and Structures* vol. 84, pp. 541–548.

OJIO, T., CAREY, C., OBRIEN, E., DOHERTY, C. and TAYLOR, S. (2015): Contactless Bridge Weigh-in-Motion, in: *Journal of Bridge Engineering* January.

PAN, C.-D., YU, L. and LIU, H.-L. (2017): Identification of moving vehicle forces on bridge structures via moving average Tikhonov regularization, in: *Smart Materials and Structures* vol. 26.8.

PEETERS, B. and DE ROECK, G. (1999): Reference-based stochastic subspace identification for output-only modal analysis, in: *Mechanical Systems and Signal Processing* vol. 13.6, pp. 855–878.

PETERSEN, C. and WERKLE, H. (2017): *Dynamik der Baukonstruktionen*, 2nd ed., Wiesbaden: Springer Vieweg, p. 1748.

PHILLIPS, D. L. (1962): A Technique for the Numerical Solution of Certain Integral Equations of the First Kind, in: *Journal of the Association for Cimputing Machinery* vol. 9, pp. 84–97.

PIERACCINI, M. (2013): Monitoring of Civil Infrastructures by Interferometric Radar: A Review, in: *The Scientific World Journal* vol. 2013.

PIERACCINI, M., ATZENI, C. and SPINELLI, P. (2007): Static and dynamic testing of bridges through microwave interferometry, in: *NDT&E International* vol. 40, pp. 208–214.

PIERACCINI, M., FRATINI, M., PARRINI, F., ATZENI, C. and GIANI, B. (2008): Interferometric radar vs. accelerometer for dynamic monitoring of large structures: An experimental comparison, in: *NDT&E International* vol. 41, pp. 258–264.

PINKAEW, T. (2006): Identification of vehicle axle loads from bridge responses using updated static component technique, in: *Engineering Structures* vol. 28, pp. 1599–1608.

PINKAEW, T. and ASNACHINDA, P. (2007): Experimental study on the identification of dynamic axle loads of moving vehicles from the bending moments of bridges, in: *Engineering Structures* vol. 29, pp. 2282–2293.

PIRCHER, H., STADLER, C., GLATZL, J. and SEITZ, P. (2009): Dynamische Berechnung von Eisenbahnbrücken im Zuge von Hochgeschwindigkeitsstrecken - Ein alternatives Verfahren zur Simulation von Zugüberfahrten, in: *Bautechnik* vol. 86.1, pp. 1–13.

POWELL, M. J. D. (1970): "A New Algorithm for Unconstrained Optimization", in: *Nonlinear Programming*, ed. by J. B. ROSE, O. L. MANGASARIAN and K. RITTER, New York, USA: Academic Press, pp. 31–65.

QIAO, B., ZHANG, X., GAO, J. and CHEN, X. (2016): Impact-force sparse reconstruction from highly incomplete and inaccurate measurements, in: *Journal of Sound and Vibration* vol. 376, pp. 72–94.

QIAO, B., ZHANG, X., GAO, J., LIU, R. and CHEN, X. (2017): Sparse deconvolution for the large-scale ill-posed inverse problem of impact force reconstruction, in: *Mechanical Systems and Signal Processing* vol. 83, pp. 93–115.

QIAO, B., ZHANG, X., LUO, X. and CHEN, X. (2015): A force identification method using cubic B-spline scaling functions, in: *Journal of Sound and Vibration* vol. 337, pp. 28–44.

QU, W., WANG, Y. and YONGLIN, P. (2011): Multi-axle moving train loads identification on simply supported bridge by using simulated annealing genetic algorithm, in: *International Journal of Structural Stability and Dynamics* vol. 11.1, pp. 57–71.

QUOOS, V. (1996): *Fiktive Verkehrslastmodelle für Haupt- und Nebenstrecken der DB AG*, tech. rep., Magdeburg: DB AG Geschäftsbereich Netz, NGT.

QUOOS, V., GEISSLER, K., SEDLACEK, G. and HENSEN, W. (2000): Zum Restnutzungsdauernachweis für stählerne Eisenbahnbrücken, in: *Stahlbau* vol. 69.1, pp. 10–24.

RACIC, V., BROWNJOHN, J. and PAVIC, A. (2010): Reproduction and application of human bouncing and jumping forces from visual marker data, in: *Journal of Sound and Vibration* vol. 329.16, pp. 3397–3416.

RACIC, V., PAVIĆ, A. and BROWNJOHN, J. (2009): Experimental identification and analytical modelling of human walking forces: Literature review, in: *Journal of Sound and Vibration* vol. 326, pp. 1–49.

RACIC, V., PAVIĆ, A., MARK, J. and BROWNJOHN, W. (2013): Modern facilities for experimental measurement of dynamic loads induced by humans: A literature review, in: *Shock and Vibration* vol. 20, pp. 53–67.

RAUERT, T., BIGELOW, H., HOFFMEISTER, B. and FELDMANN, M. (2010): On the prediction of the interaction effect caused by continuous ballast on filler beam railway bridges by experimentally supported numerical studies, in: *Engineering Structures* vol. 32.12, pp. 3981–3988.

RAUERT, T. (2011): Zum Einfluss baulicher Randbedingungen auf das dynamische Verhalten von WIB-Eisenbahnbrücken, PhD Thesis, RWTH Aachen.

REBELO, C., SIMÕES DA SILVA, L., RIGUEIRO, C. and PIRCHER, M. (2008): Dynamic behaviour of twin single-span ballasted railway viaducts - Field measurements and modal identification, in: *Engineering Structures* vol. 30.9, pp. 2460–2469.

REITERER, M. and BRUSCHETINI-AMBRO, S.-Z. (2019): Dynamik von Eisenbahnbrücken - Diskrepanz zwischen Messung und Berechnung, in: *Bauingenieur - D-A-CH Mitteilungsblatt* vol. 94, pp. 9–21.

REVI, D. A., ALVAREZ, A. M., WALSH, C. J., ROSSI, S. M. M. D. and AWAD, L. N. (2020): Indirect measurement of anterior-posterior ground reaction forces using a minimal set of wearable inertial sensors: from healthy to hemiparetic walking, in: *Journal of Neuroengineering and Rehabilitation* vol. 17.82, pp. 1–13.

REYNDERS, E. and DE ROECK, G. (2008): Reference-based combined deterministic-stochastic subspace identification for experimental and operational modal analysis, in: *Mechanical Systems and Signal Processing* vol. 22.3, pp. 617–637.

REZAYAT, A., NASSIRI, V., DE PAUW, B., ERTVELDT, J., VANLANDUIT, S. and GUILLAUME, P. (2016): Identification of dynamic forces using group-sparsity in frequency domain, in: *Mechanical Systems and Signal Processing* vol. 70-71, pp. 756–768.

RICHTER, M. (2015): *Inverse Probleme. Grundlage, Theorie und Anwendungsbeispiele*, 1st ed., Berlin und Heidelberg, Germany: Springer Spektrum, p. 128.

RIGUEIRO, C., REBELO, C. and SIMÕES DA SILVA, L. (2010): Influence of ballast models in the dynamic response of railway viaducts, in: *Journal of Sound and Vibration* vol. 329.15, pp. 3030–3040.

RÖDELSPERGER, S., LÄUFER, G., GERSTENECKER, C. and BECKER, M. (2010): Monitoring of displacements with ground-based microwave interferometry: IBIS-S and IBIS-L, in: *Journal of Applied Geodesy* vol. 4, pp. 41–54.

SANCHEZ, J. and BENAROYA, H. (2014): Review of force reconstruction techniques, in: *Journal of Sound and Vibration* vol. 333.14, pp. 2999–3018.

SANTAMARINA, J. C. and FRATTA, D. (2005): *Discrete Signals and Inverse Problems - An Introduction for Engineers*, Chichester, UK: John Wiley & Sons, Ltd, p. 350.

SCHILL, F. J. (2018): Überwachung von Tragwerken mit Profilscannern, PhD Thesis, Technische Universität Darmstadt, Germany.

SCHILL, F. and EICHHORN, A. (2019a): Deformation Monitoring of Railway Bridges with a Profile Laser Scanner, in: *ZFV* vol. 144.2, pp. 109–118.

SCHILL, F., SVIRIDOVA, A. and EICHHORN, A. (2019b): "Deformation monitoring of noise barriers with profile laser scanning", in: *4th Joint International Symposium on Deformation Monitoring (JIDSM)*, Athens, Greece.

SCHNEIDER, J., BECKER, M., FIRUS, A., PULLAMTHARA, J. J. and DRASS, M. (2015): "Identifikation dynamischer Strukturparameter von Eisenbahnbrücken mittels terrestrischer Mikrowelleninterferometrie", in: *8. Symposium Experimentelle Untersuchungen von Baukonstruktionen*, ed. by M. CURBACH, H. OPITZ and S. SCHEERER, Dresden, pp. 47–60.

SCHULTZ, G. A., SCHNABEL, R. B. and BYRD, R. H. (1985): A Family of Trust-Region-Based Algorithms for Unconstrained Minimization with Strong Global Convergence Properties, in: *SIAM Journal on Numerical Analysis* vol. 22.1, pp. 47–67.

SCHUMACHER, A. (2013): *Optimierung mechanischer Strukturen*, 2nd ed., Berlin, Heidelberg: Springer, p. 306.

SEILER, C. and HÜTTNER, S. (2004): Ein einheitliches Modell zur Beschreibung von Fußgängerlasten für verschiedene Bewegungsarten – Theorie, experimentelle Messungen und praktische Anwendung, in: *Bauingenieur* vol. 79, pp. 483–796.

SHAHABPOOR, E. and PAVIC, A. (2018): Estimation of vertical walking ground reaction force in real-life environments using single IMU sensor, in: *Journal of Biomechanics* vol. 79, pp. 181–190.

SHAHABPOOR, E., PAVIĆ, A. and RACIC, V. (2016): Interaction between Walking Humans and Structures in Vertical Direction: A Literature Review, in: *Shock and Vibration* vol. 2016, pp. 1–22.

SNYDER, R. E. and MOSES, F. (1986): Application of In-Motion Weighing Using Instrumented Bridges, in: *Transportation Reseach Record* vol. 1072, pp. 71–78.

SORENSEN, D. (1982): Newton's method with a model trust region modification, in: *SIAM Journal on Numerical Analysis* vol. 19.2, pp. 409–426.

STOLLWITZER, A. and FINK, J. (2020): Analyse des dynamischen Schotteroberbauverhaltens anhand einer Großversuchsanlage mit vertikaler Kraftanregung, in: *Bauingenieur* vol. 95, pp. 12–21.

TAYLOR, J. (2001): *Ultra-wideband radar technology*, Boca Raton, Florida, USA: CRC Press LLC.

THE MATHWORKS INC. (2019): MATLAB v. 2019b, Natick, Massachusetts.

TIKHONOV, A. N. (1963): Solution of incorrectly formulated problems and the regularization method, in: *Soviet Mathematics Doklady* vol. 4, pp. 1035–1038.

TIKHONOV, A. N., GONCHARSKY, A. V., STEPANOV, V. V. and YAGOLA, A. G. (1995): *Numerical Methods for the Solution of Ill-Posed Problems*, ed. by M. HAZEWINKEL, 1st ed., Moscow, Russia: Springer-Science+Business Media, B.V.

TRUJILLO, D. M. (1978): Application of dynamic programming to the general inverse problem, in: *International Journal for Numerical Methods in Engineering* vol. 12, pp. 613–624.

UHL, T. (2007): The inverse identification problem and its technical application, in: *Archive of Applied Mechanics* vol. 77, pp. 325–337.

USLU, A. A., SANLITURK, K. Y. and GÜL, M. (2007): "Force estimation using vibration data", in: *Inter-Noise 2007*, Istanbul, Turkey, pp. 2961–2969.

VAN HAWERMEIREN, J., VAN NIMMEN, K., VANWANSEELE, B. and VAN DEN BROECK, P. (2020): "Contact force reconstruction on vibrating structures", in: *Proceedings of the 11th International Conference on Structural Dynamics 2020*, ed. by M. PAPADRAKAKIS, M. FRAGIADAKIS and C. PAPADIMITRIOU, Athens, pp. 1854–1865.

VAN LOO, H. and ŽNIDARIČ, A. (2019): *Guide for Users of Weigh-In-Motion*, ISWIM, International Society for Weigh-in-Motion.

VAN NIMMEN, K., LOMBAERT, G., JONKERS, I. and DE ROECK, G. (2014): Characterisation of walking loads by 3D inertial motion tracking, in: *Journal of Sound and Vibration* vol. 333.20, pp. 5212–5226.

VAN NIMMEN, K., LOMBAERT, G., ROECK, G. de and VAN DEN BROECK, P. (2017): The impact of vertical human-structure interaction on the response of footbridges to pedestrian excitation, in: *Journal of Sound and Vibration* vol. 402, pp. 104–121.

VAN NIMMEN, K., ZHAO, G., SEYFARTH, A. and VAN DEN BROECK, P. (2018): A Robust Methodology for the Reconstruction of the Vertical Pedestrian-Induced Load from the Registered Body Motion, in: *Vibration* vol. 2, pp. 250–268.

VAN DEN BROECK, P., VAN HAWERMEIREN, J. and VAN NIMMEN, K. (2020): "An open access benchmark dataset on pedestrian-induced vibrations collected on the Eeklo footbridge", in: *Proceedings of the 11th International Conference on Structural Dynamics 2020*, ed. by M. PAPADRAKAKIS, M. FRAGIADAKIS and C. PAPADIMITRIOU, Athens, pp. 1866–1873.

VOSPERNIG, M. and REITERER, M. (2020): Evaluierung der dynamischen Systemeigenschaften von einfeldrigen Stahlbeton-Eisenbahnbrücken, in: *Beton- und Stahlbetonbau* vol. 115.6, pp. 424–437.

VOSPERNIG, M., WAGNER, R. and GASPARIN, G. (2016): "Condition Monitoring of a Weak Railway Bridge by Axel Load and Structure Resistance Evaluation", in: *8th European Workshop on Structural Health Monitoring (EWSHM 2016)*, Bilbao, pp. 1–8.

WAMBACQ, J., MAES, K., REZAYAT, A., GUILLAUME, P. and LOMBAERT, G. (2019): Localization of dynamic forces on structures with an interior point method using group sparsity, in: *Mechanical Systems and Signal Processing* vol. 115, pp. 593–606.

WANG, Y. and QU, W. (2011): Moving Train Loads Identification on a Continuous Steel Truss Girder by Using Dynamic Displacement Influence Line Method, in: *International Journal of Steel Structures* vol. 11.2, pp. 109–115.

WERKLE, H., FIRUS, A. and FRANCKE, W. (2016): Einfluss personeninduzierter Deckenschwingungen auf die Messgenauigkeit von Kraftmessplatten in Ganglaboren, in: *Bauingenieur* vol. 91, pp. 110–119.

WERKLE, H. (2021): *Finite Elements in Structural Analysis - Theoretical Concepts and Modeling Procedures in Statics and Dynamics of Structures*, Cham, Switzerland: Springer.

Wu, S. Q. and Law, S. S. (2010): Moving force identification based on stochastic finite element model, in: *Engineering Structures* vol. 32.4, pp. 1016–1027.

Wu, S. Q. and Law, S. S. (2012): Statistical moving load identification including uncertainty, in: *Probabilistic Engineering Mechanics* vol. 29, pp. 70–78.

Yau, J., Martínez-Rodrigo, M. and Doménech, A. (2019): An equivalent additional damping approach to assess vehicle-bridge interaction for train-induced vibration of short-span railway bridges, in: *Engineering Structures* vol. 188.January, pp. 469–479.

Yu, L. and Chan, T. H. T. (2004): Identification of Multi-Axle Vehicle Loads on Bridges, in: *Journal of Vibration and Acoustics* vol. 1, pp. 17–26.

Yu, L., Chan, T. H. T. and Zhu, J.-H. (2008a): A MOM-based algorithm for moving force identification: Part I - Theory and numerical simulation, in: *Structural Engineering and Mechanics* vol. 29.2, pp. 135–154.

Yu, L., Chan, T. H. T. and Zhu, J.-H. (2008b): A MOM-based algorithm for moving force identification: Part II - Experiment and comparative studies, in: *Structural Engineering and Mechanics* vol. 29.2, pp. 155–169.

Yu, L and Chan, T. H. T. (2003): Moving force identification based on the frequency – time domain method, in: *Journal of Sound and Vibration* vol. 261, pp. 329–349.

Yu, L. and Chan, T. H. T. (2007): Recent research on identification of moving loads on bridges, in: *Journal of Sound and Vibration* vol. 305.1-2, pp. 3–21.

Yu, Y., Cai, C. S. and Deng, L. (2016): "State-of-the-art review on bridge weigh-in-motion technology", in: *Proceedings of the ICE: bridge Engineering*, vol. 19, 9, pp. 1–17.

Zabel, V. and Brehm, M. (2008): Das dynamische Verhalten von Eisenbahnbrücken mit kurzer Spannweite - numerische und experimentelle Untersuchungen, in: *Bauingenieur - D-A-CH Mitteilungsblatt* vol. 83, pp. 9–14.

Zhang, E., Antoni, J. and Feissel, P. (2012): Bayesian force reconstruction with an uncertain model, in: *Journal of Sound and Vibration* vol. 331.4, pp. 798–814.

Zhang, N., Xia, H. and Guo, W. (2008): Vehicle-bridge interaction analysis under high-speed trains, in: *Journal of Sound and Vibration* vol. 309, pp. 407–425.

Zhu, T., Xiao, S. and Yang, G. (2014): Force identification in time domain based on dynamic programming, in: *Applied Mathematics and Computation* vol. 235, pp. 226–234.

Zhu, X. Q. and Law, S. S. (2000): Identification of vehicle axle loads from bridge dynamic responses, in: *Journal of Sound and Vibration* vol. 236.4, pp. 705–724.

Zhu, X. Q. and Law, S. S. (2001): Orthogonal function in moving loads identification on a multi-span bridge, in: *Journal of Sound and Vibration* vol. 245.2, pp. 329–345.

Zhu, X. Q. and Law, S. S. (2002a): Moving Loads Identification Through Regularization, in: *Journal of Engineering Mechanics* vol. 128.9, pp. 989–1000.

Zhu, X. Q. and Law, S. S. (2002b): Practical aspects in moving load identification, in: *Journal of Sound and Vibration* vol. 258.1, pp. 123–146.

ZHU, X. Q. and LAW, S. S. (2006a): Moving load identification on multi-span continuous bridges with elastic bearings, in: *Mechanical Systems and Signal Processing* vol. 20, pp. 1759–1782.

ZHU, X. Q. and LAW, S. S. (2007): Damage Detection in Simply Supported Concrete Bridge Structure Under Moving Vehicular Loads, in: *Journal of Vibration and Acoustics* vol. 129.1, pp. 58–65.

ZHU, X. Q. and LAW, S. S. (2016): Recent developments in inverse problems of vehicle – bridge interaction dynamics, in: *Journal of Civil Structural Health Monitoring* vol. 6, pp. 107–128.

ZHU, X. Q., LAW, S. S. and BU, J. Q. (2006b): A State Space Formulation for Moving Loads Identification, in: *Journal of Vibration and Acoustics* vol. 128, pp. 509–520.

ZIENKIEWICZ, O. C. (2005): *The Finite Element Method: Its Basis and Fundamentals*, 6th ed., Oxford, UK: Elsevier Butterworth-Heinemann.

ŽIVANOVIĆ, S. (2016): Modelling Human Actions on Lightweight Structures: Experimental and Numerical Developments, in: *MATEC Web of Conferences* February.

ŽIVANOVIĆ, S., PAVIĆ, A. and REYNOLDS, P. (2005): Vibration serviceability of footbridges under human-induced excitation: a literature review, in: *Journal of Sound and Vibration* vol. 279.1-2, pp. 1–74.

ŽNIDARIČ, A., KALIN, J., KRESLIN, M., FAVAI, P. and KOLAKOWSKI, P. (2016): Railway bridge Weigh-in-Motion system, in: *Transportation Research Procedia* vol. 14, pp. 4010–4019.

Standards and technical guidelines

ASTM E1318 (2017): Standard Specification for Highway Weigh-In-Motion (WIM) Systems with User Requirements and Test Methods.

DIN EN 1991-2 (2010): Eurocode 1: Einwirkungen auf Tragwerke – Teil 2: Verkehrslasten auf Brücken; Deutsche Fassung EN 1991-2:2003 + AC:2010.

DIN EN 1992-1-1 (2010): Eurocode 2, Design of concrete structures – Part 1-1: General rules and rules for buildings; German version EN 1992-1-1:2004 + AC:2010.

DIN EN 1993-1-9 (2009): Eurocode 3, Design of steel structures – Part 1-9: Fatigue; German version EN 1993-1-9:2005 + AC:2009.

ÖBB INFRASTRUKTUR (2011): Richtlinie für die dynamische Berechnung von Eisenbahnbrücken Ausgabe : 2011-02-01.

RIL 804 (2013): Eisenbahnbrücken (und sonstige Ingenieurbauwerke) planen, bauen und instand halten (DB Netz AG).

RIL 805 (2010): Tragsicherheit bestehender Eisenbahnbrücken (DB Netz AG).

Table of appendices

A. Firus, *A Contribution to Moving Force Identification in Bridge Dynamics*, Mechanik, Werkstoffe und Konstruktion im Bauwesen 65, https://doi.org/10.1007/978-3-658-39838-5

A Derivations of solutions for inverse problems

A.1 Derivation of the normal equations

In the following, the normal equations are derived starting from the minimization problem given in Eq. 4.5:

$$\begin{aligned}\min_{\boldsymbol{x}} \mathcal{F}(\boldsymbol{x}) = \|\boldsymbol{r}(\boldsymbol{x})\|_2^2 = \boldsymbol{r}^\top(\boldsymbol{x}) \cdot \boldsymbol{r}(\boldsymbol{x}) = \\ \boldsymbol{x}^\top \boldsymbol{A}^\top \boldsymbol{A}\boldsymbol{x} - \boldsymbol{x}^\top \boldsymbol{A}^\top \boldsymbol{b} - \boldsymbol{b}^\top \boldsymbol{A}\boldsymbol{x} + \boldsymbol{b}^\top \boldsymbol{b}\end{aligned} \tag{A.1}$$

The first derivative (gradient) of Eq. A.1 with respect to $\boldsymbol{x}$ yields:

$$\nabla \mathcal{F}(\boldsymbol{x}) = 2\boldsymbol{A}^\top \boldsymbol{A}\boldsymbol{x} - 2\boldsymbol{A}^\top \boldsymbol{b} \tag{A.2}$$

The minimizer of Eq. A.1 satisfies $\nabla f(\boldsymbol{x}) = \boldsymbol{0}$. This results in:

$$2\boldsymbol{A}^\top \boldsymbol{A}\boldsymbol{x} - 2\boldsymbol{A}^\top \boldsymbol{b} = \boldsymbol{0} \tag{A.3}$$

Dividing Eq. A.3 by 2 and rearranging the terms leads to the so called normal equations:

$$\boldsymbol{A}^\top \boldsymbol{A}\boldsymbol{x} = \boldsymbol{A}^\top \boldsymbol{b} \tag{A.4}$$

The solution of the normal equations Eq. A.4 can be achieved as follows:

$$\boldsymbol{x} = \underbrace{\left(\boldsymbol{A}^\top \boldsymbol{A}\right)^{-1} \boldsymbol{A}^\top}_{\boldsymbol{A}^\dagger} \boldsymbol{b} \tag{A.5}$$

The matrix:

$$\boldsymbol{A}^\dagger = \left(\boldsymbol{A}^\top \boldsymbol{A}\right)^{-1} \boldsymbol{A}^\top \quad \text{for } m \geq n \tag{A.6}$$

A. Firus, *A Contribution to Moving Force Identification in Bridge Dynamics*, Mechanik, Werkstoffe und Konstruktion im Bauwesen 65, https://doi.org/10.1007/978-3-658-39838-5

is called the pseudoinverse or the Moore-Penrose generalized inverse. The solution to the full-rank overdetermined least squares problem $\boldsymbol{Ax} = \boldsymbol{b}$ is:

$$\boldsymbol{x} = \boldsymbol{A}^{\dagger}\boldsymbol{b} \tag{A.7}$$

The fulfilling of the normal equations is necessary and sufficient for finding the minimizer $\boldsymbol{x}$ of the minimization problem described by Eq. A.1 (RICHTER, 2015). This is due to the convexity of the linear least square problems[1].

For the sake of completeness, the expression for the pseudoinverse matrix of a underdetermined equation system ($m < n$) is given by:

$$\boldsymbol{A}^{\dagger} = \boldsymbol{A}^{\top}\left(\boldsymbol{A}\boldsymbol{A}^{\top}\right)^{-1} \quad \text{for } m < n \tag{A.8}$$

The differences between the expressions in Eq. A.6 and Eq. A.8 occur do the a slight different approach in their derivation, which is needed due to the compatibility issues of the matrix operations related to the dimensions of the matrix $\boldsymbol{A}$ ($m < n$ in Eq. A.8 instead of $m \geq n$ in Eq. A.6) (FORD, 2015).

A.2 Solution of inverse problems using Tikhonov regularization

In the following, the Tikhonov regularization will be briefly described from the mathematical point of view. Let the linear system:

$$\boldsymbol{Ax} + \boldsymbol{e} = \boldsymbol{b} \tag{A.9}$$

be overdetermined, where $\boldsymbol{A} \in \mathbb{R}^{m,n}$ $(m \geq n)$, $\boldsymbol{b}$, $\boldsymbol{e} \in \mathbb{R}^{m}$ and $\boldsymbol{x} \in \mathbb{R}^{n}$. The coefficient matrix $\boldsymbol{A}$ and the vector of the right side (measurements) $\boldsymbol{b}$ are given, while $\boldsymbol{e}$ represents an assumed random perturbation (noise) vector. Considering the noise perturbation of the problem, it is expected that the direct solution of the least squares problem:

$$\min_{\boldsymbol{x}} \mathcal{F}(\boldsymbol{x}) = \|\boldsymbol{Ax} - \boldsymbol{b}\|_2^2 \tag{A.10}$$

leads to a solution, which is dominated by contributions from data errors (e.g. noise) or rounding errors, i.e. the problem is ill-conditioned (GOLUB et al., 1997). In order to achieve a more meaningful approximate solution of the linear system, the Tikhonov regularization may be employed. It improves the condition of the inverse problem by replacing the original

[1] The least squares problems are always convex, since the Hessian matrix $2\boldsymbol{A}^{\top}\boldsymbol{A}$ (obtained by derivation of the gradient expression with respect to $\boldsymbol{x}$) is always positive semidefinite. This can be easily proven: $2\boldsymbol{y}^{\top}\boldsymbol{A}^{\top}\boldsymbol{A}\boldsymbol{y} = 2\left(\boldsymbol{A}^{\top}\boldsymbol{y}\right)^{\top}\left(\boldsymbol{A}^{\top}\boldsymbol{y}\right) = 2\|\boldsymbol{A}^{\top}\boldsymbol{y}\|_2^2 \geq 0$

problem (Eq. A.10) with the following least squares problem, which is expectedly less sensitive to perturbations:

$$\min_{\boldsymbol{x}} \mathcal{F}(\boldsymbol{x},\alpha) = \|\boldsymbol{A}\boldsymbol{x}-\boldsymbol{b}\|_2^2 + \alpha\,\|\boldsymbol{\Gamma}\boldsymbol{x}\|_2^2 \tag{A.11}$$

where $\alpha \geq 0$ is the regularization parameter, which defines the amount of regularization introduced into the problem. $\boldsymbol{\Gamma}$ denotes the Tikhonov matrix, which is used to describe a priori information on the solution (HANSEN, 1998).

In order to allow a better mathematical manipulation, Eq. A.11 can be rewritten as:

$$\min_{\boldsymbol{x}} \mathcal{F}(\boldsymbol{x}) = \boldsymbol{x}^\top\boldsymbol{A}^\top\boldsymbol{A}\boldsymbol{x} - \boldsymbol{x}^\top\boldsymbol{A}^\top\boldsymbol{b} - \boldsymbol{b}^\top\boldsymbol{A}\boldsymbol{x} + \boldsymbol{b}^\top\boldsymbol{b} + \alpha\cdot\boldsymbol{x}^\top\boldsymbol{\Gamma}^\top\boldsymbol{\Gamma}\boldsymbol{x} \tag{A.12}$$

In a similar way as for the unregularized problem (cf. Eq. A.1 to Eq. A.4), the gradient of $\mathcal{F}(\boldsymbol{x},\alpha)$ is determined and set to zero:

$$\nabla\mathcal{F}(\boldsymbol{x}) = 2\boldsymbol{A}^\top\boldsymbol{A}\boldsymbol{x} - 2\boldsymbol{A}^\top\boldsymbol{b} + 2\cdot\alpha\cdot\boldsymbol{\Gamma}^\top\boldsymbol{\Gamma}\boldsymbol{x} = \boldsymbol{0} \tag{A.13}$$

This results in:

$$\left(\boldsymbol{A}^\top\boldsymbol{A} + \alpha\boldsymbol{\Gamma}^\top\boldsymbol{\Gamma}\right)\boldsymbol{x} = \boldsymbol{A}^\top\boldsymbol{b} \tag{A.14}$$

For any $\alpha > 0$ and under the assumption that $\left(\boldsymbol{A}^\top\boldsymbol{A} + \alpha\boldsymbol{\Gamma}^\top\boldsymbol{\Gamma}\right)$ is invertible, the equation system in Eq. A.15 delivers an unique solution:

$$\boldsymbol{x} = \left(\boldsymbol{A}^\top\boldsymbol{A} + \alpha\boldsymbol{\Gamma}^\top\boldsymbol{\Gamma}\right)^{-1}\boldsymbol{A}^\top\boldsymbol{b} \tag{A.15}$$

A.3 Trust region solution using iterative approach

The approaches for solving the trust region subproblem (Eq. 4.24) discussed in Section 4.4.3.2 to 4.4.3.4 only deliver approximate solutions. They are sufficient to ensure a good convergence behavior with justifiable implementation and computation costs. For the sake of completeness, an iterative solution method, which delivers a (nearly) exact solution of Eq. 4.24, is briefly introduced in the following (HEBDEN, 1973).

The main idea of the iterative algorithm is to find the value of λ, for which the solution of Eq. 4.24 satisfies Eq. 4.26, so that:

$$\boldsymbol{p}^{(k+1)}(\lambda) = -(\tilde{\boldsymbol{H}}^{(k)} + \lambda\boldsymbol{I})^{-1}\nabla\mathcal{F}(\boldsymbol{x}^{(k)}) \tag{A.16}$$

This leads – if λ is high enough to ensure a positive definite $\tilde{\boldsymbol{H}}$ and under consideration of Eq. 4.27a – to the following one-dimensional root-finding problem:

$$\left\|\boldsymbol{p}^{(k+1)}(\lambda)\right\|_2 = \Delta^{(k)} \tag{A.17}$$

The roots of Eq. A.17 can be found using the iterative root-finding Newton's method with a slightly reformulated problem, which provides a more convenient numerical behavior (NOCEDAL et al., 2006; MORE, 1977; HEBDEN, 1973):

$$\phi(\lambda) = \frac{1}{\Delta^{(k)}} - \frac{1}{\|\boldsymbol{p}^{(k+1)}(\lambda)\|_2} \tag{A.18}$$

whereas there is exactly one appropriate $\lambda^{*(k)}$, which satisfies Eq. 4.26 and Eq. 4.27. It is the unique solution in the interval $[-\lambda_1^{(k)}, \infty]$, whereas $\lambda_1^{(k)}$ represents the first eigenvalue of $\tilde{\boldsymbol{H}}^{(k)}$. It is to be noted that $\boldsymbol{p}^{(k+1)}(\lambda)$ can be expressed using an eigendecomposition of $\tilde{\boldsymbol{H}}^{(k)}$, in order to facilitate the manipulation of the computation (MORÉ et al., 1983; NOCEDAL et al., 2006). The iteration scheme (index l) of the root-finding method is:

$$\lambda^{(k,l+1)} = \lambda^{(k,l)} - \frac{\phi(\lambda^{(k,l)})}{\phi'(\lambda^{(k,l)})} \tag{A.19}$$

The iterative method for solving the trust region subproblem delivers (nearly) exact results. However, an additional iterative process (index l) is necessary to be solved in each iteration step of the trust region optimization (index k), which leads to increased computational effort. Therefore, it is recommended to be only used if the problem size is relatively small (n is not too large).

B Additional force identification results for the Schmutter bridge

B.1 Re-built responses for different passages

The proposed MFI identification method was applied to a total of five passages of the ICE 4 over the Schmutter bridge. The re-built responses for the passages no. 27 and 45 have already been presented in Section 7.6.3.6. For the sake of completeness, the corresponding results for the passages no. 13, 29 and 49 are illustrated in the following.

Passage no. 13:

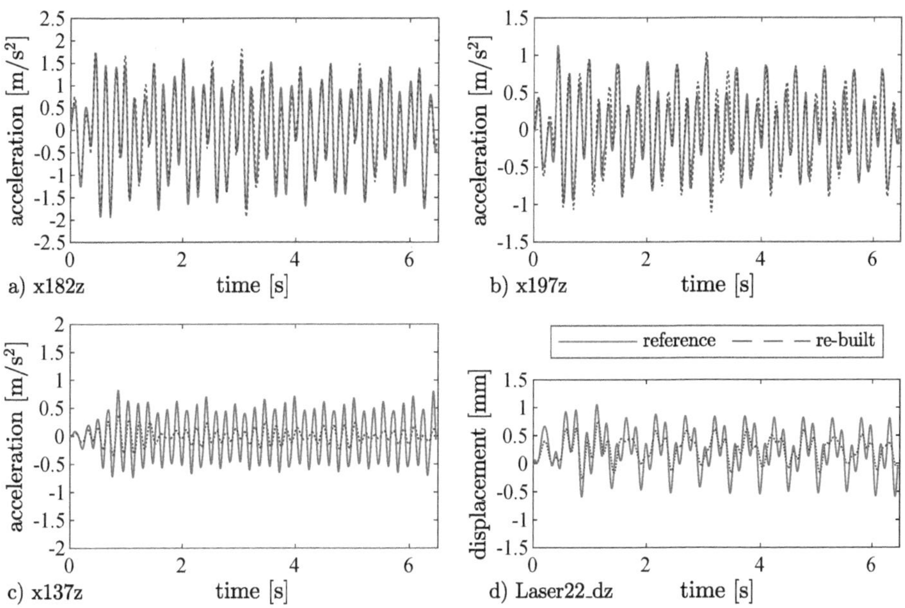

Figure B.1 Reference and re-built time histories at "unmeasured" locations for the passage no. 13 and measurement configuration 1 under consideration of the refined weighting parameters of the penalty functions: a) acceleration at MP x182z; b) acceleration at MP x197z; c) acceleration at MP x137z; d) displacement at Laser22_dz

A. Firus, *A Contribution to Moving Force Identification in Bridge Dynamics*, Mechanik, Werkstoffe und Konstruktion im Bauwesen 65, https://doi.org/10.1007/978-3-658-39838-5

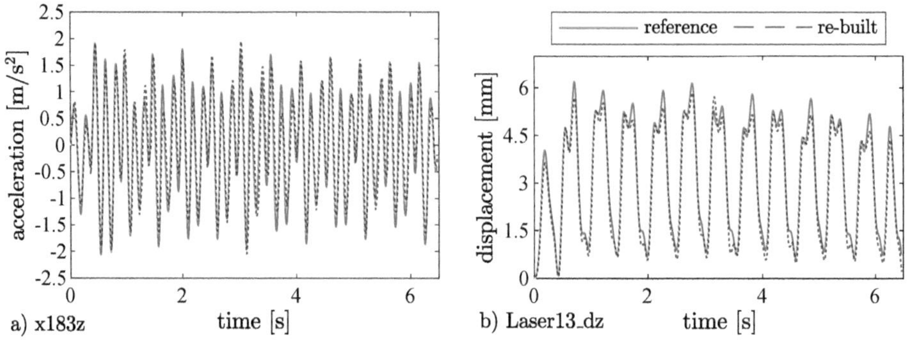

Figure B.2 Reference and re-built time histories at measured locations for the passage no. 13 and measurement configuration 1 under consideration of refined weighting parameters of the penalty functions: a) acceleration at MP x183z; b) displacement at MP Laser13_dz

Passage no. 29:

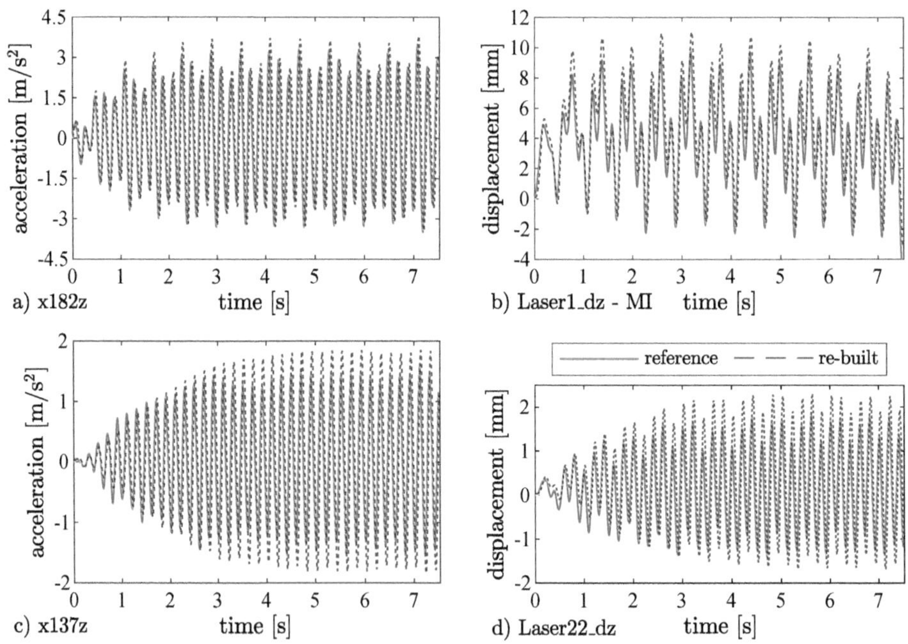

Figure B.3 Reference and re-built time histories at "unmeasured" locations for the passage no. 29 and measurement configuration 1 under consideration of the refined weighting parameters of the penalty functions: a) acceleration at MP x182z; b) displacement at MP Laser1_dz (measured using the microwave interferometer); c) acceleration at MP x137z; d) displacement at Laser22_dz

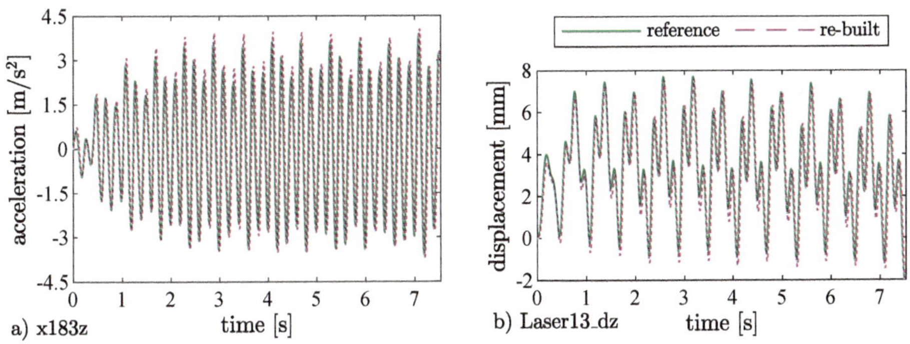

Figure B.4 Reference and re-built time histories at measured locations for the passage no. 29 and measurement configuration 1 under consideration of refined weighting parameters of the penalty functions: a) acceleration at MP x183z; b) displacement at MP Laser13_dz

Passage no. 49:

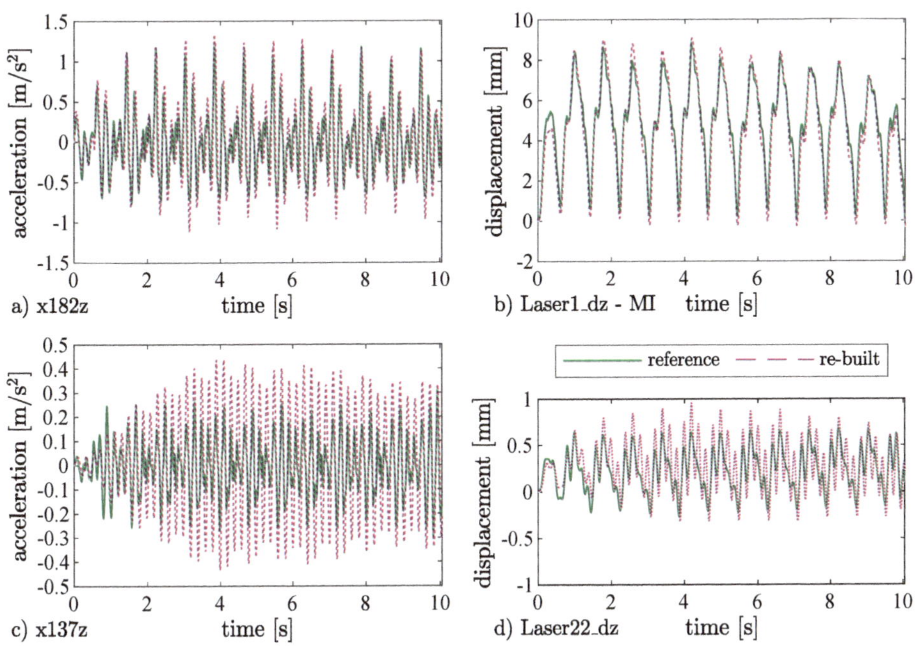

Figure B.5 Reference and re-built time histories at "unmeasured" locations for the passage no. 49 and measurement configuration 1 under consideration of the refined weighting parameters of the penalty functions: a) acceleration at MP x182z; b) displacement at MP Laser1_dz (measured using the microwave interferometer); c) acceleration at MP x137z; d) displacement at Laser22_dz

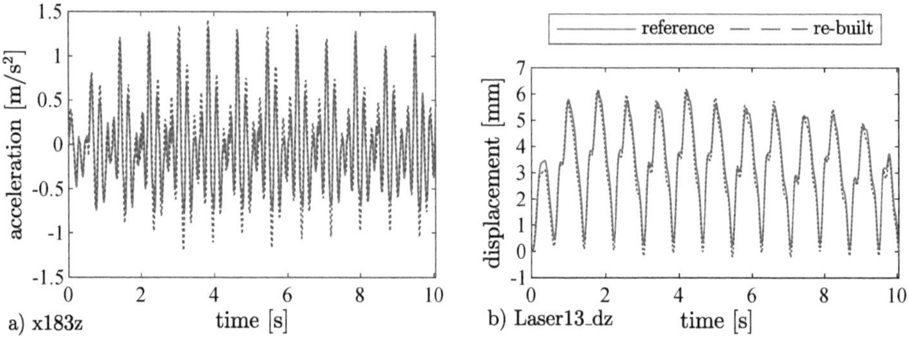

Figure B.6 Reference and re-built time histories at measured locations for the passage no. 49 and measurement configuration 1 under consideration of refined weighting parameters of the penalty functions: a) acceleration at MP x183z; b) displacement at MP Laser13_dz

B.2 Results for different passages using unrefined weighting parameters

The force identification results obtained under consideration of the unrefined set of weighting parameters (cf. Tab. 7.4) are shown in the following:

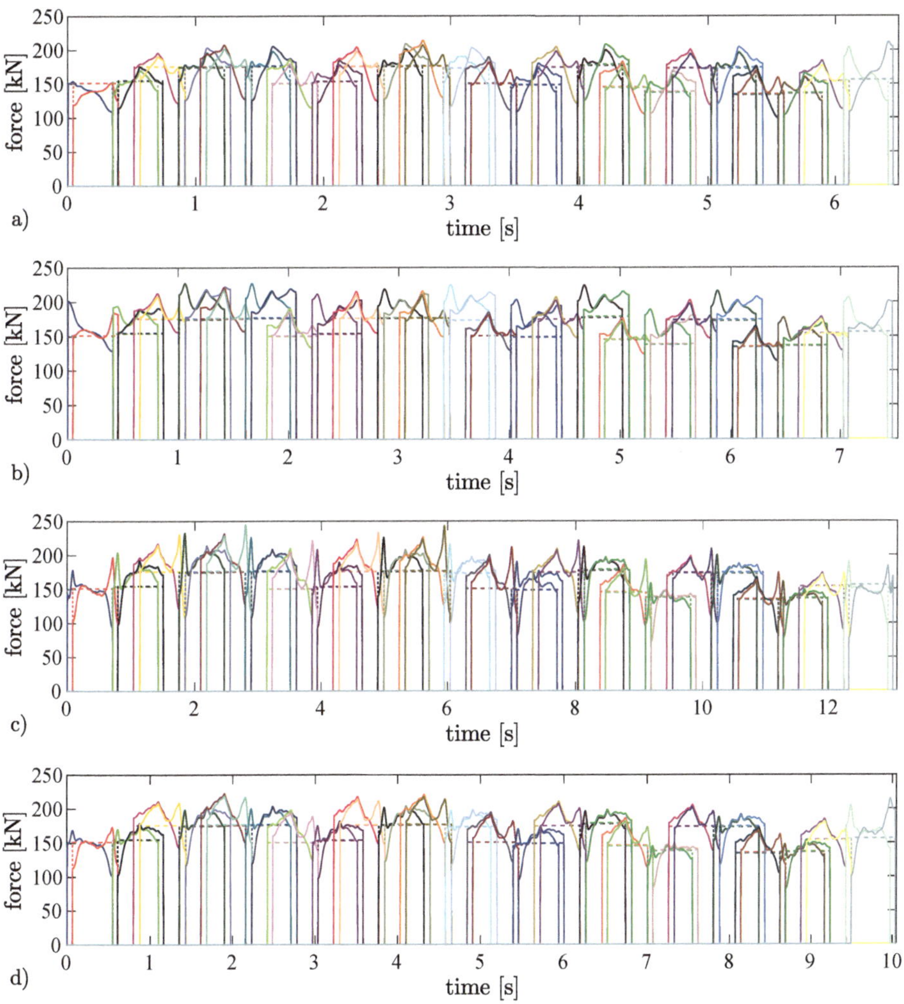

Figure B.7 Identified dynamic force time histories and reference static axle loads for different regular passages obtained under consideration of the unrefined weighting parameters of the penalty functions $w_{p1} = 5 \cdot 10^{-7}$ and $w_{p2} = 1 \cdot 10^{-2}$: a) passage no. 13 (199 km/h); b) passage no. 29 (172 km/h); c) passage no. 45 (99 km/h); d) passage no. 49 (128 km/h)

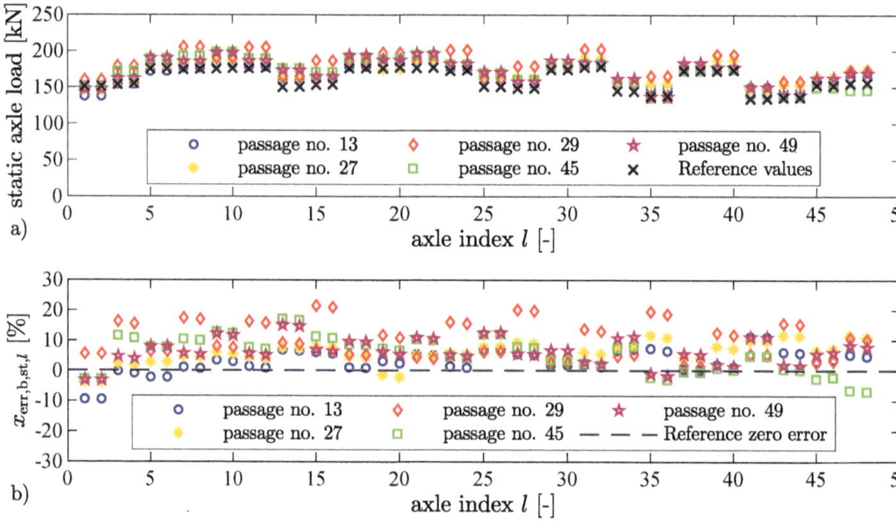

Figure B.8 Comparison of the identified static axle loads with the reference measured values for different regular passages under consideration of the unrefined weighting parameters of the penalty functions ($w_{\mathrm{p1}} = 5 \cdot 10^{-7}$ and $w_{\mathrm{p2}} = 1 \cdot 10^{-2}$): a) comparison of the force values; b) relative errors of the identified static forces with respect to the corresponding measured results

B.3 Results for different measurement set-ups

The force identification results obtained for the measurement configurations 1 to 12 (cf. Tab. 7.5 and Fig. 7.64) are illustrated in the following:

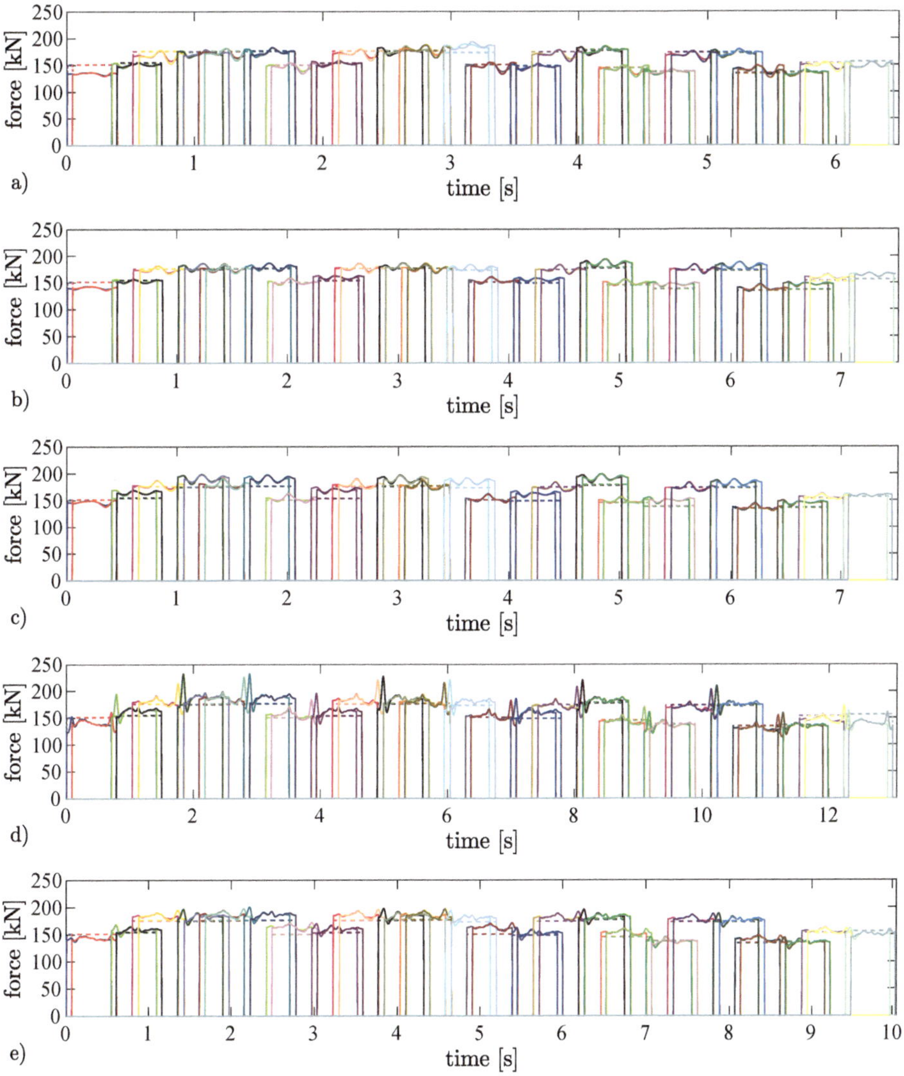

Figure B.9 Identified dynamic force time histories and reference static axle loads for measurement configuration 1: a) passage no. 13; b) passage no. 27; c) passage no. 29; d) passage no. 45; e) passage no. 49

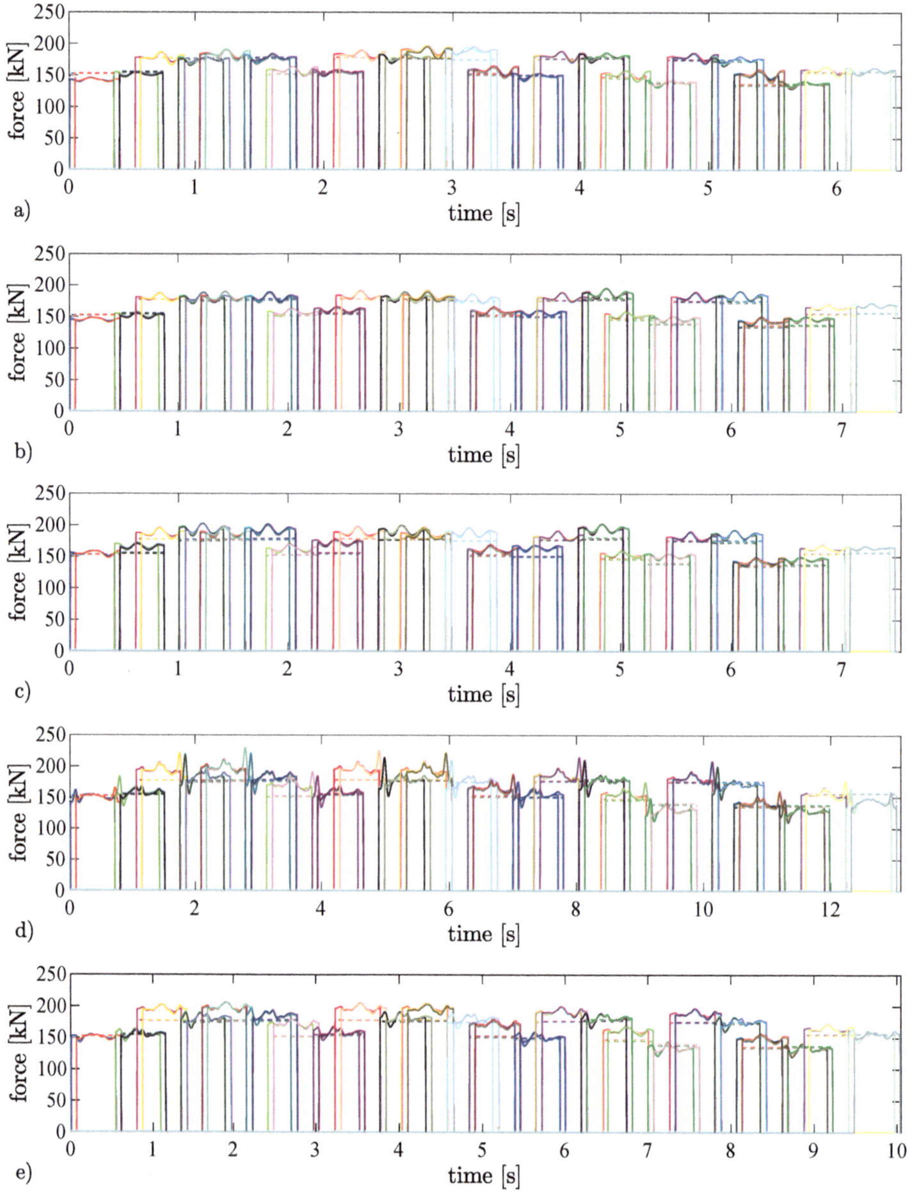

Figure B.10 Identified dynamic force time histories and reference static axle loads for measurement configuration 2: a) passage no. 13; b) passage no. 27; c) passage no. 29; d) passage no. 45; e) passage no. 49

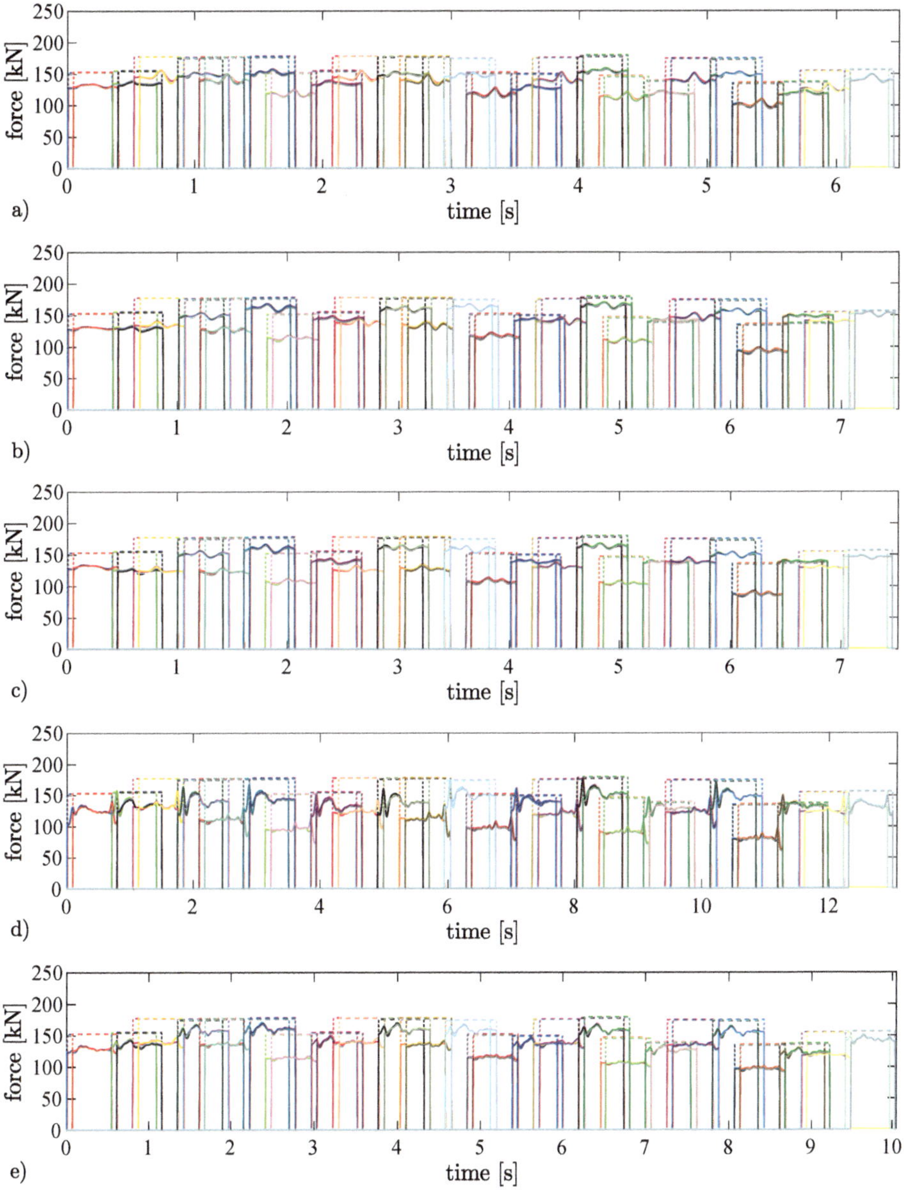

Figure B.11 Identified dynamic force time histories and reference static axle loads for measurement configuration 3: a) passage no. 13; b) passage no. 27; c) passage no. 29; d) passage no. 45; e) passage no. 49

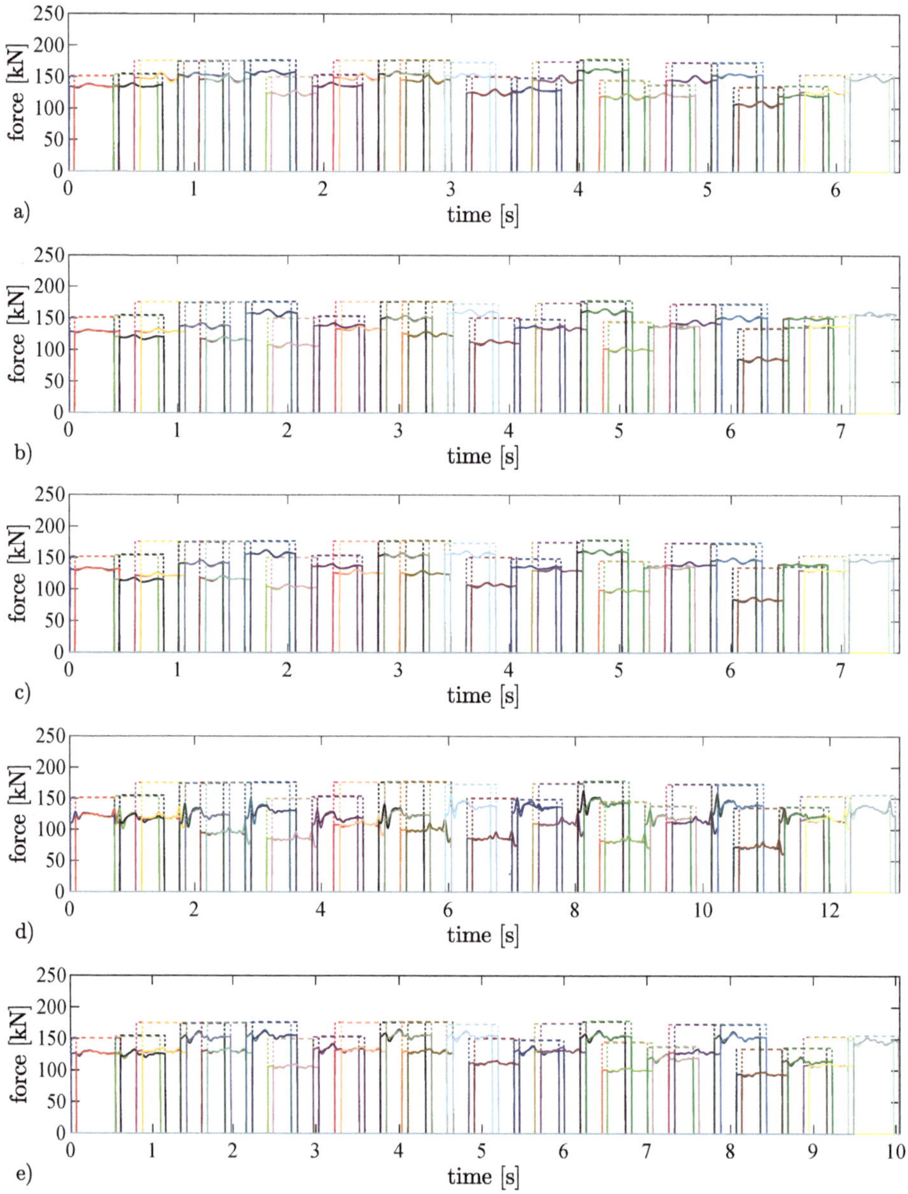

Figure B.12 Identified dynamic force time histories and reference static axle loads for measurement configuration 4: a) passage no. 13; b) passage no. 27; c) passage no. 29; d) passage no. 45; e) passage no. 49

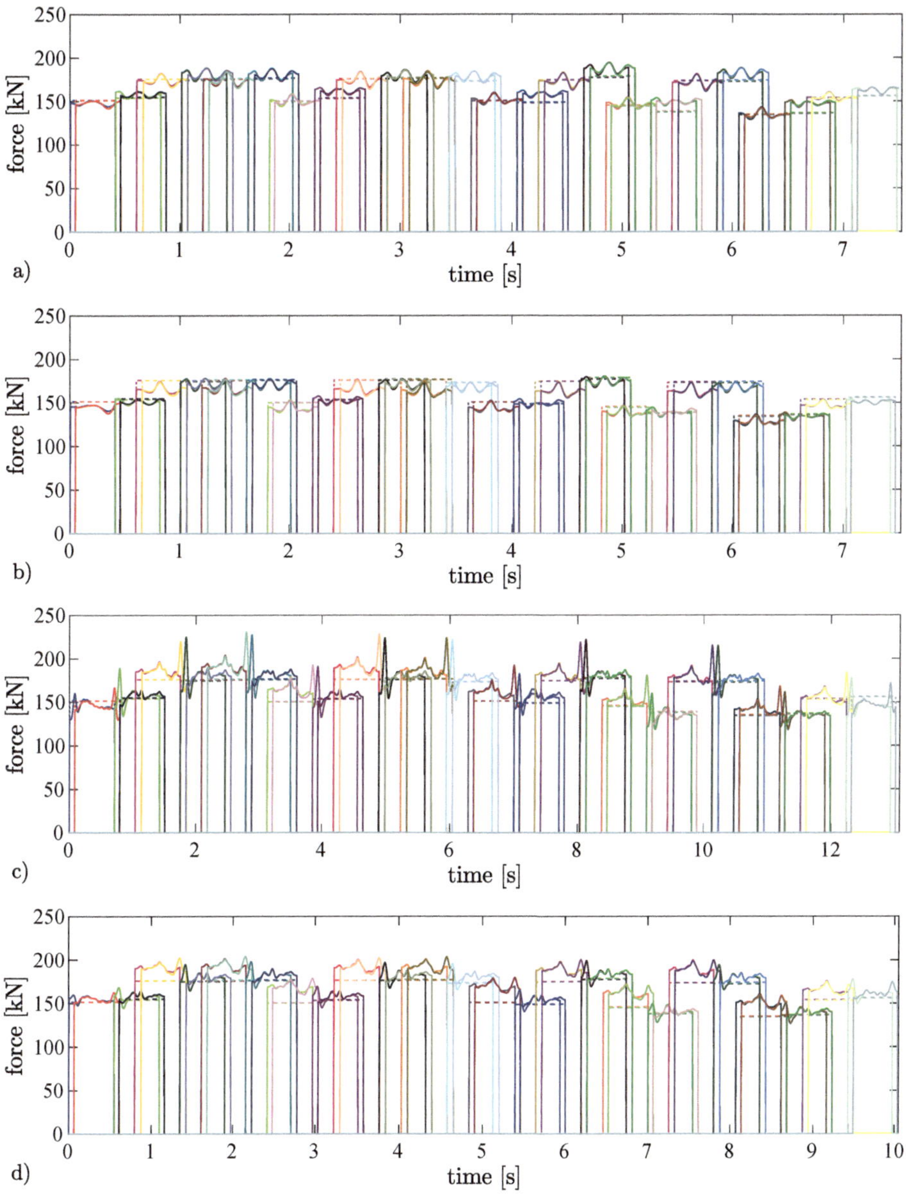

Figure B.13 Identified dynamic force time histories and reference static axle loads for measurement configuration 5: a) passage no. 27; b) passage no. 29; c) passage no. 45; d) passage no. 49

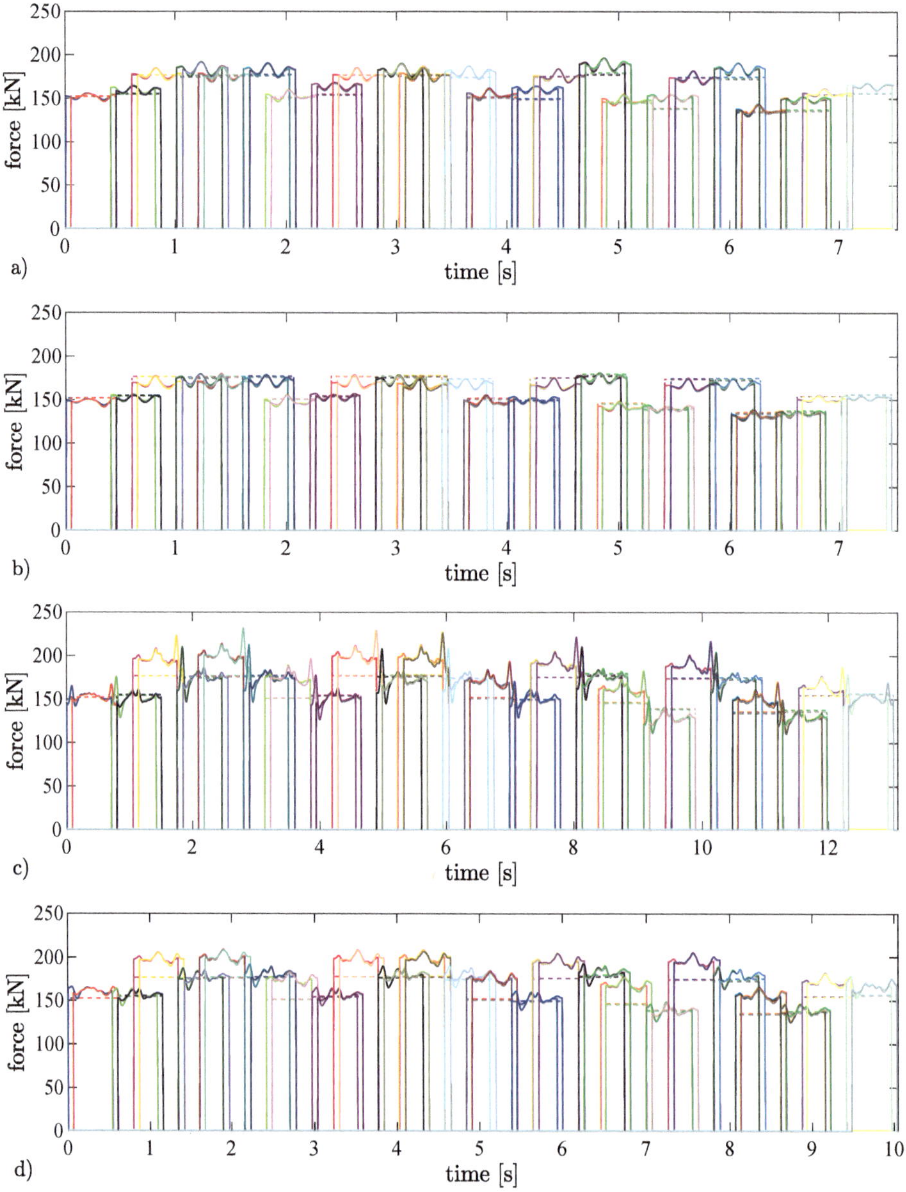

Figure B.14 Identified dynamic force time histories and reference static axle loads for measurement configuration 6: a) passage no. 27; b) passage no. 29; c) passage no. 45; d) passage no. 49

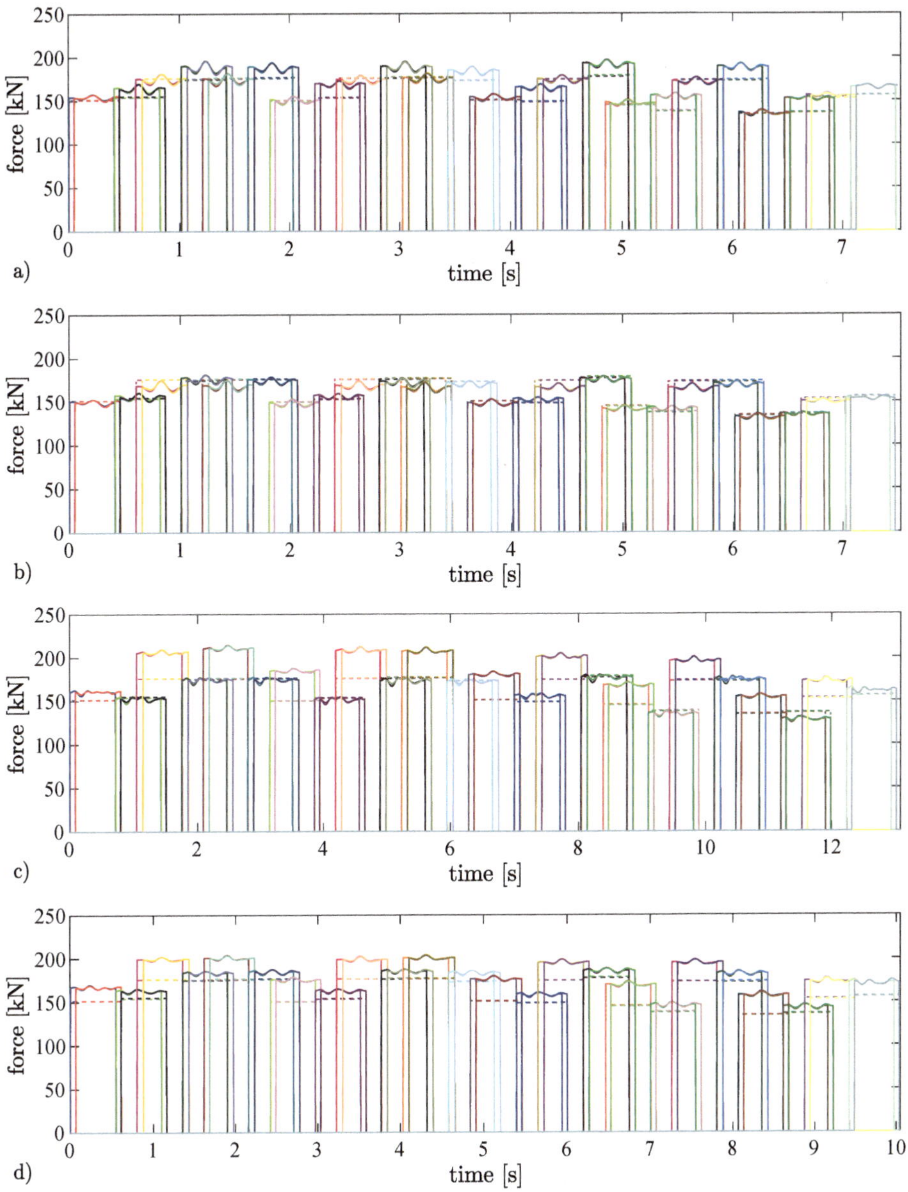

Figure B.15 Identified dynamic force time histories and reference static axle loads for measurement configuration 7: a) passage no. 27; b) passage no. 29; c) passage no. 45; d) passage no. 49

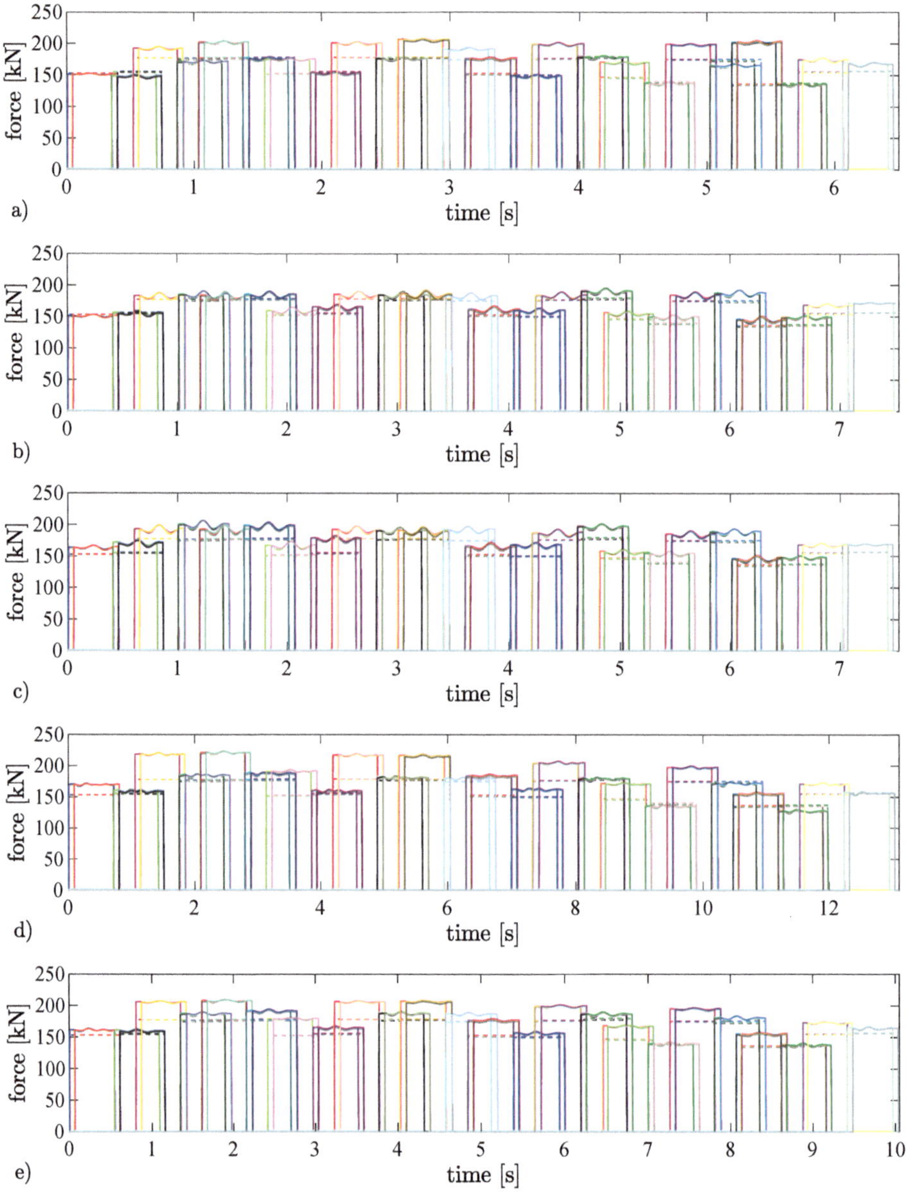

Figure B.16 Identified dynamic force time histories and reference static axle loads for measurement configuration 8: a) passage no. 13; b) passage no. 27; c) passage no. 29; d) passage no. 45; e) passage no. 49

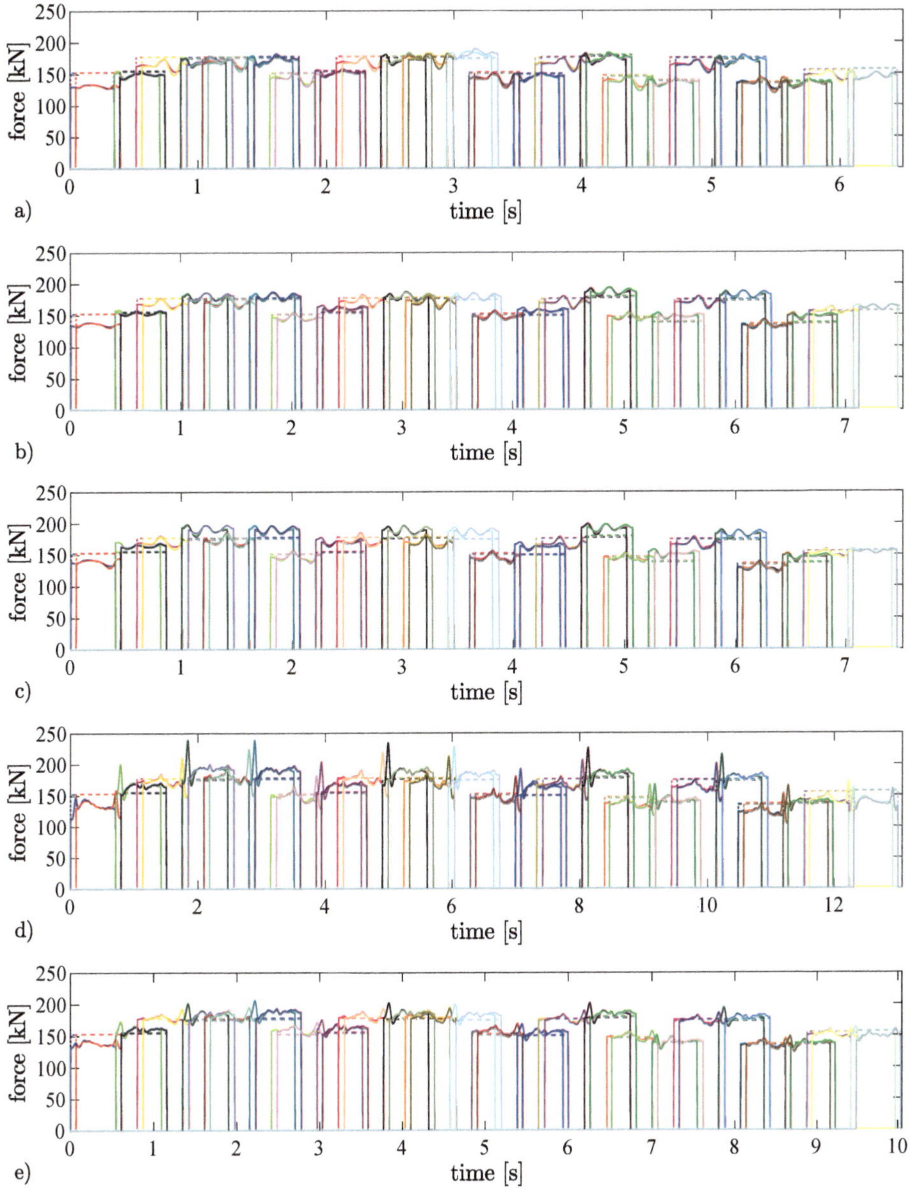

Figure B.17 Identified dynamic force time histories and reference static axle loads for measurement configuration 9: a) passage no. 13; b) passage no. 27; c) passage no. 29; d) passage no. 45; e) passage no. 49

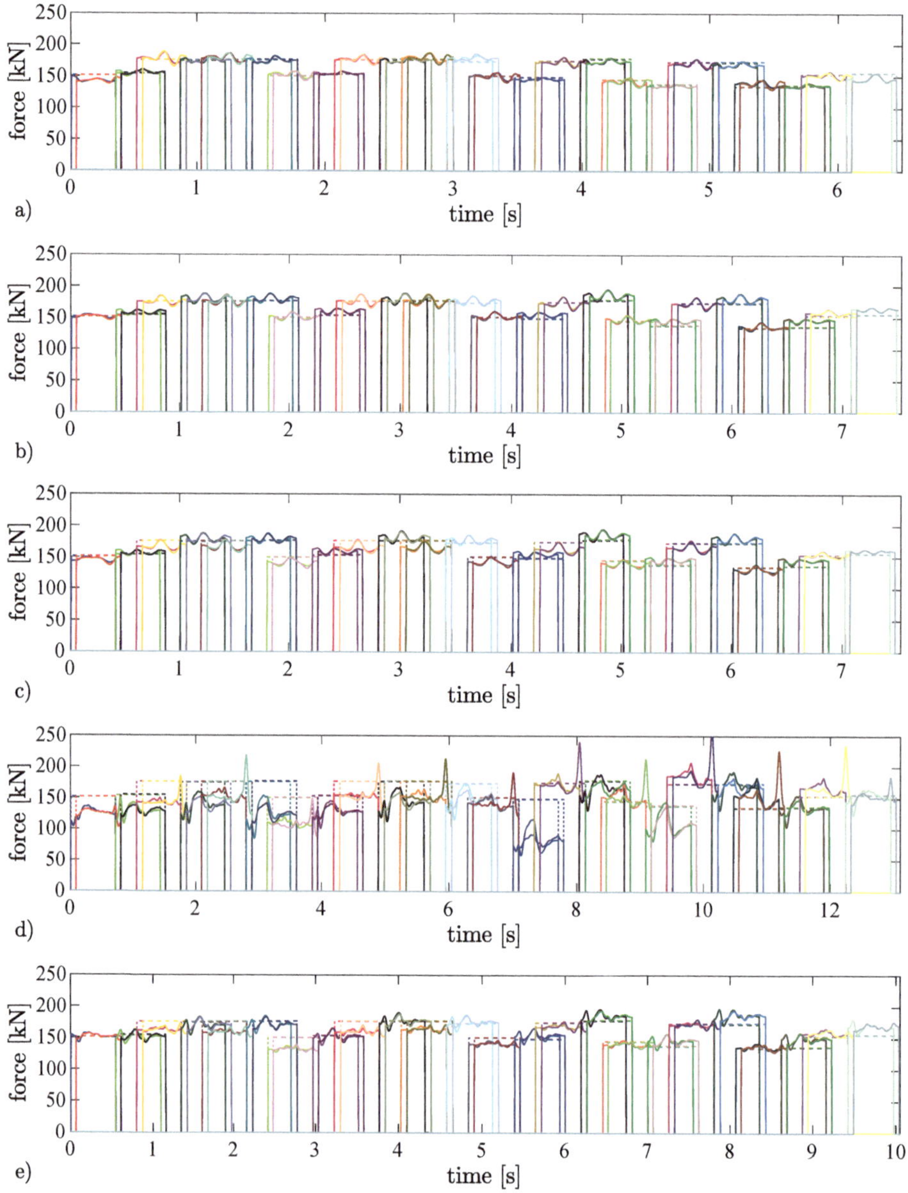

Figure B.18 Identified dynamic force time histories and reference static axle loads for measurement configuration 10: a) passage no. 13; b) passage no. 27; c) passage no. 29; d) passage no. 45; e) passage no. 49

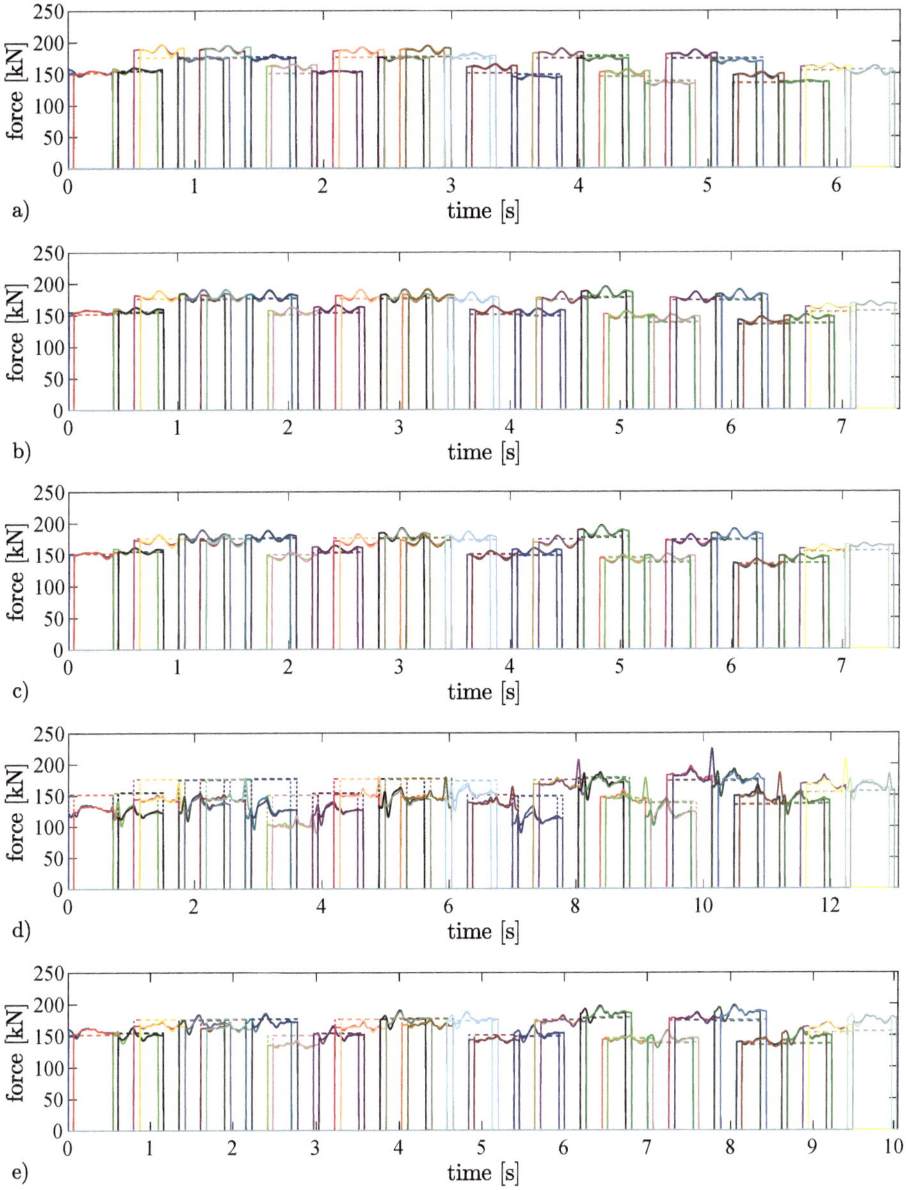

Figure B.19 Identified dynamic force time histories and reference static axle loads for measurement configuration 11: a) passage no. 13; b) passage no. 27; c) passage no. 29; d) passage no. 45; e) passage no. 49

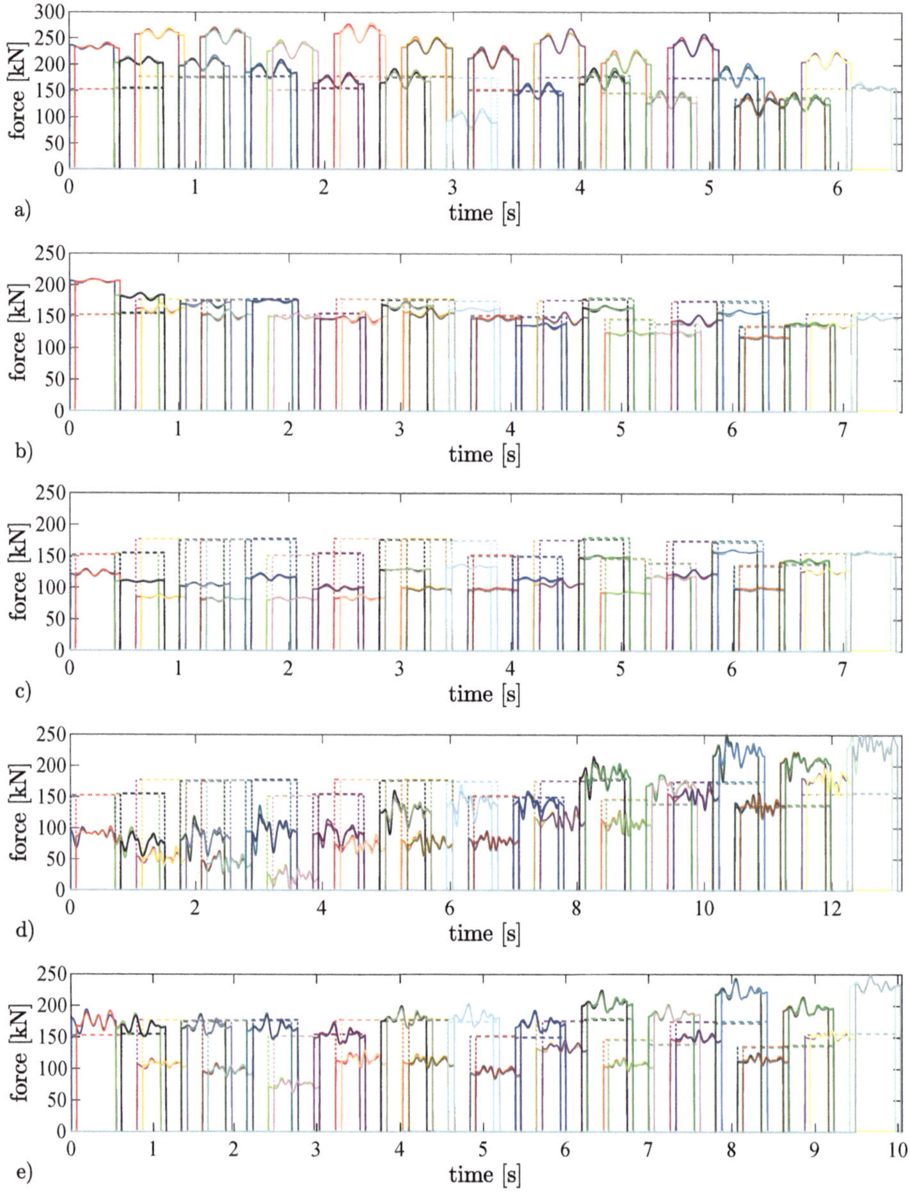

Figure B.20 Identified dynamic force time histories and reference static axle loads for measurement configuration 12: a) passage no. 13; b) passage no. 27; c) passage no. 29; d) passage no. 45; e) passage no. 49

B.4 Results for different damping ratios

The force identification results obtained for the different damping ratios (cf. Tab. 7.6) are illustrated in the following:

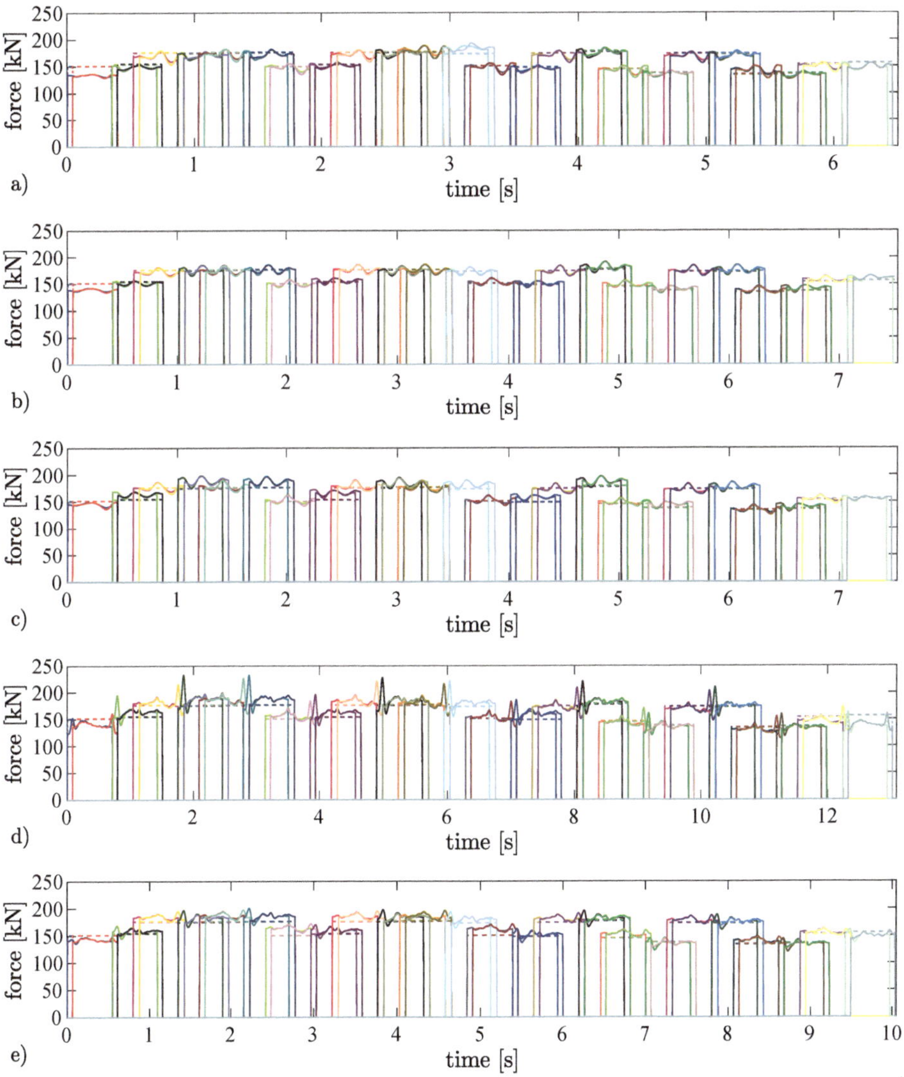

Figure B.21 Identified dynamic force time histories and reference static axle loads of the Schmutter bridge for a damping ratio $\xi = 1.74$ %: a) passage no. 13; b) passage no. 27; c) passage no. 29; d) passage no. 45; e) passage no. 49

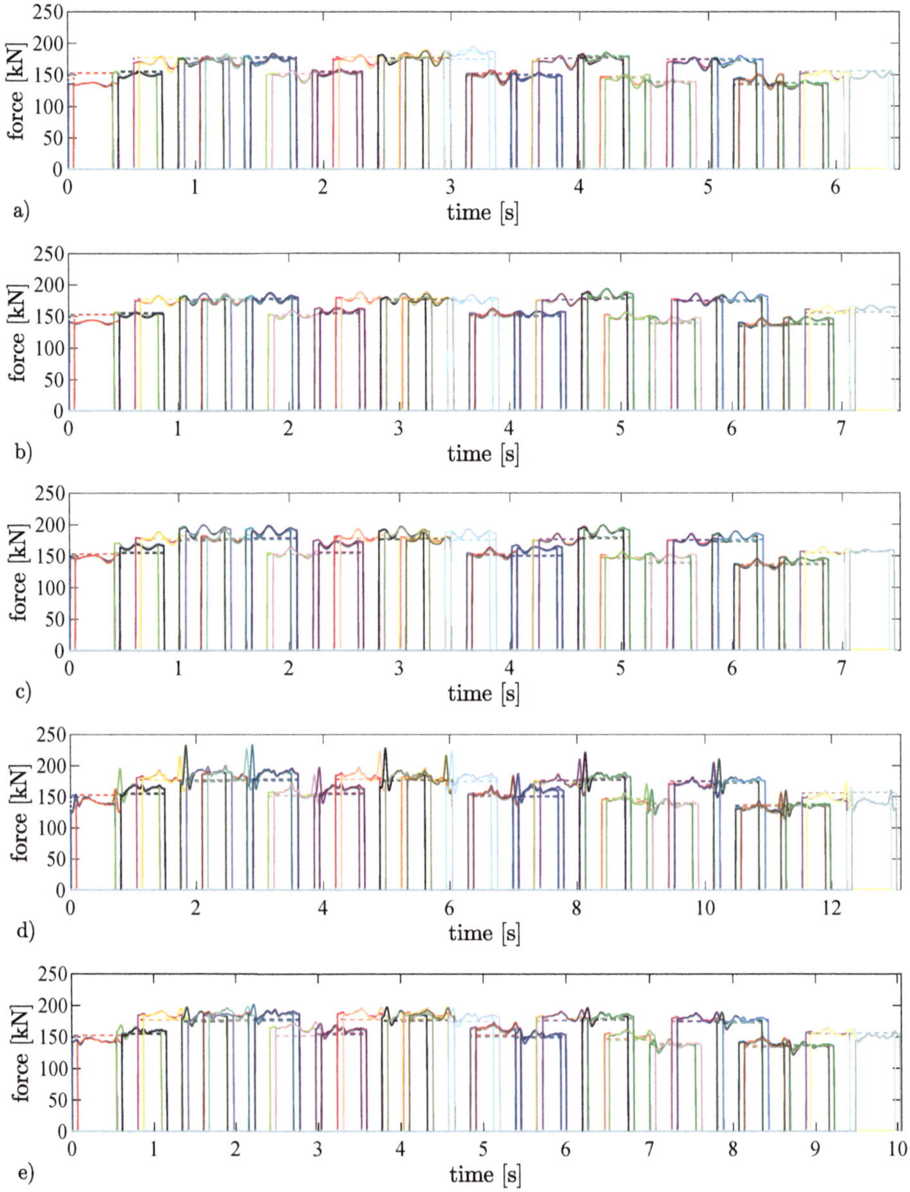

Figure B.22 Identified dynamic force time histories and reference static axle loads of the Schmutter bridge for a damping ratio $\xi = 2.24$ %: a) passage no. 13; b) passage no. 27; c) passage no. 29; d) passage no. 45; e) passage no. 49

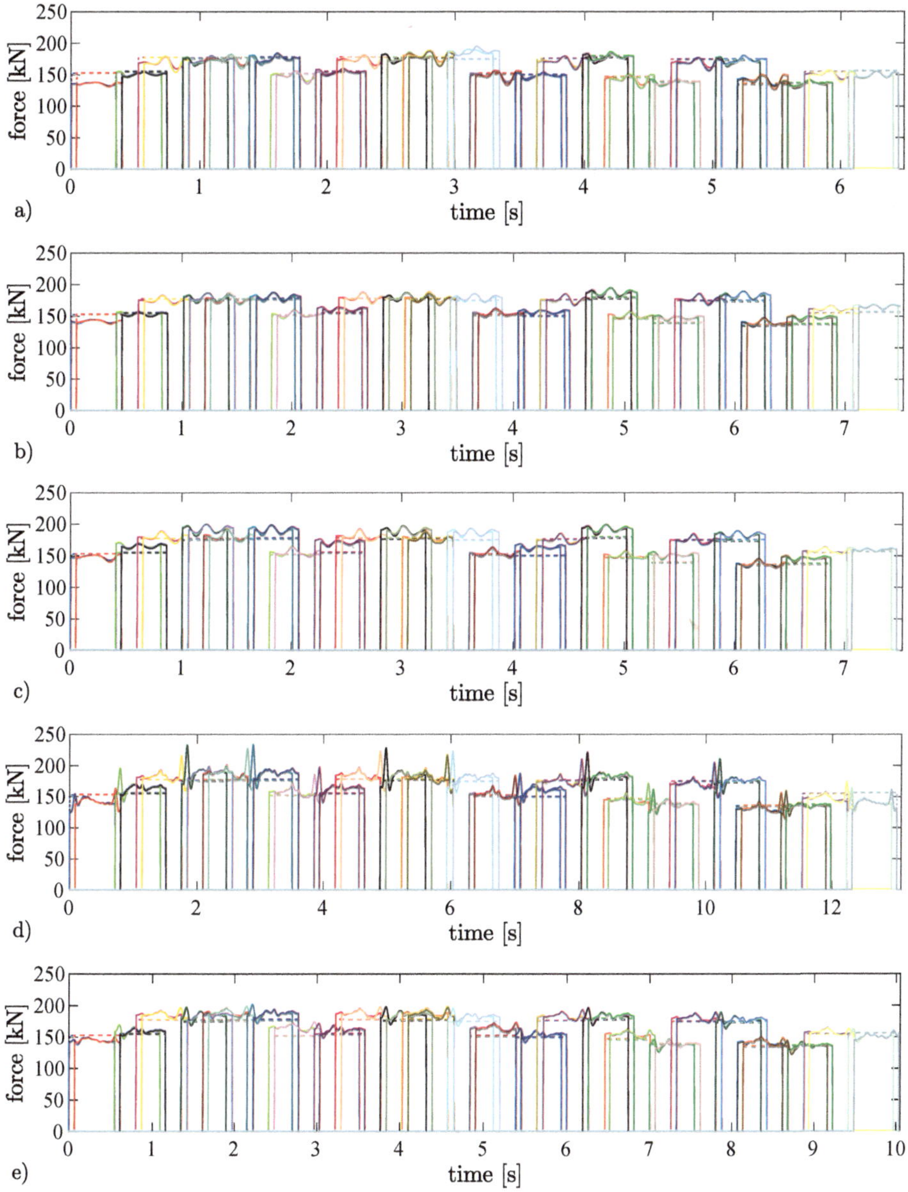

Figure B.23 Identified dynamic force time histories and reference static axle loads of the Schmutter bridge for a damping ratio $\xi = 2.74$ %: a) passage no. 13; b) passage no. 27; c) passage no. 29; d) passage no. 45; e) passage no. 49

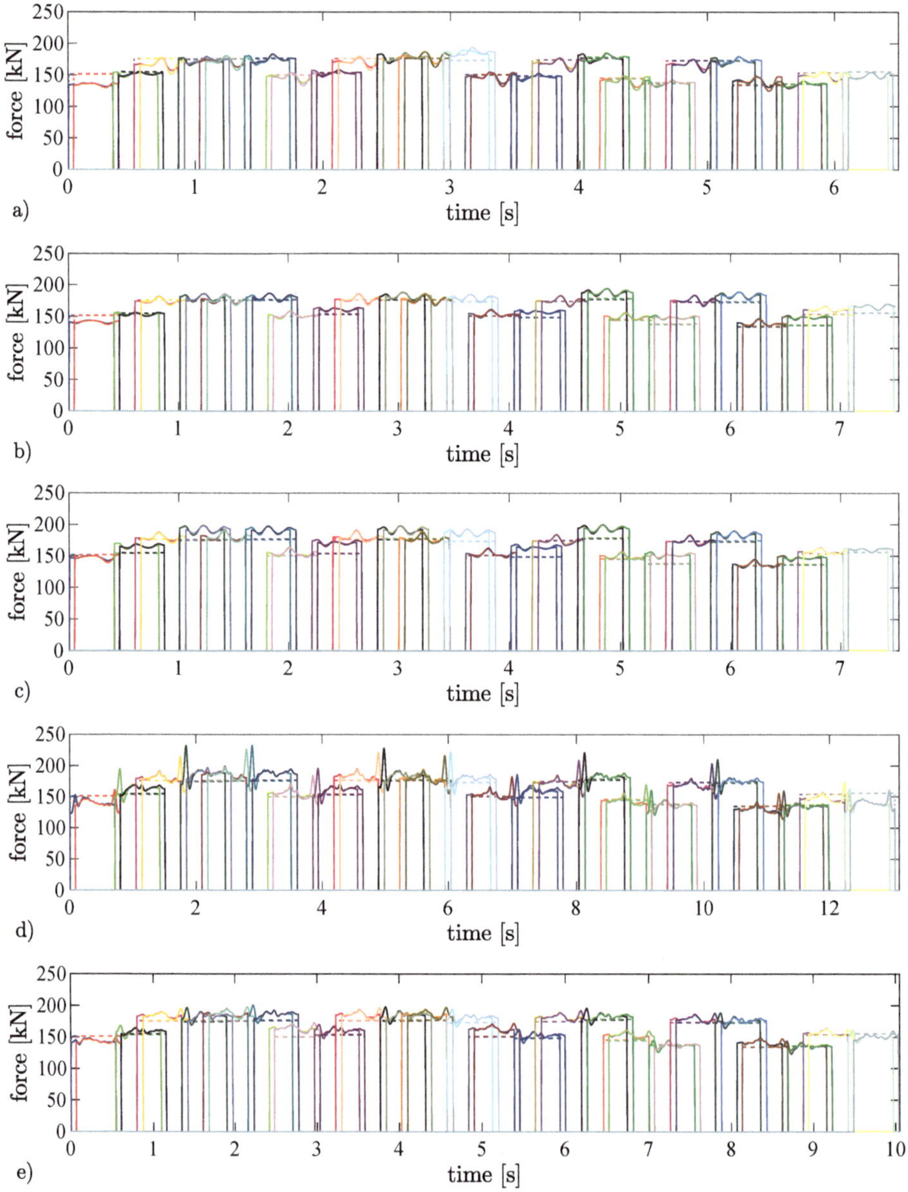

Figure B.24 Identified dynamic force time histories and reference static axle loads of the Schmutter bridge for a damping ratio $\xi = 3.24$ %: a) passage no. 13; b) passage no. 27; c) passage no. 29; d) passage no. 45; e) passage no. 49

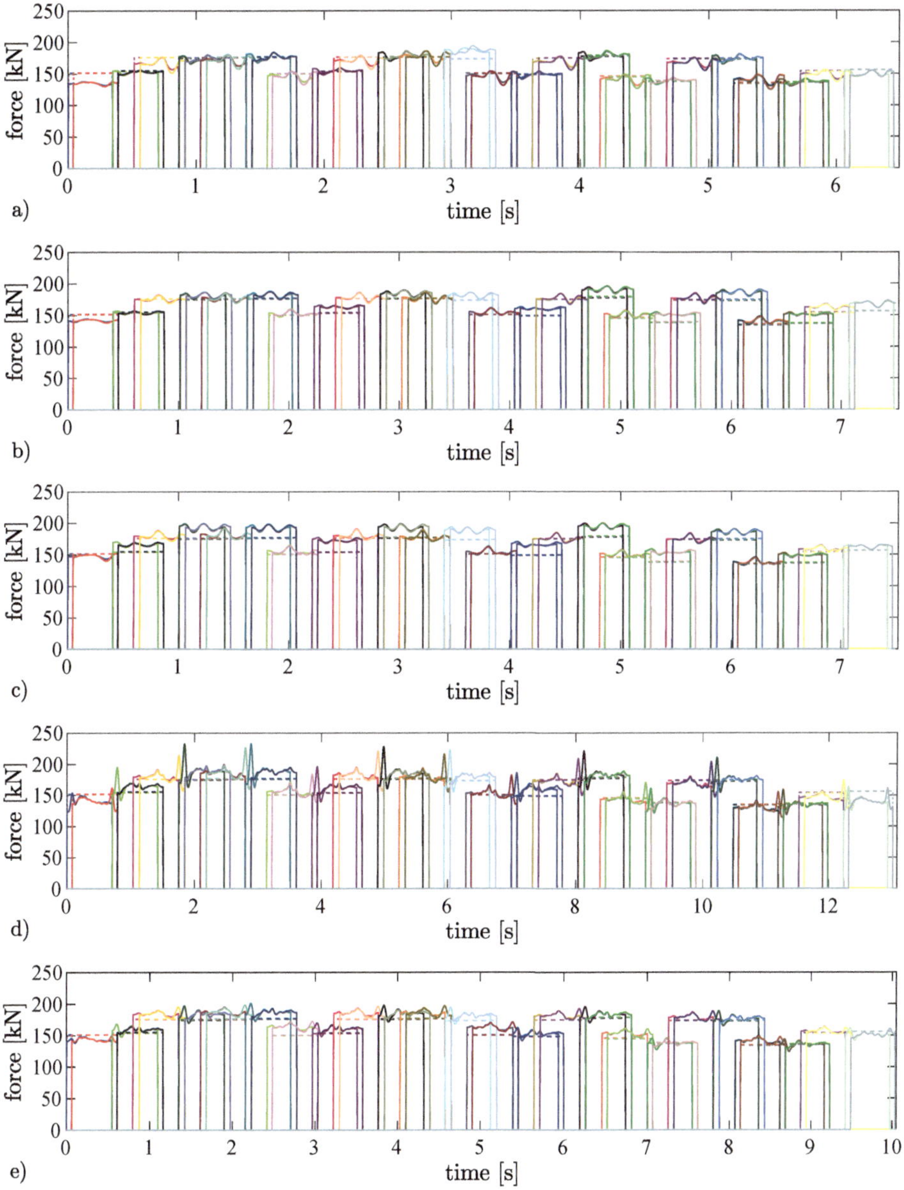

Figure B.25 Identified dynamic force time histories and reference static axle loads of the Schmutter bridge for a damping ratio $\xi = 3.74$ %: a) passage no. 13; b) passage no. 27; c) passage no. 29; d) passage no. 45; e) passage no. 49

B.5 Results for different number of modes

The force identification results obtained for the different number of modes considered in the inverse analysis (cf. Tab. 7.7) are illustrated in the following:

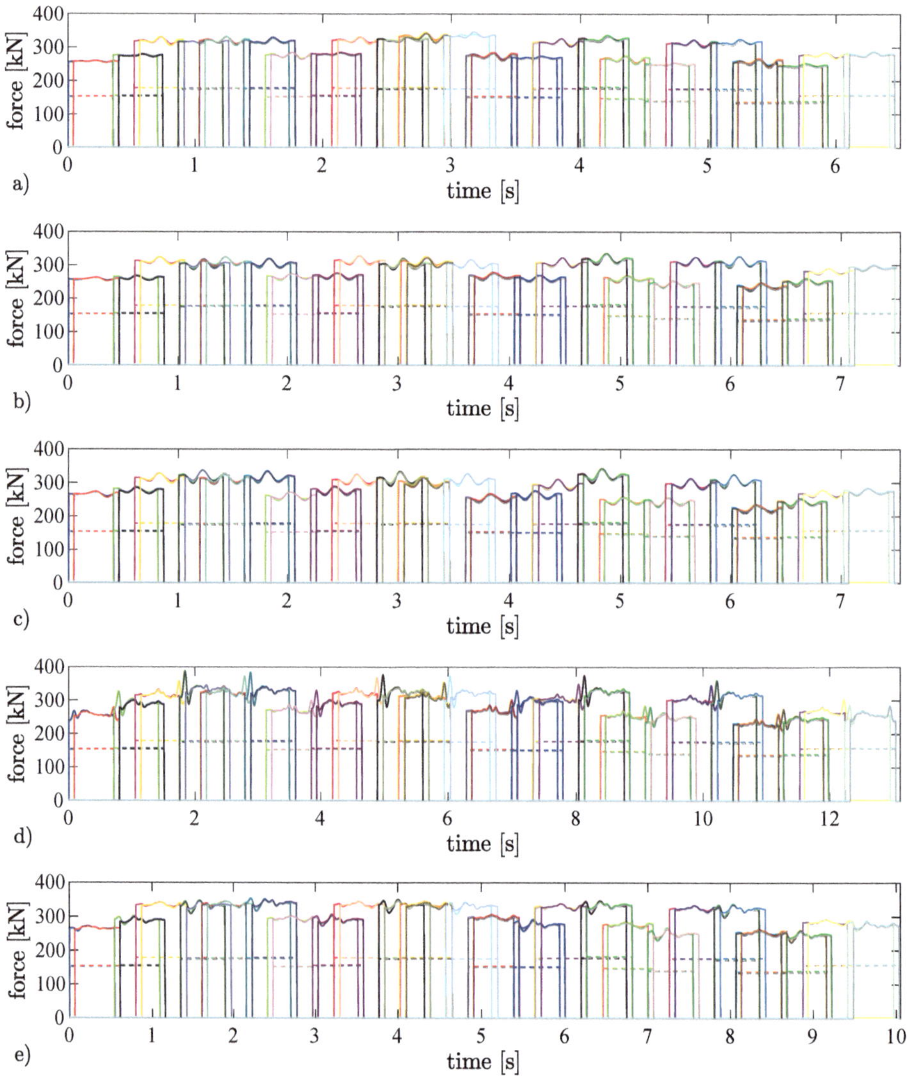

Figure B.26 Identified dynamic force time histories and reference static axle loads of the Schmutter bridge obtained under consideration of 1 mode: a) passage no. 13; b) passage no. 27; c) passage no. 29; d) passage no. 45; e) passage no. 49

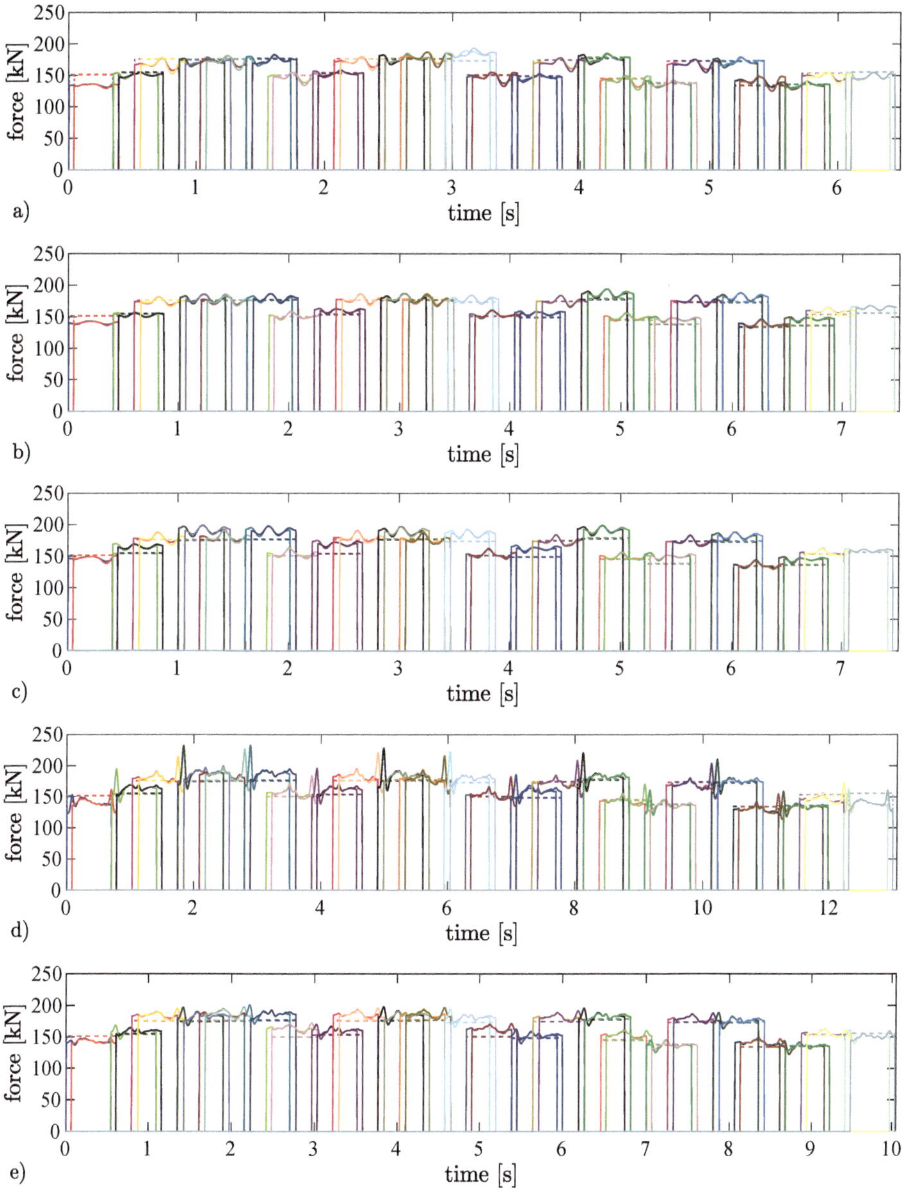

Figure B.27 Identified dynamic force time histories and reference static axle loads of the Schmutter bridge obtained under consideration of 2 modes: a) passage no. 13; b) passage no. 27; c) passage no. 29; d) passage no. 45; e) passage no. 49

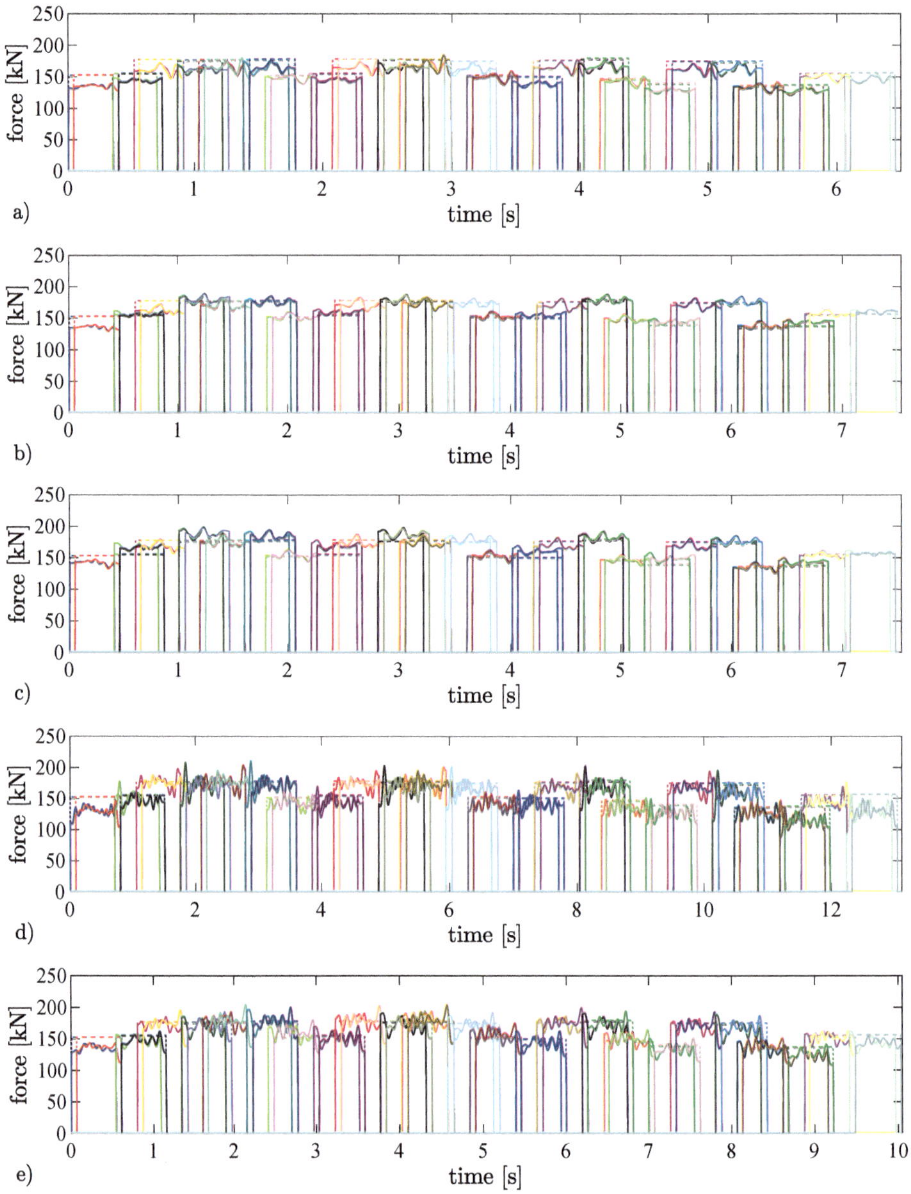

Figure B.28 Identified dynamic force time histories and reference static axle loads of the Schmutter bridge obtained under consideration of 3 modes: a) passage no. 13; b) passage no. 27; c) passage no. 29; d) passage no. 45; e) passage no. 49

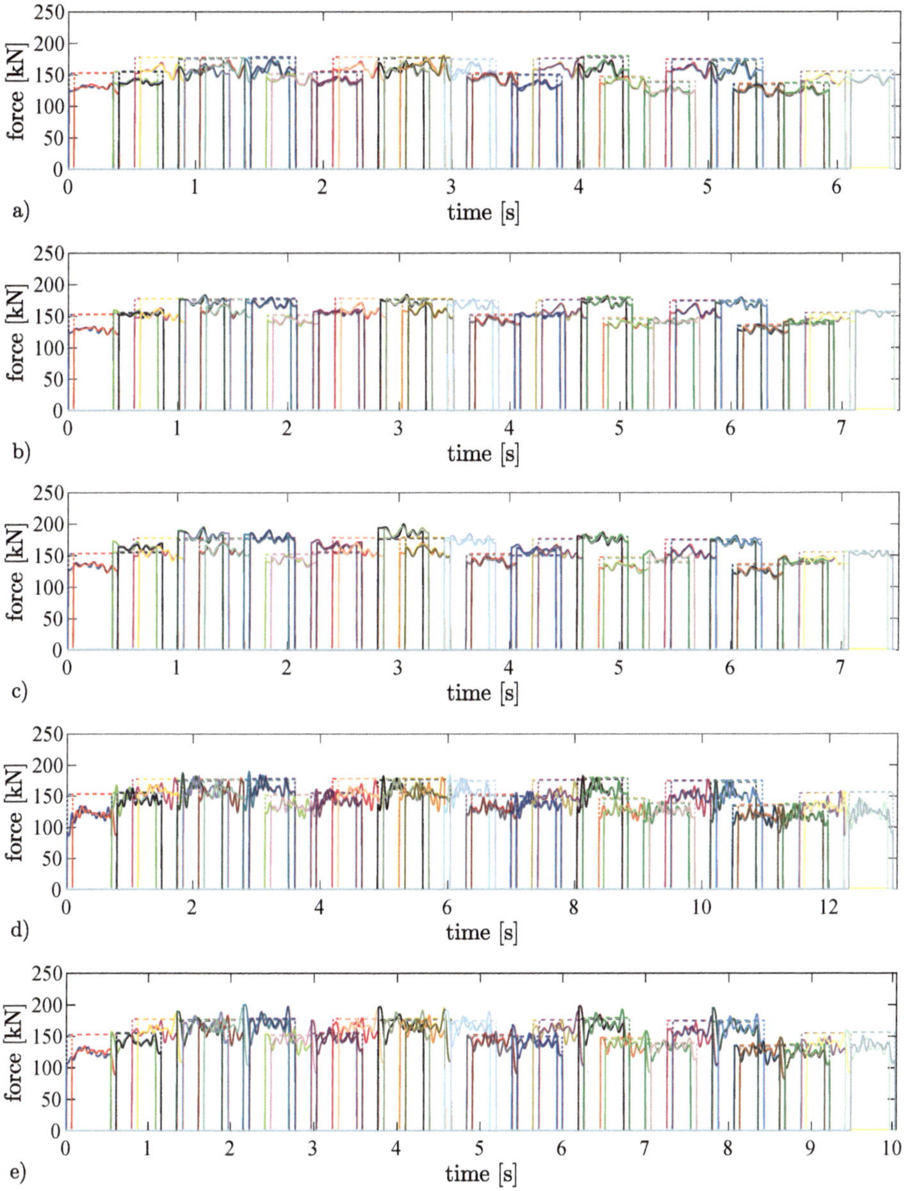

Figure B.29 Identified dynamic force time histories and reference static axle loads of the Schmutter bridge obtained under consideration of 4 modes: a) passage no. 13; b) passage no. 27; c) passage no. 29; d) passage no. 45; e) passage no. 49

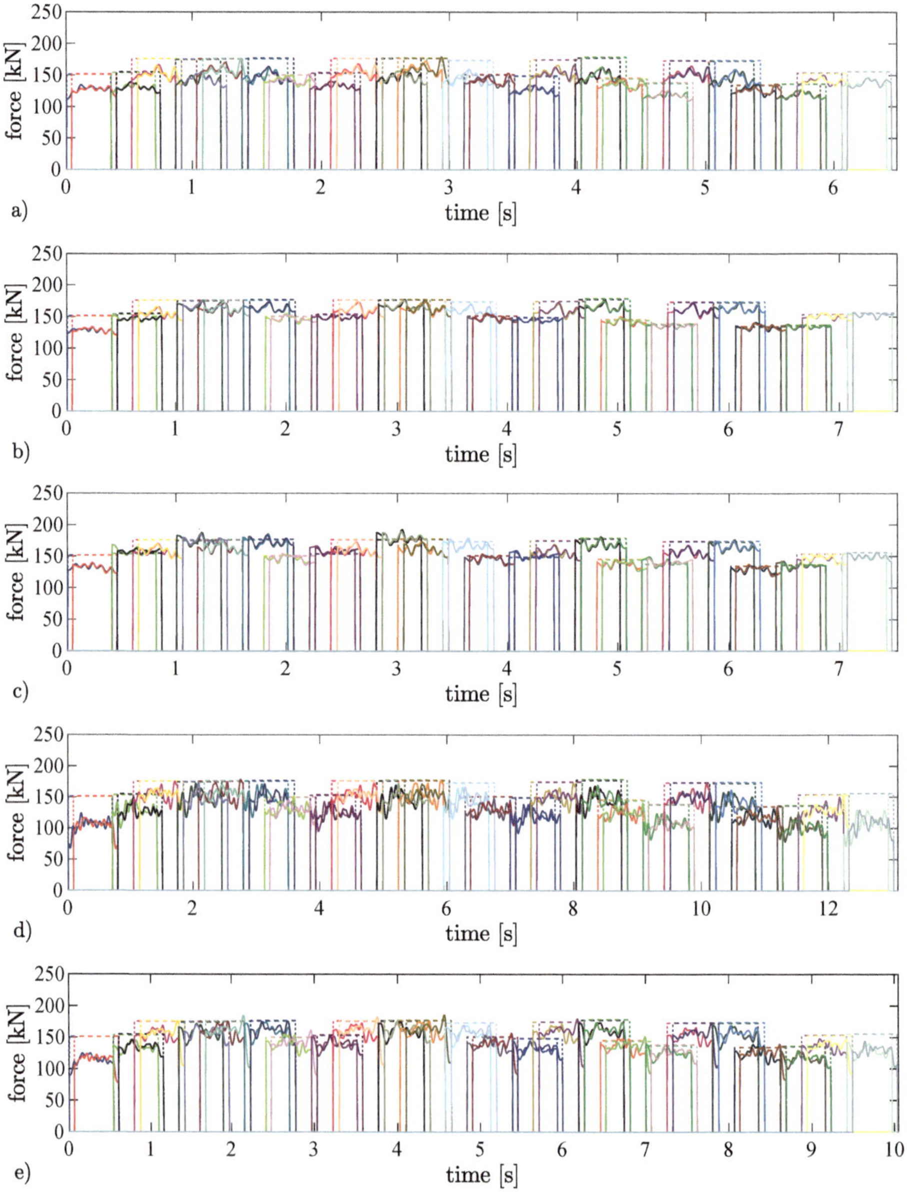

Figure B.30 Identified dynamic force time histories and reference static axle loads of the Schmutter bridge obtained under consideration of 5 modes: a) passage no. 13; b) passage no. 27; c) passage no. 29; d) passage no. 45; e) passage no. 49

B.6 Results for a perturbed speed

The force identification results obtained under consideration of perturbed vehicle speeds (cf. Tab. 7.8) are illustrated in the following:

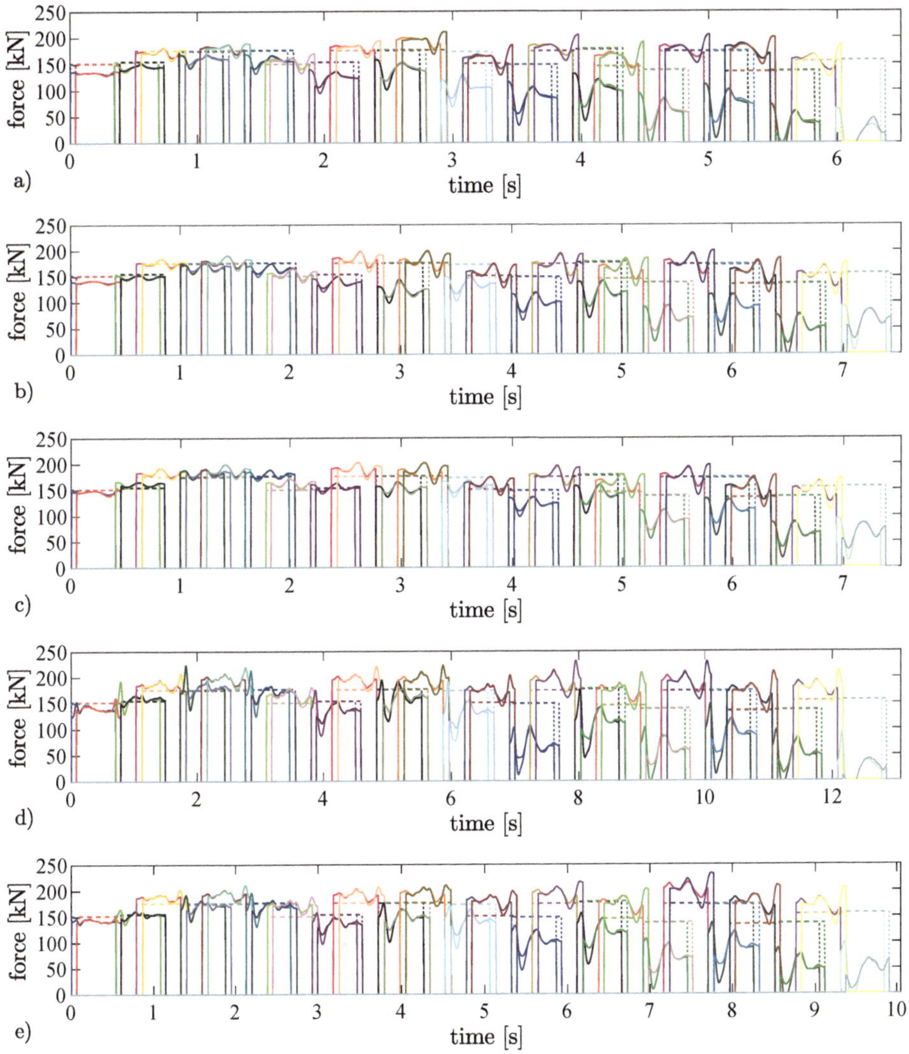

Figure B.31 Identified dynamic force time histories and reference static axle loads of the Schmutter bridge obtained under consideration of a perturbed speed: a) passage no. 13 (201.84 km/h); b) passage no. 27 (173.06 km/h); c) passage no. 29 (174.30 km/h); d) passage no. 45 (100.03 km/h); e) passage no. 49 (129.88 km/h)

Printed by Printforce, the Netherlands